Sustainability of Integrated Water Resources Management

Shimelis Gebriye Setegn
Maria Concepcion Donoso
Editors

Sustainability of Integrated Water Resources Management

Water Governance, Climate and Ecohydrology

Springer

Editors
Shimelis Gebriye Setegn
Department of Environmental and Occupational Health and Global Water for Sustainability Program (GLOWS)
Florida International University
Miami, FL, USA

Maria Concepcion Donoso
Global Water for Sustainability Program
Florida International University
Miami, FL, USA

ISBN 978-3-319-12193-2 ISBN 978-3-319-12194-9 (eBook)
DOI 10.1007/978-3-319-12194-9

Library of Congress Control Number: 2015939723

Springer Cham Heidelberg New York Dordrecht London

Printed on acid-free paper

Springer International Publishing AG Switzerland is part of Springer Science+Business Media (www.springer.com)

Foreword

All around the world, people are feeling enormous pressures from the challenges of managing water resources. Climate change is affecting the global water cycle, leading to irregular rainfall, more floods and more droughts. Water availability has been reduced due to mining of groundwater, pollution and abstraction from upstream water sources. Expanding cities and accelerating economic activity are increasing the demand for energy and food and creating further unsustainable pressures on water resources. Competition for land, water and food is threatening to exacerbate poverty, hunger and environmental deterioration.

Addressing the complex inter-linkages effectively requires an integrated framework that leverages the synergies among access to water and sanitation, education and health, equality and women's empowerment, energy security, food and nutrition, climate, biodiversity and ecosystems, governance and the rule of law. The Millennium Development Goals included the target of halving, by 2015, the proportion of people without sustainable access to safe drinking water and sanitation. Member States of the United Nations in 2010 explicitly recognized the human right to water and sanitation and acknowledged that clean drinking water and sanitation are essential to the realization of all human rights. Building on the achievements and the lessons learned thus far, the new sustainable development agenda will be adopted in September 2015 to set the world on a path to sustainable development.

The new sustainable development agenda will build on the work of the Open Working Group established after the Rio+20 Conference in 2012. It conducted an unprecedented transparent and inclusive process, open to the participation of all stakeholders, including civil society, businesses, academia, local authorities, parliamentarians and citizens. The outcome of the Open Working Group contains a proposal for the Sustainable Development Goals (SDG), which will be the main basis for the final set of negotiations leading to the world leaders' Summit in September 2015.

The new agenda will be strongly human-development focused while addressing the planetary boundaries, economic growth and social inclusion in an integrated manner. It will reflect universality, integration and transformative change, backed by accountability supporting the data revolution.

During the deliberations of the Open Working Group, member states acknowledged that water is at the core of sustainable development, as water and sanitation are central to the achievement of many other development goals and play a vital role in economic growth and poverty eradication. The relevance of sustainable management of watersheds and other water-related ecosystems has also been acknowledged. The Integrated Water Resources Management (IWRM) perspective is reflected in the proposal for an SDG on "Ensuring Availability and sustainable management of water and sanitation for all", with a target to implement IWRM at all levels by 2030.

This book will contribute to providing science-based evidence on the ways of better implementing its goals and targets, monitoring progress through appropriate indicators and data, and the means for implementation. The text addresses some of the theoretical, practical and political issues encompassed by IWRM in a comprehensive and multifaceted way. It includes the households, local, country and regional perspectives on IWRM, linking experiences and evidence with the major global challenges. The book covers a wide range of factors that ultimately influence the effective exercise of the human rights to water and sanitation in a sustainable way, including the links with governance and conflicts.

I believe this publication will help better inform the post-2015 development agenda by bridging the gap between evidence and policy making, between science and programs, between academics, policy makers, civil society, businesses and communities.

Special Advisor of the UN Secretary-General on Post-2015 Development Planning, United Nations
Amina J. Mohammed

Contents

Contributors

Abou Amani UNESCO, Nairobi Regional Office for Eastern Africa, Nairobi, Kenya

Elizabeth P. Anderson School of Environment, Arts and Society, Florida International University, Miami, FL, USA

Jorge Arriaga-Medina Red del Agua UNAM, Universidad Nacional Autónoma de México, Coyoacán, CP, Mexico

Maria Catalina Bosch Global Water for Sustainability, Florida International University, North Miami, FL, USA

Luc Descroix LTHE, Bâtiment OSUG-B, Domaine universitaire, Grenoble cedex 09, France

Robert Dessouassi Niger Basin Authority Executive Secretariat, Niamey, Niger

Naresh Devineni Columbia Water Center, Columbia University, New York, NY, USA

Department of Civil Engineering, The City College of New York, New York, NY, USA

Malinali Domínguez-Mares Red del Agua UNAM, Universidad Nacional Autónoma de México, Mexico, CP, Mexico

Maria Concepcion Donoso Global Water for Sustainability Program, Florida International University, Miami, FL, USA

Marcela Duhne-Ramírez Laboratoire d'Etude des Transferts en Hydrologie et Environnement, LTHE, Grenoble, France

Jochen Eckart Patel College of Global Sustainability, University of South Florida, Tampa, FL, USA

Diana del Cisne Encalada-Jumbo Department of Economics, UTPL, Loja, Ecuador

C. Gay-García Research Program in Climate Change (PINCC), National Autonomous University of Mexico (UNAM), Mexico City, Mexico

Kebreab Ghebremichael Patel College of Global Sustainability, University of South Florida, Tampa, FL, USA

Fernando González-Villarreal Universidad Nacional Autónoma de México, Mexico, CP, Mexico

Orville Grey Department of Earth and Environment, Florida International University, Miami, FL, USA

Alfonso Gutiérrez-López Centro de Investigaciones del Agua, CIAQ, Universidad Autónoma de Querétaro, Col. Las Campanas, México

Lauren Herwehe School of Geography and Development, University of Arizona, Tucson, AZ, USA

Juan Bautista Justo Universidad Nacional del Comahue, Neuquén, Argentina

Krishna Khatri Patel College of Global Sustainability, University of South Florida, Tampa, FL, USA

Mathew Kilanski Department of Geological Sciences, Earth and Energy Resources, University of Texas, Austin, TX, USA

Jenna Kromann Department of Geological Sciences, Earth and Energy Resources, University of Texas, Austin, TX, USA

Upmanu Lall Columbia Water Center, Columbia University, New York, NY, USA

Belize Lane Department of Land, Air and Water Resources, University of California, Davis, CA, USA

Claudia Cecilia Lardizabal Department Francisco Morazan, National Autonomous University of Honduras, Tegucigalpa, Honduras

Cecilia Lartigue Programme for Management, Use and Reuse of Water (PUMAGUA), National Autonomous University of Mexico, Mexico City, Mexico

Thierry Lebel LTHE, Bâtiment OSUG-B, Domaine universitaire, Grenoble cedex 09, France

Ryan H. Lee Arid Lands Resource Sciences, University of Arizona, Tucson, AZ, USA

David MacPhee Department of Mechanical Engineering, San Diego State University, San Diego, CA, USA

Alexandra Manueles Department of Environment and Development, Zamorano University, Francisco Morazan, Honduras

Liber Martín CONICET/Universidad Nacional de Cuyo, Mendoza, Argentina

Michael E. McClain UNESCO-IHE Institute of Water Education, Delft, The Netherlands

Denisse McLean Department of Environment and Development, Zamorano University, Francisco Morazan, Honduras

Assefa M. Melesse Department of Environmental and Occupational Health and Global Water for Sustainability Program, GLOWS, Florida International University, Miami, FL, USA

Luis Metzger National Service of Meteorology and Hydrology, Lima, Perú

Ulla Mörtberg Division of Land and Water Resources Engineering, KTH Royal Institute of Technology, Stockholm, Sweden

I. Navarro-González Institute of Engineering, National Autonomous University of Mexico (UNAM), Mexico City, Mexico

J.E. Ospina-Noreña Faculty of Agricultural Sciences, Department of Agronomy, National University of Colombia, Campus Bogota, Bogotá, Colombia

Adwoa Paintsil Water Resources Commission, Accra, Ghana

Janaina Camile Pasqual International Center of Hydroinformatic, Foz do Iguassu, Brazil

International Center of Renewable Energy-Biogas, Foz do Iguassu, Brazil

Pontifical Catholic University of Parana, Curitiba, Brazil

R.T. Montes-Rojas General Coordination of Adaptation to Climate Change, National Institute of Ecology and Climate Change, Mexico City, Mexico

C. Rueda-Abad Research Program in Climate Change (PINCC), National Autonomous University of Mexico (UNAM), Mexico City, Mexico

Israel Ruiz-González Centro de Investigaciones del Agua, CIAQ, Universidad Autónoma de Querétaro, Santiago de Querétaro, Mexico

Tess A. Russo Department of Geosciences, The Pennsylvania State University, University Park, PA, USA

Columbia Water Center, Columbia University, New York, NY, USA

Amartya K. Saha Global Water for Sustainability (GLOWS), Department of Earth and Environment, Florida International University, North Miami, FL, USA

Partha Pratim Saha School of Environmental Sciences, Charles Sturt University, Wagga Wagga, NSW, Australia

Arie Sanders Department of Environment and Development, Zamorano University, Francisco Morazan, Honduras

Christopher A. Scott Udall Center for Studies in Public Policy and School of Geography and Development, University of Arizona, Tucson, AZ, USA

Shimelis Gebriye Setegn Department of Environmental and Occupational Health and Global Water for Sustainability Program (GLOWS), Florida International University, Miami, FL, USA

Ryan B. Stoa Florida International University, College of Law and Global Water for Sustainability Program – GLOWS, Miami, FL, USA

José Alberto Tejada-Guibert Global Waters for Sustainability Program – GLOWS, Florida International University, Miami, FL, USA

Selome M. Tessema Division of Land and Water Resources Engineering, KTH Royal Institute of Technology, Stockholm, Sweden

Seneshaw Tsegaye Patel College of Global Sustainability, University of South Florida, Tampa, FL, USA

Kala Vairavamoorthy Patel College of Global Sustainability, University of South Florida, Tampa, FL, USA

Rafael Val-Segura PUMAGUA, Universidad Nacional Autónoma de México, Coyoacán, CP, Mexico

Armando Sánchez-Vargas Institut for Economics Research, UNAM, Mexico City, Mexico

Dale Webber Department of Life Sciences, Faculty of Pure and Applied Sciences, The University of West Indies, Kingston, Jamaica

Ketema Zeleke Graham Centre for Agricultural Innovation, Charles Sturt University, Wagga Wagga, NSW, Australia

School of Agricultural and Wine Sciences, Charles Sturt University, Wagga Wagga, NSW, Australia

Chapter 1
Introduction: Sustainability of Integrated Water Resources Management (IWRM)

Shimelis Gebriye Setegn

1.1 Overview

Water is essential for life, ecosystems, and social and economic development. We depend on a reliable, clean supply of drinking water to sustain our health. Water is also needed for agriculture, energy production, navigation, recreation, and manufacturing. Its exploitation and use must be well planned and managed in a sustainable manner. Water availability has been reduced due to periodic droughts, overconsumption of surface and groundwater resources, and pollution and climate change. Population increase, fast growth of cities, and accelerating economic activity are increasing the demand for water, energy, and food and creating further pressures on water resources. In many developing countries, the lack of adequate, clean, and safe water, pollution of aquatic environments, and the mismanagement of natural resources are still major causes of environmental health problem and mortality. Irregular rainfall, more floods, and droughts are becoming more frequent events in different parts of the world.

The crucial importance of water to the various aspects of human health, development, and well-being has led to specific objectives concerning water and the support to each of the eight millennium development goals (MDGs), established by the UN in the year 2000. With a human population and water demand that are continuing to grow, the management of water resources will become of vital importance. Moreover, sustainable freshwater resource management will need to be included in future development plans and implementations. The sustainability of integrated water resources management (IWRM) in the face of climate variability and change is an important issue when planning and/or developing policies that

S.G. Setegn (✉)
Department of Environmental and Occupational Health and Global Water for Sustainability Program (GLOWS), Florida International University, 11200 SW 8th Street, Miami, FL 33199, USA
e-mail: ssetegn@fiu.edu

S.G. Setegn, M.C. Donoso (eds.), *Sustainability of Integrated Water Resources Management*, DOI 10.1007/978-3-319-12194-9_1

consider the impact of climate change, ecohydrology, and water governance in the context of a more holistic approach to ensure sustainable management of water resources. Sustainable IWRM is more about processes, and more should be done to articulate the most essential IWRM components that ensure the ongoing IWRM sustainability efforts. Hence this book addresses the importance of integrated water resources management toward achieving water, energy, and food security. It addresses appropriate means of managing the scarce surface and water resources in the face of climate change and increased population pressure and high water demand. The book also addresses the question of how to define and measure the sustainability of IWRM. Main topics covered in this book include global prospective of IWRM; allocation of environmental flows in IWRM; ecohydrology, water resources, and environmental sustainability; climate change and IWRM; IWRM and water governance including social, economic, public health, and cultural aspects; climate change resiliency actions related to water resource management sustainability; and tools in support of sustainability for IWRM.

This book will be of interest to researchers, practitioners, water resource managers, policy- and decision-makers, donors, international institutions, governmental and nongovernmental organizations, educators, as well as graduate and undergraduate students. It is a useful reference for integrated water resources management (IWRM), ecohydrology, climate change impact and adaptations, water governance, environmental flows, geographic information system and modeling tools, water and energy nexus, and related topics.

1.2 Integrated Water Resources Management: Global Perspective

IWRM, as the Global Water Partnership defined, is the process of promoting the coordinated development and management of water, land, and related resources, to maximize the resultant economic and social welfare in an equitable manner without compromising the sustainability of vital ecosystems. IWRM aims to support countries in their efforts to tackle specific water challenges, e.g., water scarcity, waterborne diseases, floods, droughts, and access to water and sanitation, and thus sustain their development to achieve the goals such as poverty alleviation, food security, economic growth, and ecological conservation. IWRM is a comprehensive, participatory planning and implementation tool for managing and developing water resources, ensuring the protection of ecosystems for future generations.

Efforts to achieve the millennium development goals involve planning and action in water resource development, management, and use. Better management and development of water resources through IWRM approach has been recognized in 2002 during the Johannesburg World Summit on Sustainable Development with the Summit urging all countries "to develop IWRM and water efficiency plans by 2005." IWRM is a constantly evolving subject, and its development and application have received intense attention and contributions from many parties, including national authorities, international and intergovernmental bodies, and academic and nongovernmental organizations.

In this book we have addressed several issues of IWRM with special emphasis to African, Latin American, and global perspective of IWRM.

1.3 Ecohydrology, Water Resources, and Environmental Sustainability

A defining characteristic of integrated water resources management (IWRM) is its commitment to balance socioeconomic development of water resources with environmental sustainability. This is articulated in the definition of IWRM by the Global Water Partnership (GWP 2000) and is being adopted in new water policies and legislation worldwide (UNEP 2012). A major component of environmental sustainability in water resource development is the explicit allocation of water to meet ecosystem needs. This environmental water allocation is commonly referred to as an environmental flow. The most widely accepted definition of environmental flows is "the quantity, timing, and quality of water flows required to sustain freshwater and estuarine ecosystems and the human livelihoods and well-being that depend on these ecosystems" (Brisbane Declaration 2007). Ecosystem services, or the beneficial roles that forests and wetlands have on water availability and quality, are being increasingly recognized worldwide. Forests and wetlands store water during the rains, promote groundwater recharge, and feed streams and springs in the dry season. Harnessing this inherent capacity of ecosystems to maintain water quality and to regulate hydrology is then the logical way to manage water resources sustainably and affordably. Understanding the links between different ecosystems in a catchment and local/regional hydrology enables restoration and maintenance of the ecosystems along with the services they provide.

This book consists of different issues on how environmental flow and ecohydrology play significant roles for sustainable management of water resources. The main topics addressed in this issue include allocation of environmental flows in IWRM, understanding and maintaining ecosystem services for IWRM, and application of ecohydrology in IWRM.

1.4 Climate Change and Integrated Water Resources Management (IWRM)

A reliable and clean supply of drinking water is necessary to sustain human health. Water is also needed for agriculture, energy production, navigation, recreation, and manufacturing. These demands place pressures on water resources that are likely to be exacerbated by climate change. In many areas, climate change is likely to reduce surface and groundwater resources, accompanied by increasing water demand. A major effect of climate change is likely to be alterations in hydrologic cycles and changes in water availability. Increased evaporation, combined with changes in precipitation, has the potential to affect runoff, the frequency and intensity of floods and droughts, soil moisture, and available water for irrigation and hydroelectric generation.

The Intergovernmental Panel on Climate Change (IPCC 2007) findings suggest that developing countries will be more vulnerable to climate change due to their economic, climatic, and geographic settings. According to IPCC (2007) report, the population at risk of increased water stress in Africa is projected to be between 75 and 250 and 350 and 600 million people by the 2020s and 2050s, respectively. Moreover, yields from rain-fed agriculture could be reduced by up to 50 %, in countries which depend mainly on rain-fed agriculture.

In some areas, climate change increases runoff, flooding, or sea level rise. Changes in the amount of rainfall during storms provide evidence that the water cycle is already changing. Setegn et al. (2011) investigated how changes in temperature and precipitation might translate into changes in stream flows and other hydrological components using downscaled outputs from four climate models.

This book consists of topics on sustainability of water resources in tropical regions in the face of climate change, sustainable development and integrated water resources management, water resources variability due to climate change in Mexico, and sustainable management of floods and extreme events.

1.5 IWRM and Water Governance

Meeting the millennium development goals for water and sanitation in the next decade will require substantial economic resources, sustainable technological solutions, and courageous political will. The challenge is to mobilize the political will to implement water resource development programs which cater in an equitable manner for the various demands on water. A great number of governments and international organizations have launched water-related programs and interventions all over the world as an effective way to improve people's health and welfare. But the challenges to overcome the impacts will be very high. An integrated approach should be designed to decrease the alarming impact of water quality, chemical impurities, and other water pollutions.

IWRM is not just about managing physical resources; it also requires and promotes the positive changes in water governance regarding the enabling environment, institutional roles, and management instruments. IWRM systems should, therefore, not only be responsive to changes among its development process, for example, between projected goals and decision-makers' willingness, but also be capable of adapting to new economic, social, and environmental conditions and to changing human values over a long-term implementation.

The major environmental issues of concern to policy-makers are the increased vulnerability of groundwater quality and the sustainability of natural resources for future generations. To understand the sustainability of the natural resources such as water in general, one needs to understand the impact of future land use changes on the natural resources. Climate change is predicted to negatively alter global and basin hydrologic cycles, stream flows, and water availability (IPCC 2007; Setegn et al. 2011). IWRM is viewed as the water management and governance paradigm best suited for "securing water for people" while reconciling economic efficiency, social equity, and environmental sustainability (Global Water Partnership 2000). Without proper water governance, there is likely to be increased competition for water between sectors and an escalation of water crises of various kinds, triggering emergencies in a range of water-dependent sectors.

In this book we have addressed several water governance issues in the area of water resource management for sustainable environmental public health, climate change and sustainable water access, lessons for treaty formation and development, and basin comanagement plans: a Participative Approach to Water Governance; IWRM and vertical integration across local-, meso-, and macroscale institutions; and the environmental regulatory shift and its impact on water resources management in Latin America.

1.6 Climate Change Resiliency Actions Related to Water Resources Management Sustainability

The most important impacts of climate change will be exerted on water resources and water management systems, reflecting the importance of water resources to social development. Extreme events, linked to climate change, might affect the quantity and quality of water available in rivers, lakes, and underground reservoirs which, in turn, might generate water scarcity at the household level, affecting people's well-being (Bates et al. 2008). These impacts will affect strongly populations with lack of financial resources to implement adaptation plans. Water is also at the heart of adaptation to climate change, serving as the crucial link between the climate system, human society, and the environment.

Adaptation to the global climate change and variability is considered a cornerstone for the application of IWRM for it to be truly effective and sustainable. The adaptive nature of IWRM is considered a good platform to incorporate climate change adaptation. The major driving instrument for international efforts in climate change adaptation is the United Nations Framework Convention on Climate Change (UNFCCC), an international environmental treaty negotiated at the Rio 1992 UN Conference on Environment and Development that entered into force in 1994.

According to EPA, managing water resources will likely become more challenging with projected climate changes and anticipated population and economic growth. In many areas, climate change is likely to increase water demand while shrinking water supplies. This shifting balance would challenge water managers to simultaneously meet the needs of growing communities, sensitive ecosystems, farmers, ranchers, energy producers, and manufacturers. Freshwater resources along the coasts face risks from sea level rise. As the sea rises, saltwater moves into freshwater areas. The impacts of climate change on water availability and water quality will affect many sectors, including energy production, infrastructure, human health, agriculture, and ecosystems. Adaptation to climate change is closely linked to water and its role in sustainable development. Various necessary adaptation measures that deal with climate variability and build upon existing land and water management practices have the potential to create resilience to climate change and to enhance water security and thus directly contribute to development. Adaptation to climate change is urgent. Water plays a pivotal role in it, but the political world has yet to recognize this notion (UN 2010).

As the time limit for the millennium development goals (MDGs) draws to a close in 2015, the global community is taking stock of how it can move toward a sustainable future. A global goal for water and associated targets would build on the MDGs and redouble efforts to develop water supplies and sanitation services for human needs.

In this book several issues were covered in the area of climate change mitigation and adaptation and the role of international ocean and freshwater agreements, sustainable development and the importance of water resources, efficient use of water resources for sustainability and land use, and climate change impact on the coastal zones.

References

Bates BC, Kundzewicz ZW, Wu S, Palutikof PJ et al (2008) Climate change and water: technical paper of the Intergovernmental Panel on Climate Change. IPCC Secretariat, Geneva. 210 pp

Brisbane Declaration (2007) The Brisbane Declaration: environmental flows are essential for freshwater ecosystem health and human well-being. In: 10th international river symposium, 3–6 September 2007, Brisbane. http://www.conservationgateway.org/ConservationPractices/Freshwater/EnvironmentalFlows/MethodsandTools/ELOHA/Pages/Brisbane-Declaration.aspx

GWP (Global Water Partnership) (2000) Integrated Water Resources Management. GWP TAC Background Paper #4. http://www.gwpforum.org/gwp/library/TACNO4.PDF

Intergovernmental Panel on Climate Change (IPCC) (2007) In: Parry ML et al (eds) Climate change 2007: impacts, adaptation, and vulnerability contribution of Working Group II to the third assessment report of the Intergovernmental Panel on Climate Change. Cambridge University Press, Cambridge

Setegn SG, Rayner D, Melesse AM, Dargahi B, Srinivasan R (2011) Impact of climate change on the hydroclimatology of Lake Tana Basin, Ethiopia. Water Resour Res 47:W04511. doi:10.1029/2010WR009248. (Citation 26)

UNEP (UN Environment Program) (2012) The UN-water status report on the application of integrated approaches to water resources management

UN-Water Policy Brief climate change adaptation: the pivotal role of water 2010 http://www.epa.gov/climatechange/impacts-adaptation/water.html

Part I
Integrated Water Resources Management (IWRM): Global Perspectives

Chapter 2
Integrated Water Resources Management in Latin America and the Caribbean

Maria Concepcion Donoso and Maria Catalina Bosch

Abstract In this chapter, we present an overview of selected Integrated Water Resources Management (IWRM) schemes, legislation, policies, plans, and governance structures designed and implemented by the countries of the region of Latin America and the Caribbean (LAC). Conceptual reasons required a brief introduction on the inspiring ideas of IWRM, stress having been made on the justification of the concept as reflected in its "integrated" note. Such preamble should enable the reader to assess the orthodoxy of the IWRM schemes predominant in the Region as well as the national enabling instruments, measures, and policies enumerated in the second part of the paper, as compared with the IWRM theoretical tenets. Also mentioned is the coordinating, financial, and advisory role of international organizations in response to the limitations of countries to resolve transboundary water issues and in some cases challenges created in federal states by multiple (national, state/provincial, municipal) jurisdictions. Reference is additionally made to the endemic LAC issue of mismatch between abstract legal instruments and actual implementation, as an additional criterion for the reader to judge the value of the actions pursued in this area by the individual nations. The main part of this chapter includes country-by-country available information on the schemes and instruments in force in the Region. The final section of the chapter concludes with findings and an overall assessment of the Region's achievements and margins for improvement going forward.

Keywords IWRM • Sustainability • Water management • Latin America • The Caribbean

---

M.C. Donoso (✉) • M.C. Bosch
Global Water for Sustainability Program, Florida International University,
3000 NE 151st Street, AC1- 267, Miami, FL 33181, USA
e-mail: mcdonoso@fiu.edu; catabosch81@hotmail.com

S.G. Setegn, M.C. Donoso (eds.), *Sustainability of Integrated Water Resources Management*, DOI 10.1007/978-3-319-12194-9_2

2.1 IWRM Drivers, Definition, and Justification

2.1.1 Drivers

The adoption of an Integrated Water Resources Management (IWRM) approach is today recommended as a response to crucial national development—or, more dramatically, survival—requirements, including mitigation of climate variations, such as extreme meteorological phenomena and environmental risks, particularly as it relates to freshwater and coastal areas and in realms such as those of sustainable development; conflicts around water use and ownership (with their potential for in-country, regional, and international commotions); achievement of a range of United Nations (UN) Millennium Development Goals -MDGs-, including those on water and energy, fight against hunger, rural women empowerment; as well as transgenerational responsibilities and coordination needs (UN Millennium Project 2005). As the global community is undertaking the challenge to define the SDGs (Sustainable Development Goals), the need for a holistic integrated approach to water management is becoming ever more evident. In this context, it is no surprise that following the 2012 UN Conference on Sustainable Development (Rio+20),[1] a set of SDGs were proposed by an Open Working Group established by the UN, among which Goal 6 specifies "ensure availability and sustainable management of water and sanitation for all"(UN 2014).

2.1.2 Definition

Among the several definitions related to IWRM, two of them are mostly accepted in the LAC region. The first one was put forward by the Global Water Partnership (GWP) which states that IWRM consists of "a process which promotes the co-ordinated development and management of water, land and related resources, in order to maximize the resultant economic and social welfare in an equitable manner without compromising the sustainability of vital ecosystems"(Agarwal et al. 1999). Similarly accepted is the definition of Integrated River Basin Management (IRBM) presented more recently by the United Nations Educational, Scientific and Cultural Organization (UNESCO) International Hydrological Programme (IHP) which refers to IRBM as "a holistic approach that seeks to integrate the management of the physical environment within that of the broader socio-economic and political framework. The river basin approach seeks to focus on implementing IWRM principles on the basis of better coordination amongst operating and water management entities within a river basin, with a focus on allocating and delivering reliable water-dependent services in an equitable manner" (UNESCO 2009).

[1] The UN Conference on Sustainable Development (Rio+20) was held in Rio de Janeiro, Brazil, on June 20–22, 2012.

2.1.3 Justification

Attune with the definitions accepted, this essay contends that:

- *The integral nature of the challenges faced by water resources management requires a response that also needs to be holistic in nature,*
  and
- *Such response should consist of sound management—a term which clearly denotes the fact that vital goods like water resources, rather than being rightfully 'owned' by a given generation—in the classic legal meaning of discretional use -, enjoyment and arbitrary disposition powers on one's property—should be sustainably administered, and "legated" unimpaired and, if possible, enhanced, to future generations.*

Such opinion is tacitly recognized in and supported by the IWRM Dublin Principle N° 1, in that it proclaims: "Freshwater is a finite and vulnerable resource essential to sustain life, development, and the environment"; one that requires a minimum level of institutional capacity for carrying out "IWRM-inspired principles" (GWP 2012).

Important to note is the fact that IWRM "is not a scientific theory that needs to be proved or disproved by scholars. Rather, it is a set of common sense suggestions as to what makes up important management aspects" (Hassing et al. 2009). Stemming from this caveat is also the fact that the above IWRM definition is an "open-end" statement—one which can be indefinitely enriched by further common sense contributions like those we suggest in the next section of this paper. One should therefore agree with the contention that IWRM "is not just another Water Plan" (Lenton 2005a, b).

2.1.4 IWRM in Latin America and the Caribbean

This section introduces the main IWRM-inspired or IWRM-consistent enabling instruments as enshrined or ratified by the LAC region's States. Different criteria have been proposed in the literature to facilitate useful comparisons and assessments of the effective implementation of IWRM (UNEP 2002, 2010). In making such comparisons among countries, it is important to define the national level of capacity to adopt the IWRM concept, among other criteria. Some methodologies being applied (e.g., Hassing et al 2009) relate to a developed numerical scale for the weight of the IWRM principles at national levels. The following criteria have been used for this type of approach as comparative indicators of IWRM success:

- Participation of stakeholders in water management
- River basin management approach/plan implementation
- Existence of a water law
- Finance contribution by users to water management

- Polluter pays principle enacted
- User pays principle enacted
- Role of women in water management
- Separation of water management and service provision
- Water-use efficiency
- Private sector involvement

Other valuation criteria that could be complementary to those cited above, as based on the "common sense" standard mentioned in the previous section, are, namely:

- Contribution to economic growth implicit in the IWRM scheme
- Ease to obtain water resources
- Ease to reduce the pressure for water resources (as caused, for example, by consumption of commodities with larger water footprints)
- Socioeconomic equity, including recognition of indigenous peoples' ancestral water rights and customary water laws, and conflict mediation venues and procedures
- Facilitation of international cooperation
- Opening of information channels among stakeholders
- Availability of IWRM monitoring tools
- Sensible and fair tradeoffs among various objectives
- Country "ownership" of the IWRM action sets

Among the most recent efforts aimed to define the success or sustainability of water resources management schemes and processes, it is worth mentioning the work being implemented by the Florida International University (FIU) Global Water for Sustainability (GLOWS) program. As part of the execution of the agreement between FIU and the US Agency for International Development (USAID) E3 Water Office, since early 2012, the GLOWS program is engaged in an initiative to develop and apply measurement frameworks to gauge the sustainability of both water-related services and water resources management programs in the context of its own and other's efforts. As part of this initiative, GLOWS is working to create and test a measurement system to track core elements contributing to "sustainability" of water resources and their management. In this context, the objective of the new Water Resources Management Sustainability Index Tool (WRM SIT) being developed is to monitor core elements contributing to sustainable management of water resources in support of USAID's own sector programming as well as that of the development community at large. The WRM SIT is based on a "pillar" conceptual model, where indicators are organized under general factor categories of sustainability. The categories and indicators together reflect a systemic understanding and Theory of Change around what drives Water Resources Management sustainability. Five categories of sustainability have been identified that cover the spectrum of factors shown by evidence and experience to be most critical for ensuring sustainable water resources management over time. These categories are referenced to governance, financial, technical, environmental, and socioeconomic factors. The measurement of sustainability is defined through the application of a rigorous methodology and data collection procedure followed by a process of validation with the participation of stakeholders at various levels (GLOWS-FIU 2014).

2.1.5 Adoption of IWRM Processes and Roadmaps

Most Latin America and the Caribbean nations have taken steps towards IWRM process implementation. However, the rate of progress in actual IWRM implementation varies among the Region's countries. Similarly, the IWRM criteria that are considered most relevant to particular countries for attaining successful IWRM vary depending mostly on sociocultural factors, level of economic growth, and the overall enabling environment. The following examples illustrate the above statement and in some cases highlight innovative approaches being proposed or implemented by countries of the Region.

Within *Grenada*, broad stakeholder involvement is recognized as crucial and critical for successful IWRM. The importance of participatory processes as it relates to the implementation of specific actions requiring an integral approach to water resources management is captured in the statement: "achieving the MDGs in the water sector is a shared responsibility involving multiple and mutually dependent stakeholders from various sectoral and institutional backgrounds, such as ministries, public agencies, sub-national authorities and private actors including citizens and not-for-profit organizations. But such actors sometimes have conflicting priorities and interests, which may create obstacles for adopting convergent targets" (Akhmouch 2012).

For *Argentina*, securing Federal Government concurrence is fundamental, given the shared nature of the country's Provincial and Federal authorities management and financial responsibilities for a range of public services, including provision of water and sanitation facilities and infrastructure development. Another interesting aspect related to water management in this country refers to the provision in the Constitution (Constitución de la Nación Argentina 1994) in terms of transboundary waters treaties or agreements. The Federal Government is vested the authority to enter into treaties related to transboundary waters with neighboring countries, but it is for the provinces to define the mechanisms to implement these.

In *Colombia*, fostering water resources sustainable management is highlighted by the country's National Development Plans (NDPs) since its 2006–2010 version. A relevant virtue of this country's model is its emphasis in transparency. Water allocation experience in the San Felipe catchment area is a case in point: As stated by Quiroga (1997), "although all stakeholders are not involved in the allocations [in such project], the transparency of the process allows users to challenge others' claims to abstraction and the allocations before they are finalized. . .." This example also underlines implicitly the importance of having reliable data on which to make informed decisions.

Another interesting case study refers to the small Central American country of *Nicaragua*. This nation took a fundamental step in IWRM by developing an Environmental Action Plan under the General Law on National Waters (2007). The actual progress ratio is, nonetheless, questionable, as contended by the International Food Policy Research Institute, which claims the legal provisions enacted remain *dead letter* (Novo and Garrido 2010). These authors substantiate such contention in the following terms: "The water law and sector regulations have not been harmonized and local environmental plans have not been updated based on the law."

As for *Uruguay*, the advance of IWRM has been supported by the Inter-American Development Bank (IDB) through Loan UR-1076 considered to be an IWRM-contributing operation as it entails: 1. "mainstreaming of climate change adaptation to the IWRM process in Uruguay, through including its principles into the National Plan of IWRM, which is currently being developed"; 2. "reducing vulnerability to drought and flood events and prevention of health problems derived from hydrological conditions by defining climate change impacts affecting the water resources along the country as well as identifying adaptation measures for such impacts"; and 3. "coordinating the policies definition and the water resources management with the rest of sector policies" (Inter-American Development Bank 1998).

In *Bolivia*, an innovative *multi-criteria decision analysis* (MCDA) began to be explored in 2010, as a step towards IWRM, in the Lake Poopo area, one of the country's most underdeveloped regions (Calizaya et al. 2010). This approach involves the use of an Integrated Water Sustainability Index (IWSI) and considers stakeholder participation as well as an institutional arrangement structure (Lund University 2009).

Chile, however, provides an example of progress towards the three "E"s of integrated water resources management (IWRM) referenced by the Global Water Partnership (1999), namely, "economic efficiency," "social equity," and "environmental sustainability." As referred by Williams and Carriger (2006), the country has attempted to create a modus vivendi—and, arguably, a synergy—between economic growth and water sustainability. These scholars contend that "in Chile, development has placed additional pressure on the environment in general, and on water resources in particular. Over the two decades studied by a recent review, the use of wells in agriculture has increased sixfold, the use of wells for drinking water fourfold, and, during the last decade, 40 aquifers have been closed to new concessions. This said, environmental sustainability in Chile has actually improved in recent years. This was largely the result of factors outside the water sector, including economic growth, which has provided the financial and technological resources needed to bring about environmental improvements, and an ideological shift, which resulted in greater attention to social and environmental issues on the part of the Government and Chile's citizens. *Improvements in water-use efficiency* have been considerable, especially in those areas linked to exports. In some cases, these were a side effect of the drive to produce higher-quality products for the international market" (Williams and Carriger 2006). It is important to note that the enabling environment is unique in Chile in comparison to the rest of LAC. Water use in this country is granted by the Direccion General de Aguas (DGA) based on availability of water resources. Productive use is a priority and to an extent mandatory. Water concession recipients who fail to use the allocated volume of water may lose their privilege over the use of water. Exclusions apply when water is used for environmental conservation.

We cannot thus see any two countries where IWRM has been adopted in order to attain the very same objectives. Such feature confirms the assertion quoted in the first section of this chapter that IWRM is ultimately an expression of *common sense*, rather than a dogmatic formula to be rigidly enforced throughout the Region ignoring the unique characteristics and needs of each country. Such ductility adds to the worth of the IWRM concept by encouraging the development of novel legal and enabling formulae in response to supervening local circumstances.

2.2 IWRM Contributions to the MDGs' Process

Poverty-, Hunger-, Primary Education-, Women's *Social Capital-*, Child Mortality-, Diseases-, and Environmental Quality MDGs are patently associated with better IWRM. Such reality was recognized already in April 2005 at the Annual Meeting of the Inter-American Development Bank (IDB) Assembly of Governors and the Inter-American Investment Corporation (ICC, an arm of the IDB Group). The participants of the referred venue arrived to the "global consensus that IWRM is crucial in the quest to achieve the MDGs, not only in water and sanitation but in other areas, which include eradicating extreme poverty and hunger; reducing child mortality; improving maternal health combating major diseases; and improving environmental sustainability" (Inter-American Development Bank 2005).

Although several LAC countries have specifically recognized in their Constitutions and/or regulatory codes their inhabitants' right to a healthy and ecologically sound and sustainable environment, *Costa Rica* is one of the Region's nations where such outcome has been most thoroughly considered and decisively furthered through concrete actions, in particular through the development and implementation of the 2002 Integrated Water Resources Management Strategy (IWRMS). This strategy, for which the Costa Rican Ministry of Environment and Energy, with the technical cooperation of the IDB, holds responsibility, attempts to respond to the following questions:

- What are the critical problems?
- What is the vision of the society and of the Government and how to address these?
- What is the strategy required for the water resources to become the driver of the country's sustainable development?

The Costa Rica IWRM strategy intends "to develop a proper Country Concept and find the way to reach the rationality and sustainable water management, to contribute to achieve the Country development goals and the international commitments" (Villalta Fernández 2005) and is explicitly linked with the country's water policy, its current development plan, and a future version thereof. In turn, the strategy addresses a number of issues critical for the successful implementation of IWRM, including data availability, coordination, decentralization, IWRM legal, financial and institutional structures, and a whole range of water quality issues. This universe of interconnected elements, and its explicit quest for modernization, makes the Costa Rican case an—so far unique—interesting paradigm of the application of IWRM in the Region.

2.3 In-Country Enabling Environment for Centralization or Decentralization of WRM

This specific area of analysis responds to the need to avoid contradictory water resources management policies at national/state/municipal levels in different countries with particular government systems. We will be presenting a sample of country cases where the dialectics and concurring jurisdictions among such management levels are apparent.

Argentina chose to address the issue of conservation and management of natural resources and thus of water resources management in particular, at a constitutional level. In its 1994 Constitution (Section 41), Argentina adopted two convergent mechanisms, mandating that "The Nation shall regulate the minimum protection standards, and the provinces those necessary to reinforce them, without altering their local jurisdictions," that is, assigning the Federal Government the main coordinating responsibilities and counterbalancing such authority by prohibiting the central authorities to impair the provinces' jurisdictions over the resources in their respective geographical boundaries.

Mexico created the Water National Commission (Comision Nacional de Aguas) CONAGUA, an autonomous branch of the Secretary of Environment and Natural Resources (SEMARNAT, for its Spanish acronym). In spite of its affiliation to such particular entity, SEMARNAT, CONAGUA has global coordination and managerial powers for all of the public agencies in the country (those having national, state, or municipal jurisdictions), and its technical council includes representatives of a range of top governmental bodies.

Brazil, a country where water resources responsibilities are particularly fragmented, created an entity—the National Water Agency (ANA, for its Portuguese acronym)—technically ascribed to the Ministry of the Environment. This institution "plays a number of management and co-ordination roles, and consists of ten functional superintendences with implementing and administrative functions. Providing a managerial structure, an authority and the means to implement and co-ordinate the National Water Law, ANA has brought a general improvement of water resources in Brazil" (Akhmouch 2012).

A similar mechanism was set up in *Peru* in 2008 through the National Water Authority (ANA, for its Spanish acronym), with the explicit mission to "recognize and assure the economic, social, and environmental values of water and to involve all levels of government and the civil society." Peru is one of the countries where international aid with a transnational focus has made a contribution towards IWRM. The Global Water for Sustainability (GLOWS) Program, led by Florida International University (FIU), implemented a paradigmatic IWRM project for the Peruvian-Ecuadorian Pastaza river basin (GLOWS-FIU 2010; Nissim 2012). Community-based fisheries management, formation of permanent participatory structures, reducing water contamination in the basin's oil-producing area, and transboundary lesson-learning efforts were some of the major components of this GLOWS IWRM effort.

A different special case is that of *Cuba*, where original centralization of water and sanitation services provision was attenuated through the establishment of provincial delegations under the umbrella of the National Institute for Water Resources in pursuance of a series of financial, data gathering, registration, and environmental objectives (García Fernández 2006). These provincial entities are to carry specific IWRM duties, the implementation of which is hampered by economic constraints.

Also, *Nicaragua* is a particular case, in that water resources matters coordination takes place within the smallest unitary local government entity—rather than at federal level—as a response to the need created by the large number of municipalities (151 of them). Similar considerations apply to *Panama* and the *Dominican Republic*. *Guatemala*, particularly, stands out in terms of the governmental top level of the body created to perform IWRM functions, namely, the Gabinete Específico del Agua (GEA) (Specific Water Cabinet), established in August 2008. The GEA is chaired by the Vice President of the Republic and comprises ten Ministries and seven State Secretariats.

Mention should also be made to the fact that interagency and/or inter-sectorial planning coordination is supplemented, in some LAC countries, by coordination within particular areas of Government and within/among economic sectors. Remarkable, in such respect, is the Permanent Forum on Agricultural Development, in Brazil.

Finally, an overview of the distribution of responsibilities among water policy-making agencies in LAC, as presented by Akhmouch (2012), shows the diversities in numbers of authorities involved in both *policy making* and *regulatory powers* at central government level as well as an even more pointed disparity at subnational level. The number of line ministries, departments, public agencies, and other entities operating at the national level charged with regulatory tasks, for instance, oscillates from 4 to 13. In terms of the enabling environment at the water sector subnational level, the countries of the Region widely differ in a spectrum of criteria, including:

- Governance system (unitary, federal, or quasi-federal systems)
- Type of involvement (central "dominant" role, joint role with Central Government, no competence)

- Water resources management schemes in place
- Nature of water supply (domestic) provision (private, governmental, communal)
- Budget for water resources sector development and conservation

Take for instance *Honduras*, where municipalities, inter-municipal bodies, and water-specific bodies are responsible for water resources management and water supply, or Mexico, where four different policy-making instances are involved.

Finally, in assessing the scope of IWRM in LAC, it is of interest to denote the important advisory role of international organizations in the spheres of coordination and harmonization in areas such as data gathering and scholarly work, facilitation of peaceful solution to conflicts, including water-related conflicts, international law development, especially in the context of treaty development, and sharing best practices, among others.

2.4 Federal Versus Unitary Government Systems Harmonization

IWRM description and assessment efforts would not be complete without considering the constitutional structure of the particular country being considered. Such factor is a determinant in matters such as central versus local power sharing (with the financial authority balance in water resources often making a difference between IWRM success and failure). Moreover, the matter is made even more complex by the fact that sometimes the real-world nature of a particular state as federal or unitary cannot be determined by the letter of constitutional provisions, with political realities prevailing. A case in point is that of *Mexico*, as highlighted by Feldman (2014), who contrasts (a) the theoretical nature of the country as a federal state from its inception as an independent entity; (b) the long-lasting marked prevalence of the Central Government in spite of the constituents' will; and (c) the contemporary evolution of the country, whereby the federal nature of Mexico's Government is becoming more akin to its constitutional tenets, because of a shift in the balance of *political* power, rather than as a result of abstract *legal* considerations.

The impact on IWRM of the Federal/Unitary configuration of the Government is examined in more detail in another chapter of this book by comparing the cases of Argentina and Uruguay and identifying the areas where the constitutional arrangements of either country explain the differences in terms of water-related human rights, environmental protection and sustainability, etc. Particular attention is paid to the legal relevance of international conventions—an area where the Constitutions of Uruguay and Argentina differ (Bosch and Donoso 2015).

2.5 Public Participation in IWRM in LAC

The discussion of this issue might well be started by quoting the following proposition in a document on this matter of the Inter-American Development Bank:

> Public participation in decision-making and due regard for the environment and the economic, political and social realities of each country will be a common denominator. (García 1998)

Such statement, entirely consistent with Dublin Principle II, however, appears to contrast with the findings presented in a publication sponsored by the Organisation for Economic Co-operation and Development (OECD), which characterizes the LAC water resources scenario as one of "accountability gap," with "absence of monitoring and evaluation of outcomes" and "limited citizen participation" (OECD 2011). Akhmouch (2012) followed the methodology introduced in the OECD (2011) survey which covered 17 countries, two of which were from the LAC region, namely, Chile and Mexico. This author's stern observations from his own survey of 13 LAC countries deserve being quoted in length:

> **The accountability gap is . . . considered an important obstacle to** inclusive water policy in more than 90 % of LAC countries surveyed. Generally, the main issues relate to a lack of public concern and low involvement of water users' associations in policy making. Indeed, limited citizens' participation was pointed out as an important gap in more than two-thirds of countries surveyed. But challenges related to the evaluation of water policies at central and sub-national level are also crucial to approach the accountability gap. Inadequate monitoring, reporting, sharing and dissemination of water policy performance also prevent policy coherence at horizontal and vertical levels. Periodic assessment of progress toward established policy goals is vital for understanding whether the applied efforts are effective and for adjusting policy where necessary. But feasibility is often limited due to considerations of political, financial and capacity, and this complicates the implementation of central government decisions at the subnational level. In the absence of monitoring and evaluation of water policy outcomes were considered important obstacles to water policy implementation at the territorial level in almost all LAC countries surveyed (11 out of 13).

The above described scenario for the LAC region can be presented as a convergence of several negative factors, where inadequate assessment of prevailing IWRM situation and tendencies, the deficit of the institutional capacity, limited budgetary and financial assets, and other factors important to the dimensions of IWRM is compounded by the convergent limited public eagerness to be informed and actively participate in IWRM, no matter how vital water resources are in a transgenerational human perspective, achievement of development goals, and enjoyment of fundamental human rights.

The share of responsibility of government and civil society structures seems to be confined, under the above analysis, to (a) feeble or nonexistent ongoing monitoring and evaluation of results of IWRM processes/schemes, (b) inadequate educational efforts aimed at actively encouraging people to participate in water management processes, and (c) inadequate distribution of IWRM responsibilities among different levels of government, a dimension that can also be termed as *faulty interfacing*. Added to the above is the need, as discussed in another chapter of this work, to keep in mind the complex nature of water management, where matters such as riparian and coastal waters and ecological and climatic problems are acute.

Still, other considerations can be inferred, or at least surmised, from the above strikingly strong quotation. One is the "risk," for political actors, to encourage active and ongoing public participation and defeat the (improper) rulers' dislike for democratic control. Finally, the problem could also be considered an aspect of larger cultural dimensions, which is the subject matter of the following section of this paper.

2.6 The IWRM's Cultural Dimension in LAC

As we have just contended, the immediate causes of the public participation deficit in the region of Latin America and the Caribbean can be traced to a wider and more convoluted root issue—one worth considering given the fact that academic consensus recognizes IWRM success as contingent to vibrant public participation.

Participation inadequacies are noticeable in the cultural-wide regional landscape. Although an attempt to detect its root causes would bring us beyond the down-to-earth approach of this paper, in that they would require an in-depth examination of a complex historical background where historians are far from having come to a consensus. However, one needs to admit that water resources-related stakeholder groups, to begin with, are clearly distinct and disparate and their mutual relations could often be contentious. Except in those particular cases where obvious "economic-only" competitions are involved, the LAC region, similar to other parts of the world, bears the weight of centuries-long cultural differences (and sometimes enmities) with water-related disputes being but one of the components. Language pluralism, although not a major issue in most of the Region, could also in some specific cases be a cultural divide. Bolivia, having been officially renamed a *Plurinational State*, is only too eloquent a fact in such order of ideas.

These brief remarks, added to the existence of border disputes that remain unresolved several centuries after their birth (the failure of the Tordesillas Treaty being a case in point), compounded by intermittent armed conflicts and political feuds involving a specific number of Latin American countries, seem to support the thesis that economic factors and/or mismanagement add to deeper cultural issues. The brighter point of this panorama, in our opinion, is that increasing positive efforts aimed to mitigate cultural and socioeconomic differences would certainly contribute to assuaging the faulty participation deficit.

However, despite the cultural diversity present in LAC, it does not constitute a major element hampering the success of IWRM in the Region. Contrary to other regions of the world, there are more similarities than differences among countries in Latin America or island states in the Caribbean. The fact that there are only two official languages (Spanish and Portuguese) common to more than 95 % of the Latin American countries and one language (English) common to more than 80 % of the Caribbean countries largely facilitates the interaction and dialogue within and among countries in relation to water resources management issues.

2.7 IWRM Gender Connotations in LAC: Towards a "New Attitude"

Although gender matters cannot be isolated from cultural ones, describing them as cultural in nature, as sometimes contended, seems questionable, thus deserving an independent section in this paper. In fact, one can claim the role of women in water management matters is supracultural. Aureli and Brelet (2004) describe the subject of "Women and Water" as "an ethical issue." They persuasively demonstrate a range of matters of fundamental human behavior interest which are tightly connected with—and in part dependent upon—the role of women in the area of water resources, namely, environmental sustainability vs. degradation; health (contingent as it is to fresh water availability); women access to formal education versus excessively time-consuming water-securing efforts; social waste of women value as a source of environmental knowledge; gross inequities in work retribution; women's increased exposure to violence; as well as a poverty-perpetuating factor in poor countries, regions, or communities.

May we add that like all rights, those of women in the water realm have the counterpart of other parties' duties, with *ethics* being the link between both elements. This means that social tolerance for unethical behavior against women, including in water-related matters, cannot but erode the moral stamina of the society at large.

From the purely economic standpoint, reduction of the time spent by women in securing water supply is relevant in having their energies available for other roles as domestic food producers or providers to the family economy. Reduced health and caregiving burdens brought about by improved water services give women more time for productive endeavors, adult education, empowerment activities, and leisure. The World Health Organization, in turn, has highlighted the social relevance of women health as it relates to the wellbeing of their offsprings and the future generations, in general.

In Latin America, as a whole, the role of women in decision-making positions in society is rapidly increasing. Although, still a majority of high level positions are under the responsibility of men, in most LAC countries, women are rapidly advancing in more permanent medium level positions. This also holds in the water sector, both for private and public entities. In particular, in the water sciences and academic sector, a wave of young professionals are competing and successfully engaging in key positions such as deans, department heads, program directors, etc. This trend was already detected by Donoso et al. (1998) in her analysis of the participation of women in water-related positions in academia and research as well as in the overall water sector in LAC. Notwithstanding this positive regional trend of inclusion of women in the IWRM process, there is still much to do, in particular in guaranteeing the participation of the most poor and vulnerable (less educated) female stakeholders of the basins.

2.8 Concluding Remarks

IWRM has a relevant role to perform in Latin America and the Caribbean, given the Region's vast, but not unlimited, water resources; the nature of its economic development model, where agriculture will continue to be "an actor of increasing importance in the global food supply"; and its "enormous potential to contribute to its own and global food security and sustainable and equitable development" (FAO 2010). The later statement is endorsed by the undisputable fact of the sheer reality that with less than 10 % of the total world population (population close to 600 million people), LAC has 23 % of the world's arable lands, 31 % of its water resources, 23 % of its forests, and 46 % of its tropical forests (FAO 2010).

In parallel, safe water (and sanitation) services are basic to attain sustainable development, and such accepted ethic and legal tenet should be the point of departure for any attempt to implement a fair and mutually accepted framework to manage water resources within each country and regionally. Such efforts are also needed—to a lesser but growing extent prompted by globalization—in the international arena. This is particularly true in terms of water resources, where issues (e.g., pollution, overexploitation, and others) are transboundary in nature.

The Region's territorial vastness and its nations' disparities in areas such as legal system backgrounds—civil law vs. common law traditions; consuetudinary law and customary procedures and traditional courts vs. written codes; constitutional frames (unitary vs. federal structures); centralization vs. decentralization models covering a range of choices mostly dictated by unique historic and cultural backdrops; as well as the various development levels of the Region's economies—require a measure of coordination and harmonization, in reference to the use and conservation on natural resources, which can be attained through integrated management. In terms of water resources, such optional coordination schemes are referred to as IWRM.

The "I" (integrated) element in the IWRM concept is also valuable in that it fosters an integral vision and integral solutions to problems that are intrinsically interconnected, while the notion that *management*, rather ownership, is the key for adequate water resources regimes of use and conservation. Limited powers and transgenerational obligations such as those exerted by administrators are the reverse of classic *jus abutendi* theoretically enjoyed by the typical individual property owner.

Another feature of IWRM is its proclaimed nature of *common sense* product. This attribute is relevant in that malleable as it is by definition, IWRM allows each LAC country to choose the management instruments that best fit its current needs as adapted to its unique circumstances, fine-tuning them as new factors step in. In addition, IWRM *common sense* quality should help mitigate a widespread error in the approach of developing countries to their development problem: the illusion that problems get resolved by merely reforming the laws based on abstract theories.

In view of the above, the crucial challenge, for the countries of the Region, is to closely monitor the ongoing consistency between the IWRM legal texts adopted and the national water resources needs in evolution. Worth considering is the fact that keen efforts need to be devoted to operationalization of the adopted model (Biswas 2004).

Despite the advance of IWRM in LAC compared to other regions, efforts are still required to improve key elements such as public participation and gender equality in the framework of cultural diversity and sustainable economic growth. As we shape the millennium challenges for the LAC region, full engagement of the countries in appropriate IWRM schemes is expected to be a key factor in attaining a sustainable future.

References

Agarwal A, de los Ángeles MS, Bhatia R et al (1999) Integrated water resources management. GWP TAC background paper, Sweden, p 71. ISSN: 1403-5324

Akhmouch A (2012) Water governance in Latin America and the Caribbean: a multi-level approach. OECD regional development working papers, 2012/04, OECD Publishing, Paris. http://dx.doi.org/10.1787/5k9crzqk3ttj-en

Aureli A, Brelet C (2004) Women and water: an ethical issue. Essay no. 4, water and ethics series. UNESCO International Hydrological Programme and World Commission on the Ethics of Scientific Knowledge and Technology, Paris

Biswas A (2004) Integrated water resources management: a reassessment. Water Int 29(2):248–256

Bosch MC, Donoso MC (2015) Environmental provisions in the constitutions of Uruguay and Argentina affecting water resource management. In: Setegn S, Donoso M (eds) Sustainability of integrated water resources management (IWRM) in the face of climate variability and change. Springer, Germany (in press)

Calizaya A, Meixner O, Berndtsson L et al (2010) Multi-criteria decision analysis (MCDA) for integrated water resources management (IWRM) in the Lake Poopo Basin, Bolivia. Lund University, Lund, Sweden. ISBN 0920-4741

Constitución de la Nación Argentina (1994) http://leyes-ar.com/constitucion_nacional.htm

Donoso MC, Bakkum A, Troetsch M (1998) Woman and water in humid tropics. In: Tortajada C (ed) Women and water management. Oxford India Publication, New Delhi, p 231

Feldman D (2014) Constitutional law. Enclopædia Britannica. http://www.britannica.com/EBchecked/topic/134322/constitutional-law/22076/Classifying-states-as-federal-or-unitary

FAO (2010) Food and Agriculture Organization working paper LAC/03/10. FAO, Rome

García L (1998) Technical study on IWRM for the Inter-American Development Bank IWRM strategy. http://www.pnuma.org/agua-miaac/PERIODISTAS%20BOLIVIA/MATERIAL%20ADICIONAL/BIBLIOGRAFIA-WEBGRAFIA/Agua/Integrate%20water.pdf

García Fernández JM (2006) Experiencias cubanas en la institucionalización del manejo integrado de cuencas. Voluntad Hidráulica 98:15–28

Global Water Partnership (1999) The Dublin principles for water as reflected in a comparative assessment of institutional and legal arrangements for IWRM. Background paper no 3. http://www.gwp.org/en/The-Challenge/What-is-IWRM/IWRM-Principles/

Global Water Partnership (2012) IWRM principles, 2012. http://www.gwp.org/en/The-Challenge/What-is-IWRM/IWRM-Principles/

GLOWS-FIU (2010) Integrated water resources management in the Pastaza River Basin, Ecuador & Peru. Global Water for Sustainability Program – Florida International University. Program Report, Miami

GLOWS-FIU (2014) Water resources management sustainability index tool (WRM SIT): Volume I – Manual and Volume II – Implementation resources. Global Water for Sustainability Program, Florida international University – documents under revision

Hassing J, Ipsen N, Jønch Clausen T et al (2009) Water in a changing world. Integrated water resources management (IWRM) in action. UNESCO and UN world water assessment programme. Publication series. World water assessment paper no. 3. UNESCO, Paris, France. ISBN 978-92-3-104114-3

Inter-American Development Bank (1998) Strategy for integrated water resources management. Document GN-1908-4 as approved by the IADB Board of Directors on May 27, 1998 and published in December 1998 (No. ENV-125)
Inter-American Development Bank (2005) Water and the Millennium Development Goals investments needs in Latin America and the Caribbean, Washington, DC. http://idbdocs.iadb.org/wsdocs/getdocument.aspx?docnum=959175
Lenton R (2005a) Water resources management and the Millennium Development Goals. UN millennium project task force on water and sanitation technical committee, Global Water Partnership, Okinawa. http://idbdocs.iadb.org/wsdocs/getdocument.aspx?docnum=973151
Lenton R (2005b) Water and the Millennium Development Goals: its contribution to development. Paper submitted at Panel 1, Inter-American Development Bank Seminar, Okinawa. http://www.iadb.org/en/topics/water-sanitation/water-and-the-millennium-development-goals-its-contribution-to-development,1539.html
Ley General de Aguas Nacionales – Nicaragua (2007) http://legislacion.asamblea.gob.ni/Normaweb.nsf/%28%24All%29/C0C1931F74480A55062573760075BD4B
Lund University (2009) Water resources management efforts for best water allocation in the Lake Poopo basin, Bolivia. Lund University Publications. https://lup.lub.lu.se/search/publication/1487891
Nissim R (2012) Global water for sustainability: delivering clean water around the world. FIU Magazine, August, Miami
Novo P, Garrido A (2010) The new Nicaraguan water law in context: institutions and challenges for water management and governance. IFPRI discussion paper 01005, International Food Policy Research Institute. http://www.ifpri.org/sites/default/files/publications/ifpridp01005.pdf
Organization for Cooperation and Economic Development (OECD) (2011) Water governance in OECD countries: a multi-level approach – highlights. OECD Publishing, Paris. ISBN 9789264119277. http://www.oecd.org/governance/regional-policy/48885867.pdf
Quiroga ER (1997) Rural water distribution and management: the case of San Felipe water supply utility. Universidad del Valle, Institute Cinara, Cali, Colombia (unpublished)
UN Millennium Project (2005) Investing in development: a practical plan to achieve the Millennium Development Goals. Interview. UNDP
UNESCO (2009) IWRM guidelines at River Basin level, Part I, Principles. UNESCO-IHP, Paris
United Nations (2014) Introduction to the proposal of the Open Working Group for sustainable development goals. 13th Session of the OWG, July 19, 2014. http://sustainabledevelopment.un.org/?menu=1300
United Nations Environment Programme (2002) Economic instruments for environmental protection. Division of Technology, Industry & Economics, Economics & Trade Branch, UNEP briefs on economics, trade and sustainable development, information and policy tools from the United Nations Environment Programme
United Nations Environment Programme (2010) Integrated Water Resources Management Programme, Key features, lessons learned and recommendations from implementation of the IWRM 2005 Target. UNEP Collaborating Centre on Water and Environment. http://www.unepdhi.org/~/media/Microsite_UNEPDHI/Publications/documents/unep_DHI/UNEP-DHI_lessons_learned.ashx
Villalta Fernández RA (2005) Water resources management contribution for the development and accomplishment of the Millennium's Development Goals (MDGs). Costa Rican Institute of Aqueducts and Sewage Systems, Document presented at the Inter-American Development Bank Seminar, Okinawa, Japan, April 6. http://idbdocs.iadb.org/wsdocs/getdocument.aspx?docnum=973143
Williams S, Carriger S (2006) Water and sustainable development: lessons from Chile. Technical committee policy brief no. 2. http://www.gwp.org/Global/GWP-CACENA_Files/en/pdf/policy_brief_2_chile.pdf

Chapter 3
Integrated Water Resources Management: African Perspectives

Abou Amani, Robert Dessouassi, and Adwoa Paintsil

Abstract The importance of water resources for sustainable development for African countries has been recognised at the highest level of the African Heads of States through the Sharm el-Sheikh declarations calling for the equitable and sustainable use, and promotion of integrated management and development, of national and shared water resources in Africa. Integrated Water Resources Management (IWRM) was adopted since the adoption of the Africa Water Vision 2025 and the Johannesburg Declaration on Sustainable Development (2002), which called for the preparation and implementation of IWRM plans. The crucial role of water resources for socio-economic development in Africa led to the creation of African Ministers' Council on Water (AMCOW) as the African voice and leadership on water and sanitation in Africa with the key objective of mobilising countries, regional institutions and partners to address in a sustainable and coherence way the water challenges in Africa.

The shared nature of water resources in Africa with more than 80 shared river basins and lakes and at least 60 transboundary aquifers systems led to the establishment of various entities at national, subregional and regional levels for the promotion of sustainable water resources management through IWRM. This paper presents an overview of the status of IWRM in Africa based on the different published documents including institutional, knowledge base and capacity building aspects and includes the main characteristics of water resources and related challenges. The paper calls for the implementation of a true IWRM in Africa considering climate change, integrating surface and groundwater, cultural and water quality issues.

The paper presents also a summary of two case studies on the implementation of IWRM. The first case study is on a large transboundary basin, the Niger basin

A. Amani (✉)
UNESCO, Nairobi Regional Office for Eastern Africa, Nairobi, Kenya
e-mail: a.amani@unesco.org

R. Dessouassi
Niger Basin Authority Executive Secretariat, Niamey, Niger
e-mail: Dessouassi2003@yahoo.fr

A. Paintsil
Water Resources Commission, Accra, Ghana
e-mail: himapaintsil@yahoo.com

S.G. Setegn, M.C. Donoso (eds.), *Sustainability of Integrated Water Resources Management*, DOI 10.1007/978-3-319-12194-9_3

shared by nine countries, which have adopted a shared vision through a participatory approach and the preparation of a long-term Investment Development Plan for the basin, based on a comprehensive analysis of the challenges facing the basin. The second case study is on the Densu basin, a national basin in Ghana and one of the most stressed in the country, where pollution is one of the key challenges. The application of the IWRM concept has increased awareness and led to the mobilisation of key stakeholders to contribute to the reduction in the level of pollution of the basin.

Keywords African Water Vision • Shared water resources • Climate change • IWRM stakeholders in Africa • Water quality • Flooding • Case studies Niger • Densu and Mono basins

3.1 Introduction

Water is a basic need for all humans, but circumstances limiting its availability, access and quality are perpetuated by various social, cultural aspects and policies. The African continent is endowed with abundant water resources, but these are often not mobilised to meet the needs of its peoples. Water has been recognised at the highest level of the Heads of States as a crucial resource for the sustainable development in Africa. Water resource cut across all the Millennium Development Goals (MDGs) and as highlighted by the United Nations Secretary General (UNSG) advisory board on water and sanitation, water accounts for around one third to the achievement of MDGs. Water resources in Africa have many challenges. The main challenge of African countries is to provide safe drinking water to their population. Unfortunately about 344 million of people (35 %) do not have access to safe drinking water, and about 584 million of people (59 %) do not have access to improved sanitation (UNEP 2010). Water is life, but water can also be a source of death. The lack of access to clean water and good sanitation has led to high prevalence of water-related and waterborne diseases affecting children with high rate of mortality for children under five in sub-Saharan Africa. Following the adoption by the UN General Assembly of a resolution elevating access to drinking water and sanitation as human right, African countries should revisit their policies.

Since the International Environment Conference in Dublin in 1992, the science of Integrated Water Resources Management (IWRM) has been accepted by the international community and many countries as the best approach to sustainable management of water resources. The Global Water Partnership (GWP) (2000) defines IWRM as "a process which promotes the coordinated development and management of water, land and related resources, in order to maximize the resultant economic and social welfare in an equitable manner without compromising the sustainability of vital ecosystems".

IWRM is a comprehensive, participatory planning and implementation tool for managing and developing water resources, ensuring the protection of ecosystems for future generations. Efforts to achieve the MDGs must involve planning and action in water resources development, management and use.

Better management and development of water resources through IWRM approach has been recognised in 2002 during the Johannesburg World Summit on Sustainable Development with the summit urged all countries "to develop IWRM and water efficiency plans by 2005".

The African countries need to pursue and adapt in line with the Africa Water Vision 2025 (UNECA 2000) more sustainable approaches to the use and management of water resources through a true Integrated Water Resources Management approach where water quantity and quality, surface and groundwater, climate change and cultural dimension are fully integrated.

In Africa, the concept and the application of IWRM have been promoted through various programmes on capacity building and the establishment of GWP components as non-governmental organisation at subregional and national levels. The paper presents an overview of IWRM in Africa with a focus on two case studies/examples among many which need to be shared worldwide. It includes a brief on water resources in Africa, their characteristics and related challenges and how important IWRM is for a sustainable management of water resources in Africa. The situation of IWRM in Africa including the key stakeholders and players is followed by a call for a true IWRM approach in Africa.

3.2 Overview of Africa Water Resources

The African continent is endowed with abundant water resources with an estimate of 3,931 km^3 as total renewable water resources representing 9 % of the world's total renewable water resources (UNEP 2010). This places the continent as the second driest continent due to the important arid and semiarid areas covering around half of the continent. In Africa, the total renewable water resources by country and per capita are extremely variable. For the whole continent, the available water resource per capita is estimated at about 4,000 m^3 in 2010, and at country scale, it goes from more than 10,000 m^3/hab/year for some humid countries to less than 500 m^3/hab/year for some arid and semiarid countries (UNEP 2010). The geographical landscape, the climate dimension and the population distribution should be considered when analysing figures on the African countries renewable water resources.

Water resources in Africa are mainly concentrated within large river basins such as the Congo, Nile, the Niger, Senegal, Zambezi and Volta basins. The water resources comprise various rivers, perennial and non-perennial, wetlands, lakes, dams, swamp and aquifers. In Africa, the majority of water resources both surface and groundwater are shared among countries with 80 international transboundary rivers/lakes systems and at least 60 international shared aquifers systems. Figures 3.1 and 3.2 present, respectively, the transboundary river basins and shared aquifers systems in Africa.

Fig. 3.1 River and lake basins in Africa

The key water challenges identified within the Africa Water Vision 2025 and the recent publication on Africa Water Atlas by UNEP are still relevant. The ten key water challenges from the Africa Water Vision 2025:

- Ensuring that all have sustainable access to safe and adequate water supply and sanitation services to meet basic needs
- Ensuring that water does not become the limiting factor in food and energy security
- Ensuring that water for sustaining the environment and life-supporting ecosystems is adequate in quantity and quality
- Reforming water resources institutions to establish good governance and an enabling environment for sustainable management of national and transboundary water basins and for securing regional cooperation on water quantity and water quality issues
- Securing and retaining skilled and motivated water professionals

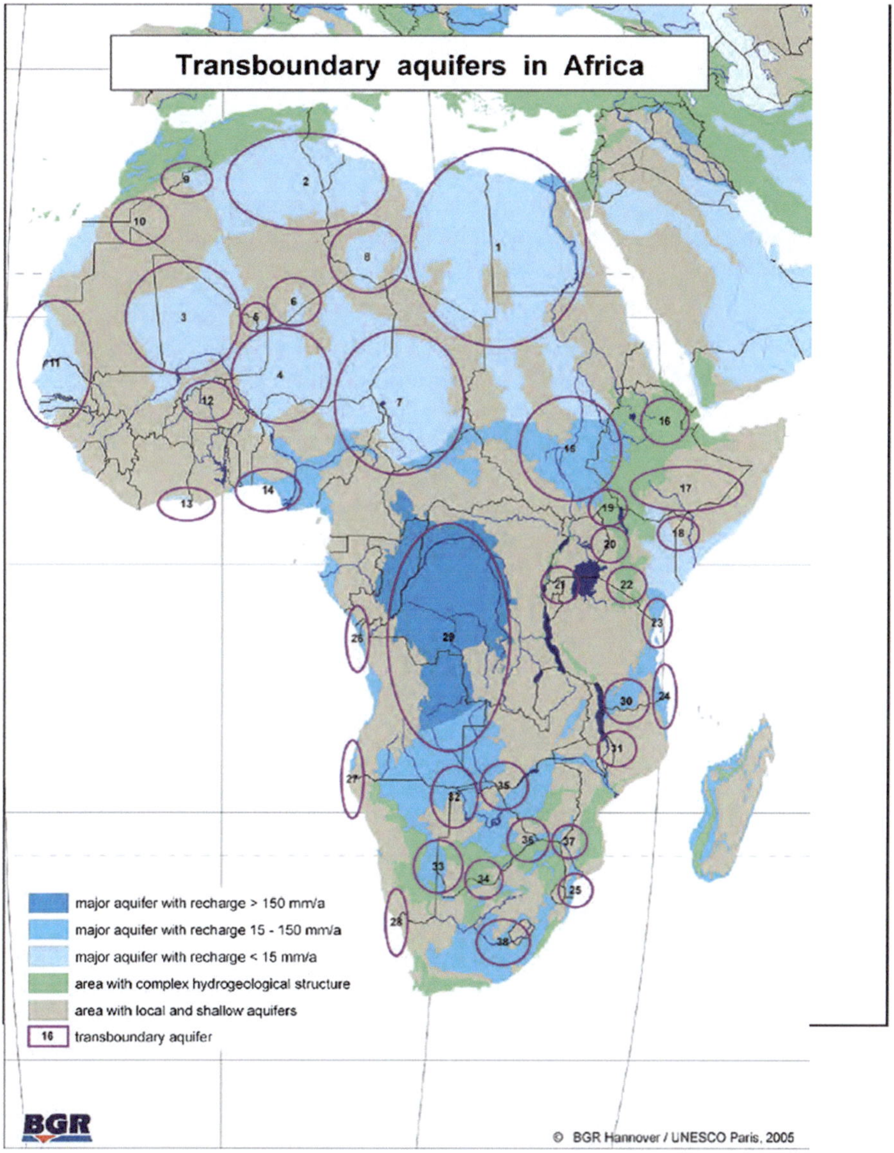

Fig. 3.2 Transboundary aquifers in Africa in 2005

- Developing effective systems and capacity for research and development in water and for the collection, assessment and dissemination of data and information on water resources
- Developing effective and reliable strategies for coping with climate variability and change, growing water scarcity and the disappearance of water bodies

- Reversing growing man-made water quantity and quality problems, such as overexploitation of renewable and non-renewable water resources, and the pollution and degradation of watersheds and ecosystems
- Achieving sustainable financing for investments in water supply, sanitation, irrigation, hydropower and other uses and for the development, protection and restoration of national and transboundary water resources
- Mobilising political will, creating awareness and securing commitment among all with regard to water issues, including appropriate gender and youth involvement

The African Water Vision 2025 calls for "an Africa where there is an adequate and sustainable and management of water resources for poverty alleviation, socio-economic development, regional cooperation, and the environment". The achievement of the vision should require a holistic approach and a clear understanding of the strategic role of water for development and the necessity for its sustainable management for addressing increasing needs. The decision makers who are in charge of putting in place policies on water resources and taking decisions which have a direct or indirect impact on water resources must be aware and be informed on the key challenges facing water management in Africa and the different options for a sustainable management of the water resources of the continent.

3.3 Importance of IWRM for Africa

Sub-Saharan Africa is facing increasingly critical challenges in using and managing its water resources in a sustainable way and for its sustainable development. In addition to environmental threats including climate change, water resources in Africa are threatened by the growing population, rapid urbanisation and economic development and pollution. For a better management of water resources, it is crucial for all the stakeholders using the resources of the same basin to come together and have a common understanding of the challenges facing the basin in order to collectively agree on the best way to the management of the basin. Despite the international recognition of the Integrated Water Resources Management approach as the best approach for a sustainable management of water resources, the main reasons for adopting IWRM in Africa for the achievement of the Africa Water Vision 2025 and beyond are obvious and include among others:

- The fact that the majority of water resources, both surface and groundwater, are shared among countries which necessitate cooperation and participation
- The increasing pressure on water resources to satisfy the different needs for a growing population and socio-economic development which requires a holistic approach
- The threat of climate variability and change with Africa as the most vulnerable leading to a more reinforced hydrological cycle with increasing extremes both drought and floods

3.3.1 Shared Water Systems

In Africa, the international transboundary river basin systems (Fig. 3.1) cover 63 % of the continent area and represent almost 93 % of the African total surface water for 77 % of Africa's population. Africa comprises 17 large international transboundary river basins having more than 100,000 km^2 representing one third of the total world large basins. Due to the geographical landscape, river basin systems don't obey to countries international boundaries which lead to the interdependency of some countries with water flowing from countries (upstream) to countries (downstream). The water interdependency of some countries is around or greater than 90 % (case of Niger, Mauritania, etc.), meaning that only 10 % of their total surface water resources are produced inside their countries, and 90 % are coming from outside their international borders. These countries cannot have any sustainable and peaceful management of their water resources without cooperating with countries upstream and downstream (NEPAD 2005).

Regarding groundwater which represents in Africa for more than 75 % as source for water supply for rural and urban areas and particularly in arid and semiarid areas, the exact number of international aquifers shared among countries is not known as it is the case for river basin systems. The main reason is that knowledge base and quality hydrological data are very limited and fragmented. At the beginning of the UNESCO Internationally Shared Aquifer Resources Management (ISARM) programme, 38 international aquifer systems have been identified (UNESCO 2005). By putting in place subregional working groups where in each group experts from the different countries were brought together to share and analyse their national available hydrological data, the programme was able to identify now at least 60 international transboundary aquifers in Africa (Fig. 3.2). With an improved knowledge base, this number will certainly increase.

The transboundary nature of majority of water resources in Africa requires cooperation at basin level which is essential in order to anticipate potential conflicts or to transform potential conflict into cooperation as it has been well documented by the UNESCO project PCCP (From Potential Conflict to Potential Cooperation) which conducted two case studies in Africa for the Mono (UNESCO 2008) and Icomati basins (Vas, Alvaro Carmo and Pieter van der Zaag 2003).

3.3.2 Increasing Pressure for the Satisfaction of Various Needs

Even though Africa has an average 4,000 m^3/hab/year of renewable water resources, unfortunately more than 300 million of the population do not have access to clean drinking water, and around 600 million do not have access to sanitation. Also many African countries particularly those in the arid and semiarid regions with limited resources are still struggling with food insecurity and malnutrition in terms of quantity to face the needs of their growing population. In total, it is estimated that African countries are importing food for the level of 17 billion USD every year.

In Africa, only around 5 % of the renewable water resources are mobilised, and only 7 % of the hydropower potential has been mobilised. Also less than 10 % of arable irrigable land has been mobilised. It is clear that for any sustainable socio-economic development in Africa and for the achievement of MDGs and beyond, African countries should mobilise significantly and manage sustainably their water resources both surface and groundwater.

With the growing of the African population and the continued development of African countries, there will be more and more pressure on the use of water resources which can lead to degradation of its quality posing serious risks to human and ecosystem health. It is therefore indispensable to adapt a holistic and integrated approach for a sustainable sharing and peaceful management of water resources within African countries. The initial estimation cost for the implementation of the Africa Water Vision 2025 was around 20 billion USD per year over 20 years. These figures were revised to around 50 billion per year by taking into account the cost of investment on related hydraulic infrastructures (AMCOW, AfDB 2010).

3.3.3 Climate Variability and Change Including Hydrological Extremes

The fourth IPCC report published in 2007 has clearly identified Africa and small islands as the most vulnerable to climate change. The main reasons are the great climate variability, the high dependency of many African countries economies to climate and poverty reducing adaptation capacities. Due to the changing of climate conditions, many African countries have been facing the consequences of hydrological extremes (floods, droughts). Floods have become recurrent phenomena in many African countries. Drought is also regularly observed, and it constitutes a threat to the achievement of food security in many countries particularly those in arid and semiarid areas like in the Sahel and Eastern Africa. Even though the level of impacts of climate change on water resources in Africa is not well known, there is a great likelihood that the hydrological cycle will change accompanied with an increase of the frequency and the intensity of extreme events, floods and droughts. From the fourth IPCC report (IPCC 2007), the main climate change impacts in Africa will be among others: agriculture production and food security are likely to be compromised in many countries; water stress is likely to be aggravated in many countries; low-lying lands are likely to be inundated; changes in a variety of ecosystems will be faster; and human health could be negatively impacted. The potential impact of climate change in the hydrological cycle in Africa will need more strategic measures for a better governance and management of water resources in Africa.

Climate variability and change and its impacts on water resources have regional dimension in Africa, and any country stand-alone solutions should not be sustainable and optimum if not considered within a regional perspective. Even though it is well known that droughts have a regional behaviour and impacts case in the Sahel and in the Horn of Africa requesting regional interventions, many African regions have been dealing regularly with transboundary flooding, for examples, within the Niger, Volta, Zambezi and Limpopo basins among others. This situation requires regional intervention for a proper prevention, mitigation and management of transboundary floods phenomena.

3.4 IWRM Situation in Africa

3.4.1 AMCOW Leadership

The creation of the African Ministers Committee on Water (AMCOW) in 2002 and its recognition as a technical specialised committee of the African Union (AU) and the different ministerial and Heads of States declarations related to water in Africa (Ethekwini Ministerial 2008, Tunis Ministerial declaration 2008 and Sharm el-Sheikh Heads of States declaration 2009) are clear signs for a high political commitment level on water resources in Africa. The African Water Vision 2025 has clearly highlighted the central role of sustainable management of water resources for the achievement of the different development objectives in Africa and recognised IWRM as the approach to be used.

Recognising the critical importance of water resources for sustainable development, MDG achievements and socio-economic growth in Africa, despite adopting IWRM as a general approach for water management in Africa, AMCOW has taken forward the IWRM concept during the first Africa Water Week into the concept of water for growth and development. The African context as far as water is concerned goes beyond the water management only; it should embed the socio-economic development in order to mobilise water resources (with a current mobilising rate of less than 5 %) to boost development in Africa.

As clearly indicated in the AMCOW work plan, it is now the time for implementation by pursuing the Africa water agenda as articulated by the Africa Water Vision 2025 and putting into actions the Sharm el-Sheikh declaration of Heads of States and the Tunis and eThekwini Ministerial declarations. The AMCOW work plan which is an overall framework put in place a set of actions to be undertaken by AMCOW, regional entities including among other Regional Economic Communities (RECs) and river basin organisations and national governments in the following seven areas: water infrastructure for economic growth; managing transboundary water resources; meeting the sanitation, hygiene and water MDG gap; global changes and risk management; governance and management; finance and education; knowledge and capacity development.

3.4.2 Other Major Actors

Many efforts have been ongoing for the promotion of IWRM in sub-Saharan Africa by various organisations at regional, subregional and national levels (Amani and Nguetora 2003). The Global Water Partnership has been instrumental in that regard with the establishment of four subregional components, namely, GWP West, Central, Southern and Eastern Africa. Each subregional component has been assisting countries to establish their national water partnerships. More than 30 national partnerships have been created for the promotion of IWRM at countries level.

The creation of the Africa Groundwater Commission to assist guiding countries for the development and sustainable management of groundwater resource in the continent is a good opportunity to fully integrate groundwater issues within IWRM in Africa. UNESCO and UNEP have been instrumental by supporting AMCOW on the creation of the Commission and have been providing advice and support for its establishment.

It is also important to mention the role of Regional Economic Communities in the promotion of IWRM within their respective subregions. In the Southern Africa Development Community (SADC) subregion, a protocol on watercourse was signed by the different countries considering the principle of IWRM and leading to the creation of river basin organisations in the subregion. The SADC subregion has a strategic water action plan implemented by the Secretariat through a water division. In the Economic Community of West African States (ECOWAS) subregion, there is a full regional centre in charge of coordinating water issues with the subregion through a water resources regional action plan. In the Central Africa subregion, the countries of Economic Community of Central African States (ECCAS) have been in the process of the preparation of a regional IWRM action plan and the establishment of a regional water coordination centre. In the IGAD, through the inland water programme, a structure should be put in place dealing with issue of water resources for the subregion.

The river basin organisations in Africa constitute a key pillar for the implementation of IWRM which considers the basin as the appropriate level for addressing water issues in a holistic way. Even though around 22 river basins have treaties providing cooperation framework to all or limited number of countries sharing the basins, the river basin organisations with operational administrative structure are limited in number. The main river basin organisations in Africa are (UNECA 2000) for Congo basin, CICOS (International Commission of Congo-Oubangui-Sangha); for Gambia basin, OMVG (Organisation for the Management of Gambia River); for the Lake Chad hydrological basin, LCBC (Lake Chad Basin Commission); for the Lake Victoria hydrological basin, LVBC (Lake Victoria Basin Commission); for the Limpopo basin, LIMPCOM (Limpopo Watercourse Commission); for the Niger basin, NBA (Niger Basin Authority); for the Okavango basin, OKACOM (Permanent Okavango River Basin Water Commission); for the Orange basin, ORASECOM (Orange-Senqu River Commission); for the Senegal basin, OMVS

(Organisation for the Development of Senegal river); for the Volta basin, VBA (Volta Basin Authority); and for the Zambezi basin, ZAMCOM (Zambezi Watercourse Commission). It is worth to mention that for the Nile basin, the NBI (Nile Basin Initiative), even though is not a river basin organisation per se like the others, has contributed to promoting cooperation among the countries sharing the Nile basin.

The African Network of Basin Organisations (ANBO) created in 2002 is another important actor having the overall objective of promoting cooperation for the management of transboundary river basins within the spirit of IWRM.

Networks for capacity building on IWRM have been also established through CapNet, a global capacity building network on IWRM contributing to the establishment of subregional networks in Africa such as WaterNet in Southern and Eastern Africa, WANet in West Africa and NileNet for the Nile basin. It is worth to mention that IWRM is taught in various water training centres, universities and centres of excellence in Africa contributing to its promotion.

All these different actors, AMCOW, ANBO, river basin organisations, the different Regional Economic Communities, the networks of the Global Water Partnership in Africa and the IWRM capacity building networks, the countries and various water-related training centres, have contributed to promote and root IWRM within the management of water resources in Africa at local, national and transboundary levels.

3.4.3 Status of IWRM Application in Africa

In 2012, a comprehensive status report was published on the application of integrated approaches to water resources management in Africa. The report was based on the detailed responses of 40 African countries to a survey on key IWRM elements and detailed interviews for 10 countries. The key IWRM elements considered included among others the enabling environment (policies, laws and plans), governance and institutional framework (institutional framework, stakeholders participation, capacity building), applying the management instruments and financing infrastructure and water management. The report provided a very rich information on IWRM status in African countries with key messages on the status of applying integrated approaches to water resources management in Africa. Selected messages are indicated below (AMCOW 2012):

- Seventy six percent of reporting African countries are implementing national water laws, and 44 % are implementing plans based on the application of integrated approaches as stated in Agenda 21 and described in the Africa Water Vision 2025.
- Countries with improved enabling environment on water resources management are more likely to have improved governance and institutions as well as to progress faster with infrastructure development and financing.

- Some countries reported on good progress on financing for water resources infrastructures. Generally though financing of water management is poorly addressed and not well appreciated.
- Progress with development of and implementation of transboundary agreements is one of the most advanced elements of water resources management involving 77 % of reporting African countries.
- Progress with instituting water management instruments has lagged behind compared to other elements of IWRM. Progress has been observed primarily in those countries with improved enabling environment and institutions.
- Floods, drought and water pollution are the greatest threats to water resources in Africa as reported by many countries, and a great deal of efforts are needed to address these issues.
- Concerns over institutional capacity constraints and no evident responses with capacity development programmes in many countries.

In general, there has been a good progress on the application of integrated approaches to water resources management in Africa. The existence of various actors providing the leadership in the promotion of IWRM and calling for the creation of new river basin organisations and the strengthening of the existing ones will definitively provide the momentum for a faster progress.

3.5 Needs for a True IWRM in Africa

Despite the good progress, there is need for the promotion of a more integrated IWRM towards water for economic growth and development in Africa. Even though things have started changing, from our experiences on the many interventions on the promotion of IWRM by different actors, we got the feeling that the focus was on a very limited elements of IWRM such as participation, the enabling environment and surface water in quantity. The aspects dealing with the real integration seemed to be neglected. For example, it is only recently that certain regional water partnerships in Africa have started considering groundwater and climate change issues within their advocacies. The following key issues should receive more attention for a better application of IWRM in Africa with integration at the centre.

3.5.1 Integrating Water Quantity and Quality

It is important not to forget that water quantity and quality go together when it comes to water resources management. Pollution on water resources is contributing to diminish the quantity of water available for the different uses. In many African countries, the quality of water resources is not well monitored. Quantity and quality should have equal consideration within national water management plan. Some river basin organisations have started considering seriously water quality issues within the creation of river basin observatories which will contribute to improve the knowledge base on water quality and lead a real integration of quantity and quality.

3.5.2 Integrating Surface and Groundwater

The promotion of IWRM in Africa has initially focused essentially on surface water with river basin organisations. Groundwater was not clearly considered in an integrated way with surface water. Despite that groundwater is a major source for water supply in rural and urban in many countries, groundwater is not well known like surface water. Also many river basin organisations have had surface water (rivers) as their priorities. Things have started changing with some river basin organisations fully considering surface and groundwater within their water resources management plans and also regional water partnerships entities did now consider groundwater as important as surface water within their awareness raising and advocacy on the promotion of IWRM. A bigger effort should be considered on improving the knowledge base on groundwater for a better integration of surface and groundwater resources within IWRM in Africa. The ongoing UNESCO Internationally Shared Aquifer Resources Management (ISARM) programme and the operationalisation of the Africa Groundwater Commission will contribute to improve the knowledge base on shared aquifer systems in Africa.

3.5.3 Integrating Climate Change Dimension and Related Disasters

The Integrated Water Resources Management has been promoted as a good adaptation measure to climate change and its impacts on water resources (Niasse et al. 2004; ENDA 2007). The impacts of climate change with increasing or decreasing of water resources and the reinforcement of hydrological extremes, drought and floods will need a holistic and integrated approach for a better management of water resources in Africa. Therefore, climate change and related impacts on water resources including adaptation actions should be fully considered within IWRM plan. Floods and droughts should be mitigated in an integrated way at basin level with the need for a comprehensive plan for transboundary water system. For example, for transboundary river basin, transboundary floods and droughts should be addressed in more upstream/downstream integration.

3.5.4 Integrating Cultural Dimensions

There is an abundant literature on IWRM, in which cultural dimensions of water management are usually not clearly articulated. UNESCO introduced the theme

Water and Cultural Diversity into the international discourse in 2000, as a way of mainstreaming cultural diversity into IWRM. This recognised the fact that sustainable solutions to water-related problems must reflect cultural dimensions of people's interactions with water. Culture is therefore recognised as a powerful and harmonising aspect of water resources management. In 2007, UNESCO launched the *Water and Cultural Diversity Project* to accelerate comprehensive assessment of existing research and case studies and to foster interdisciplinary and multicultural research partnerships. Despite this progress, cultural practices have not been effectively integrated into training and management programmes.

The cultural dimension of water and the place of culture for a better management of water resources have not been well captured within the different training materials on IWRM in Africa. Many studies have shown the importance of water in our different cultures and the place of culture for the conservation of water. There is a need to mainstreaming cultural diversity within IWRM national action plan of the different countries particularly when it comes to the implementation.

3.5.5 Data as an Investment and Foundation

Good quality data is needed to prepare and implement a comprehensive IWRM plan integrating quantity and quality, surface water and groundwater, climate change and related disasters and cultural dimension. Data on climate, surface and groundwater resources availability, water quality and demand are the prerequisite and foundation for any sustainable management of water system. Since the results of the sub-Saharan Hydrological Assessment (HASSA) in the 1990s, indicating the advanced deterioration of water-related networks (climate and hydrological networks) in various countries and basins in Africa, the problem has unfortunately become worst. It is critical to address the issue of data in Africa, if not it will hampered all planned water investment and management projects. Rehabilitation of climatic and hydrological network should be considered as an investment and not a waste of money as it is still considered because of lack of sensitization to many policy makers of Ministry in charge of water affairs. Even though, remote sensing data and geoinformation can contribute to complement and fill the gap in some basins, good quality raw data will still be needed.

3.6 Case Studies

Many examples are available in Africa where principles of IWRM have adopted within different national water policies and strategies and within subregion water strategy and river basin strategies. Two case examples/studies are presented to illustrate concrete application of the principles of IWRM and to what extent the adoption of IWRM led to address in a holistic and better way of water resources challenges. The first case study is on the Niger basin through an intensive participatory process involving the nine countries sharing the basin and which led to the adoption of the shared vision and the preparation and starting implementation of an investment regional programme for the whole basin. The second case study/example concerns the Densu basin in Ghana where following the water act and adoption of a national IWRM strategy that basin was considered as the first pilot for using IWRM approach to address challenges mainly pollution facing the basin.

3.6.1 Niger Basin Shared Vision

3.6.1.1 Overview of the Niger Basin

The Niger basin has an area of about 2,100,000 km^2 and it is shared among the following countries: Guinea, Mali, Ivory Coast, Burkina Faso, Benin, Niger, Algeria, Nigeria, Chad and Cameroon (Fig. 3.1 with Niger basin indicated in West Africa). The hydrological active part of the basin has an estimated area of 1,500,000 km^2 which doesn't include Algeria part which is inactive. The Niger River has around 4,200 km of length putting it as the third in Africa and ninth in world. The source of the river is from the mountain of Fouta Jalon in Guinea. Based on the hydrological conditions, the basin is generally divided in four part: the higher Niger from the source of the river to the entry of the second part of the basin called the inner delta which is a large important and rich floodplain ecosystem area in Mali and the third portion of the basin called the medium Niger goes from the outlet of the inner delta at Ansongo in Mali to the entry of the river in Nigeria. The last portion of the basin called the lower Niger is from the Niger and Nigeria border to the delta including the Benue tributary. The nine countries sharing the active hydrological part of the basin created the Niger River Authority (NBA) in 1980 to promote coordination and cooperation for the management of the resources of the basin. The Niger Basin Authority replaced the River Niger Commission (RNC) which was created in 1964. Following the financial crisis faced by NBA, the convention was revised in 1987 in order to redefine the NBA objectives so as to make it more operational and promote an integrated and sustainable management of water resources in the basin with the ultimate objective of meeting the socio-economic needs of the populations. The new agreement assigns the following specific objectives to NBA:

- Harmonise and coordinate national policies to develop the Niger basin's water resources.
- Take part in development planning by preparing an integrated development plan for the basin.
- Promote and participate to the design and operation of infrastructure and joint interest projects.

- Ensure surveillance and control of all forms of navigation on the river, its tributaries and sub-tributaries in conformance with the Act of Niamey.
- Take part in the formulation of applications for support services and funding for the studies and works necessary to develop the basin's resources.

3.6.1.2 The Main Challenges Facing the Basin

Despite this reform, the NBA continued to face difficulties in the performance of its assigned tasks, which led to insufficient tangible products to improve the socio-economic conditions of the basin's populations. Three reasons for this functional problem (beyond administrative, technical and financial problems) can be emphasised: (1) an unsuitable legal and institutional framework, (2) the absence of Integrated Water Resources Management (IWRM) policy at transboundary level and (3) the disparity of the national legal and institutional frameworks and subregional instruments, which are either barely compatible or incompatible with NBA objectives. The solutions to these functioning problems are very important institutional issues for the Basin and will enable asset management, operational management, tactical management and the sharing of the benefits from the infrastructure and works related to water resources.

Apart the institutional issue, the basin has been facing various hydrological, environmental and socio-economic challenges. During the diagnosis phase, 14 thematic and cross-cutting areas were considered, and for each one, a list of two to five issues has been identified. The identified issues are grouped into two categories: issues related to the ecosystem conservation of the basin and issues related to the development of socio-economic infrastructures. In the category of the ecosystem conservation of the basin, the priority issues identified are:

- Develop knowledge about water resources and their management.
- Prevention of water pollution.
- Wetland and biodiversity conservation.
- Catchment area protection and development.

Regarding the development of socio-economic infrastructures, the priority issues identified are:

- Improving the efficiency of existent infrastructures.
- Define the combination(s) of large infrastructure and management instruction for equitable and sustainable irrigation, minimise negative impacts on environment and optimise the hydropower production and navigation.
- Identify mitigation measures in areas subject to the negative impacts of the infrastructure/works.
- Identify development action in the parts of the river basin that are not along the river.

3.6.1.3 The Process Towards a Niger Shared Vision

The Heads of State and Government of the Niger Basin Authority, during their sixth and seventh Summit meetings held in Bamako (Mali) in December 2000 and Abuja (Nigeria) in February 2002 respectively, requested that NBA should develop a "Clear and Shared Vision" with a Sustainable Development Action Plan (SDAP) coupled with an investment programme and projects. When the Heads of State and Government signed the Paris Declaration in April 2004, it was a reiteration of the same request, recognising the need for all to adopt Integrated Water Resources Management principles. At their meeting in May 2005, the Council of Ministers of the NBA described the SDAP as being a strategic document to define and guide the "shared vision" process conducted by all the Niger basin countries by means of integrated management of their water resources and the associated ecosystems in order to improve the living standards and the prosperity of their populations.

The formulation of the strategy should be guided by the statement of the shared vision adopted by the extraordinary session of the NBA Council of Ministers held in Abuja in May 2005: "The River Niger Basin, a common space of sustainable development through an integrated management of water resources and related ecosystems for an enhancement of the living conditions and the prosperity of the populations by 2025". The top priority concerns and potential action of the SDAP are (1) the conservation of the basin's ecosystems, (2) the development of socio-economic infrastructure and (3) capacity building and stakeholder involvement. These priority concerns are all intricately linked to sustainable development which cannot exist without them.

The development of the SDAP (NBA 2007) was done in two (2) phases: the first phase consisted in the preparation of the diagnostic assessment and the second related to the development of a master plan for the development and management of the basin. The diagnostic assessment presents the situation report and the trends observed, the development constraints and opportunities and the other international, regional and subregional initiatives and programmes, while the master plan deals with the orientations and principles for a sustainable development, the justifications and development priorities throughout the basin, the analysis and the hierarchical classification of the priorities and the institutional aspects of the implementation of the SDAP.

The preparation of the strategic development plan and adoption of a new institutional framework for NBA went through an intensive participatory process involving the representatives of countries sharing the basin and civil societies and key stakeholders having interest and depending on the resources of the basin. The diagnosis assessment was the first fundamental step which led to the identification of the major key challenges facing the basin through various national consultations, national and regional workshops. In total, 14 sectoral themes and cross-cutting themes were analysed during the diagnosis phase. The themes are agriculture, livestock, fisheries and aquaculture, energy, mining, forestry, trade and industry, tourism, drinking water and sanitation, health, transport, water resources and catchment management, environment and biodiversity and human dynamic and land use. The analysis for each sectoral and cross-cutting theme was first been conducted in each of the nine countries of the basin during a national validation workshop in each country. From the results of the nine national diagnosis assessments, a regional diagnosis analysis was produced and validated during a regional workshop comprising representatives of the countries, civil societies and other key stakeholders. The institutional framework was addressed as a prerequisite indispensable for the success of the implementation of the shared vision.

For the preparation of the strategic development plan based on the different diagnosis assessment, similar approach was used with the preparation of a national strategic plan in each country from the priorities of the country and the identified national challenges to be addressed. The national document was validated during a national consultation in each country. Then the national strategic documents were harmonised and integrated in a coherence way to produce the global strategic development plan for the whole basin. The process of the shared vision and the preparation of the basin-wide strategic development plan were accompanied by a dedicated team at the Secretariat of NBA and the political support with regular councils of Minister in charge of NBA matters approving step by step the different documents produced.

3.6.1.4 Main Outcomes of the Shared Vision

The main outcomes of the whole process are:

- A revamped Niger Basin Authority with adapted institutional framework including the creation of Niger Basin Observatory, creation of a consultation and participation framework including civil society and the creation of a water charter of the basin, consultative panel on large infrastructures
- A strategic development programme for the whole basin

The priority areas of the SDAP are (1) *the conservation of the ecosystems of the basin*, (2) *the development of socio-economic infrastructure* and (3) *capacity building for the actors*. The SDAP aims at the following objectives:

1. To formulate an action plan (diagnosis and master plan for the development and management of the basin) to support the sustainable development of the River Niger basin
2. To give a concrete content to the principles of the "Paris Declaration" taking especially into account the geopolitical aspect, the priorities of the member countries and the subsidiarity principle
3. To translate the shared vision by 2025 in concrete actions so as to combat poverty, protect the environment of the River Niger basin and reinforce cooperation among the NBA member countries

4. To ensure a responsible and sustainable involvement of the civil society and the private actors in the NBA member countries in the implementation of the shared vision

The SDAP document integrates several actions which are carried out by other programmes and projects undergoing execution and/or planning. The actions integrated in the SDAP document concerned the following ongoing available strategic plans, programmes and projects: master plan of the silting control programme in the River Niger basin (AfDB-NBA), the strategic actions plan of the project "Reversing Land and Water Degradation Trends" (GEF-NBA), the programme development of Water Resources and Sustainable Management of Ecosystems (WB-NBA) and the Niger HYCOS Project (FDA-NBA).

The main lesson learned, like in any other river basin organisation having impact, is that the political will at the highest level of Heads of States and the commitment of the council of Ministers is indispensable for such kind of transformation happening with the NBA and the Niger basin.

3.6.2 Addressing Water Quality Challenges Within the Densu Basin

Following the water sector reforms in Ghana, the Water Resources Commission was established by Act 522 of 1996 with the mandate of regulating and managing the sustainable utilisation water resources. The strategy and concept of IWRM was adopted by the Commission. The Commission also has the responsibility of the coordination of the government water and related policies. River basins were adopted as the unit for management and not political or administrative boundaries. Based on a national diagnosis of key challenges facing the different basins in the country, it was found that the Densu basin was one of the most stressed urban basins with pollution as the main challenge. This was the reason why the Water Resources Commission (WRC) in 2001 selected the Densu River basin as the first pilot basin where the complete approach of IWRM would be rolled out.

3.6.2.1 The Densu Basin

The Densu River basin as indicated in Fig. 3.3 is located at southern part of the country. It has an area of 2,488 km^2 and a population of 460,270 estimated in 2000 leading to a population density above the national average (77 persons per km^2). Agriculture plays an important role. Water is mainly used within the basin for domestic, industrial and irrigation purposes through four reservoirs. Among all, the Weija Dam is the largest one located at the lower course of the river. The river is very important for Accra where over half million people from the capital have their potable water coming from the river.

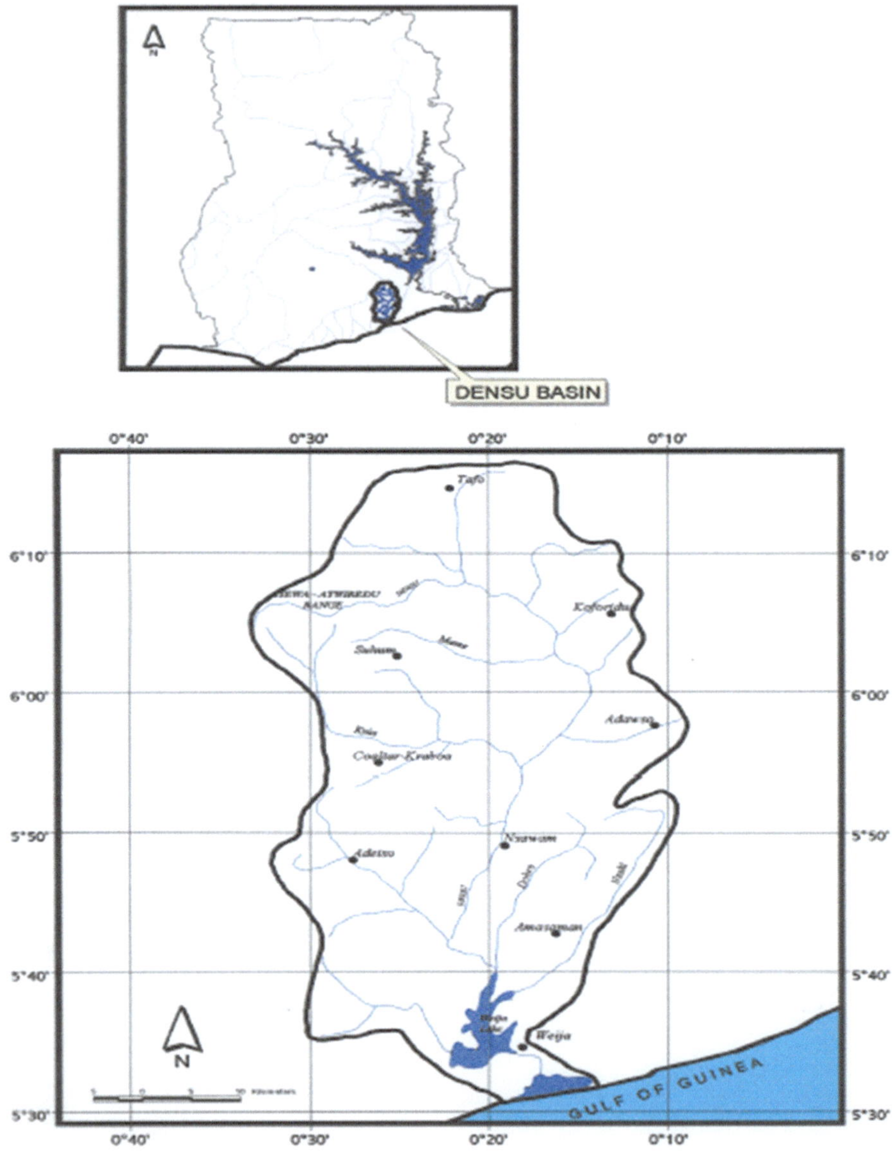

Fig. 3.3 The Densu basin in Ghana

The main challenges identified by the Water Resources Commission (WRC) within the Densu basin during its national diagnosis of different basins in the country were prioritised as (WRI 2000): water shortage, water pollution and improper land use, lack of comprehensive institutional and legal framework, inadequate reliable management information and data for water resources management, water weeds, flooding, high salinity of groundwater and high iron concentrations in groundwater.

3.6.2.2 The Pollution in the Densu Basin

Pollution of the Densu River basin was widespread and identified causes include dumping of solid waste and discharges of untreated urban domestic and industrial wastes, leaching of agrochemicals from commercial farms, as well as improper fishing practices such as the use of chemicals and rotten tree trunks. Furthermore faecal pollution is widespread in the Densu basin due to poor waste management practices and inadequate sanitary and waste disposal facilities leading to disposal of domestic sewage and garbage into the river directly. In one of the larger townships, Nsawam, an estimated 33 metric tons of waste are generated annually with a large amount being disposed off on the fringes of the town close to the river. This waste contaminated surface water sources through runoff. In the Weija Lake, high nutrient levels enhance the proliferation of algae thereby increasing water treatment costs. Persistent algal blooms were recorded in the Weija Lake and as high as 60–140 mg/l (Ansa-Asare 1996; Ansa-Asare 1998). Unfavourable odour and taste resulting from the pollution persist, even after treatment, and it costs much more for water treatment at Weija as compared to other treatment facilities in the country.

The problems of the Densu River have resulted also from the fast uncontrolled urbanisation and uncoordinated developments within the river catchment. Improper land use has also led to erosion and siltation of the river channel and consequently flooding and intermittent water shortages in the Densu basin. Risk associated with pollution of the Densu River can be summarised as follows: algal toxicity, health hazards, loss of livelihoods, loss of biodiversity, high water treatment cost and water use conflicts.

3.6.2.3 Interventions Using IWRM Approach

Despite the well-known deterioration of the water quality of the basin and some interventions by NGOs, but probably due to lack of coordination and political will, the deterioration continued. In 2001, the Water Resources Commission decided to revamp the process of addressing the pollution challenge by introducing the Integrated Water Resources Management approach within the Densu basin (Asamoah et al. 2008). A Densu basin board was established in 2003 comprising all interested parties and key stakeholders within the basin in order to address the management of the basin in a sustainable and holistic way. The board is composed of representatives from five major district assemblies within the basin, the regional coordinating councils under which Densu basin falls, various decentralised government departments, the Environmental Protection Agency, NGO and the traditional authority. For the restoration of the ecological health of the basin, the board has been implementing the following activities: capacity building, public awareness campaign programmes, monitoring of the catchment from upstream to downstream in a systematic manner and collaboration with the major relevant local government institutions in the basin on proper waste management practices.

Regarding the capacity building, various training workshops on IWRM have been conducted for the board members and for key stakeholders including Members of Parliament, Regional and Sector Ministries and other political leaders. Members of the board were strongly sensitised on the issues facing the basin with a tour from upstream to downstream. The success of the training component on IWRM at local level within the basin led some districts adding IWRM activities within their development plans. Concrete outcomes of the capacity building led to the relocation of waste dumping sites of some municipal assemblies such as the Akwapim South Municipal Assembly relocated their refuse dumping site from the banks of the river at Nsawam Bridge. The Suhum/Kraboa/Coaltar District Assembly has also relocated its dumping site to avoid polluting the river.

In conjunction with the capacity building programme, an aggressive public awareness and sensitization programmes took place with periodic radio programmes to sensitive local communities leading to the appreciation of many of the members of the district assemblies on the pollution which is the effects of their activities on water bodies. Children within the various schools in the different districts were sensitised. The consultative and networking efforts among members of the board and key stakeholders led to the higher commitment and will to address the pollution issues within the basin.

After few years of interventions, positives outcomes were observed on the decline of water pollution within the basin. The water quality of the river has been monitored from four sites: Potroase, which is at the source of the river, Mangoase and Nsawam at the midstream and Weija at the downstream since 2005. A water quality index (WQI) was used for the assessment. The Ghana water quality index is a translation of more technical data into a simplified description of the state of water quality. The index is thus a systematic way of interpreting measurements of ambient water quality parameters (dissolved oxygen, biochemical oxygen demand, ammonia, faecal coliform, pH, nitrate, phosphate, suspended solids, electrical conductivity and temperature) that have been checked against natural or desirable conditions. The index is based on the attainment of water quality objectives. The objectives are, safe limits, set by the WRC to protect the most sensitive uses of water.

The index classifies water quality into four categories: good, fairly good, poor and grossly polluted. Each category describes the state of water quality compared to the objectives that usually represent the natural state. The index thus indicates the degree to which the natural water quality is affected by human activity. Results showed that the quality of the river has improved from the source to the Weija reservoir. However, in 2007 there was a slight decline in water quality partly to the construction of a major road, which attracted many hawkers and attending problems. In all the best results were obtained in 2006, as compared to 2005 and 2007. Potroase which is at the source of the river remained in Class I and is described as good, unpolluted in 2006 and 2007. Mangoase had better water quality than

Nsawam, which recorded the worst in the results and was classified as poor quality. Nsawam is a highly populated capital of the Akwapim South Municipal Assembly and is a commercial town and as such receives many people in transit. In 2007 the river was dredged and could be the cause of a decline in water quality. At the Weija Lake, which is the downstream, the river has been impounded and showed its capacity at recovering from pollution.

Recent monitoring from 2010 to 2012 showed that the results had declined due to illegal mining activities within the basin. These activities have been a great challenge in almost all the river basins in Ghana. The government having noticed the threat to water sources has instituted an inter-ministerial committee to flash out all the illegal miners. With this intervention, the water quality will continue to improve.

3.7 Conclusion

The integrated water resources management has been mainstreamed within the Africa Water Vision 2025, and AMCOW has adopted its concept with a more development dimension towards socio-economic growth and development in Africa. The crucial role of water for sustainable development in Africa has been recognised by the African Heads of States through the Sharm el-Sheikh declarations calling for an equitable and sustainable use, and promotion of integrated management and development, of national and shared water resources in Africa. The shared nature of the majority of water resources in Africa with 80 transboundary rivers/lakes systems and at least 60 transboundary aquifers combined with natural threat of regional dimension such as climate variability and change calls for the application of integrated approaches to a sustainable and peaceful management of water resources in Africa.

Since the introduction and adoption of IWRM as the best approach for a sustainable management of water resources, numerous stakeholders have been contributing to the promotion and effective application of IWRM in Africa. The key stakeholders include among others, AMCOW, network of Global Water Partnership in Africa comprising subregional and national water partnerships, Regional Economic Communities, river basin organisations, network of capacity building and water-related training institutions. Even though the application of IWRM is a time-consuming process, there was global progress in Africa on the application of integrated approaches to water resources management in Africa as indicated by the AMCOW 2012 status report.

The good momentum gained so far on the promotion of IWRM in Africa by various stakeholders leading to its progressive integration within national, regional and transboundary water resources policies, strategies and plans should continue to be intensified.

Based on the case of Niger basin and other basin such as the Senegal basin, the application of IWRM for a large transboundary basin is a very challenging process which needs a strong political will at the level of Heads of States and Ministers of the countries sharing the basin and a strong commitment of professionals in the different countries and at the Secretariat of the river basin for conducting the process.

For a national basin, the commitment of different stakeholders having direct or indirect interest within the basin is the key and sustaining efforts that should be maintained like in the Densu basin where the water quality of the Densu basin will continue to improve if the interventions are sustained in the future.

An IWRM at basin scale, integrating surface water and groundwater quantity and quality, climate change and related hydrological extremes, upstream/downstream, cultural issues and putting the issue of quality data and information system as its foundation, should be the way forward in Africa.

References

Amani A, Nguetora M (2003) Towards an integrated water resources management in West Africa: status and challenges. Agrhymet Info 5(1), 1st trimester 2003:5–10 (in French)

AMCOW (2012) Status report on the application of integrated approaches to water resources management in Africa. AMCOW, Abuja

AMCOW, AfDB (2010) Africa regional paper: bridging divides in Africa's water security: an agenda to implement existing political commitments. 5th World Water Forum Secretariat, Libadiye Caddesi No:54, Küçükçamlıca-Üsküdar, Istanbul/Turkey

Ansa-Asare OD (1996) Environmental impact of the production of exportable pineapples – a case study of the Densu Basin. Ghana J Chem 2(1):1–7

Ansa-Asare O.D. and Asante K.A. (1998) A comparative study of the nutrient status of two reservoirs in south-east Ghana. Lakes and Reservoirs: Research and Management, 3:205–217 pp

Asamoah I, Jacobi S, Alfa, B (2008) Integrated water resources management plan for Densu River Basin, Ghana. In: 33rd WEDC international conference, Accra, Ghana, 2008

ENDA (2007) Climate change adaptation and water resources in West Africa. Writeshop report, 21–24 February 2007, Dakar, Senegal

Ethekwini Ministerial Declaration (2008) http://www.wsp.org/sites/wsp.org/files/publications/eThekwiniAfricaSan.pdf

GWP (2000) Integrated water resources management, TAC background paper no. 4. Global Water Partnership, Sweeden

IPCC (2007) Africa: climate change 2007: impacts, adaptation and vulnerability, Working group II, fourth report. Cambridge University Press, Cambridge

Johannesburg Declaration on Sustainable Development (2002) http://www.joburg.org.za/pdfs/johannesburgdeclaration.pdf

NBA (2007) Sustainable development action plan of the Niger Basin. Synthesis report. NBA (Niger Basin Authority), Niamey, Niger

NEPAD (2005) NEPAD Short-Term Action Plan (STAP) for transboundary water resources. Framework for implementation. NEPAD, Johannnesburg, South Africa

Niasse M, Afouda A, Amani A (2004) Reduce the vulnerability of West Africa face to the impacts of climate variability and change on water resources, humid zones and desertification. Elements for a regional strategy for preparedness and adaptation. UICN, Gland, Suisse and Cambridge, Royaume-Uni, xviii + 71pp

UNECA (2000) Transboundary river/Lake Basin water development in Africa. Prospects, problems and achievements. UNECA (2000) Africa water vision 2025. UNECA, Addis Ababa, Ethiopia

UNEP (2010) Africa water atlas. Division of Early Warning and Assessment (DEWA).United Nations Environment Programme (UNEP), Nairobi

UNESCO (2008) PCCP case study on the Mono Basin shared between Togo and Benin (Unpublished). UNESCO, Paris, France

UNESCO (2005) Transboundary aquifers in Africa, Water Sciences Division, Headquarters Paris, France

Sharm El-Shekh Heads of States Declaration on water and sanitation (2008) http://www.au.int/en/sites/default/files/ASSEMBLY_EN_30_JUNE_1_JULY_2008_AUC_ELEVENTH_ORDINARY_SESSION_DECISIONS_DECLARATIONS_%20TRIBUTE_RESOLUTION.pdf

Tunis Ministerial declaration (2008) http://www.afdb.org/fileadmin/uploads/afdb/Documents/Generic-Documents/First%20African%20Water%20Week%20Tunis%20Ministerial%20Declaration%20-%20English.pdf

Vas AC, Pieter van der Zaag (2003) Sharing the incomati waters: cooperation and competition in the balance, UNESCO/IHP/WWAP, IHP-VI/Technical documents in hydrology, PC-CP series No. 14 (SC-2003/WS/46). UNESCO, Paris, France

Water Research Institute (WRI) Accra (2000) Water resources management problems identification, analysis and prioritisation study; carried out for the Water Resources Commission (WRC). WRI, Accra, Ghana

Chapter 4
A Paradigm Shift in Urban Water Management: An Imperative to Achieve Sustainability

Kala Vairavamoorthy, Jochen Eckart, Seneshaw Tsegaye, Kebreab Ghebremichael, and Krishna Khatri

Abstract With increasing global change pressures (such as urbanization and climate change) and existing unsustainable factors and risks inherent to conventional urban water management, cities in the future will experience difficulties in efficiently managing scarcer and less reliable water resources. In order to meet these challenges, there needs to be a paradigm shift to integrated urban water management (IUWM). This paradigm shift is based on several key concepts including resilience of urban water systems to global change pressures, interventions over the entire urban water cycle, reconsideration of the way water is used (and reused), and greater application of natural systems for treatment. This chapter will present an integrated framework supporting IUWM and its key principles.

Keywords Integrated urban water management • Integrated urban water frameworks • IUWM principles

4.1 Introduction

It has become obvious that the current practice of urban water management is not sustainable and the need to adopt innovative approaches is evident. With increasing global change pressures and existing unsustainable factors, cities in the future will experience difficulties in efficiently managing scarcer and less reliable water resources. Nevertheless, there are great opportunities to implement innovative approaches in emerging urban areas, which may allow direct implementation of radically different water system configurations. The primary objective of this chapter is to present the paradigm shift toward integrated urban water management and its contribution to the development of resilient, livable, and healthy cities.

K. Vairavamoorthy (✉) • J. Eckart • S. Tsegaye • K. Ghebremichael • K. Khatri
Patel College of Global Sustainability, University of South Florida, 4202 E Fowler Avenue CGS 101, Tampa, FL 33620, USA
e-mail: airavk@usf.edu1

S.G. Setegn, M.C. Donoso (eds.), *Sustainability of Integrated Water Resources Management*, DOI 10.1007/978-3-319-12194-9_4

4.2 Challenges and Opportunities of Urban Water in Cities of the Future

4.2.1 Urban Water Challenges

It is widely accepted that one of the major challenges of the twenty-first century is to provide a safe water supply and basic sanitation for all. Safe water supply and reliable sanitation are fundamental to a community's health and development. Water and sanitation are fundamental requirements in the fight against poverty, hunger, and child mortality and in achieving gender equality. Providing adequate water supply and sanitation, particularly in urban areas, is a challenging task for governments throughout the world. Already, half of the world's population lives in cities which have very inadequate infrastructure and only limited resources to address water and wastewater management in an efficient and sustainable way. Because of these widespread infrastructure problems, almost 85 % of all wastewater is discharged directly into surface water bodies without treatment. This creates a critical health problem, restricts development, and increases poverty through higher health care costs and lost labor productivity (Corcoran et al. 2010).

Improving water supply and management will be even more difficult due to predicted dramatic global changes. Climate change is predicted to alter precipitation patterns and their variability, affecting the availability of water. In developing countries, technological and financial constraints make it difficult to maintain and upgrade the infrastructure necessary to deliver quality water to all sectors. This situation is exacerbated by population growth, urbanization, and industrial activities that dramatically increase water use and wastewater discharge. The global population is expected to exceed nine billion by 2050 (UN 2010). The urban population is projected to rise, nearly doubling from the current 3.4 billion to 6.4 billion by 2050, with the number of people living in slums rising even faster, from 1.0 to 1.4 billion in just a decade (UN 2010). Water scarcity, coupled with inadequate infrastructure and management of wastewater, is at the heart of the water and sanitation crisis. Current models of urban water management and their corresponding infrastructure have already failed or are on the verge of collapse from the perspective of cost-effectiveness, performance, and sustainability.

4.2.2 Urbanization Offers Great Opportunities

Urbanization, while challenging, also offers opportunities to implement a new paradigm for urban water management. This is particularly the case in many of the emerging towns and villages in developing countries. Pilgrim (2007) reported that for every large town, there are an estimated ten small towns, and this is expected to increase fourfold in the next 30 years. Because these emerging urban areas often don't have mature infrastructure and lack governance structures or existing urban planning, they provide real opportunities to implement innovative solutions for water and sanitation challenges. New development plans in these emerging areas could allow direct implementation of radically different system configurations where surface water, groundwater, and storm water are combined as potential sources, where source separation of wastes is applied to implement reclamation schemes (wastewater recycling, nutrient and energy recovery schemes) and where mixed land use development that promotes cascading water uses between domestic, industry, and agriculture sectors is considered. Although these emerging areas present the potential to do things differently, the window of opportunity to create a more sustainable pathway is relatively small (5–15 years). Quick action is needed if one is going to create a paradigm shift (World Bank 2009).

Compared to emerging towns, existing cities provide limited opportunities to rethink urban water management as the built environment already exists, and they often have locked-in, legacy infrastructure operated by silo institutions. In these cities, a paradigm shift will be difficult. However, there are opportunities to apply new approaches for water supply and sanitation even in growing existing cities, by "ring-fencing" the infrastructure of the existing city (instead of extending it further into the growing areas) and then considering the growing areas as independent clusters that can be developed according to the integrated paradigm (as proposed for emerging cities).

4.2.3 A Paradigm Shift Toward Integrated Urban Water Management

Cities are now faced with difficult future strategic decisions – do they continue business as usual, following a conventional technical, institutional, and economic approach for water and sanitation; do they tinker, where they follow the conventional approaches while trying to optimize and fine-tune them; or do they look for a new paradigm. Such a new paradigm of integrated urban water management intervenes over the entire urban water cycle and provides security through diversifying water sources, reconsiders the way water is used (and reused), understands wastewater as a valuable resource, and builds governance structures that cover the entire urban water cycle and water and sanitation systems resilient in the face of global change pressures.

In contrast to traditional water management systems, integrated urban water management (IUWM) develops efficient, flexible urban water systems by adopting a holistic view of all components of the urban water cycle (water supply, sanitation, storm water management), in the context of the wider watershed. The concept of IUWM developed from the broader framework of integrated water resource management (IWRM) which gained attention in the 1990s and has recently begun to include urban water issues. Both approaches embrace comparable core principles. Both IUWM and IWRM foster a participatory planning approach including all stakeholders affected by and interested in water management, and both want to overcome the limitations of the narrow disciplinary water management. Within this general agreement, however, there are practical differences between the two approaches. While IWRM encompasses a whole catchment, IUWM works within the boundaries of a city as a major decision-making unit while considering the interactions with the wider catchment. Because of the difference in scales, IWRM is more focused on the integration of the different water use sectors (agriculture, domestic, industry, and ecosystem), while IUWM aims to integrate the different parts and flows of the urban water cycle (water supply, sanitation, and drainage). This difference in scale dictates that the main tasks of both approaches are also different. While IWRM manages the equitable allocation of water resources between different users in a catchment, IUWM manages urban water management issues such as water sources, water demand management, provision of water supply, sanitation and drainage service, flood protection, etc.

Since 2000, the concept of IUWM has been successfully applied on city scale in order to integrate all aspects of the urban water cycle. The first applications of IUWM are documented from Australia (Mitchell 2004), Brazil (Braga 2000), and Singapore (PUB 2010). The research project "SWITCH" verified the concept of IUWM based on research and demonstration activities in eight cities around the globe (Howe et al. 2011). Recently, the IUWM approach was applied to African cities (Jacobsen et al. 2012). These examples illustrate that IUWM can successfully contribute to water security and can help to improve the provision of water and sanitation services. The discussion in this paper will focus on the principles of IUWM, frameworks and models for IUWM, and the documentation of best practices on IUWM.

4.3 Principles of Integrated Urban Water Management

4.3.1 Water Security Through Diversity

One of the major challenges in meeting the future water and sanitation goals is servicing more people with less water. This requires us to critically look into water use practices and to develop strategies that maximize water use efficiency. Balancing the demands for water between the various sectors will need to be accompanied by the use of new and alternative resources provided by emerging technologies. The integrated framework proposed here provides alternative water sources and facilitates the diversification of the water resources on which cities rely. By increasing diversity of sources, IUWM increases the security of the water supply and reduces cities' vulnerability to extreme events.

The integrated framework fosters reuse strategies that reduce water demand by using wastewater as a valuable resource. Advanced treatment technologies such as membranes or natural treatment systems facilitate the reuse and recycling of wastewater, enabling gray water recycling, closing the black water loop, and enabling the recycling of wastewater for drinking water purposes. Membrane technologies have become a particularly useful way to treat wastewater to a high standard and can reduce freshwater demand. Urban water systems designed with distributed clusters, each with a semi-central system of advanced treatment technologies, can reduce freshwater demand by up to 80 % by substitution with reclaimed water (Bieker et al. 2010; Otterpohl et al. 2003).

The decentralized infrastructure of distributed clusters is the best way to exploit alternative water sources. In this approach, locally available sources such as rainwater/storm water, local groundwater, and reclaimed wastewater become potential sources of water to offset the freshwater demand from the central water supply system. The use of local resources should be complemented with storage innovations such as aquifer recharge and recovery technologies.

In addition to recycling wastewater, IUWM uses water multiple times by first cascading it from higher- to lower-quality needs (e.g., using household gray water for irrigation) and then reclaiming water and returning it to the supply side of the infrastructure. The cascading use of water facilitates a reuse of water with less treatment effort. The integrated framework requires knowledge of water quality necessary for all applications so that water quality can be matched to water use. This system also requires a multiple barrier approach and water safety plans to safeguard water quality from the source to the tap.

While it is crucial to develop alternative sources of urban water, IUWM includes strategies to improve the end use efficiency as well as system efficiency tailored for low-income countries. For example, reducing the high leakage rate of the water supply system, a common problem in many cities in low-income countries, provides a huge opportunity to improve the system efficiency. Advanced leak detection technologies in combination with techniques for decoupling highly chaotic and interconnected networks can improve leakage management. Compared to other options, leakage management and water conservation are some of the cheapest and the largest sources of water available within cities (Haddad and Lindner 2001). Considering a conservative figure of 35 % average water loss, the World Bank estimates that utilities lose approximately 26.7 billion m^3 every year. Reducing this amount by half would enable to supply an additional 90 million people in low-income countries (Kingdom et al. 2006).

4.3.2 The Need to Consider Wastewater as a Resource

One of the major opportunities to addressing the global sanitation challenge is to change the way people think about wastewater – we should stop viewing it as a burden and begin to see it as a resource. While wastewater can be a new source of water as we have seen, it can also be a source of energy and valuable nutrients. The integrated framework proposed here suggests that entrepreneurs interested in the recovery of these resources could be new driving forces for the provision of sanitation.

Wastewater is often grossly undervalued as a potential resource. All too frequently, wastewater is ignored and left to drain away. Smart and sustained investment in wastewater management can capture more economic advantage of every drop of wastewater. For instance, by using advanced treatment technologies, an integrated system can effectively make use of fecal sludge, produce clean energy, reclaim valuable nutrients, produce bioplastics, etc. (Bieker et al. 2010). Besides reclaiming valuable resources, the new paradigm also reduces the wastewater discharge to the environment and the associated negative effects on water resources.

Innovative use of wastewater can convert current liabilities (e.g., energy required for wastewater treatment) into assets (e.g., energy from wastewater treatment). In Africa, it is expected that by 2030, biomass resources will replace 30 % of imported oil, provide 20 % of transportation fuels, and produce 5 % of electric power demand and 25 % of chemical needs (Fenton 2011). Being a finite resource, phosphorus has taken center stage in recent years, and current global reserves may be depleted in 50–100 years (Cornel et al. 2011). Its demand for sustained food production is projected to increase, and hence alternative and more sustainable sources such as recovery from urine and wastewater are becoming increasingly important. Treating phosphorus as a finite resource shifts the management paradigm from mitigating a noxious substance (due to its negative impacts on aquatic ecology by eutrophication) to recovering and reusing a precious element.

With proper management, wastewater can become not only a resource but a source of new livelihoods. The potentials of cascading use of water and the use of reclaimed water for urban water are already described in the section above. In addition, wastewater reuse in agriculture can provide benefits to farmers in conserving freshwater resources, improving soil integrity, preventing discharge to surface and groundwater waters, and improving economic efficiency. Reclaimed water can also play an important role in ecosystem restoration.

New technologies that promote wastewater as a resource are crucial, but they must be tailored for the unique conditions that exist in low-income countries. Some of the technologies appropriate for low-income countries include low-cost membrane filtration systems such as membrane bioreactors, hybrid systems of natural and advanced treatment, microbial fuel cells, electrochemical processes, and source separation of different waste streams. These game-changing technologies can lead to new ways of providing water and sanitation and are instrumental for the new paradigm.

Clearly, there is a need to provide a platform for the development of new and innovative technologies and management practices to promote wastewater as resource. The potential outcome could be wise investments in wastewater management generating significant returns, and instead of being a source of problems, well-managed wastewater will be a positive addition to the environment which in turn will lead to improved food security, health, and therefore economy.

4.3.3 IUWM Can Be a Useful Approach to Develop Resilient Water Supply and Sanitation Systems

New water and sanitation systems must be sensitive to the uncertainties created by global change. The resilience of the water and sanitation system will depend on the relations among the natural, the built, and the human environment and on the ability of this complex system to adapt to unexpected change. To reduce the future regret, new systems must be designed to be robust and adaptable against future uncertainties (Khatri and Vairavamoorhty 2009). Thus, the challenge facing the city of the future is to develop robust yet flexible systems that can respond to inevitable shocks related to future global change pressures.

A more modular approach to water and sanitation can create flexible systems that are characterized by their ability to cope with uncertainties and have the capability to adapt to new, different, or changing requirement. Modular diversity exponentially increases the amount of possible configurations of urban water systems that can be achieved from a given set of inputs (complex adaptive systems). For example, in relation to storm water management, small-scale decentralized system such as sustainable urban drainage systems (SUDS) or low-impact development (LID) has the ability to respond more flexibly to changes in boundary conditions (Eckart et al. 2010). In relation to sanitation, decentralized systems provide modular diversity and natural treatment processes (that use the natural capacities of soil and vegetation to absorb and retain water and to take up, transform, or otherwise treat pollutants) and have a great deal of adaptability to almost all conceivable applications and improved renewal and readjustment opportunities (essential for flexibility). These flexible systems are characterized by their ability to cope with uncertainties and hence have the capability to adapt to new or changing requirements (Ashley et al. 2007).

4.3.4 IUWM Framework and Diverse Scales

Because the IUWM framework captures the interactions between the various components of the urban water system, it can maximize the opportunities and minimize the threats to urban water management (Mitchell 2004; SWITCH 2011). For example, experience has shown that poor sanitation increases the pollution of potential water sources and treated drinking water. An integrated framework exposes these negative interactions. In addition to exposing some of the negative impacts, an integrated framework also highlights positive interactions such as the potential for providing more people with water and sanitation services while using less resource. For example, it highlights opportunities for reuse and recycling and the potential of alternative sources for water such as storm water.

The ability of the integrated framework to capture these system interactions stems from the way it can be applied at the neighborhood/cluster scale, at the city scale, and at the catchment scale. Because it can consider different scales, the integrated system can assess the full range of threats and opportunities for urban water management (see Fig. 4.1 for the nested structure of the different frameworks). Optimal solutions to water and sanitation challenges are likely to be a combination of interventions across these scales.

Neighborhood/Cluster Scale This scale allows water and other resource flows to be described between the various components within neighborhoods. It allows negative interactions between components to be articulated (e.g., cross-contamination of treated water by dysfunctional or poor sanitation systems; impact of septic tanks, pit latrines, etc., on groundwater sources and receiving water bodies; etc.) and helps to identify the root causes of many of the threats to urban water management. In addition, this scale allows identifying positive interactions between components (e.g., opportunities for addressing competing needs for water and other resources by exploring and maximizing recycling, reuse of water, and recovery of energy and nutrients) and facilitates the identification of opportunities. Figure 4.2 presents an integrated framework for water systems in low-income neighborhoods in developing countries. It illustrates typical elements such as onsite sanitation and negative interactions such as the cross-contamination.

Urban/City Scale This scale describes water and other resource flows between different neighborhoods and clusters within an urban/city space. For example, negative interactions on the city scale include lack of drainage provision in one cluster (e.g., a slum) impacting the performance of drainage in the entire city and lack of sanitation in one cluster (e.g., slum) degrading the quality of water sources in other clusters. Positive interactions that could be articulated at this scale include cascading use of water between clusters (e.g., wastewater from one cluster being used for urban agriculture in another cluster) and arguments for integrated infrastructure provision among clusters (i.e., all communities including low-income groups and slums), where service provision for the entire city benefits both individual clusters and the greater city (i.e., an integrated drainage network that includes all clusters makes more sense than one that intentionally avoids some clusters (i.e., slum clusters)). Figure 4.3 presents a diagram of the flows between clusters which illustrate the abovementioned positive and negative interactions.

The Catchment Scale This scale allows an integrated system to capture the interactions between cities and the catchments in which they are embedded. Activities at the catchment scale determine a city's access to adequate quality water and also provide flood protection. Upstream changes in land use patterns or water allocation may change the local hydrology and available water resources and can result in the necessity of watershed protection plans or water allocation strategies (Gleick 2009; Anderson and Iyaduri 2003). On the other hand, the city's impact on the watershed has to be considered. This may refer to the efficient use of the water resources within cities as well as impact of cities on downstream uses through the discharge of wastewater and storm water. The link between IUWM and IWRM at catchment scale should be strengthened to facilitate an equitable allocation of water resources

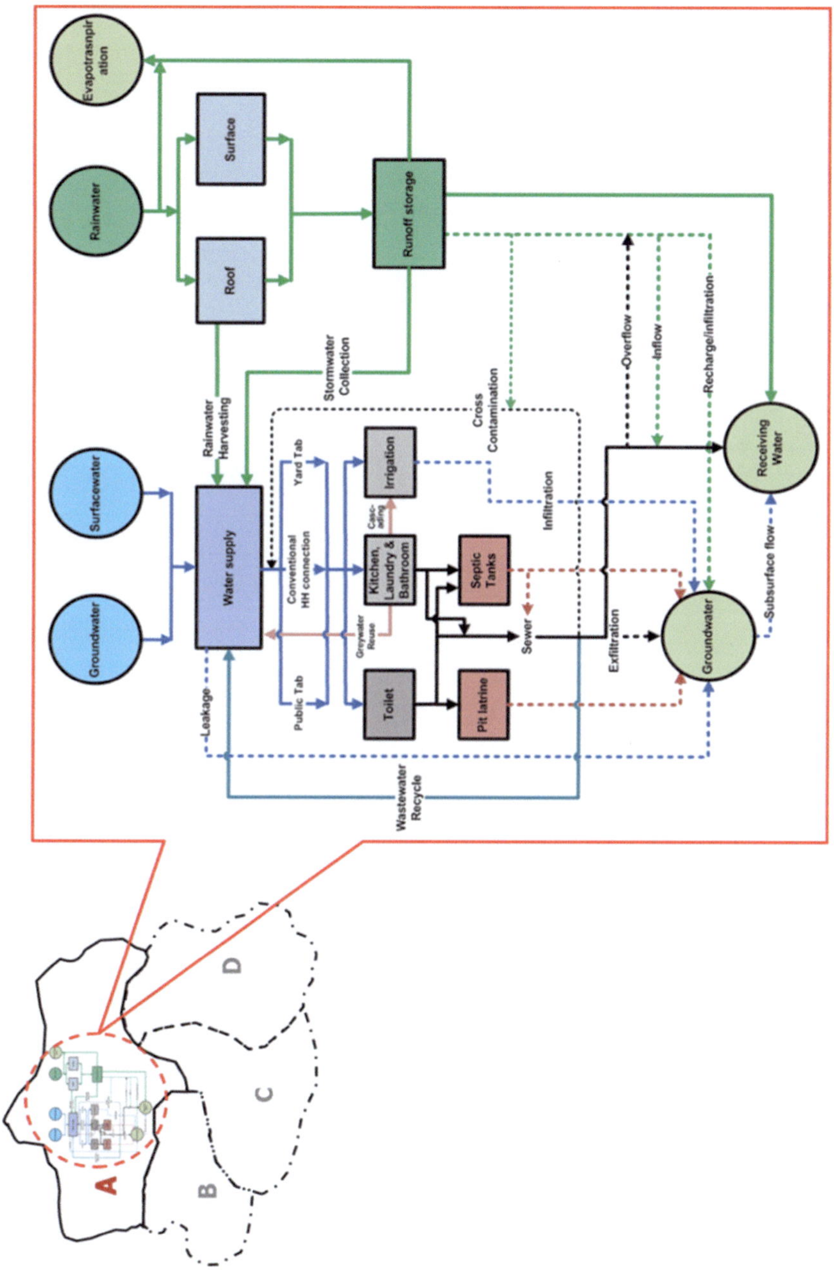

Fig. 4.1 Nested structure of the integrated urban water framework (Vairavamoorthy et al. 2012)

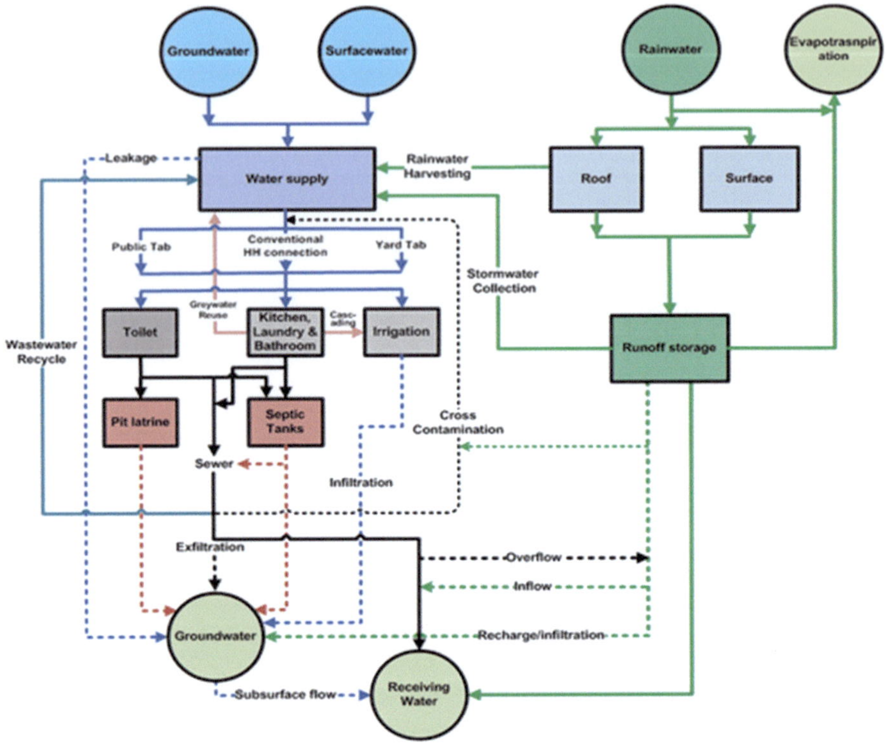

Fig. 4.2 Integrated urban water framework for low-income urban areas (Vairavamoorthy et al. 2012)

between the different sectors. Both IWRM and IUWM contribute to the vision of improved water security and climate resilience.

The integrated framework facilitates a structured and integrated analysis of strategies to improve water and sanitation provision. By improving the understanding of the highly complex interactions across scales and between the different parts of the urban water cycle, IUWM provides a framework for integrated and adaptive decision-making.

4.3.5 Institutional Mapping and the Need for Appropriate Governance Structures

While the multiple scales of IUWM provide a framework for integrated and adaptive decision-making, poor governance and dysfunctional institutional arrangements frequently make this integrated process impossible. Existing policy

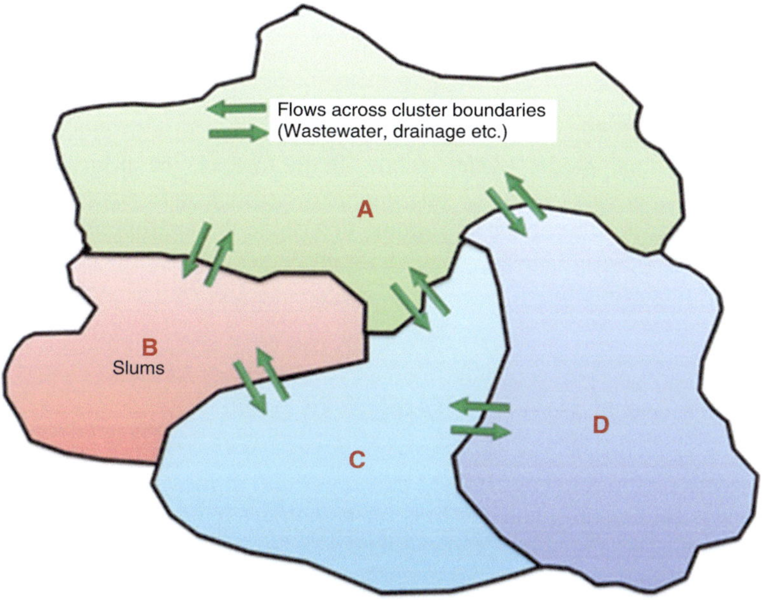

Fig. 4.3 Schematic depicting flows between neighborhoods/clusters in an urban area (Vairavamoorthy et al. 2012)

and regulations developed in isolated institutional structures will impede integration at a city scale. Sound policy, regulatory, and monitoring systems are the first requirements to achieve the targeted performance by public as well as private sector operators (Lenton and Wright 2004). At the same time, institutions that provide water and sanitation services need technical, financial, managerial, and social capacity that is lacking in many parts of the low-income countries. Integrated decision-making and management require reforms that strengthen institutional leadership, capacity, and oversight, make regulation transparent, provide adequate water finance, and build private sector partnerships and accountability to consumers.

Operationalizing the integrated framework described above presents a complex array of administrative, political, institutional, social, and economic challenges in cities (SWITCH 2011). Putting IUWM in place requires new governance policy and regulation to facilitate the innovative approaches (e.g., decentralized, semi-centralized systems setup), technologies (enhancing reduce, recycle, and reuse opportunities), cost-effective models, and stakeholder participation in the entire urban water cycle.

The contours of the new institutional framework necessary for IUWM have been identified through the technique of institutional mapping. Institutional mapping is a strategy that explores the space for institutional reform and organizational change. By describing the relations between all stakeholders in the provision and consumption of water and sanitation, an institutional map articulates users' needs and role in decision-making processes and identifies issues related to governance and the opportunities and constraints to achieving integrated urban water management (SWITCH 2011). Institutional mapping does not assume a prescribed governance model (i.e., either need to be centralized or decentralized or devolution model) or dictate whether private, public, or public-private partnerships are the best institutional and financial model (Kelay et al. 2006). Instead, the principle of good governance (Lockwood et al. 2010) and participatory approaches (which are more important for urban poor and slums) determine the appropriate institutional and governance structure most conducive for implementing the integrated model of urban water management.

4.4 The Case of IUWM Strategies in Arua

Although the key principles of IUWM may seem straightforward, its application on the ground is challenging. The new integrated approach to urban water management has been applied in a recently prepared feasibility study for Arua, Uganda, funded by the World Bank.

Arua is a rapidly emerging town located in northern Uganda. It is experiencing a critical shortage of water. The main water source, the Enyau River, is affected by the increasing water demands of upstream users, exacerbating the low flow conditions during the dry season. The current water supply is not sufficient to meet the existing demand, and with an estimated population growth of up to 200 % in the next 20 years, the water supply problem will only get worse. In addition to the water shortage problem, Arua also lacks adequate sanitation provisions, with dysfunctional pit latrines, open defecation, and untreated wastewater posing both health risks and water pollution risks (Jacobsen et al. 2012).

In order to cope with these challenges, a feasibility study for future water supply and sanitation was developed applying the integrated framework. The feasibility study proposed that in Arua surface water, groundwater, artificial aquifer recharge, and recycled wastewater (gray and black) were considered as potential water sources, resulting in increased water security. The feasibility report concluded that building a decentralized system for wastewater recycling using innovative options such as decentralized wastewater treatment system (DEWATS) and soil aquifer treatment (SAT) can both improve sanitation and generate additional water sources.

Decentralized systems were also recommended for Arua because they lower energy demand and reduce operational and maintenance costs, making them especially well suited to conditions in sub-Saharan Africa. The IUWM strategy also proposed the development of a strong watershed protection plan where the needs and wishes of all upstream and downstream stakeholders in the watershed are considered.

It is possible that this IUWM strategy could provide sufficient water resources to meet the increasing demand in the next 20 years. Allocation of the different water resources is prioritized from a cost-benefit perspective. The feasibility study estimates that for Arua, the average unit cost for the proposed IUWM scenario is US $0.57 per cubic meter, while the unit cost for the traditional approach of using water from conventional surface water sources (20 km away from the city) is US$0.74 per cubic meter.

4.5 Conclusions

We need to recognize that global change pressures will challenge our ability to manage urban water in the city of the future. We all live in a rapidly changing environment. The thinking behind much of urban planning today predates these changes, and the time has come to think fresh! We cannot continue investing in water infrastructure that is unsuited to future societal needs. At the same time, we have to find new ways to accommodate more people, with more needs, with the same quantity of water. All this has to be achieved while reducing our ecological footprint. This complicated challenge calls for a real paradigm shift in urban water management. In emerging urban areas in developing countries where water infrastructure is still in its infancy, there is a unique but fleeting opportunity to implement radically different urban water systems based on the principles of IUWM. The innovative approach described here exposes the potential of alternative water sources to meet the growing demand and enables the prioritizing of different water resources from a cost-benefit perspective. The Arua case study demonstrated that IUWM is a powerful approach to managing freshwater and wastewater (including storm water), as it provides the potential to satisfy the water needs of communities at the lowest cost while minimizing adverse environmental and social impacts.

References

Anderson J, Iyaduri R (2003) Integrated urban water planning: big picture planning is good for wallet and the environment. Water Sci Technol 47(7–8):19–23, IWA Publishing

Ashley R, Blanksby J, Cashman A, Jack L, Wright G, Packman J, Fewtrell L, Poole T, Maksimovic C (2007) Adaptable urban drainage: addressing change in intensity, occurrence and uncertainty of stormwater (AUDACIOUS). Built Environ 33:70–84

Bieker S, Cornel P, Wagner M (2010) Semicentralised supply and treatment systems: integrated infrastructure solutions for fast growing urban areas. Water Sci Technol 61(11):2905–2913

Braga BPF Jr (2000) The management of urban water conflicts in the Metropolitan Region of Sao Paulo. Water Int 25(2):208–213

Corcoran E, Nellemann C, Baker E, Bos R, Osborn D, Savelli H (eds) (2010) Sick water? The central role of wastewater management in sustainable development. A rapid response assessment. United Nations Environment Programme, UN-HABITAT, GRID-Arendal, Intl Atomic Energy Agency, Arendal

Cornel P, Meda A, Bieker S (2011) Wastewater as a source of energy, nutrients, and service water. In: Wilderer P (ed) Treatise on water science, vol 4. Academic, Oxford, pp 337–375
Eckart J, Sieker H, Vairavamoorthy K (2010) Flexible urban drainage systems. In: Proceedings of the water convention at Singapore International Water Week 2010
Fenton DL (2011) Biomass feedstock from MSW backbone for the biorefining industry. Missouri Recycling Association (MORA) annual conference, June 8, 2011
Gleick PH (2009) Doing more with less: improving water use efficiency nationwide. Southwest Hydrol 8(1):20–21
Haddad M, Lindner K (2001) Sustainable water demand management versus developing new and additional water in the Middle East: a critical review. Water Policy 3(2):143–163
Howe CA, Vairavamoorthy K, van der Steen NP (eds) (2011) SWITCH sustainable water management in the city of the future. Report- findings from the SWITCH project. Retrieved from http://www.switchurbanwater.eu/
Jacobsen M, Webster M, Vairavamoorthy K (2012) The future of water in African cities: why waste water? World Bank Publication, Washington, DC
Kelay T, Chenoweth J, Fife-Schwa C (2006) Trend report trend report on consumer trends, cross-cutting issues across Europe TECHNEAU, p 46
Khatri K, Vairavamoorthy K (2009) Water demand forecasting for the city of the future against the uncertainties and the global change pressures: case of Birmingham. Pro. EWRI/ASCE 2009 May (17–21), Kansas, USA
Kingdom B, Liemberger R, Marin P (2006) The challenge of reducing non-revenue water (NRW) in developing countries. The World Bank, Washington, DC
Lenton R, Wright A (2004) Interim report of task force 7 on water and sanitation, Millennium Project. www.unmillenniumproject.org
Lockwood M, Davidson J, Curtis A, Stratford E, Griffith R (2010) Governance principles for natural resource management. Soc Nat Res Int J 23(10):986–1001
Mitchell VG (2004) Integrated urban water management. A review of Australian practice. CSIRO and AWA report CMIT-2004-075
Otterpohl R, Braun U, Oldenburg M (2003) Innovative technologies for decentralised wastewater management in urban and peri-urban areas. In: IWA 5th specialized conference on small water and wastewater treatment systems, Istanbul, Turkey, 24–26 September 2002
Pilgrim NR (2007) Water working notes: principles of town water supply and sanitation, part 1: water supply. Water supply and sanitation sector board of the infrastructure network. World Bank Group, Washington, DC
PUB (2010) PUB Annual report. http://www.pub.gov.sg/annualreport2010/
SWITCH (2011) Training kit module 1 SWITCH, strategic planning-preparing for the future. http://www.switchtraining.eu/modules/module-1/
United Nations (2010) World population prospects. The 2010 revision. United Nations, New York
Vairavamoorthy K, Eckart J, Ghebremichael K, Khatri K, Tsegaye S, Kizito F, Mutikanga H, Rabaça J (2012) Final report – integrated urban water management for Mbale, Uganda, Prepared for the World Bank, April 2012
World Bank (2009) World development report 2009: reshaping economic geography. World Bank, Washington, DC

Chapter 5
Integrated Water Resources Management (IWRM) in a Changing World

José Alberto Tejada-Guibert

Abstract Integrated water resources management (IWRM) is an approach enjoying wide international acceptance and is considered a contribution to sustainable development. The governments of the vast majority of countries are in the process of implementing it or preparing to do so with the support of intergovernmental organizations and of numerous NGOs. However, the vision of what IWRM comprises and where it leads might not be clear to stakeholders and practitioners. This chapter reviews the concepts and evolution of IWRM; it points out that the historical and even the formal precursors go farther back than described in usual treatises on the matter. The chapter focuses on the related intergovernmental processes, primarily those in the UN system, as they have proved of fundamental importance in lending thrust to IWRM. It examines the principles and definitions of IWRM and points out that the IWRM platform is considered by many authors and practitioners not a blueprint, but more of a guide or philosophy that has to be adapted to the particular settings and needs – that is, IWRM could be a normative guide but not an implementation plan. The status of the adoption and application of IWRM at an international scale is examined, and the most common objections to IWRM are noted and discussed. The practical utility of IWRM as a provider of a common platform for discussing water resources management issues and sharing experiences is noted.

Keywords Water resources management • IWRM • Sustainability • Sustainable development • Water security • Dublin principles • Water-energy-food nexus • Drivers

---

J.A. Tejada-Guibert (✉)
Global Waters for Sustainability Program – GLOWS, Florida International University, Miami, FL, USA
e-mail: jatejada@globalwaters.net

S.G. Setegn, M.C. Donoso (eds.), *Sustainability of Integrated Water Resources Management*, DOI 10.1007/978-3-319-12194-9_5

5.1 Introduction

Integrated water resources management (IWRM) as a central concept to guide comprehensive water resources management at various administrative levels has received the ample and general acceptance of the responsible water resources management bodies and practitioners. IWRM aims at achieving optimal benefits of the use of water resources to society normally within a river basin as the natural planning unit and capable to being applied to larger regional or national levels (transboundary waters are a special case).

IWRM is a constantly evolving subject, and its development and application have received intense attention and contributions from many parties, including national authorities, international and intergovernmental bodies, and academic and nongovernmental organizations. The idea of IWRM is not recent; the United Nations Conference on Environment and Development (UNCED) (Rio 1992) is viewed as a significant milestone, though the concept itself may be traced much farther back. The fact that we are already in the follow-up phase of the United Nations Conference on Sustainable Development – also denominated Rio+20 (Rio 2012) – and at the end of implementation of the Millennium Development Goals when the post-2015 Sustainable Development Goals are being debated and shaped means that it is an appropriate time to take a scrutinizing look at IWRM. It is the major water resources management planning instrument, and it needs to reflect interlinkages with other development drivers to constitute an effective contribution to the process of sustainable development.

This review of the role and implication of the adoption and application of IWRM at this point is all the more relevant given the overall process of changes in the world – many of them in the realm of the unpredictable in the environmental, social, and economic dimensions that profoundly affects the setting for development planning. This process of change obeys in a significant measure the denominated "global changes," of which climate change is but one.

This chapter examines the current status of IWRM – recent developments and relevant concepts in order to give a current perspective, particularly in the context of related international initiatives, with a specific emphasis on the pursuit of sustainability. Sustainability, though firmly rooted in the modern notion of development, has proved to be a most elusive concept to define and apply and to be linked with metrics in practice. This chapter attempts to emphasize the evolving and flexible nature of IWRM that while not an exact science, it is an extraordinarily valuable guiding concept that can be continually enriched and adapted reflecting sound science and the appropriate social and economic principles and even the moral ideals for the benefit of all stakeholders.

This chapter, in addition to a general discussion of the concept of IWRM with an overview of the historical background, addresses some particular topics that have gained currency or significance recently. The discussion of specific points is succinct, as many of the topics are treated in depth in other chapters of the book.

5.2 Water Links to Development

The vital nature of water for life and development is clear. It is linked strongly to food, health, energy, trade and all walks of life, and the economy. It is likewise central to cultural and social aspects of humanity. However, the freshwater resources of the world are subject to increasing pressures due to a multitude of factors, including population growth, and the accelerated process of change under uncertainty. These are critical challenges that nations and society must confront in order to ensure not only a livable but a fulfilling and sustained environment. Thus, water plays an essential role in the development of nations – more precisely in economic development, which through policy interventions aims to improve the economic and social well-being of people – that is, a higher-order objective than just economic growth.

There are many drivers at global and local scale affecting water, and these need to be understood and to be taken into account explicitly in order to foresee future actions and consequences. These include (UN WWAP 2009):

- Demographics: population dynamics (growth, age distribution, urbanization, and migration) create pressures on freshwater resources through increased water demands (reflecting also changed patterns of consumption) and pollution.
- Economic: growth and changes in the global economy greatly impact on water resources and their use – globalization, global economic crisis, and international trade resulting in "virtual" water transfers – particularly related to agricultural products.
- Technical innovation: technological innovation affects all walks of water management from data recovery, water treatment, etc.
- Climate change: CC affects sharply on the hydrologic cycle, variation, and variability, in time and in space, a compounded effect when coupled with sea level rise. Extreme events (floods and droughts) may be more acute and frequent in various parts of the world.
- Land use change: changes in the natural landscape associated with population dynamics (migration, urbanization), agricultural expansion, etc. This generally means greater water demand and a degradation of the natural habitat.

Figure 5.1 illustrates the key drivers and causal links affecting water stress and sustainability and human well-being. The complexity of the interactions is evident; it is not a static picture, but it is dynamic. The speed of change of these relationships can be highly variable, and there are various degrees of uncertainty; the information available to the water manager and decision maker must be appropriate corresponding to the scale and nature of the area and the timescale.

Institutional constraint in the management of water, an often entrenched top-down sectorial approaches to water resources management, and a lack of emphasis on improving management of existing resources but rather on expanding to new sources lead to a suboptimal and uncoordinated management of the resource. There is now a growing recognition that global interdependencies will

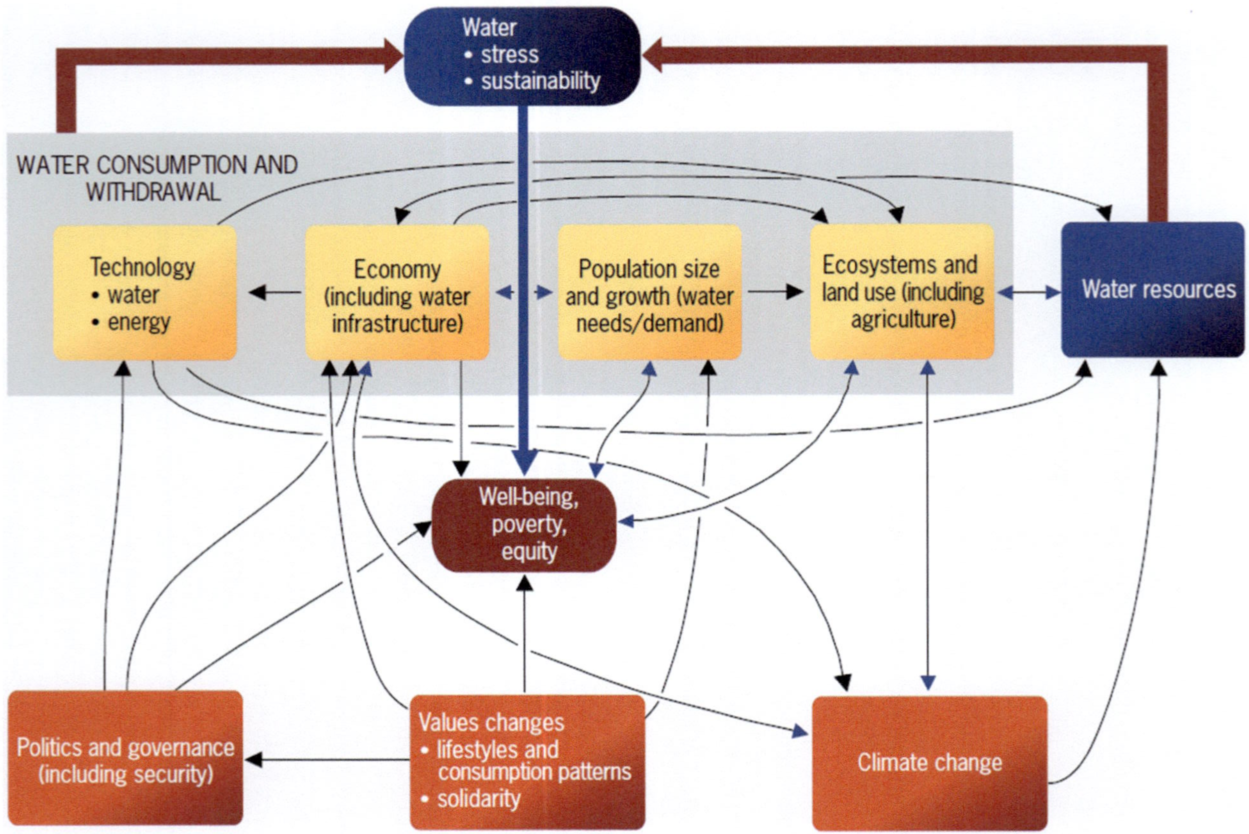

Fig. 5.1 Key drivers and causal links affecting water suess and sustainability and human well-being (Source: Gallopín (2012) as cited in UN WWAP 2012a, Fig. 9.1, Vol. 1, p. 267, Used by permission of UNESCO)

increasingly be manifested through water. Consequently, sector-based decision-making would need to be replaced with a wider, more functional framework reflecting all significant impinging factors (UN WWAP 2012a). These observations have a direct effect on what an effective, modern IWRM framework in tune with the times should address in a changing world.

This section points out some of the complex issues to be considered in development planning, particularly as linked to water and to the nature of ongoing change processes. The linkages of sustainable development to water resources management are elaborated in another chapter of this book.

5.3 Basics of IWRM

5.3.1 Evolution of IWRM

The concept and development of integrated water resources management (IWRM) emerged as a response of increasingly acute freshwater challenges. The accelerated process of change at global scale with its regional and local variants as referred in the preceding section calls for a water management approach to face the progressive scarcity and degradation of water resources that generate growing conflicts and competition for the resource, to safeguard the environment, and to maximize the benefits to stakeholders within the framework of a harmonious development – sustainable development. The background is presented here.

5.3.1.1 Early Development

Humanity, in the course of its existence, has dealt with the problem of water provision. A number of ancient civilizations, including Mesopotamia, China, and Egypt, were able to have substantial periods of stability based on irrigation economies. The ancient Chinese expressed it eloquently, as the combination of the Chinese characters of "river" and "dike" meant "political order" (Delli Priscoli 2013). In Europe, Spain has dealt for centuries with its climatic variability, aridity in particular, by means of water infrastructure such as the 2,000-year-old Cornalvo reservoir, which is still in operation, and by establishing institutions such as irrigation associations, some of which have existed for more than a thousand years: the Water Tribunal of Valencia has performed its function with stakeholder participation for 1,000 years. In addition, the Hydrographic Confederations (Confederaciones Hidrográficas), established in 1926, are the oldest river basin management authorities in the world (World Bank 2008).

An early example of a forerunner of integrated basin management is the Tennessee Valley Authority (TVA) established in 1933, which integrated the functions of navigation, flood control, and power production while also dealing with erosion control, recreation, public health, and welfare. It was part of President Franklin D. Roosevelt's New Deal, with the aim of overcoming the Great Depression through the construction of large public works. The TVA approach contained many elements of the current conception of IWRM: comprehensive planning of natural resource utilization combined with economic, social, and even environmental objectives (Snellen and Schrevel 2004).

A report on Integrated River Basin Development by an international panel of experts submitted to the Secretary-General of the UN in November 1957 described "the challenge that is presented by the orderly development of the rivers in the world, and the lines along which we believe the United Nations and its specialized agencies might suitable move in dealing with it" (United Nations 1958). The report also indicates integration: water infrastructure by itself does not bring development; supporting services are needed as well. This is a sectorial point of view as the integration refers to the support services to irrigation development projects, not true integration with other water uses and other development issues (Snellen and Schrevel 2004).

Also significant from the historical and practical standpoint, France in 1964, in accordance with its National Water Law of that year, established river basin agencies (agences de l'eau) in the country's six major river basins. This is generally held as an example of actual IWRM given the characteristics of a basin as administrative unit, to the strengths and the application of IWRM tools and concepts – such as stakeholder participation, economic instruments, and appropriate regulation (Mei Xie 2006).

5.3.1.2 Modern Times

Here we consider, perhaps a bit arbitrarily, that the initiation of the modern era of water resources management concepts is the point where the general conception of IWRM, as we now know it, was first suggested to the world. The first global intergovernmental conference on water (and the only such intergovernmental meetings to date), the United Nations Water Conference, Mar del Plata (1977), duly stated among its recommendations for policy, planning, and management (United Nations 1977):

> Increased attention should be paid to the integrated planning of water management. Integrated policies and legislative and administrative guidelines are needed so as to ensure a good adaptation of resources to needs and reduce, if necessary, the risk of serious supply shortages and ecological damage, to ensure public acceptance of planned water schemes and to ensure their financing. Particular consideration should be given not only to the cost-effectiveness of planned water schemes, but also to ensuring optimal social benefits of water resources use, as well as to the protection of human health and the environment as a whole.

This is one of the first formal internationally agreed statements leading to the IWRM approach and is a valid forerunner to more recent developments referred to below.

After the intergovernmental conference in Mar del Plata in 1977, the next significant development took place in the International Conference on Water and the Environment (ICWE) in Dublin, Ireland, on 26–31 January 1992. ICWE, though it was a nongovernmental conference, was attended by nearly 500 experts from many organizations, including government-designated experts and representatives of intergovernmental and nongovernmental organizations. The four Dublin principles that hold great significance for IWRM were agreed there.

The Dublin principles hold that:

1. Freshwater is a finite and vulnerable resource, essential to sustain life, development, and the environment.
2. Water development and management should be based on a participatory approach involving users, planners, and policy makers at all levels.
3. Women play a central part in the provision, management, and safeguarding of water.
4. Water has an economic value in all its competing uses and should be recognized as an economic good.

The above principles, which may be said to have nurtured the school of thought permeating modern water resources management approaches, are further elaborated below (discussion partially adapted from [Mei Xie 2006]), highlighting their link to the "pillars" of IWRM.

Principle 1 "Ecological": It recognizes the all-embracing approach that must be adopted in water resources management, "linking social and economic development with protection of natural systems" (ICWE 1992). This is put into evidence in WRM at the level of a catchment or river basin, the natural planning and administrative unit of IWRM, requiring the intersectorial integration for an effective application. The finiteness of water as a critical constraint, formerly, was conceived as essentially unlimited subject to temporal and spatial limitations; now given the growing pollution, water is even a diminishing resource in numerous locations and added difficulty to be overcome.

Principle 2 "Institutional": The participatory approach "advocates increased accountability of management institutions and full consultation and involvement of users in the planning and implementation of water projects," raising awareness of water issues among policy makers and all stakeholders. It emphasizes subsidiarity by which management decisions should be taken at the lowest appropriate level (with central government retaining regulatory and support roles).

Principle 3 "Gender": This principle calls for ensuring gender equity in water issues. Traditionally in large portions of the world, women, though responsible for the collection and transport of domestic water resources, have been excluded from the water management decision-making. Thus, IWRM processes emphasize empowering women throughout.

Principle 4 "Economic": This principle emphasizes the importance of economic tools in helping achieve efficient and equitable use of water resources. It is fundamental to acknowledge the basic right of all human beings to access to clean water and sanitation at an affordable price. Managing water as an economic good is essential to achieving financial sustainability of water service provision, via water rates at levels that ensure full cost recovery. A necessary extension is the recognition likewise of water as a social good, as the application of this principle provides the basis for the parallel goal of social equity.

The Dublin principles were a clarion call for the formulation of a solid IWRM framework. They were reflected later that year at the UN Conference on Environment and Development in Rio (1992) in the instrument adopted there by the community of nations to guide their actions toward sustainable development: Agenda 21, whose Chapter 18 (Protection of the Quality and the Supply of Freshwater Resources: Application of Integrated Approaches to the Development, Management and Use of Water Resources) is of direct application to water resources management. This agenda has guided the international and national policy decisions regarding the environment, and its validity still holds.

5.3.2 Defining IWRM

With the dynamics of change and the inherent uncertainty, it is clear that IWRM is a process that evolves continuously over time – principles change, as do technologies and procedures (UNESCO-IHP et al. 2009). There are different ways of looking at the integrative and evolving nature of IWRM. It is illustrative to cite some ways this is seen:

> [IWRM] is a step-by-step process of managing water resources in a harmonious and environmentally sustainable way by gradually uniting stakeholders and involving them in planning and decision making processes, while accounting for evolving social demands due to such changes as population growth, rising demand for environmental conservation, changes in perspectives of the cultural and economic value of water, and climate change. It is an open-ended process that evolves in a spiral manner over time as one move towards more coordinated water resources management. (UNESCO-IHP et al. 2009)

> IWRM is an empirical concept which was built up from the on-the-ground experience of practitioners.... The concept has been adopted widely by water managers, decision-makers and politicians around the world. (Hassing et al. 2009)

> Integrated water resources management is therefore a systematic process for the sustainable development, allocation and monitoring of water resource use in the context of social, economic and environmental objectives. (Cap-Net, GWP and UNDP 2005)

> IWRM should be viewed as a process rather a one-shot approach – one that is long-term and forward – moving but iterative rather than linear in nature. (GWP Toolbox website, http://www.gwp.org/ToolBox/ABOUT/IWRM-Plans/)

These various takes of the IWRM approach do not really clash but are rather complementary views from slightly differing angles, making it clear that IWRM is an evolving process and that it is not prescriptive, but rather dynamic and adaptive and that it pursues a sustainable outcome.

Though there are no formally adopted definitions of IWRM, the one formulated by the Global Water Partnership (GWP 2000) is widely used. It states:

> IWRM is a process which promotes the coordinated development and management of water, land and related resources, in order to maximize the resultant economic and social welfare in an equitable manner without compromising the sustainability of vital ecosystems.

This definition encompasses the main aspects that need to be envisaged. As in all concise definitions of a complex concept such as IWRM, various concerns can be raised regarding the precise scope of its various provisions: responsible parties, participants in the coordination, meaning of "equitable," etc. Nonetheless, it is a competent working definition of what requires to be addressed.

USAID (US Agency in International Development) offers a more detailed definition (and more "functional" according to some authors [Mei Xie 2006]):

> Integrated Water Resources Management (IWRM) is a participatory planning and implementation process, based on sound science, which brings stakeholders together to determine how to meet society's long-term needs for water and coastal resources while maintaining essential ecological services and economic benefits.

USAID tellingly further elaborates beyond its strict definition:

> IWRM helps to protect the world's environment, foster economic growth and sustainable agricultural development, promote democratic participation in governance, and improve human health. Worldwide, water policy and management are beginning to reflect the fundamentally interconnected nature of hydrological resources, and IWRM is emerging as an accepted alternative to the sector-by-sector, top-down management style that has dominated in the past. (http://www.usaid.gov/our_work/environment/water/what_is_iwrm.html)

This definition and the linked qualifications are widely encompassing, including elements stemming directly from the Dublin principles and the preceding GWP definition, plus specifics such as "coastal resources," "sustainable agricultural development," and "human health" and the significant statement that IWRM is the alternative to the usual sectorial "top-down management." The latest USAID water development strategy (USAID 2013) does not contain a new definition of IWRM, but it may be construed that some shifts of emphasis would be introduced reflecting what USAID considers key water-related development challenges now. Among them:

- Inadequate access to safe drinking water and sanitation
- Limited access to freshwater for food security
- The impact of climate change on water resources
- The significant energy requirements for water
- Water as a potential source of conflict

As indicated by Merrey (2008), most IWRM proponents accept the basic concept as articulated by GWP and USAID.

5.3.3 *Guidance for Implementation*

5.3.3.1 Global International Guidance

Putting into practice, an IWRM process requires key enabling elements as represented by the "three pillars" of IWRM (Hassing et al. 2009) depicted in Fig. 5.2:

1. Moving toward an enabling environment of appropriate policies, strategies, and legislation for sustainable water resources development and management
2. Putting in place the institutional framework through which the policies, strategies, and legislation can be implemented
3. Setting up the management instruments required by these institutions to do their job

The importance of an enabling environment (policy and laws) supported by the appropriate institutional framework at all levels plus competent management instruments is highlighted here.

However, the above representation is static – a process such as IWRM is dynamic; thus, we can envision a sequence of tasks in order to implement the various steps involved in IWRM, many times referred to the river basin as the natural planning unit. Policy making, planning, and management might be

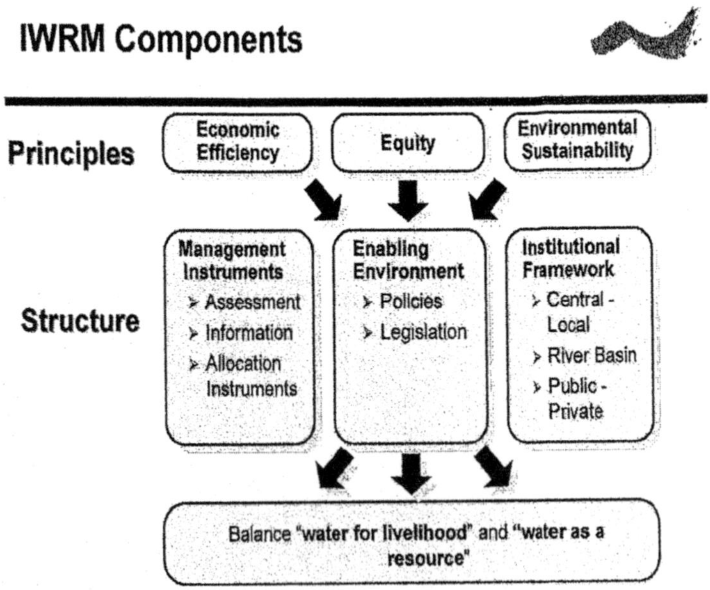

Fig. 5.2 The three pillars of IWRM: an enabling environment, an institutional framework and management instruments (Source: Hassing et al. 2009, Fig. 1, p. 4, used by permission of UNESCO)

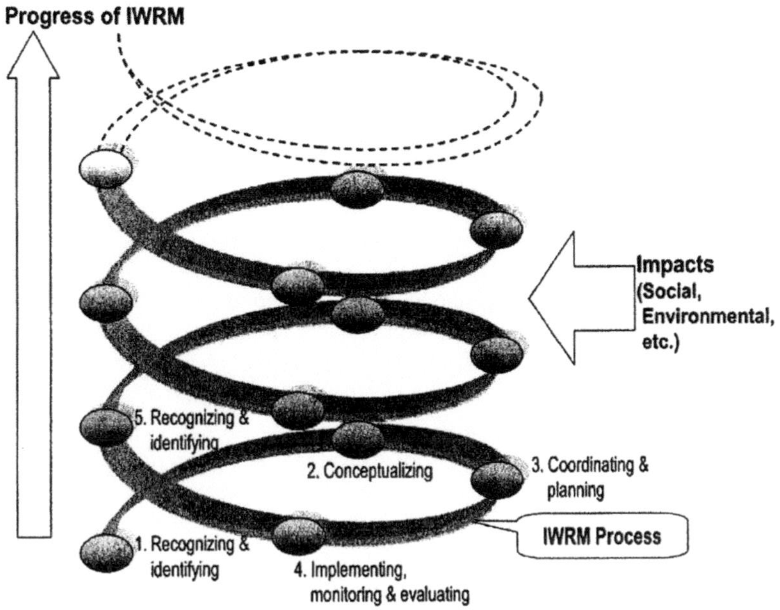

Fig. 5.3 IWRM spiral and process (Source: UNESCO-IHP et al. 2009, IWRM Guidelines at River Basin Level, Fig. 3, p. 9, used by permission of UNESCO)

considered as a series of sequential steps in basin management. The initial step normally involves establishing broad policy goals based on the national development objective (where we want to get to). From there the identification of water management issues that need to be faced can follow, allowing the formulation of the strategy to overcome the problem, selection, and implementation of a plan. Thereafter, the evaluation of the outcomes, allows learning from these outcomese in order to revise our plan to make it work better in the future (UN-Water and GWP, 2007). This is a very simplified, even ideal description of the planning and implementation cycle; there are many ways it can be interrupted, be sent down another path, and have internal loops – whether it is caused by politics, finances, natural disasters, or whichever other way reality chooses to enter. What should be clear is that there is a logical and a foreseen sequence based on solid principles.

It has also been noted that IWRM is not a one-off exercise; actually not only is the philosophy of IWRM constantly evolving but also the settings upon which it is applied undergo constant changes. This leads to the need of revisiting constantly an IWRM process and assessing its performance on the basis of a well-thought-out monitoring plan. This led UNESCO to posit IWRM as a spiral process progressing through times. Figure 5.3 represents this vision.

For the practitioners, a number of tools and publications have been made available. Among the recent guides for the application of IWRM, we can mention *IWRM Guidelines at River Basin Level* (UNESCO-IHP et al. 2009) and *A Handbook for Integrated Water Resources Management in Basins* (GWP, INBO

[International Network of Basin Organizations] 2009). Though both publications came out simultaneously, they are intentionally of a complementary nature, as both declare in their presentations. More recently, *The Handbook for Integrated Water Resources Management in Transboundary Basins of Rivers, Lakes and Aquifers* (INBO and GWP 2012) has been issued; this very welcome addition addressing shared waters is the product of the collaboration of INBO, GWP, UNESCO, the UN Economic Commission for Europe (UNECE), the Global Environment Facility (GEF), and the French Development Agency (AFD) – this by itself represents a striking example of an integrated effort.

The Global Water Partnership has likewise created an IWRM toolbox designed to support the development and application of IWRM approaches, providing IWRM practitioners with a wide range of tools and instruments to fit diverse needs. The tools fall into three main categories, replicating the three pillars of IWRM described earlier: (a) enabling environment, (b) institutional roles, and (c) management instruments. There are altogether 59 different tools available in the Toolbox. The Toolbox was originally launched in year 2000, and it is periodically renewed; Version 3.0 was launched in September 2013 (http://www.gwp.org/en/ToolBox/).

5.3.3.2 European Union

Thus far, the international accent has focused on the UN-related international efforts. Another significant initiative has been led by the European community. The European Union Water Framework Directive (WFD) (http://ec.europa.eu/environment/water/water-framework/) came into force in December 2000, launching an integrated water protection policy in Europe meant to bring about a coordinated management of waters in river basins beyond regional and national borders. The key objective of the WFD is to achieve good status for all water bodies by 2015. It has undergone some amendments since its launching. The EU's Water Framework Directive focuses on integrated water protection, seeking to achieve the following goals, among others:

- Water protection should no longer stop at administrative or national borders. Rather, a holistic approach is called for, treating the river basin as a whole, as defined by its natural boundaries.
- Water protection should no longer be treated as a purely ecological or technical problem. In the future, the related economic and social issues will also be taken into account.
- Water resources should be managed sustainably and thus secured for future generations.
- The public should be involved in the measures undertaken.
- Deterioration in the state of groundwater, surface waters, and aquatic habitats should be prevented.

These goals are compatible with a number of the main directions of IWRM and are mainly oriented to achieving and maintaining the "good status" of the resource, not being as all encompassing as the full scope of IWRM would require. The WFD from the start placed a strong emphasis on the protection of groundwater, something which has not been as evident in the IWRM formulations elsewhere in the world, though there are some explicit efforts in this direction (Kennedy et al. 2009; INBO and GWP 2012).

A recent official report on the implementation of the WFD (European Commission 2012), focusing primarily on the WFD's mandated River Basin Management Plans (RBMPs), concluded that "A lot of effort has been put into the preparation and drafting of the RBMPs. Our knowledge about the status of EU waters and the activities that influence them is better than ever before. However, the Commission's assessment shows that a more determined effort is needed to ensure achievement of WFD objectives . . .." Some concerns have been raised by civil society about the effectiveness of the WFD's mandated RBMPs (EEB 2010).

There are a number of efforts for upgrading the standards in Europe as well. Timmerman et al. (2008) examined the IWRM-related research in Europe and concluded that it already addresses many of the future challenges. However, it is also found that to enable a transition toward the adaptiveness in water management, it should lend specific focus to incorporating into the integrative concept of IWRM aspects such as equity and fairness; vulnerability and resilience of systems; development of governance and participation models in the transboundary context, taking into account the differences in water management competences and political processes between countries; and development of an improved understanding of the consequences of climate change along with new approaches to deal with uncertainty and scenario planning.

5.3.4 Status of the Practice of IWRM in the World

UN-Water was asked by the UN Commission on Sustainable Development (CSD) in 2005 to produce status reports on the progress of water resources management for its succeeding meetings. The first such report was presented to CSD for its 16th session (UN-Water 2008).

Another parallel and complementary study was carried out supported by the UN system based on a 2007 survey of 58 mostly developing countries in an attempt to establish a more detailed picture with respect to the adoption of IWRM by countries, as reported in Hassing et al. (2009). One indicator was aimed at assessing the institutional capacity for the implementation of IWRM such as policy formulation, water allocation, and water demand management. Figure 5.4 illustrates the results. Countries show the highest score for the formulation of policies and accompanying legislation, while cost recovery practices and water demand management capabilities were among the lowest. This may point out that many countries may be setting

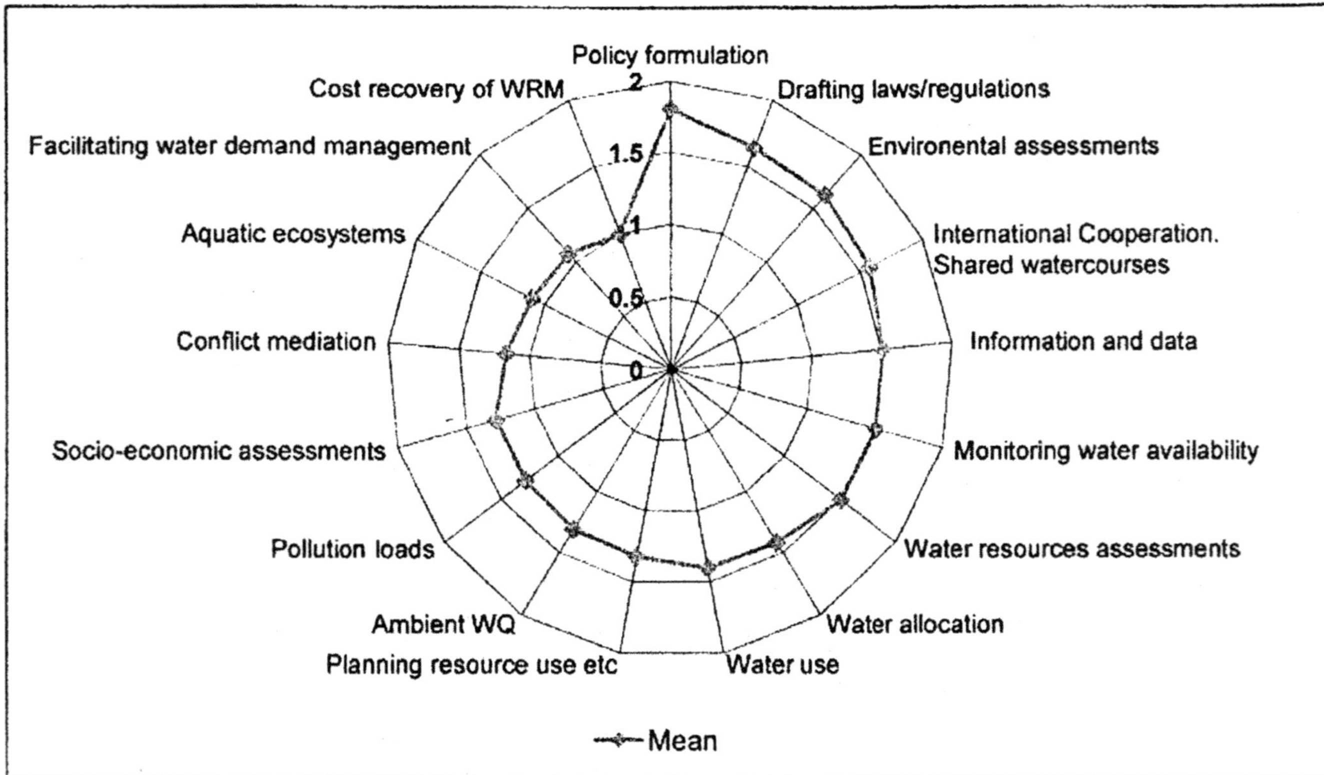

The ratings that the respondents were asked to give were: 0 = function not established, 1= function has many gaps in quality and coverage, 2 = function has some gaps in quality and coverage, 3 = function operates at realistic goal levels.

Fig. 5.4 The institutional capacity for carrying various IWRM-inspired functions (Source: Hassing et al. 2009, Fig. 2, p. 5, used by permission of UNESCO)

up the institutional structure for IWRM and they will enter a learning process to enhance the other necessary capabilities.

Thereafter, UN-Water prepared a specific status report (UN-Water 2012) for submission to the UN Conference on Sustainable Development 2012 (Rio+20). To this end, a detailed global survey was conducted in 2011, and the results were elaborated on the basis of data from 130 countries. This analysis looked at issues pertaining to the management, development, and use of freshwater resources, as well as possible outcomes and impacts of the application of integrated approaches to water resources management.

Among the key findings of the report, we cite:

- Integrated approaches to water resources management and development are critical for progress toward a green economy in the context of sustainable development and poverty eradication and for the attainment of climate adaptation and resilience.
- Since 1992, 80 % of countries have undertaken reforms to improve the enabling environment for water resources management based on the application of integrated approaches as defined in Agenda 21 adopted that year in the first Rio conference.
- 65 % of countries have developed integrated water resources management plans as called for by the Johannesburg Plan of Implementation (2002) issues at the follow-up Rio+10 conference, and 34 % report an advanced stage of implementation. However, progress appears to have slowed or even regressed in a number of developing and least developed countries after the first progress report was issued in 2008.
- Water-related risks and the competition for water resources are perceived by a majority of countries to have increased over the past 20 years.
- Though countries that have adopted integrated approaches report more advanced infrastructure development, further efforts are needed to ensure appropriate levels of coordination.
- Countries report improvements to the institutional framework together with improved policies, laws, and systems over the past 20 years. This has led to better water resources management practices bringing important socioeconomic benefits.

This report showed that while there is still a long way to go, progress toward the goal of sustainable water resources management is clearly being made.

5.4 Convergence or Divergence?

Up to this point, we have examined the IWRM process, including the historical background, the evolution of the concept, and even the status of its application internationally. There is certainly progress, and IWRM enjoys wide acceptance at the practitioner and governmental level. The associated literature is widespread, and there is work to incorporate within its framework the latest perceived challenges and internationally agreed priorities. However, it is always advisable to take an objective and critical look at platforms that are guiding significant national and international decisions. Various authors have indeed looked critically at IWRM and manifested dissatisfaction at its characteristics. Some of the common claimed objections are discussed below.

IWRM is not clearly defined. The currently accepted definitions attempt to encapsulate in brief statements views and guidance for the solution of complex and very diverse problems at widely varying scales. The generality and the vastness of the intended scope of IWRM lead to criticism that it is ill-defined from the conceptual to the procedural dimension (Biswas 2008; Merrey 2008). For instance, in the definition by GWP that was cited, questions arise about which are related resources to water and land, how does one maximize economic and social welfare in an equitable fashion, etc., and to the means to apply IWRM. A frequent observation is that while IWRM has been designed by and for water resources practitioners having a short-sighted scope, the actual playing field goes much beyond the water community and that the major development decisions are made outside the water sector.

IWRM is not implementable. Merrey (2008) states that "attempts to implement full Integrated Water Resources Management (IWRM) are doomed to failure and disappointment." Biswas (2008) also echoes this sentiment. Part of the problem derives from the alleged vagueness of the definition indicated above – no clear directions. Also, while IWRM pretends to provide the elements to resolve major water management issues, reality is becoming more complex as a consequence of intersectorial competition for the resource, of the interactive pressure from the various change drivers, of political transformations, etc., so effective solutions do not come easily. An author blames the "wicked" nature of the problem: "Wicked problems and messy situations are typified by multiple and competing goals, little scientific agreement on cause-effect relationships, limited time and resources, lack of information, and structural inequities in access to information and the distribution of political power" (Lachapelle et al. 2003). Under these conditions, normal planning procedures simply fail.

Existing formulations of IWRM do not possess the requisite flexibility. A parallel theme to the previous entry is that for IWRM to overcome its shortcomings in handling competently the full panoply of reality, it should adopt a more flexible stance or form. As it is, the platform is rigid and unbending, as it is mainly of a normative or prescriptive nature (what should be done). Merrey (2008) suggests an alternative approach to IWRM of interpretive, expedient nature that retains its inspiration from the Dublin principles and does not abandon the integrative perspective of IWRM, but requires critical and insightful thinking to formulate the problems to be solved and the operational steps to follow. Figure 5.5 illustrates the contrast between normative IWRM and adaptive water resources management.

The foundations of IWRM may not be valid everywhere. Another striking case is the claimed limited applicability of one of the basic principles of IWRM in the context of a Western European country. On the one hand and as referred earlier in

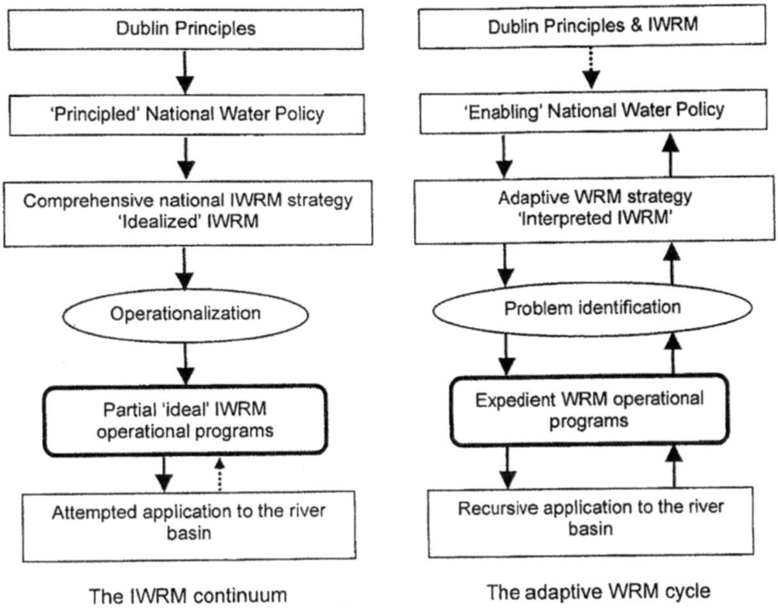

Fig. 5.5 The deployment of IWRM policy and operations – a partial ideal or expediency? (Source: Lankford et al. (2007) as cited in Merrey (2008, Fig. 2, p. 903))

Section 5.3.1.1, France set up the system of river basin agencies in 1964; these are held to be archetypical of the IWRM philosophy (formulated later) that calls for administrative units based on natural basin boundaries and that places emphasis on a participatory approach; further, the French central government has delegated a number of functions to these bodies. This is in sharp contrast with Germany that has (1) a directive government and a highly regulated water administration being organized according to administrative boundaries and (2) a management process in which relatively little participation exists. This could be judged as rigid and counterproductive were it not that water management in Germany functions efficiently and to the satisfaction of the population. Matz (2008) ascribes this result to the difference in mind-sets stemming from historical and cultural aspects; he recommends "Do not make participation a dogma." This advice contradicts Dublin Principle 2 that calls for public participation; thus, this principle could be said not to be generally applicable in Germany (participation is substituted by other mechanisms). Paradoxically, the creation of the French river agencies was inspired by the German model of the Genossenschaften, water management cooperative unions, the first of which was established in 1904 to manage the Emscher basin (Barraqué 1995) – but this model was not generalized to the rest of Germany.

The IWRM discourse has become too solemn and has been appropriated by a few (thus alien to many). The IWRM discourse has undergone in the last two decades a process of formalization, from adoption of solid principles to establishing guiding definitions and becoming an internationally accepted platform part of the sustainable development process. There was a progression from diverse initial positions reflecting specific regional and national experiences in water management and the specialized knowledge derived from scholars and professional associations to the eventual interventions of global networks aiming to create a consistent organized body of knowledge of IWRM theory to enable its application (Mukhtarov 2008). This process, in the minds of some, led to the creation of simplistic, palatable, wholesome standardized IWRM packages for wide dissemination and implementation; this resulted in some of the shortcomings and disconnects with the complexities of reality cited earlier. Furthermore, some critics hold that the IWRM rhetoric became more something resembling scriptures, elevated to a religion-like status; Delli Priscoli (2013) remarked "Are we chanting an IWRM mantra as one former Chair of the GWP once quipped?"; these perceptions need to be taken into account. Matz (2008) is particularly scathing: "IWRM is a means to an end and not a means in itself. However, the religious ferocity with which it is nowadays sometimes promoted nourishes the critical voices of its conceptual basement. It is time to review some of the ivory-tower like criteria that are used in an inflationary way."

It is clear that IWRM has generated differing views and has been controversial at times. This, in the author's opinion, is not so much a showing of weakness or ineffectiveness of IWRM, but rather evidence of a vigorous and healthy process that invites debate and distills applicable ideas and components to produce a strengthened body of knowledge with the proper qualifications. As a significant side benefit, this process has also provided a common language to the international community and a means to share water management experiences, stimulating the generation of exchanges and the upgrading of capabilities in the field engaging national, international, and intergovernmental organizations. As Merrey (2008) concludes, despite his critical observations, he "... has not advocated abandoning IWRM. IWRM continues to be shorthand for a valid model emphasizing the systematic interconnectedness within river basins. However ... the connections are not only hydrological and ecological but also – indeed primarily – political." Thus, without belaboring the point, convincing evidence has been produced that shows convergence is indeed taking place in the IWRM process.

5.5 Final Remarks

We have just made a short journey exploring the evolution of the IWRM and sustainability, the development of principles and concepts, and the implementation inroads and general performance. IWRM is an approach that by now has gained the favor of governments and that is being applied throughout the community of nations. The latest status UN report submitted to the 2012 Rio+20 conference (UN-Water 2012) gives a generally positive report about progress attained in the implementation of IWRM. Most water resources policy makers and practitioners at all levels have embraced IWRM; thus, it may be said that its adoption is following an expected path.

Despite these positive signs, some qualifications are applicable: IWRM is an approach, a philosophy if you will, that cannot be taken as a fixed prescription, but as a basis to tailor specific policies and solutions according to the varying realities: an adaptive and flexible tool. The IWRM concept serves to feed policy debates and as a common language of communication and to trade experiences between practitioners. This implies that the IWRM discourse should be considered from a nondogmatic and non-rhetorical stance. IWRM is an evolving subject in a period of over two decades; it has matured, adopting a general formulation – you can say this is the "normative" or "prescriptive" IWRM that gives way to the "adaptive" or "expedient" IWRM for effective implantation. Various issues of critical importance have come forward in recent years related to global changes, water security, the human right to water, and green economy which, depending of the basin, country, and region, shift the emphasis of the implementation goals of IWRM.

The IWRM implementation framework must be endowed with the capability to handle the emerging complexities and uncertainties and to become an integral part and support of the sustainable development process. IWRM would be largely ineffective if it remained a narrow tool within the water resources management discipline. The water-related interdependencies and interlinkages between many of the key development aspects are growing more evident; thus, water management issues cannot be resolved unilaterally by the water community. "Global interdependencies will increasingly be woven through water," declares the World Water Development Report (UN WWAP 2012b), adding that "without fully implemented (and adaptable) plans for IWRM, the 'nexus' dialogue creates a pragmatic and substantial opportunity for informed decision-making outside the 'water box', complementary to IWRM," highlighting the approach to achieve an intersectorial treatment.

The intergovernmental processes, particularly through the UN system with the support of a network of collaborating organizations, have resulted of fundamental importance to thrust forward the IWRM approach. Nonetheless, a considerable effort is still needed in positioning water in the intergovernmental arena so that its importance is properly reflected in the overall sustainable development debate. Two examples of the relegation of water are that, aside from two targets on water supply and sanitation, no prominent role was assigned to water in the Millennium Development Goals; further, up to the 19th session of the conference of the parties of the United Nations Framework Convention on Climate Change (UN FCCC 2014), there was no major focus on water. These two shortcomings mentioned above, and no doubt many more, have to be overcome; the fact that it is actually the countries that must act and for far too long what the community of water professionals have kept to themselves ("preaching to the converted") – obviously, this is a case for the freshwater community to go "out of the water box" at different levels and get the message out so that it reaches the broader international and intergovernmental venues to enable the changes. This is necessary in order to aim for truly integrated and sustainable solutions.

The necessity of having adaptiveness built into the IWRM process has been already remarked, but no matter how good the implementation plans, the institutional and professional capacity must be there for the execution. It has been noted that a regression in the application of IWRM has been observed in a number of developing and least developed countries in the last few years, underscoring the capacity-building needs; thus, a reinforced focus on this aspect must be brought to bear.

References

Barraqué B (1995) Les politiques de l'eau en Europe. Revue française de science politique 45(3):420–453. doi:10.3406/rfsp.1995.403539. http://www.persee.fr/web/revues/home/prescript/article/rfsp_0035-2950_1995_num_45_3_403539

Biswas AK (2008) Integrated Water Resources Management: is it working? Int J Water Resour Dev 24(1):5–22. doi:10.1080/07900620701871718

Cap-Net (International Network for Capacity Building in Integrated Water Resources Management), GWP (Global Water Partnership), UNDP (United Nations Development Programme) (2005) Integrated Water Resources Management plans – Training manual and operational guide. http://pawd.pbworks.com/f/IWRM+Planning-Training+Manual+Cap-Net+GWP.pdf

Delli Priscoli J (2013) Keynote address: clothing the IWRM emperor by using collaborative modeling for decision support. J Am Water Resour Assoc 49(3):609–613. doi:10.1111/jawr.12072

EEB (European Environmental Bureau) (2010) 10 years of the Water Framework Directive: a toothless tiger? A snapshot assessment of EU environmental ambitions, Brussels

European Commission (2012) Report from the commission to the European parliament and the council on the implementation of the Water Framework Directive (2000/60/EC) River Basin Management Plans, COM(2012) 670 final, Brussels. http://ec.europa.eu/environment/water/water-framework/implrep2007/index_en.htm

Gallopín GC (2012) Global water futures 2050 – Five stylized scenarios, UNESCO, Paris

GWP (Global Water Partnership) (2000) Integrated Water Resources Management, Background paper 4. GWP Technical Committee, Stockholm

GWP, INBO (International Network of Basin Organizations) (2009) A handbook for Integrated Water Resources Management in basins. Global Water Partnership, Stockholm

Hassing J, Ipsen N, Clausen TJ, Larsen H (2009) Integrated Water Resources Management in action. WWAP (World Water Assessment Programme)-Side Publication Papers- Dialogue, UNESCO, Paris

ICWE (1992) just refers to the International Conference on Water and the Environment that took place in Dublin in 1992

INBO, GWP (2012) The handbook for Integrated Water Resources Management in transboundary basins of rivers, lakes and aquifers (with collaboration of UNECE, UNESCO, GEF, EVREN and AFD). ISBN 978-91-85321-85-8

Kennedy K, Simonovic S, Tejada-Guibert A, Doria M, Martín JL (2009) IWRM implementation in basins, sub-basins and aquifers: state of the art review. International Hydrological Programme of UNESCO, WWAP Side Publications – Insights, UNESCO, Paris

Lachapelle PR, McCool SF, Patterson ME (2003) Barriers to effective natural resource planning in a "messy" world. Soc Nat Resour 16(6):473–490

Lankford B, Merrey DJ, Cour J, Hepworth N (2007) From integrated to expedient: an adaptive framework for river basin management in developing countries, IWMI research report no. 110. IWMI (International Water Management Institute), Colombo

Matz M (2008) Rethinking IWRM under cultural considerations. In: Scheuman W, Neubert S, Kipping M (eds) Water politics and development cooperation: local power plays and global governance. Springer, Berlin/Heidelberg. ISBN 978-3-540-76707-0

Mei Xie (2006) Integrated Water Resources Management (IWRM) – introduction to principles and practices, World Bank Institute, Africa Regional Workshop on IWRM, Nairobi

Merrey DJ (2008) Is normative Integrated Water Resources Management implementable? Phys Chem Earth 33(8–13):899–905. doi:10.1016/j.pce.2008.06.026, Elsevier

Mukhtarov F (2008) Intellectual history and current status of Integrated Water Resources Management: a global perspective. In: Pahl C, Kabat P, Möltgen J (eds) Adaptive and integrated water management – coping with complexity and uncertainty, Library of Congress Control Number: 2007937517. Springer, Berlin/Heidelberg. ISBN 978-3-540-75940

Snellen WB, Schrevel A (2004) IRM: for sustainable use of water; 50 years of international experience with the concept of integrated water management. Background document to the FAO/Netherlands conference on water for food and ecosystems. Alterra report 1143, The Hague/Wageningen, the Netherlands. Available at: http://edepot.wur.nl/30428. Accessed Apr 2015

Timmerman JG, Pahl-Wostl C, Möltgen J (eds) (2008) The adaptiveness of IWRM: analysing European IWRM research. IWA Publishing, London. ISBN 1843391724

UN FCCC (2014) Report of the conference of the parties on its nineteenth session, held in Warsaw from 11 to 23 November 2013, Addendum Part two: Action taken by the conference of the parties at its nineteenth session, FCCC/CP/2013/10/Add.2

UN-Water, GWP (2007) Roadmapping for advancing Integrated Water Resources Management (IWRM) Processes. Available at: http://www.unwater.org/downloads/UNW_ROADMAPPING_IWRM.pdf

UN-Water (2008) Status report on IWRM and water efficiency plans for CSD16. Available at: http://www.unwater.org/downloads/UNW_Status_Report_IWRM.pdf

UN-Water (2012) The UN-Water status report on the application of integrated approaches to water resources management, United Nations Environment Programme, Nairobi, Kenya

UNESCO-IHP (International Hydrological Programme), WWAP, NARBO (2009) IWRM at River Basin Level – guidelines, Part 1 Principles, Part 2 (Part 2-1 The guidelines for IWRM coordination; Part 2-2 The guidelines for flood management; Part 2-3 Invitation to IWRM for irrigation practitioners). UNESCO, Paris

United Nations (1958) Integrated river basin development; report by a panel of experts. Sales number 58.II.B.3 "E/3066". United Nations Publications, New York

United Nations (1977) Report of the United Nations water conference, Mar del Plata

UN WWAP (World Water Assessment Programme) (2009) Part 1 – Understanding what drives the pressures on water. In: United Nations world water development report 3 – water in a changing world, UNESCO, Paris

UN WWAP (World Water Assessment Programme) (2012a) Understanding uncertainty and risks associated with key drivers. In: United Nations world water development report 4 – managing water under uncertainty, vol 1. UNESCO, Paris

UN WWAP (World Water Assessment Programme) (2012b) Recognizing the centrality of water and its global dimensions. In: United Nations world water development report 4 – managing water under uncertainty, vol 1. UNESCO, Paris

USAID (2013) Water and development strategy 2013–2018. Washington, D.C

World Bank (2008) Practical solutions to water challenges – learning from the Spanish experience. World Bank, Washington, DC

Chapter 6
Water Resources Management and Sustainability in Mexico

Rafael Val-Segura and Jorge Arriaga-Medina

Abstract Mexico faces major challenges in water resources management. The variability in time and space of water resources, the increase in the number of users and the inequality in the consumption, the persistence of inadequate finance systems, the absence of a new water culture, and the lack of information systems and well-trained personnel in the departments and agencies dealing with water management call for the urgent application of the principles of integrated water resources management (IWRM) and frame these within a sustainable development strategy. This chapter presents some strategic guidelines for the transition from a "hydraulic" policy to a more comprehensive water policy in Mexico, mainly derived from the experience that the National Autonomous University of Mexico (UNAM) has developed through the UNAM's Water Management, Use and Reuse Program and the fundamentals of the Water Responsibility Program within the International Hydrological Program of the United Nations Educational, Scientific and Cultural Organization (UNESCO).

Keywords Integrated water resources management • Mexico • PUMAGUA • UNAM – Water Responsibility Program

6.1 Main Challenges in Water Management in Mexico

Mexico faces major challenges in water resources management. These challenges, according to the information available, point to a future water crisis. There are clear indications that the crisis is a reality in most of the Mexican territories. These need to be attended by new paradigms based on decentralization, differentiated responsibility, and an ecosystem approach that goes beyond the traditional view supported

R. Val-Segura (✉)
PUMAGUA, Universidad Nacional Autónoma de México, Torre de Ingeniería, Piso 5, Ala Norte Cubículo 3, Circuito Escolar, Ciudad Universitaria, CP 04510 Coyoacán, Mexico
e-mail: rvals@iingen.unam.mx

J. Arriaga-Medina
Red del Agua UNAM, Universidad Nacional Autónoma de México, Torre de Ingeniería, Piso 5, Ala Norte Cubículo 5, Circuito Escolar, Ciudad Universitaria, CP 04510 Coyoacán, Mexico
e-mail: jarriagam@iingen.unam.mx

S.G. Setegn, M.C. Donoso (eds.), *Sustainability of Integrated Water Resources Management*, DOI 10.1007/978-3-319-12194-9_6

by indiscriminate investment in infrastructure; otherwise, water may transcend in the coming years from an engine of development to a factor that limits it.

According to the Mexican National Population Council (CONAPO), the number of inhabitants in Mexico in 2050 will exceed 150 million (CONAPO 2013), which means an increase of more than 30 million people that will require drinking water and sanitation, food, and other goods that need water in their production processes. This growth will be faster in areas of high population density and territories where water resources are under strong anthropogenic pressure, so we can expect changes in their availability and quality as well. On the other hand, the increased temperature, decreased precipitation, and increased intensity of extreme events projected for Mexico in the second half of this century as a result of climate change will affect water availability in most of the territories – although its effects will be particularly intense in the north of the country (Martínez et al. 2010) – and increase the vulnerability of coastal dwellers, as pointed out by González-Villarreal et al. in another chapter of this book (González et al. 2014). In addition, most problems caused by the above factors will affect precisely the most vulnerable populations; hence, any policy designed should be guided by a strong equity criterion, which can only be built through the full participation of all stakeholders and sectors involved.

Making the integrated management of water resources in Mexico a reality involves, besides the understanding of the mentioned scenarios, considering the following challenges.

6.1.1 Variability in Time and Space of Water Resources

Due to its geographical location, two-thirds of Mexican territories are in an area where rainfall is scarce, reaching in some areas of the north annual values of only 169 mm. On the other hand, in the south are some elevated areas of the Grijalva basin which experience heavy rainfall with annual values of 4,200 mm. These can be counted among the high precipitation areas of the world. The spatial variability of precipitation is also combined with a very variable temporal distribution, because 68 % of the annual rains occur between June and September.

6.1.2 Coverage of Drinking Water and Sanitation

According to the information from the National Water Commission, national drinking water coverage of 91.6 % was recorded by the end of 2012, but with a clear disparity between states. While Yucatan and Colima, for example, have coverage above 98 %, the states with the highest degrees of marginalization, such as Oaxaca and Guerrero, only reach 80 % service provision. In absolute terms, this implies that about 9.5 million people do not have water supply service. Of these, 5.5 million people are located in rural areas, in locations with less than 2,500 inhabitants (CONAGUA 2011). In terms of sanitation, national coverage is estimated above 90.2 %, and again, the states of Oaxaca and Guerrero are those with the lowest coverage. Of the 11.1 million people without access to the service, 8 million are located in rural areas (CONAGUA 2012).

6.1.3 Water Quality

Despite the difficulties in assessing water quality,[1] it is estimated that about 30 % of surface waters, which provide about 30 % of the human consumption, are contaminated or heavily contaminated, and only 28.3 % have excellent quality (CONAGUA 2011). Regarding groundwater, the information is even more limited, as the federal government considers only the content of ionic salts – usually harmless to health – as a criterion for classification (Jimenez et al. 2010). According to the criteria defined in the Mexican norm NOM127-SSA1-1994, amended in 2000, 97.6 % of the supplied water is disinfected, at least with free chlorine residual. Finally, the Federal Commission for the Protection against Sanitary Risks (COFEPRIS) reported in July 2013 that 99 % of the beaches were suitable for recreational activities, while the presence of the bacterium *Enterococcus faecalis* was within the norm (SEMARNAT 2013).

6.1.4 Water and Food Safety

As in most developing countries, Mexico uses a significant amount of water for agriculture, around 76.7 % (FAO 2013). The production structure also presents severe dysfunctions, because the major producing areas are in the north, where water resources are scarce and high temperatures increase evapotranspiration. Although Mexico ranks sixth worldwide in terms of area equipped with irrigation (with 6.46 million hectares), only 18 % of agricultural production units are using this technology (CONAGUA 2011).

6.1.5 Water and Environment Preservation

There is a symbiotic relationship between water and ecosystems, since both provide to each other essential services to continue their natural cycles (Falkenmark 2003).

[1] In Mexico, different methodologies have been used to classify water quality. Recently, three major indicators have been defined: 5-day biochemical oxygen demand (BOD_5), chemical oxygen demand (COD), and [total suspended solids (TSS)].

Therefore, a change in any element of one of them will necessarily impact on the other. Although this link is becoming more widely recognized, the expansion of the agricultural frontier and livestock grazing areas, as well as the intensification of production and deforestation, has exerted the main pressure on ecosystems. As a result, nowadays only 17 % of the rainforests and 26 % of the dry forests of the country still exist (Balvanera and Cotler 2009). This has negative consequences on water resources: transforms climate and water quality, reduces groundwater recharge, and reduces the ability of ecosystems to react to extreme natural events.

6.1.6 Financial Mechanism

We can observe only three sources of finance of water system: user fees, government budgets, and subsidies and grants from donor agencies. The first has a strong disparity, as the largest user of water –agriculture – is exempt from taxes as an equity policy, while urban and industrial centers are subject to different taxes, but certainly not enough to recover the total costs of the service. The government budget specifically dedicated to drinking water and sanitation has the following composition: 49 % federal, 18.5 % state, and 12 % municipal funds (CONAGUA 2011). As we observe, the distribution presents inconsistencies according to national legislation, because the level with the greatest responsibilities has the lowest budget. A number of external institutions contribute to external financing, among them the Inter-American Development Bank, the International Bank for Reconstruction and Development, the Japan Bank for International Cooperation, and the Global Environment Facility. This external participation, while benefiting Mexico with innovation and development derived from international experiences, can increase the country's borrowing and be oriented toward non-priority areas for national development.

6.1.7 Water Governance

Mexico's water policy has replicated the most visible feature of the Mexican political system: its centrality. Under a scheme called by some as hydrocracy (Wester et al. 2009), planning, design, and implementation of water policy have been defined from the agendas and interests established by a bureaucracy with high resistance to profound institutional changes and a lack of experience in relation to the sustainability agenda. Important steps have been taken to ensure greater participation of users and local authorities at different levels, for example, with the creation of boards, commissions, and committees at basin level and groundwater technical committees and beaches, however, these schemes are still limited.

6.1.8 New Water Culture

Since the creation of the National Water Commission in 1989, the generation of a new water culture has been one of its main objectives. However, the strategies for developing this new culture have been partial and short-term. Among the problems that can be identified are (1) lack of professional staff, (2) transmission of knowledge generated, (3) no follow-up, (4) focus mainly in urban areas, (5) competition between departments within agencies, (6) legal and economic uncertainty, and (7) failure to review the regulatory framework (Perevochtchikova 2012). Despite the advances that have been reached in the subject by some NGOs and universities, their work has not been incorporated into government strategies.

6.2 Integrated Water Resources Management as a Sustainability Paradigm

From the 1970s, in particular since the1972 Stockholm Conference on the establishment of the United Nations Program for the Environment, the long road to a new paradigm focused on sustainability and the recognition of water as an essential element for humanity and ecosystems development has begun. However, the integration of the integrated water resources management (IWRM) principles on the national legislation and effective programs of action has been slow and with variable results. Mexico is an example of this situation.

IWRM is recognized as a process that promotes the coordinated development and management of water, land, and related resources, in order to maximize the resultant economic and social welfare in an equitable manner without compromising the sustainability of vital ecosystems (González-Villarreal et al. 2000). Associated with a shift in the cultural paradigm, IWRM considers the integrated element as the coordination between the hydrological, economic, social, and environmental perspectives and reflects the multidimensional, multisectorial, and multiregional approach in water management, also influenced by interests and agendas, which must be addressed in a coordinated way by the different institutions involved (González-Villarreal et al. 2006). This concept is not seen as an end but as a pathway to achieve sustainable development objectives, as shown in the diagram below (Fig. 6.1).

It was not until 1992, at the Conference on Water Resources and Environment, when a broad group of experts met to define some guidelines that enable more effective implementation of IWRM (Andrade 2004). Results of this debate were the four principles below:

1. Freshwater is a finite and vulnerable resource, essential to sustain life, development, and the environment. Given its importance for life, water management demands a comprehensive approach that is able to balance social and economic development with the natural cycles, and it considers water and man as components of one system. Similarly, the basin is recognized as a basic planning unit.

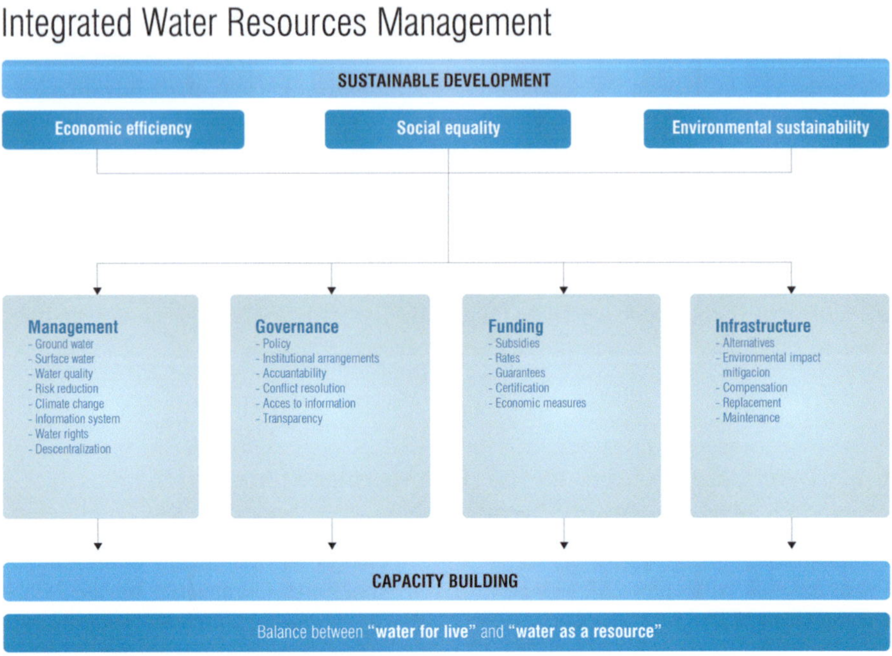

Fig. 6.1 Integrated water resources management process

2. Water development and management should be based on a participatory approach, involving users, planners, and policy-makers at all levels. Hierarchical and centralized systems in water management are recognized as contrary to the principles of sustainability; hence, the need to promote new schemes in which all actors and sectors involved in resource management should participate in the design, planning, execution, and evaluation of water-related projects. This requires a process of decentralization and a more informed society that is involved in transparency and accountability.
3. Women play a central part in the provision, management, and safeguarding of water. The role of women in water management is recognized at all levels, and the processes of inclusion are emphasized. This principle also calls for establishing statistical information systems and programs that promote gender equality.
4. Water has an economic value in all its competing uses and should be recognized as an economic good. This principle recognizes that many failures in water resources management are the result of treating them as free goods, which has encouraged overuse and waste. This principle does not compete with the basic right of humans and ecosystems to access water in the quantity and quality necessary to continue with its basic activities. Seeing water as an economic good doesn't mean necessarily to establish charging criterion, but pay attention to its total value (González-Villarreal and Solanes 1999).

The creation of these guiding principles meant a breakthrough in the transition to a new paradigm of sustainability in water management. This model proposes that water supply, sewerage and sanitation, and other services traditionally treated as building infrastructure under standardized schemes should be viewed as part of an extended water cycle, in which man also has an important place. The characteristics of this alternative are summarized in the following table, made by Pinkham (1999) (Table 6.1):

Table 6.1 Old and emerging paradigm for water systems

The old paradigm	The emerging paradigm
Human waste is a nuisance. It is to be disposed of after minimum required treatment to reduce its harmful properties	Human waste is a resource. It should be captured and processed effectively and put to use nourishing land and crops
Stormwater is a mistake. Convey stormwater away from urban areas as rapidly as possible	Stormwater is a resource. Harvest stormwater as a water supply, and infiltrate or retain it to support urban aquifers, waterways, and vegetation
Build to demand. It is necessary to build more capacity as demand increases	Manage demand. Demand management opportunities are real and increasing. Take advantage of all cost-effective options before increasing infrastructure capacity
Demand is a matter of quantity. The amount of water required or produced by water end-users is the only end-use parameter relevant to infrastructure choices. Treat all supply-side water to potable standards, and collect all wastewater from treatment in one system	Demand is multifaceted. Infrastructure choices should match the varying characteristics of water required or produced by different end-users: quantity, quality (biological, chemical, physical), level of reliability, etc.
One use (throughput). Water follows a one-way path from supply, to a single use, to treatment and disposal to the environment	Reuse and reclamation. Water can be used multiple times, by cascading it from higher- to lower-quality needs (e.g., using household graywater for irrigation) and by reclamation treatment for return to the supply side of the infrastructure
Gray infrastructure. The only things we call infrastructure are made of concrete, metal, and plastic	Green infrastructure. Besides pipes and treatment plants, infrastructure includes the natural capacities of soil and vegetation to absorb and treat water
Integration by accident. Water supply, stormwater, and wastewater systems may be managed by the same agency as a matter of local historic happenstance. Physically, however, the systems should be separated	Physical and institutional integration by design. Important linkages can be made between physical infrastructures for water supply, stormwater, and wastewater management. Realizing the benefits of integration requires highly coordinated management
Collaboration = public relations. Approach other agencies, and the public approval of pre-chosen solutions is required	Collaboration = engagement. Enlist other agencies and the public in the search for effective, multi-benefits solutions
Bigger/centralized is better. Larger systems, especially treatment plants, attain economies of scale	Small/ decentralized is possible, often desirable. Small scale systems are effective and can be economic, especially when diseconomies of scale in conventional distribution/collection networks are considered

From this new paradigm and the IWRM guiding principles, decision-makers worldwide decided at the Earth Summit (1992) to take action. The task has not been easy in Mexico, because there isn't a scenario in which all stakeholders can come together. The government has to accept its limitations in the public administration and recognize that not all capabilities are in the government. On the other hand, there is a need to recognize that society can't replace the government's actions, but also can't take a passive approach in relation to the day-to-day challenges (Carabias and Landa 2005).

The importance of this concept was not recognized in Mexico until 2004, when there were some reforms to the National Water Act, which is the most important tool for water management in Mexico, only superseded by the Constitution. The incorporation of IWRM to legislation, however, has failed to make adjustments in the institutional arrangement and legal framework. Restructuring the distribution of power between the authorities and matching them with their fiscal and operational possibilities; encouraging society's participation through rules, mechanisms, and clearly defined scope; and ensuring the incorporation of ecosystem vision in the design and implementation of water management policies are among the actions required to move toward a most comprehensive water policy.

Complications experienced by Mexico in the implementation of IWRM principles reflect, in addition to the lack of political will, the difficulties which have been experienced by almost all countries. In India, for example, the ministries that have a direct relationship with the water sector, such as agriculture, rural development, and environment, decided to commit to implement the IWRM principles and incorporate them into their plans and programs. The result, however, has been a duplication of efforts in a basin and sometimes conflicts between ministries. The opposite case is represented by the Netherlands, where environmental management and water resources management are institutionally separate, but conflicts still exist (Saravanan et al. 2009). Those cases are examples that it's necessary to put the principles of IWRM into action, not only in legal frameworks. An institutional change is a condition sine qua non in the sustainability pathway.

Even in countries with long experience in water management and with a more comprehensive regulatory framework, like Australia and South Africa, the implementation of IWRM has not been easy. While in Australia the legislation in some states only has been promoted by small groups or has not been applied at all, in South Africa, whose National Water Act is considered one of the most advanced in the reform toward IWRM, the objectives of equity, efficiency, and sustainability have been insufficient because of the different ways that actors apply their principles and how goals and objectives should be achieved (Saravanan et al. 2009).

Although there are serious difficulties in the implementation of its principles, IWRM should be recognized as the most advanced framework to formulate specific proposals. As Fernando González-Villarreal and Solanes (1999) point out: "the important thing is to focus on IWRM to find solutions to our problems and not, as it sometimes seems, to search for problems that fit a preconceived solution." This means to recognize regional differences and encourage each of the actors involved in water management to assume their responsibilities according to their place in the system. Examples of successful IWRM have been contributed by the National Autonomous University of Mexico (UNAM) through the UNAM's Water Management, Use and Reuse Program.

6.3 The UNAM's Contributions Toward IWRM

The National Autonomous University of Mexico has been characterized by its social and deep concern for environmental issues, particularly those related to water. From the UNAM's perspective, effective solutions for problems related to water use and conservation can only be achieved from interdisciplinary debate and the application of actions for the benefit of humans and ecosystems. Given the inherent complexity of the subject, the integration of a knowledge network that could link all the people that are interested in water was the best choice, in order to create viable and appropriate mechanisms to analyze and solve water problems at all levels. This gave rise to the UNAM's Water Network (RAUNAM).

In the framework of the IV World Water Forum, held in Mexico in 2006, 26 university departments decided to create RAUNAM. Also a seed fund was established for the development of multidisciplinary projects, which contribute to the UNAM's Water Management, Use and Reuse Program, known as PUMAGUA. Both projects respond to the growing concerns of the university authorities that the academic and scientific sector participates in the solution of water problems in Mexico City. An effective way to do that was through a program that was able to integrate IWRM principles at all campuses around the country, by initiating actions at the Central University City Campus (González-Villarreal and Domínguez 2012).

The Central University City Campus (CU) has an area of over 750 acres, in which are located 411 buildings and 237.3 acres corresponding to the Pedregal de San Angel Ecological Reserve (REPSA), which is characterized by its high level of endemic species. In 2007, an area of the campus was added to the list of World Heritage of the United Nations Educational, Scientific and Cultural Organization (UNESCO). This space is attended regularly by more than 132,000 people, and about 20,000 people visit daily. Due to its high concentration of cultural spaces, in 2010 alone, more than 1,600,000 people attended concerts, dance exhibitions, workshops, and conferences. This does not include the number of participants of the popular football matches held at the University Stadium (UNAM 2013). The large number of visitors raises water demand considerably.

Initiating actions at the Central University City Campus, PUMAGUA became active in 2008 with the development of a diagnosis of the situation and use of water resources. The result was that water supplied had good quality and obey the current standards, although there were irregularities in the concentration of free residual chlorine. Moreover, of 100 l/s extracted from three wells, 30 were used for human consumption, 20 for irrigation, and 50 were lost in leaks. It was also found that 15% of bathrooms had leaks, while 12% were out of service. There were 26 treatment plants BRAIN type whose volume processed did not meet the quality standard of the tributaries, working well below its design flow. Only one third of the green areas (50 ha) were irrigated with recycled water, and there were no detailed plans for the building or for the entire campus. Finally, from a survey of the university community, it was found that they did not perceive the problem of water waste and showed no clear interest of efficient actions.

Diagnostic results allowed PUMAGUA staff to focus their aim to implement a management, water use and reuse program at the UNAM with the participation of the entire community. To make this real, PUMAGUA have set three goals: (1) reduce the consumption of drinking water by 50%, (2) improved water quality for human use and consumption to make it drinkable and treated water to meet Mexican standards, and (3) encourage community participation in responsible water use. The activities performed to achieve these goals are organized into three areas:

- Hydraulic Balance – Responsible for designing and implementing actions to significantly reduce the water supply. Includes measurement and real-time monitoring of supply, leakage detection, segmentation and control of pressure in the distribution network, replacement of bathrooms, and replacement of intense water use vegetation with native plants, which is subject strictly to natural rainfall regimes.
- Water Quality – This area analyzes real-time water quality for human consumption and reuse. It checks whether water meets the Official Mexican Standards and monitors other microbiological parameters that are not covered by Mexican law but can also cause some adverse effects on health.
- Communication/Participation – Work to engage the community in the program. This area promotes responsible use of water and acceptance of the implemented measures by users and shares the information through social networks and media (PUMAGUA 2013).

By the end of 2012, PUMAGUA had achieved considerable results. Water extraction was reduced by 23% in the three wells. This was possible by the detection of leaks in the main supply networks within buildings, replacement of bathroom furniture, change of vegetation with native plants as those in the REPSA, and active community participation.

CU campus has an automatic disinfection system in the three wells, complying with the international standard for human drinking water. This allows to certificate that all people can drink safe water throughout the CU campus. Also, it implies a reduction in the consumption of bottled water and, consequently, the substantial decrease in solid waste generation.

The campus also has a flow measurement system in real time including electromagnetic water meters on the main network and supply wells, level sensors in tanks and distribution outlets, and water micrometers in buildings. 203 water meters have been installed so far in various campus entities.

PUMAGUA is developing a platform for real-time monitoring that will provide alerts about leaks inside buildings and deficiencies in water quality both for use and consumption. This platform will be consulted via the Internet to see both the quantity and the quality of the water consumed in the CU campus. This tool will also display the water consumption of the UNAM campus, generating a greater sense of responsibility for the conservation of the resource.

PUMAGUA has rehabilitated treatment plants that can comply with Mexican norm NOM-003-SEMARNAT-1997 for reuse on gardens. Among these actions, it is important to highlight the rehabilitation of the wastewater treatment plant Cerro del Agua, because it has the most advanced technology in Latin America. PUMAGUA has monitored water quality to comply with the Mexican standard, in addition to complementary microbiological parameters monitored within the program.

The program involves more than 110 units, which apply at least one of the five "PUMAGUA actions": installing water meters, replacing bathroom furniture, planting native vegetation, attending workshops given by PUMAGUA, and disseminating information. There is a water observatory that people can access via the Internet. It shows indicators of quantity, quality, and participation. Also, a person can see if there are leak problems, report the loss of flow within the unit, see the water quality, and find out the days when the standard for human use were not met and consumption in terms of residual free chlorine concentration.

In relation to the participation of all stakeholders in the efficient use of water, PUMAGUA has coordinated trainings on the following topics: installation and maintenance of water meters, monitoring of bathroom furniture, repairing of leaks, and security measures to prevent vandalism. It has also carried out workshops for gardeners with the aim of promoting efficient irrigation. PUMAGUA has also done literary and photography contests, festivals, research projects, conferences, workshops, and training of human resources in order to involve the community.

Development of human resources is a key activity for PUMAGUA. Since the start of the program, more than 15,000 students have been involved in at least one of the activities designed by PUMAGUA, and more than 250 research proposals on water use at the UNAM have been developed. Additionally, the work done by the program has been presented to a wider community through scientific publications, national media (television, radio, and newspapers), conferences, and briefings.

Thanks to the success of PUMAGUA, its principles have been recognized by the International Hydrological Program of UNESCO as essential for IWRM, even giving rise to the Water Responsibility Program (Val et al. 2011). The main concept behind this program refers to the commitment to involve different actors of society in water management, in order to maintain water availability and quality to be sufficient, safe, acceptable, accessible, and affordable to ensure the human right to water and to preserve at the same time the ecosystem functions (Val et al. 2012).

This commitment is applied through various activities and practices. It means social participation in quality preservation and the promotion of optimal savings within the basin in which the stakeholders are living.

The program is developed at the microscale, corresponding to individual actions, and at a macroscale, it comprises the institutions responsible for water management and productive sectors, which are in constant interaction, assigning each one them a different degree of responsibility, as we can see in the table below (Fig. 6.2).

Fig. 6.2 Actors involved in Water Responsibility Program

The Water Responsibility Program is still under construction. To reach its goals, four phases have been identified, which will allow for more efficient water management in Latin America. These are:

Phase 1. It consists in generating a Water Responsibility Index that is based on the participation of different stakeholders. It allows to know the way people use water and to gather their perception about the problems and challenges within each sector. Through this indicator, it is possible to define the responsibility degree that corresponds to each one of them, in order to determine the type of action to be carried out in each case.

Phase 2. Program implementation through:

- Diagnosis of the situation that exists in Latin America and the Caribbean, in terms of rules and legal frameworks.
- Creation and implementation of the Water Responsibility Program tailored to the different stakeholders.
- Final evaluation

Phase 3. Planning and implementing adjustments to the standards through a virtual forum.

Phase 4. Publication of manuals to guide the implementation of the Water Responsibility Program under various scenarios.

So far, the program has completed Phase 1, with the development of a proposal of indicators by stakeholders. They generate scores that are between zero and one, one being the expression of more water responsibility (Val et al. 2013). At the end, a summation of the components is done by each sector and the total water responsibility level is obtained. Indicators are designed for effective implementation at different stakeholders, as they can be applied to very limited spaces, as a school or a municipality, to a basin and countries as a whole. The proposed indicators should be discussed in broader forums to incorporate as many realities as possible, but they are a first approach to make real IWRM at different levels.

6.4 Final Thoughts

IWRM has been presented as a gradual approach for the solution of problems for structural changes in a way that is economically efficient, socially equitable, environmentally sustainable, and politically feasible and acceptable to all citizens (Oswald et al. 2011). Making IWRM a reality is urgent; otherwise, the scenarios of degradation and resource depletion can transform water from an engine for development to a constant source of conflicts.

Despite the efforts made in Mexico, incorporating the principles of IWRM in Mexican law has proven not sufficient for transforming them into comprehensive and long-term plans and programs. The 2030 Water Agenda proposed by the previous government administration as a general framework for the transition to a medium-term water policy did not reach the expected results because it failed in the definition of concrete proposals by sectors and stakeholders and the establishment of differentiated responsibilities and was not supported by the academic sector. The start of a new government and the requirement to formulate a new water law open important opportunities for sustainable water management in Mexico. However, the new administration must move beyond the entrenched hydrocracy and incorporate the needs and visions of other social sectors, especially the academic, who have demonstrated through the generation of science and technology that their contributions can't be neglected.

The pathway to decentralization in water management in Mexico has been slow but has allowed some progress. Without intending to replace the obligations that the government has on water resources, other players must take more proactive actions according to their position in the system. The experiences of PUMAGUA and Water Responsibility Program have shown that participation of all under comprehensive frameworks is necessary for achieving sustainable development. These efforts should not remain isolated, but should be replicated at different scales. To achieve these goals, political will is substantial.

References

Andrade A (2004) Lineamientos para la aplicación del enfoque ecosistémico a la gestión integral del recurso hídrico. PNUMA, Mexico City

Balvanera P, Cotler E (2009) Estado y tendencia de los servicios ecosistémicos. In: José Sarukhán (coordinator). Capital Natural de México. CONABIO, Mexico

Carabias J, Landa R (2005) Agua, medio ambiente y sociedad. Hacia la gestión integral de los recursos hídricos en México. UNAM-COLMEX-Fundación Gonzalo Río Arronte, Mexico City

CONAGUA (2011) Estadísticas del agua en México. SEMARNAT, Mexico City

CONAGUA (2012) Situación del Subsector Agua Potable, Alcantarillado y Saneamiento. SEMARNAT, Mexico City

CONAPO (2013) Proyecciones de la Población. SEGOB. http://www.conapo.gob.mx/es/CONAPO/Proyecciones_de_la_Poblacion_2010-2050. Accessed 1 August de 2013

Falkenmark M (2003) Water management and ecosystems: living with change. GWP, Stockholm
FAO (2013) Aquastat. FAO. http://www.fao.org/nr/water/aquastat/data/query/results.html. Accesed 1 Aug 2013
González-Villarreal F, Domínguez M (2012) In Perevochtchikova M (coordinator). Cultura del agua en México. Conceptualización y vulnerabilidad social. UNAM-PINCC-RAUNAM-Miguel Ángel Porrúa, Mexico City
González-Villarreal F, Solanes M (1999) The Dublin principles for water as reflected in a comparative assessment of institutional and legal arrangements for integrated water resources management. GWP, Stockholm
González-Villarreal F et al (2000) Manejo integrado de recursos hídricos. GWP, Stockholm
González-Villarreal F et al (2006) Orientaciones estratégicas. Propuestas para el manejo del agua en México. UNAM, Mexico City
González-Villarreal F et al (2014) Sustainability of water resources in tropical regions in the face of climate change. In: Setegn SG, Donoso PI (eds) Sustainability of integrated water resources management water governance, climate and ecohydrology. Springer, Miami
Jiménez B et al (2010) Calidad. In: Jiménez B et al (eds) El agua en México: cauces y encauces. Academia Mexicana de Ciencias-CONAGUA, Mexico City
Martínez P et al (2010) Efectos del cambio climático en los recursos hídricos. In: Jiménez B et al (eds) El agua en México: cauces y encauces. Academia Mexicana de Ciencias-CONAGUA, Mexico City
Oswald U et al (2011) Manejo integral en cuencas hidrológicas. Multidisciplina y multiinstitucionalidad como paradigmas de acción. In: Oswald U (Coordinator). Retos de la investigación del agua en México. CRIM-UNAM, Cuernavaca
Perevochtchikova M (2012) Nueva cultura del agua en México. In Perevochtchikova M (coordinator) Cultura del agua en México. Conceptualización y vulnerabilidad social. UNAM-PINCC-RAUNAM-Miguel Ángel Porrúa, Mexico
Pinkham R (1999) 21st Century water systems: Scenarios visions and drivers. http://www.rmi.org/Knowledge-Center/Library/W99-21_21stCenturyWater. Accessed 2 July 2013
PUMAGUA (2013) Informe PUMAGUA 2011–2012. UNAM, Mexico City
Saravanan VS et al (2009) Critical review of integrated water resources management: moving beyond polarised discourse. Nat Res Forum 33:76–86
SEMARNAT (2013) Resultados de calidad de agua de mar. SEMARNAT. http://www.semarnat.gob.mx/playas/playas_limpias/Paginas/resultados.aspx. Accessed 6 July 2013
UNAM (2013) Campus central de la Ciudad Universitaria Patrimonio Mundial. UNAM, Mexico City
Val R et al (2011) Programa de responsabilidad hídrica. Programa Hidrológico Internacional-UNESCO, Mexico City
Val R et al (2012) Programa de responsabilidad hídrica. Fase I Documento guía. Programa Hidrológico Internacional-UNESCO, Mexico City
Val R et al (2013) Programa de responsabilidad hídrica. Fase I Propuesta de indicadores. ProgramaHidrológicoInternacional-UNESCO, Mexico City
Wester P et al (2009) The hydraulic mission and the Mexican hydrocracy: regulating and reforming the flows of water and power. Water Altern 2(3):395–415

Part II
Echohydrology, Water Resources and Environmental Sustainability

Chapter 7
The Gap Between Best Practice and Actual Practice in the Allocation of Environmental Flows in Integrated Water Resources Management

Michael E. McClain and Elizabeth P. Anderson

Abstract A major component of environmental sustainability in water resource development is the explicit allocation of water to meet ecosystem needs. This environmental water allocation is commonly referred to as an environmental flow, which is the main subject of this chapter. A shift towards more consideration of water needs of ecosystems/environment in Central and South America has been more irregular, with some countries increasingly articulating and prioritizing these needs (e.g., Costa Rica and Colombia) and others not. The situation is similar in Africa, where ambitious new water policies with substantial attention to environmental protection have appeared in Eastern and Southern Africa (McClain et al., Int J Water Resour Dev 29(4):650–665, 2013) and Asia, where China stands out as a globally important country undergoing rapid change in its outlook towards environmental flows (Wang et al., Ecol Appl 21:163–174, 2009). In this chapter, we explore the status of environmental flow science and practice around the world, focusing on the gap that exists between environmental flow levels suggested by aquatic scientists and those actually protected in water regulations. With a wealth of science and different technologies to make use of, some of the most difficult challenges in applying best environmental flow practices lie in the governance processes and equitable allocation among water users and the environment. This brings us back to the promise of IWRM itself as a process to facilitate integration of these factors in a highly participatory fashion. In this chapter, we have endeavored to summarize the promise and highlight the current challenges of environmental flow assessment and implementation to enable the protection of ecosystems in the process of IWRM.

M.E. McClain (✉)
UNESCO-IHE Institute of Water Education, 2611 DA Delft, The Netherlands
e-mail: m.mcclain@unesco-ihe.org

E.P. Anderson
School of Environment, Arts and Society, Florida International University, Modesto Maidique Campus, ECS 486, Miami, FL, USA
e-mail: epanders@fiu.edu

S.G. Setegn, M.C. Donoso (eds.), *Sustainability of Integrated Water Resources Management*, DOI 10.1007/978-3-319-12194-9_7

Keywords Environmental flows • IWRM biodiversity • Freshwater ecosystems • Ecosystem services

7.1 Introduction

A defining characteristic of integrated water resources management (IWRM) is its commitment to balance socioeconomic development of water resources with environmental sustainability. This is articulated in the definition of IWRM by the Global Water Partnership (GWP 2000) and is being adopted in new water policies and legislation worldwide (UNEP 2012). A major component of environmental sustainability in water resource development is the explicit allocation of water to meet ecosystem needs. This environmental water allocation is commonly referred to as an environmental flow, which is the main subject of this chapter.

The 2012 UN report on the Application of Integrated Approaches to Water Resources Management (UNEP 2012) describes the priority given by countries around the world to water for ecosystems as well as the changes in priorities over the past 20 years. Not surprisingly the national situations vary by level of development. More than 50 % of countries ranked "very high" on the UN's Human Development Index (HDI) reported that water for ecosystems/environment was a high (or even highest) priority (UNEP 2012). Medium-level HDI countries reported a similar level of priority, but a smaller proportion (32 %) of low-level HDI countries gave high priority for water for ecosystems/environment. In fact, 32 % of low-level HDI countries also ranked water for ecosystems/environment as a low priority; the remaining 35 % ranked it as medium priority. In terms of changing perceptions, more than 60 % of very high-level HDI countries reported that the priority of water for ecosystems/environment had increased in the past 20 years, while slightly more than 30 % of low-level HDI countries reported the same (UNEP 2012). Despite the understandable variability among countries, the general picture (as reported) is promising for ecosystems and the environment, and changing perceptions seem to be moving (gradually) in a favorable direction.

These changes are taking place through the development and implementation of new policies and laws. A prime example is the European Union (EU) Water Framework Directive (WFD 2000), which includes an ambitious ecological goal of allowing only small deviations from natural conditions in the biology and water quality of freshwater bodies (classified as "good" condition). This new directive has stimulated a large amount of coordinated research and improved practices in the 28 EU member countries (as of 2013). In North America, the move toward greater implementation of IWRM and environmental protection has occurred largely at the level of individual US and Mexican States and Canadian Provinces. The changes have been especially large in more arid regions of the US West and Northern Mexico where over-development and utilization of water resources have caused sometimes severe environmental degradation (USACE 2012). A shift toward more consideration of water needs of ecosystems/environment in Central and South

America has been more irregular, with some countries increasingly articulating and prioritizing these needs (e.g., Costa Rica and Colombia) and others not. The situation is similar in Africa, where ambitious new water policies with substantial attention to environmental protection have appeared in Eastern and Southern Africa (McClain et al. 2013), and Asia, where China stands out as a globally important country undergoing rapid change in its outlook toward environmental flows (Wang et al. 2009).

While the global outlook in terms of acknowledged priorities and enabling policies for environmental flows is generally favorable, implementation on the ground has not kept pace. To implement policy priorities, water resources managers must first quantify the flow regime required to meet ecosystem water needs and then protect these flows in basin water allocation plans and the operational plans of water infrastructure like dams. When there are competing uses for water, however, a strong scientific and socioeconomic basis often is required to justify significant (e.g., >10 % of mean available flow) allocations of water to the environment. Otherwise, ecosystems usually receive only minimal flows that are more a reflection of the amount of water other users are willing to forego rather than the amount needed to maintain ecosystem health. Furthermore, in many countries the scientific information that is needed to determine the quantity, quality, and timing of flows needed to sustain ecosystems is limited.

In this chapter, we explore the status of environmental flow science and practice around the world, focusing on the gap that exists between environmental flow levels suggested by aquatic scientists and those actually protected in water regulations. We begin with a brief history of environmental flow science and a summary of best practices for environmental flow assessment. We then look at the actual levels of environmental flows being protected in water resources management regulations of select countries around the world. There is a sizable gap between the flow levels recommended by best practices and those protected in most regulations. We end by considering the consequences of this gap in practice and make a few suggestions for possible ways forward.

7.2 Definition, History, and Early Approaches

Today the most widely accepted definition of environmental flows is "the quantity, timing, and quality of water flows required to sustain freshwater and estuarine ecosystems and the human livelihoods and well-being that depend on these ecosystems" (Brisbane Declaration 2007). This comprehensive definition has evolved over many years as the development of environmental flow science and practice is closely tied to the development and use of water resources. The earliest interests in the low-flow characteristics of managed rivers date from the mid-1800s and relate to the self-purifying properties of flowing water. At that time, surface waters were prioritized over groundwater for domestic supply, and even the smallest flows were generally considered sufficient to protect public health (Chandler 1873).

When it became clear that the self-purifying capacity of rivers could be overwhelmed by excessive wastewater discharges, engineers developed interest in calculating the capacity of streams and rivers to receive and assimilate wastes. Because river levels vary over wide ranges, engineers sought to identify minimum flow levels that would provide dependable dilution factors in their calculations. Any flows above these minimums would provide extra dilution, but minimal flows set a base level below which dilution capacity should rarely fall. By the early to mid-twentieth century, Europe and North America had accumulated river flow records of sufficient length (>20 years) to enable detailed statistical analyses of flow characteristics, and hydrologists began to formulate indices describing the frequency of different low flows, including probability distributions of flow levels of given magnitudes and durations (1, 7, 14, 30, etc. days) (Gumbel 1954; Matalas 1963). Sanitation engineers selected certain of these indices to use in their dilution calculations. This approach to meeting water quality objectives in rivers receiving wastewater discharges continues today. For example, under the US Clean Water Act, the lowest mean 7-day flow level in a 10-year period (7Q10) is used as the basis for calculating the minimal dilution capacity of receiving waters (USEPA 1997). This figure is treated as a "critical low flow" indicating when a river is most susceptible to pollution levels exceeding water quality criteria. It is not a recommended flow but instead a critical threshold that should not be crossed. It is also not an approach that considers the ecology of streams and rivers.

Interest in environmental flows to protect the ecological condition of streams and rivers emerged in the 1950s in response to growing water withdrawals for hydropower and irrigation projects. This emergence is well documented in the western USA, where fishery biologists became concerned by the increasing number of dewatered river reaches downstream of newly constructed water projects (Orsborn and Allman 1976). Both the projects and the fishery biologists were mainly attached to State and Federal Agencies, which facilitated an interaction among agencies and a somewhat coordinated approach to the issue. Emphasis was initially placed on the determination of minimum flow levels that should be released from dams and diversions during the late summer season, which was the dry season and the time when most reaches were dewatered. Research extending over more than 20 years revealed that healthy fish populations were maintained in rivers having mean dry season flows in the range of 30–60 % of the mean annual flow (Trihey and Stalnaker 1985). This was based on the quantification of flow levels necessary to maintain the multiple habitats (defined mainly in terms of depth, velocity, and substrate material) required by riverine species.

Increasing knowledge of the relationships between flow levels and the ecological condition of rivers led to the formulation of incremental methodologies for assessing environmental flows such as the widely used Tennant method, which relates changes in the general condition of river ecosystems to changes in baseflow (Tennant 1976). The methodology describes the progressive worsening of river conditions as baseflows drop beneath the annual mean flow and highlights the need for different baseflow levels during different seasons and occasional high flows for channel maintenance. Tennant also stressed that 10 % of mean annual discharge should be the lowest instantaneous flow ever recommended, not because it provided for good ecological conditions but because anything less would likely cause "catastrophic degradation to fish and wildlife resources and harm both aquatic and riparian environments" (Tennant 1976). Flows equivalent to 10 % of the annual mean were found to provide only the minimal requirements for short-term survival of species.

Incremental methodologies like that of Tennant represented an important early advancement in the practice of environmental flows because they offered a range of options and illustrated the trade-offs involved when deciding allocations of water between ecological and other uses. They also highlighted the importance of seasonal variability in flows and the need for occasional high flow events. From the beginning, environmental flow scientists have stressed that single "minimum flow" values are unsatisfactory for sustaining ecosystems in river corridors (Stalnaker 1990).

7.3 Current Best Practices in Environmental Flow Assessment: Holistic Approaches

During the past 30 years, environmental flow science and practice have continued to improve in step with advances in ecohydrological sciences, new consideration of social-ecological systems, and adoption of principles of integrated water resources management. Environmental flow science and practice are also spreading internationally and are now on the political agenda in many countries around the world (Moore 2004; Le Quesne et al. 2010). From a technical standpoint, there have been significant advances in flow regime analysis, reach-scale habitat modeling, basin-scale runoff modeling, and basin-scale water planning systems. In the mid-1990s, environmental flow science adopted a natural flow regime paradigm, which identified five interrelated characteristics of a river's flow regime that are important for maintaining ecological integrity, namely, the magnitude, frequency, duration, timing, and rate of change of flows (Poff et al. 1997). This was accompanied by the identification of many new ecologically relevant flow indices and the development of new computational tools to quickly calculate these indices and compare their values before and after withdrawals or regulation by water projects (Richter et al. 1996). Even earlier, in the mid-1980s, new reach-scale hydraulic assessment tools enabled the integrated measure of both habitat quantity and quality for fish as a function of river flow levels (Milhous et al. 1984). With this information, habitat suitability criteria could be specified for key indicator species at different times of the year and in different stages of growth (Bovee 1986). New developments in basin-scale runoff modeling enabled the simulation of flow regimes across time and space, allowing environmental flow scientists to both reconstruct past flow regimes and project future flow regimes under different development scenarios (Arnold et al. 1998). These models are also useful in ungauged river basins, especially when satellite-derived data can be used to drive them (Tobin and Bennett 2009; Dessu and Melesse 2012). Finally, new modeling tools for water systems incorporate environmental flow recommendations directly into basin planning efforts, allowing for detailed analysis and optimization of water allocations among multiple uses and ecosystem protection (Yates et al. 2005).

Use of these new and multidisciplinary tools is ideally coordinated within a larger environmental flow assessment (EFA), which provides a structured process of collaborative analysis and flow setting among multiple scientists and stakeholders. A number of EFA frameworks have been developed over the past 20 years, and a few have emerged as best practices recognized around the world. The building block methodology (BBM) (King et al. 2000) was one of the first and is currently one of the most widely used frameworks. It is an objective-based approach involving a team of physical, biological, and social scientists working together to determine the flow requirements to either maintain a river in its current ecological condition or transition it to another (usually better) condition. Building blocks consist of monthly base flows during normal and drought years, high or flood flows required for specific ecological or channel-forming processes, and any other flow component judged important for achieving the ecological objectives set for the river. The BBM results in a specific set of flow recommendations that can be applied in water allocation plans or infrastructure operational plans. A closely related framework, the Downstream Response to Imposed Flow Transformation (DRIFT) methodology (King et al. 2003) shares many elements with the BBM but, rather than producing a fixed recommendation for application, produces a set of scenarios for water managers to consider when setting environmental flow levels. DRIFT also seeks to place environmental flows in a larger macro-economic context to facilitate better decision-making at basin scales. BBM and DRIFT are well-established best practices and have been widely applied.

The newest, and still somewhat experimental, EFA frameworks seek to assess environmental flows for many rivers at the same time over larger basins or regions. The ecological limits of hydrologic alteration (ELOHA) framework (Poff et al. 2010) was the first of these and serves as the basis for even newer frameworks adapted from it (Pahl-Wostl et al. 2013). The essential innovations of ELOHA are the classification of river sections of the assessment area into ecologically relevant flow regime types and the development of flow alteration-ecological response relationships for each flow regime type. With this information, required flow characteristics to maintain desired ecological conditions can be simultaneously assessed across the entire basin or region being examined. ELOHA also recognizes a social process to define acceptable ecological conditions based on societal values and management needs. The ecological condition defined by this social process becomes the target for the ecohydrological processes.

While current best practices call for holistic environmental flow assessments within structured frameworks and employing the latest analytical and modeling tools, these practices are applied in only a small proportion of environmental flow efforts today. One explanation for this is that most environmental flow regulations around the world, where they exist, do not insist on best practices in the assessment of flow requirements. This is partially attributable to a lack of understanding of the socioeconomic costs and benefits of maintaining environmental flows, a lack of political will, and limited legal, institutional, and monitoring arrangements (Moore 2004). Limitations in technical and financial resources are also factors, as these best practices are considerably more demanding than simple hydrological indices. In the absence of holistic EFAs, current best practice calls for application of the precautionary principle and recommends minimal alteration of hydrological regimes (Richter et al. 2012), but the reality is quite different.

7.4 Current Reality of Environmental Flow Regulations and Implementation

The widespread adoption of integrated water resources management has committed countries around the world to sustain vital ecosystems even while they continue to develop their water resources (UNEP 2012). Environmental flow science over the past 60 years indicates that in order to maintain the ecological structure and function of riverine ecosystems, different flow levels are required at different times of the year, including variable baseflows and occasional high flows. Yet, the majority of environmental flow regulations in place today continue to require nothing more than flatline minimum flows that lack any ecological considerations. Simple and invariable indices like Q95, 7Q10, and %MAR continue to dominate regulations and application (Tharme 2003). The case of Brazil is illustrative here. Responsibilities for water regulation and management are divided between federal and state authorities in Brazil, and rules vary accordingly. The Brazilian National Water Agency recommends an environmental flow level of 70 % of Q95, with allowances for variations among regions of the country (Santos and Cunha 2013). Among the states' agencies, rules range from 50 % of Q95 to 90 % of Q90 and 20–100 % of 7Q10. These figures may appear arbitrary out of the local context, and the full rationale for their selection is not clear, but what is clear is that there is no example of a requirement for variable flows throughout the year or for holistic EFA methodologies that consider ecological needs explicitly (Santos and Cunha 2013). Increasing attention has been devoted to consideration of ecological needs in recent years, but these efforts have as yet not impacted existing legislation (Sarmento 2007; Santos and Cunha 2013). In many cases, these hydrological indices are remnants of earlier times when minimum flows were required for dilution of sanitary wastes rather than ecological protection. There are even cases (e.g., countries of the former Soviet Union) where minimum flows continue to be called sanitary flows or where the terminology has recently changed from sanitary flow to environmental/ecological flow but with no change in accompanying flow levels (Abbasov and Smakhtin 2009).

Even in countries that have recently revised their environmental flow regulations, best practices have received limited attention. In 2013, the Ministry of Public

Works and the Ministry of Environment in Chile approved new legislation establishing a hydrology-based environmental flow requirement for the issuing of all new water rights (RoC 2012). The environmental flow is to be set at 20 % of the mean monthly flow (based on 25+ years of hydrological data), with a maximum allowable environmental flow of 20 % of the mean annual discharge. In other words, in months when 20 % of the mean monthly flow falls below 20 % of the mean annual flow, the monthly figure is enforced. However, in months when 20 % of mean monthly flow exceeds 20 % of mean annual flow, the annual figure is enforced. This caps the enforceable environmental flow at 20 % of the annual mean. Under special circumstances, such as in high conservation value rivers, the President of Chile may request a higher environmental flow requirement, but even this request may not exceed 40 % of mean annual flow and cannot infringe upon any established water rights.

An example of emerging recognition of more detailed ecological requirements in environmental flow rules may be found in Austria. In 2010, as part of its reformulation of national laws and policies to conform with the European Water Framework Directive, the Austrian Ministry of Life issued a new Quality Objective Ordinance on the Ecological Status of Surface Waters, which included new rules for the quantification of environmental flows (Ministry of Life 2014). Under the ordinance, environmental flows are recognized as having at least two components: a permanent minimum flow rate and a dynamic flow rate. The permanent minimum flow rate represents the lowest flow that should be allowed. It is based on the lowest daily flow level ever recorded in the river reach under consideration and is expressed as a proportion 33–50 % of the mean annual daily low flow throughout the flow record. The dynamic flow builds upon the minimum flow and includes dynamic components required to maintain (a) the seasonal character of the natural bed-sediment relocation and thus substrate composition, (b) sufficient current/flow in times of spawning migrations, (c) different habitat demands of individual age classes of key organisms during different times of the year, and (d) oxygen and thermal conditions which are typical of the water body. The resulting environmental flow recommendation is thus a combined minimum and dynamic flow regime that meets the listed requirements. Unfortunately implementation of this new rule is proceeding very slowly, as most installed infrastructure continues to operate under old rules that allowed for flows below historical daily low flows.

Another current approach that is incorporating ecological considerations comes from the State of Connecticut (CT) in the USA. At the end of 2011, the Department of Energy and Environmental Protection issued new Stream Flow Standards and Regulations (www.ct.gov/deep/streamflow). For Class 1 rivers in CT, which are those that have not yet been developed and maintain unaltered biological communities, no modification of natural flow regimes is allowed except in times of emergency (e.g., water required for firefighting). For Class 2 rivers, which are those with minimally altered biological communities, a flow equivalent to 75 % of the natural flow must be maintained in the river, and this value is to be adjusted every 2 weeks to reflect natural variations in the flow regime. For Class 3 rivers, with moderately altered biological communities, environmental flows are set according to the bioperiods of certain aquatic species. The rules recognize bioperiods related to overwintering, habitat formation, spawning, and rearing and growth. These vary somewhat depending on fish species, including salmonids, resident trout, and clupeids (herrings, shads, and sardines). Different hydrologically indexed minimum continuous flow levels ranging from Q99 to Q50 are prescribed during the different bioperiods, presumably in relation to the flow requirements to meet ecological needs in the given rivers. Adjustments in the operational plans of all permitted infrastructure (dams) must comply with the new requirements within 10 years.

One of the most advanced examples of new environmental flow legislation comes from the State of Texas in the USA. In 2007, the Texas Legislature passed Senate Bill 3, which requires that environmental flow recommendations and standards be established for all Texas river basins and estuaries. Environmental flows are to be based on the best available scientific data and to be guided by the goal of ensuring a sound ecological environment, which is defined as "an ecological environment that: supports a healthy diversity of fish and other aquatic life; sustains a full complement of important species; provides for all major habitat types including rivers and streams, reservoirs, and estuaries; sustains key ecosystem processes; and maintains water quality adequate for aquatic life" (TCEQ 2012).

The Texas approach is holistic and blends scientific and stakeholder inputs through a process anchored by key committees. At the State level, an Environmental Flows Advisory Group, consisting of nine members from different branches of government, oversees the entire process. In each of the river basins and bays under consideration, there is also a Stakeholder Committee and an Expert Science Committee. The stakeholder committees include representatives of all major interest groups, including agricultural water users, recreational water users, municipalities, industrial water users, electricity generation, commercial fisherman, public interest groups, regional water planning groups, groundwater conservation districts, river authorities, and other environmental interest groups. The Expert Science Committee reports to the Stakeholder Committee and has responsibility for conducting environmental flow analyses and developing an environmental flow regime recommendation based on the best available science.

Texas does not have a fixed methodology for estimating environmental flows but certainly adheres to holistic principles. In addition to the meeting the goal of a "sound ecological environment," environmental flow analyses must also consider other public interests and relevant factors, include "unappropriated water" set-asides considering human water needs, and include procedures to implement adjustments for permits or water right amendments. The environmental flow recommendations themselves, however, are hydrology based and must consist of a schedule of flow quantities for each season of the year (spring, summer, fall, winter). For each a subsistence flow is specified, which is the amount to protect against deleterious water quality conditions and to provide for the basic survival of species. A base flow is also prescribed to represent normal conditions between rainfall events that provide a range of habitats needed by native aquatic and riparian species. Finally, high flow pulses are prescribed to provide cues in the life cycles of native species and to maintain the physical characteristics of river channels.

The movement, however gradual, toward more ecologically relevant requirements for environmental flow setting is a positive trend, but a very large gap remains in most parts of the world between the environmental flow regime that would be recommended by best practices and the regime that is actually protected. Dam operators and water allocation managers have grown accustomed to flatline minimum flows in the range of 10 % of mean annual flow, Q95, or 7Q10 and have developed their plans and even business models around these figures. The larger levels of flow required to meet the standard of ecological sustainability expected in IWRM challenge existing models of water development, reducing water availability for well established economic water uses (e.g., hydropower and agriculture) and even calling into question the feasibility of certain projects. This makes the implementation of improved practices in environmental flow setting exceedingly challenging, but there are also significant consequences of not providing sufficient environmental flows. Some of these consequences are reviewed in the next section.

7.5 The Consequences of Not Protecting Sufficient Environmental Flows

Understanding the consequences of not protecting sufficient environmental flows requires certain familiarity with what is at stake in rivers and related ecosystems. Freshwater ecosystems harbor extraordinary species richness in multiple taxonomic groups, and rivers in particular can exert considerable influence on terrestrial, estuarine, and coastal ecosystems. Additionally, the lives, livelihoods, and well-being of human populations around the world are linked to multiple freshwater ecosystem services. In this section, we provide a general overview of the global importance of freshwater ecosystems and freshwater ecosystem services. This overview is followed by a discussion of some of the examples of consequences under scenarios where no environmental flows are protected or some environmental flows are protected, considering some of the widely applied methodologies for environmental flows.

Freshwater ecosystems—used here to mean rivers, lakes, and wetlands—harbor an estimated 10 % of the species known to science, but also rank among the most understudied of ecosystems (Dudgeon et al. 2006; Naiman and Dudgeon 2011). Of the roughly 30,000 described species of fishes, 40 % are freshwater species or spend considerable parts of their life cycle in freshwaters; these figures correspond to about one-quarter of all vertebrate species. Other vertebrate species, such as mammals, reptiles, and amphibians, inhabit freshwaters as well, and when their freshwater species totals are added to those of fishes, it becomes clear that approximately one-third vertebrate species on Earth live in freshwater (Dudgeon et al. 2006). These non-fish vertebrates include species like river dolphins and otters in the mammals' category, and species like crocodiles or turtles in the reptiles' category. And many mammal species considered primarily terrestrial depend on freshwater for parts of their life cycles or for drinking water; wildebeests and hippopotamus in Africa provide examples here (McClain et al. 2014). Richness of freshwater species also includes a plethora of invertebrate and plant species. Mussels, insects, crustaceans, and snails all inhabit freshwater ecosystems, yet the knowledge of these species, and even to some extent knowledge of freshwater vertebrate species, is incomplete. There is an absence of adequate data on freshwater species, especially in tropical regions, which have received comparatively much less study than terrestrial areas (Dudgeon et al. 2006).

Important to mention here is the fact that freshwater species are not evenly distributed in space and time. Many fishes move longitudinally along river courses, laterally between rivers and floodplains, and even between fresh and saltwater areas during their life cycles. Potamadromous species live entirely within freshwater but migrate between upstream and downstream areas or river channels and floodplains for feeding or spawning; anadromous species live mainly in saltwater but migrate to freshwater for spawning; catadromous species live mainly in freshwater but migrate to brackish or saltwater for certain periods or spawning. Thus, the life cycles of these species depend on connectivity along longitudinal and lateral corridors and a certain degree of unimpeded river flow.

Several factors influence the distribution, abundance, and diversity of freshwater species, but for running water environments, like rivers, flow acts as a master variable. The quantity, quality, and timing of flows help to define the habitats available to species both spatially and temporally. A vast literature documents the linkages between parameters like water depth and velocity—both are determined by flow—and the habitat preferences of plant and animal species (see Anderson et al. 2006b; Davies et al. 2014). Flow provides longitudinal and lateral connectivity, facilitating the movement of water, matter, and organisms between upstream and downstream and river channel and floodplain areas (Pringle 2003). For aquatic species, this flow-mediated movement means access to spawning, recruitment, and foraging habitats; seasonal variations in flows, such as higher flows at the end of a spring season, can cue organisms to begin movement. For riparian species, flow can be linked to processes of seed dispersal and seedling recruitment (Rood et al. 2005).

In addition to their importance for biodiversity, freshwater ecosystems play a critical role in securing the overall well-being of human populations worldwide. Freshwater ecosystem goods and services underpin human health, are often the backbone of local economies, play a critical role in cultural practices, and provide numerous other benefits (Brauman et al. 2007). Freshwater provisioning services provide goods like freshwater, food, and building materials. An estimated one billion people depend on fish as their primary protein source, and a significant part of this fish comes from freshwater ecosystems, although solid global estimates of the number of people engaged in inland fisheries are lacking (Allan et al. 2005). Freshwater regulating and supporting services refer to functions of freshwater ecosystems in waste assimilation, nutrient cycling, and flood control, or as transportation routes; these regulating services are essentially free, and estimates of what they would cost to replace are usually exorbitantly high. In terms of cultural services, nearly all of the world's religions have traditions that are closely linked to freshwater, rites like baptism or practices of submerging people or idols in water. Rivers and lakes also play an important role in recreation, bring noteworthy aesthetic beauty to landscapes, and are a source of revenue from tourism.

Despite the importance of freshwater ecosystems, both ecologically and to human populations, they have undergone an extreme degree of alteration worldwide. For example, dams and water diversions have affected the flow of more than half of the world's large rivers (Nilsson et al. 2005). In part as a consequence of these extensive flow alterations, freshwater species are among the most threatened species on Earth, as declines of freshwater species far exceed those of terrestrial counterparts, with some estimates suggesting extinction rates in freshwater species to be five times those of terrestrial species in some areas (Ricciardi and Rasmussen 1999; Naiman and Dudgeon 2011). Among the factors frequently linked to the increasing threats to freshwater ecosystem sustainability are overexploitation, water pollution, flow modification, habitat destruction or degradation, and exotic species introductions (Dudgeon et al. 2006). Of these, at least two—flow modification and habitat destruction or degradation—are linked directly to the concept of environmental flows, in the sense that a prescribed environmental flow has an objective of mitigating these factors' negative effects. And another two—water pollution and exotic species introductions—can be exacerbated by flow alterations. Therefore providing environmental flows can indirectly help to address these challenges to freshwater ecosystem sustainability as well. Providing a limited environmental flow, but not one that ensures sustainability in line with IWRM goals, also results in consequences.

In cases where no environmental flow is provided downstream from a water withdrawal or similar flow alteration, the ecological consequences can be severe and ecosystem services will be lost. In the worst cases, a river reach may be left completely dewatered for several kilometers, either on a permanent or intermittent basis depending on the type of water diversion project. For example, many run-of-river dams, including what are typically considered small hydropower projects, operate by diverting water from a river to an off-channel turbine house and then returning water to the river downstream. In the absence of environmental flows, the reach of river between the diversion site and turbines/return to the river is permanently dewatered and therefore becomes inhospitable to aquatic biota and potentially subject to encroachment by riparian vegetation and increased algal growth. A dewatered reach disrupts connectivity between riverine environments, so that even species like migratory biota that may pass through this reach only in certain seasons are negatively affected by dewatering in terms of blocked access to spawning or feeding sites. Additionally, extreme flow reductions like dewatering have been linked to the spread of diseases in human populations, as they can affect the biodiversity of infectious agents, reservoirs, and vectors (Sala et al. 2008; Naiman and Dudgeon 2011). For water storage dams that operate with in-channel reservoirs, dewatering may be intermittent, as flows may be released downstream from these projects during periods of electricity generation or reservoir flushing. However, in the absence of an environmental flow regime, the release of flows from these projects typically has little if any resemblance to the natural hydrograph, and therefore native aquatic and riparian biota may also find the downstream river channel uninhabitable. In addition to these more hydrologic effects, releases of flows from in-channel reservoirs and even from the turbines of run-of-river dams can have consequences for the thermal, chemical, and sediment regimes of a river. Water released from a reservoir, depending on from which part of the reservoir it is released, may be much colder in temperature (or warmer, depending on seasonality) and lower in dissolved oxygen than would have been the case in the absence of the river alteration. Water released from both in-channel storage reservoirs and turbine houses of run-of-river projects can frequently be devoid of or carry very little suspended sediment; its release can then be linked to scouring of the river channel downstream.

In cases where some environmental flow is provided, the ecological consequences and related compromises for freshwater ecosystem services may be less severe than in cases where no environmental flow is provided, as described above. Nevertheless, the approach to setting environmental flows—be it purely hydrologic, based on hydrology-habitat requirements, or holistic—heavily influences the kinds and magnitude of ecological consequences of altered flows. Provision of an environmental flow that is a flatline minimum, exemplified by approaches or common statistical methodologies like Q95 or 7Q10, is linked to multiple problems for freshwater ecosystems. For example, these approaches typically safeguard only a small percentage of river flow, creating dry or drought-like conditions for longer and at different times of the year than would naturally occur. Several studies suggest that minimum flows can provide sufficient habitat for a subset of species that occur naturally, but not all species (Anderson et al. 2006a; Waters and Post 2011). Additionally, rivers are naturally dynamic systems, and the life history strategies of aquatic biota are strongly linked to the natural dynamism of rivers. For example, the survival of fishes has been shown to be either positively or negatively related to flow variability, predictability, or seasonality, depending on whether the species is an opportunistic, periodic, or equilibrium life history strategist (Mims and Olden 2012).

Provision of an environmental flow that is very species centric is preferable to a flatline minimum but not without ecological consequences. These approaches typically focus on one or a few species and often on salmonids (Anderson et al. 2006b). However, environmental flow regimes that are based solely on physical habitat suitability for target species assume that abiotic rather than biotic interactions are the main influencing factor on species survival. Additionally, these methods may not always take into account other components of freshwater or riparian ecosystems and therefore may not adequately protect other important ecological features or functions. A good example here is the importance or linkage between the hydrologic regime and the sediment and thermal regimes of a river. The magnitude, frequency, duration, and timing of flows are tightly linked to ecological processes like movement of sediments and maintenance and formation of the river channel. Flow pulses, flushing flows, channel-forming flows, and floods and their influence on sediment dynamics in turn influence aquatic and riparian biota through the role of a river's sediment regime in structuring habitat. Temperature serves as an important cue for migratory fishes, notably *Alosa* spp. (shad), for which late spring migrations are initiated in response to warming river temperatures (Hightower et al. 2012).

The damage to freshwater ecosystems, the biodiversity they harbor, and the ecosystem services they provide are clearly visible in heavily modified rivers around the world. Moreover, the currently most widespread environmental flow practices—flatline minimum flows and single-species flows—are not providing sufficient protection. The trend toward use of holistic approaches to develop environmental flows offers some hope of minimizing the ecological consequences and compromises for freshwater ecosystem services associated with flow alterations from dams and water diversions, but these best practices are applied in only a small fraction of impacted rivers. The essential challenge now is to find a way to extend the use of best practices. We end this chapter with a few reflections on possible ways forward.

7.6 The Way Forward: Moving Toward Wider Application of Best Practices

With such clear indications of the degradation of riverine ecosystems due to over-withdrawals and severe regulation of river flows, it is essential that water managers begin to apply better practices in the determination and protection of environmental flow levels. This means establishing environmental flow rules that consider specific ecological and geomorphological needs. The political enabling conditions for this move are being established by widespread acceptance of the guiding principles of integrated water resources management and reforms to national and state water policies and laws. Establishment of specific ecological quality objectives and classification of rivers is also helping to clarify and reinforce this need. Nevertheless, it is likely to be a long, geographically unequal, and phased process that will require more knowledge of environmental flow science (especially low flow ecology), improved technologies for water infrastructure and, most importantly, increased awareness and willingness among decision-makers and water managers to improve environmental flow rules and enforce them. Improved processes of stakeholder engagement and conflict resolution will also be important to negotiate the changes among different water sectors, especially hydropower and agriculture.

The situation is made more challenging by the wide gap between current science and practice. For the past 20 years, environmental flow science has accepted the prevailing paradigm of the natural flow regime (Poff et al. 1997). This has led to many advances in understanding, but it has also positioned the science at odds with water sectors that modify, or even erase, the natural flow regime. This opposition is made more extreme by setting completely unmodified (natural) flow regimes as the ideal condition and quantifying all changes as degrees of hydrological alteration from natural conditions (Richter et al. 1996). This produced a view that all diversions from natural conditions bring risk of ecological degradation, and the threshold for presumed unacceptable risk has been set quite close to the natural end of the spectrum. Richter et al. (2012) proposed a presumptive environmental flow standard of 11–20 % alteration from natural daily flow levels to achieve a moderate level of ecological protection. This language aligns well with the objectives of many new water policies, such as the EU Water Framework Directive, that set targets allowing only small deviations from natural conditions in the biology and water quality of freshwater bodies (WFD 2000). But these presumptive standards will come as a shock to other water users and even water managers who have become accustomed to alterations in flow of 90 % or more. This is an extreme gap to bridge for environmental science, especially when the current reality in implementation is much closer to the 90 % value.

In highly altered river systems and those with plans for major alterations, the current 90 %-use reality challenges environmental flow scientists to build custom flow regimes, starting from a minimal flow baseline (critical low flow) and adding flow components one by one until ecological objectives are met. Holistic frameworks such as the Building Block Methodology are well suited to this purpose, but it too relies on application of species flow requirements taken from natural systems. Environmental flow science needs to focus new research efforts on understanding the ecohydrology of regulated river systems not only as altered ecosystems but also as novel ecosystems that are functioning and adapting in new ways. This is aligned with the general concept of novel ecosystems, which focus on how ecosystems adjust to combinations of biotic stresses derived from new species combinations and abiotic stresses derived from human modification of habitats (Hobbs et al. 2006, 2009). This concept is beginning to be applied to aquatic ecosystems, especially those that are so highly altered that natural conditions no longer serve as meaningful management targets or references for explaining system function (Moyle 2013). More scientific research is needed into these dynamics to improve flow recommendations for novel systems and make as much water as possible available for other uses.

At the same time that ecohydrologists work to sustain ecosystems with less water, other water-consuming sectors must work to improve efficiently and develop innovative technologies to also do more with less. Agricultural scientists have been working to grow more crops with less water for decades, so there are many well-established techniques and technologies available (Keating et al. 2010). The pressure to do more with less has also pushed hydropower engineers to improve generation efficiencies and develop other technologies to adapt to climate change and soften the impact of hydropower projects on aquatic systems (Ardizzon et al. 2014; IPCC 2011). These technologies, combined with improved operational practice, have been consolidated into a hydropower sustainability assessment protocol that is beginning to be applied in different parts of the world (http://www.hydrosustainability.org/).

With a wealth of science and different technologies to make use of, some of the most difficult challenges in applying best environmental flow practices lie in the governance processes and equitable allocation among water users and the environment. This brings us back to the promise of IWRM itself as a process to facilitate integration of these factors in a highly participatory fashion. In this chapter, we have endeavored to summarize the promise and highlight the current challenges of environmental flow assessment and implementation to enable the protection of ecosystems in the process of IWRM. The way forward is to continue to improve our knowledge and technologies, to reduce the gap between best practice and current practice, and most importantly to continue with the difficult work of integrating the factors into functional and effective governance structures.

References

Abbasov R, Smakhtin V (2009) Introducing environmental thresholds into water withdrawal management of mountain streams in the Kura River basin, Azerbaijan. Hydrol Sci J 54(6): 1068–1078

Allan JD, Abell R, Hogan Z, Revenga C, Taylor BW, Welcomme RL, Winemiller K (2005) Overfishing of inland waters. Bioscience 55:1041–1051

Anderson EP, Pringle CM, Freeman MC (2006a) Ecological consequences of hydropower development in Central America: impacts of small dams and water diversion on neotropical stream fish assemblages. River Res Appl 22:397–411

Anderson KE, Paul AJ, McCauley E, Jackson LJ, Post JR, Nisbet RM (2006b) Instream flow needs in streams and rivers: the importance of understanding ecological dynamics. Front Ecol Environ 4:309–318

Ardizzon G, Cavazzini G, Pavesi G (2014) A new generation of small hydro and pumped-hydro power plants: advances and future challenges. Renew Sust Energ Rev 31:746–761

Arnold JG, Srinivasan R, Muttiah RS, Williams JR (1998) Large area hydrologic modeling and assessment part I: model development1. J Am Water Resour Assoc 34(1):73–89

Bovee KD (1986) Development and evaluation of habitat suitability criteria for use in the instream flow incremental methodology. National Ecology Center, Division of Wildlife and Contaminant Research, Fish and Wildlife Service, US Department of the Interior

Brauman KA, Daily GC, Duarte TK, Mooney HA (2007) The nature and value of ecosystem services: an overview highlighting hydrologic services. Annu Rev Environ Resour 32:67–98

Brisbane Declaration (2007) The Brisbane Declaration: environmental flows are essential for freshwater ecosystem health and human well-being. In: 10th international river symposium, 3–6 September 2007, Brisbane. http://www.conservationgateway.org/ConservationPractices/Freshwater/EnvironmentalFlows/MethodsandTools/ELOHA/Pages/Brisbane-Declaration.aspx

Chandler CF (1873) Report upon the sanitary chemistry of waters, and suggestions with regard to the selection of the water supply of towns and cities. Am Public Health Assoc Public Health 1:533–563

Davies PM, Naiman RJ, Warfe DM, Pettit NE, Arthington AH, Bunn SE (2014) Flow-ecology relationships: closing the loop on effective environmental flows. Mar Freshw Res 65:133–141

Dessu SB, Melesse AM (2012) Modelling the rainfall–runoff process of the Mara River basin using the Soil and Water Assessment Tool. Hydrol Process 26:4038–4049

Dudgeon D, Arthington AH, Gessner MO, Kawabata Z, Knowler DJ, Lévêque C, Naiman RJ, Prieur-Richard AH, Soto D, Stiassny ML, Sullivan CA (2006) Freshwater biodiversity: importance, threats, status, and conservation challenges. Biol Rev 81:163–182

Gumbel EJ (1954) Statistical theory of droughts. Proc Am Soc Civ Eng 80:1–19, separate 439

GWP (Global Water Partnership) (2000) Integrated Water Resources Management. GWP TAC Background Paper #4. GWP (Global Water Partnership), Stockholm. http://www.gwpforum.org/gwp/library/TACNO4.PDF

Hightower JE, Harris JE, Raabe JK, Brownell P, Drew CA (2012) A Bayesian spawning habitat suitability model for American Shad in southeastern United States rivers. J Fish Wild Manag 3:184–198

Hobbs RJ, Arico S, Aronson J, Baron JS, Bridgewater P, Cramer VA, Epstein PR, Ewel JJ, Klink CA, Lugo AE (2006) Novel ecosystems: theoretical and management aspects of the new ecological world order. Glob Ecol Biogeogr 15(1):1–7

Hobbs RJ, Higgs E, Harris JA (2009) Novel ecosystems: implications for conservation and restoration. Trends Ecol Evol 24(11):599–605

IPCC (Intergovernmental Panel on Climate Change) (2011) Summary for policymakers. In: Edenhofer O et al (eds) IPCC special report on renewable energy sources and climate change mitigation. Cambridge University Press, Cambridge

Keating BA, Carberry PS, Bindraban PS, Asseng S, Meinke H, Dixon J (2010) Eco-efficient agriculture: concepts, challenges, and opportunities. Crop Sci 50(Suppl 1):S-109–S-119

King JM, Tharme RE, de Villeers MS (eds) (2000) Environmental flow assessments for rivers: manual for the building block methodology, Water research commission report no.: TT 131/00. Freshwater Research Unit, University of Cape Town, Cape Town

King J, Brown C, Sabet H (2003) A scenario-based holistic approach to environmental flow assessments for rivers. River Res Appl 19(5–6):619–639

Le Quesne T, Kendy E, Weston D (2010) The implementation challenge: taking stock of government policies to protect and restore environmental flows. The Nature Conservancy and WWF, Godalming

Matalas NC (1963) Probability distribution of low flows, Professional paper 434-A. U.S. Geological Survey, Washington, DC

McClain ME, Kashaigili JJ, Ndomba P (2013) Environmental flow assessment as a tool for achieving environmental objectives of African water policy, with examples from East. Int J Water Resour Dev 29(4):650–665

McClain ME, Subalusky AL, Anderson EP, Dessu SB, Melesse AM, Ndomba PM, Mtamba JOD, Tamatamah RA, Mligo C (2014) Comparing flow regime, channel hydraulics, and biological communities to infer flow-ecology relationships in the Mara River of Kenya and Tanzania. Hydrol Sci J. doi:10.1080/02626667.2013.853121

Milhous RT, Wegner DL, Waddle T (1984) User's guide to the physical habitat simulation system (PHABSIM). Department of the Interior, US Fish and Wildlife Service, Washington, DC

Mims MC, Olden JD (2012) Life history theory predicts fish assemblage response to hydrologic regimes. Ecology 93:35–45

Ministry of Life (2014) http://www.bmlfuw.gv.at/wasser/wasser-oesterreich/wasserrecht_national/planung/QZVOekologieOG.html

Moore M (2004) Perceptions and interpretations of "environmental flows" and implications for future water resource management: a survey study (MSc). Linköping University, Linköping

Moyle PB (2013) Novel aquatic ecosystems: the new reality for streams in California and other Mediterranean climate regions. River Res Appl n/a-n/a

Naiman R, Dudgeon D (2011) Global alteration of freshwaters: influences on human and environmental well-being. Ecol Res 26:865–873

Nilsson C, Reidy CA, Dynesius M, Revenga C (2005) Fragmentation and flow regulation of the world's large river systems. Science 308:405–408

Orsborn JF, Allman CH (1976) Instream flow needs, vols I and II. American Fisheries Society, Bethesda, Mary, 551 and 657pp

Pahl-Wostl C, Arthington A, Bogardi J, Bunn SE, Hoff H, Lebel L, Nikitina E, Palmer M, Poff LN, Richards K, Schlüter M, Schulze R, St-Hilaire A, Tharme R, Tockner K, Tsegai D (2013) Environmental flows and water governance: managing sustainable water uses. Curr Opin Environ Sustain 5(3–4):341–351

Poff NL, Allan JD, Bain MB, Karr JR, Prestegaard KL, Richter BD, Sparks RE, Stromberg JC (1997) The natural flow regime. Bioscience 47(11):769–784

Poff NL, Richter BD, Arthington AH, Bunn SE, Naiman RJ, Kendy E, Acreman M, Apse C, Bledsoe BP, Freeman MC, Henriksen J, Jacobson RB, Kennen JG, Merritt DM, O'Keeffe JH, Olden JD, Rogers K, Tharme RE, Warner A (2010) The ecological limits of hydrologic alteration (ELOHA): a new framework for developing regional environmental flow standards. Freshw Biol 55(1):147–170

Pringle CP (2003) What is hydrologic connectivity and why is it ecologically important? Hydrol Process 17:2685–2689
RoC (Republica de Chile), Ministerio de Medio Ambiente (2012) Dicreto 14: Abrueba reglamento para la determinación del caudal ecológico mínimo, 22 May 2012. Available online at http://www.dga.cl/legistlacionynormas/normas/Reglamentos/Reglamento_Caudal_Ecologico.pdf
Ricciardi A, Rasmussen JB (1999) Extinction rates of North American freshwater fauna. Conserv Biol 13:1220–1222
Richter BD, Baumgartner JV, Powell J, Braun DP (1996) A method for assessing hydrologic alteration within ecosystems. Conserv Biol 10(4):1163–1174
Richter BD, Davis MM, Apse C, Konrad C (2012) A presumptive standard for environmental flow protection. River Res Appl 28(8):1312–1321
Rood SB, Samuelson GM, Braatne JH, Gourley CR, Hughes FM, Mahoney JM (2005) Managing river flows to restore floodplain forests. Front Ecol Environ 3(4):193–201
Sala OE, Meyerson LA, Parmesan C (eds) (2008) Biodiversity change and human health. Island Press, Washington, DC
Santos PVCJ, da Cunha AC (2013) Outorga de recursos hídricos e vazão ambiental no Brasil: perspectivas metodológicas frente ao desenvolvimento do setor hidrelétrico na Amazônia. Rev Bras Recur Hidr 18(3):81–95
Sarmento R (2007) Estado da arte da vazão ecológica no Brasil e no mundo. UNESCO/ANA/CBHSF, UNESCO
Stalnaker CB (1990) Minimum flow is a myth. In: Ecology and assessment of warm water streams. Biol Report 90(5):31–33
TCEQ (Texas Commission on Environmental Quality) (2012) Chapter 298 – Environmental flow standards for surface water subchapter C: Sabine and Neches Rivers, and Sabine Lake Bay. Available online at http://www.tceq.texas.gov/assets/public/legal/rules/rules/pdflib/298c.pdf
Tennant DL (1976) Instream flow regimes for fish, wildlife, recreation, and related environmental resources. Fisheries 1(4):6–10
Tharme RE (2003) A global perspective on environmental flow assessment: emerging trends in the development and application of environmental flow methodologies for rivers. River Res Appl 19(5–6):397–441
Tobin KJ, Bennett ME (2009) Using SWAT to model streamflow in two river basins with ground and satellite precipitation data1. J Am Water Res Assoc 45(1):253–271
Trihey EW, Stalnaker CB (1985) Evolution and application of instream flow methodologies to small hydropower developments: an overview of the issues. ed. Symposium on small hydropower and fisheries, 1985, Bethesda, MD, pp 176–183
UNEP (UN Environment Program) (2012) The UN-Water status report on the application of integrated approaches to water resources management. United Nations, New York
USACE (US Army Corps of Engineers) (2012) Water in the U.S. American West: 150 Years of Adaptive Strategies. Available at http://www.building-collaboration-for-water.org/
USEPA (US Environmental Protection Agency) (1997) Technical guidance manual for performing wasteload allocations, Book II: Streams and rivers – Part 1: Biochemical oxygen demand/dissolved oxygen and nutrients/eutrophication. EPA DOCUMENT NUMBER: EPA-823-B-97-002
Wang X, Zhang Y, James C (2009) Approaches to providing and managing environmental flows in China. Water Resour Dev 25(2):283–300
Waters A, Post DM (2011) How low can you go? Impacts of a low-flow disturbance on aquatic insect communities. Ecol Appl 21:163–174
WFD (European Water Framework Directive) (2000) Directive 2000/60/EC of the European Parliament and of the Council of 23 October 2000 establishing a framework for Community action in the field of water policy. http://eur-lex.europa.eu/LexUriServ/LexUriServ.do?uri=CELEX:32000L0060:EN:NOT
Yates D, Sieber J, Purkey D, Huber-Lee A (2005) WEAP21—a demand-, priority-, and preference-driven water planning model. Water Int 30(4):487–500

Chapter 8
Ecohydrology: Understanding and Maintaining Ecosystem Services for IWRM

Amartya K. Saha and Shimelis Gebriye Setegn

Abstract Since historical times, natural ecosystems such as forests and wetlands are known to regulate water flow and maintain water quality. The past half a century however has witnessed unplanned and rapid development with widespread ecosystem degradation. Meanwhile water treatment and supply happens on an ad hoc basis that is neither sustainable nor affordable for most communities. The revival of an ecohydrological approach is called for, with increased use of ecosystem services in water resources management. The affordable and sustainable aspects of this approach make it especially pertinent for developing countries, given the increasing challenges posed by mounting population, consumption, and climate change. This chapter describes the general links between different ecosystems, hydrology, and water quality and outlines the steps in developing an ecohydrological approach. The next chapter describes case studies that have successfully incorporated an ecohydrological approach in different realms of water resources management in the developing world.

Keywords Ecohydrology • IWRM • Ecosystem services • Watershed ecosystems • Environmental flows

8.1 Ecosystem Services: Revival of Old Practices

Water is the driving force of all nature. – Leonardo da Vinci

A.K. Saha (✉)
Global Water for Sustainability (GLOWS), Department of Earth and Environment, Florida International University, North Miami, FL 33181, USA
e-mail: asaha@fiu.edu

S.G. Setegn
Department of Environmental and Occupational Health and Global Water for Sustainability Program (GLOWS), Florida International University, 11200 SW 8th Street, Miami, FL 33199, USA
e-mail: ssetegn@fiu.edu

S.G. Setegn, M.C. Donoso (eds.), *Sustainability of Integrated Water Resources Management*, DOI 10.1007/978-3-319-12194-9_8

8.1.1 *Forests, Wetlands, and Ecosystem Services*

Water is life, as goes the ancient saying. At the same time, organisms and ecosystems markedly influence the quality, quantity, and flux of water on the landscape. Forests regulate the water cycle by storing rainwater in the canopy and forest floor, facilitating the percolation of rainwater underground, and feeding springs and streams well into the dry season. The removal of forest cover has been unequivocally linked with flash floods and soil erosion following heavy rainfall as well as springs and rivers drying up earlier than in catchments with intact forests (e.g., Munishi and Shear 2005; Giambelluca and Gerold 2011). Wetlands store large amounts of water during the rainy season and recharge downstream rivers. Wetlands are also known as the kidneys of the landscape on account of their ability to trap and remove excess nutrients in runoff that would otherwise create harmful algal blooms and other problems for aquatic ecosystems and water quality (e.g., Mitsch and Gosselink 1993). Aquatic organisms in streams feed upon and help decompose organic matter, thereby contributing to maintaining water quality.

Harnessing this inherent capacity of ecosystems to maintain water quality and to regulate hydrology is then the logical way to manage water resources sustainably and affordably across vast areas in developing countries. This awareness of the role of forests and wetlands in maintaining water quality and flux and preventing soil erosion goes back in time. Traditions such as sacred groves arose to protect forests from being cut down; patches of pristine old growth forest are still preserved to this day on account of their status as sacred groves, such as those present in various parts of Africa (Sheridan and Nyamweru 2007) and throughout India (Malhotra et al. 2001; Khan et al. 2008). Similarly, the traditional use of wetlands in Uganda centered upon water, grazing, hunting, and fishing, while in the recent past, unsustainable practices of land drainage and sand mining have happened (Iyango et al. 2012).

8.1.2 *Rising Demands, Degrading Ecosystems, and Imperiled Freshwater*

Over the past century, the growing human population along with increasing levels of consumption of natural resources has led to the degradation of forests, wetlands, lakes, rivers, and oceans worldwide. This degradation has caused major alterations in river flows, tremendous declines in water quality, and increasing uncertainty in seasonal water availability. The loss of hydrologic regulation by ecosystems has also increased the frequency and magnitude of natural disasters: unprecedented deforestation and development practices that promote runoff are directly leading to an increase in flooding frequency. In a study over 1990–2000 with data from 56 developing countries, Bradshaw et al. (2007) confirmed this relationship, with models predicting a 4–28 % increase in flood frequency accompanying a 10 %

decrease in forest area. The same decade saw over 100,000 deaths, 320 million people displaced, and a loss exceeding US$ 1,151 billion resulting from floods where deforestation played a significant part.

Accompanying the diminishing ecosystem services is an accelerating thirst for water in cities, both in direct consumption and in water utilized in manufacture of consumer goods. At the same time, in developing countries, around 770 million people lack access to adequate clean drinking water while more than 2.5 billion people lack access to sanitation (UNICEF 2013). Providing adequate safe water to meet basic human needs is a serious and a growing problem that is all the more acute in the dry season and in years with low rainfall. The extremely limited piped water supply and inadequate wastewater treatment systems persist because of limited resources and funding, an absence of effective policies, planning, management practices, regulations, and implementation. Even when funding has been available, the conventional response has been to construct large, centralized energy-intensive wastewater treatment plants (e.g., UNEP 2004). However, even the few cities in the developing world that have such physicochemical treatment plants can handle but a fraction of the daily wastewater generated, much of which is discharged untreated into rivers, creeks, and seas.

While technical solutions for point pollution control and flood control are necessary and have their place in water resources management, their enormous construction and operating expenses make them impractical to be widely applied. Hence, any water management plan that solely relies upon technical solutions is unsustainable. Besides, the lack of understanding and consideration of ecosystem services reflects a trial and error approach to water management rather than the implementation of a policy toward sustainable use of water resources (UNEP 2004). Attaining sustainability in freshwater resource use requires not only reducing pollution but also arresting the degradation of ecological processes in landscapes. A catchment-level planning and management strategy provides a coordinating framework for water supply protection, pollution prevention, and ecosystem preservation.

8.1.3 Nature as an Ally: The Ecohydrological Approach

Ecohydrology has been defined as the integrated study of ecosystems and hydrological characteristics and processes (Zalewski 2000). The linkages between ecosystem function and hydrological processes influence water dynamics and quality (e.g., Breshears 2005). In addition, ecohydrology seeks to understand anthropogenic impacts upon these linkages (Nuttle 2002). Ecohydrology is both an old and a new field (Jackson et al. 2009): old in that ecosystem services have been historically used and new in that the emergence of modern tools (such as laser spectrometry for stable isotope analysis, time domain resistivity for estimating soil moisture, and passive integrated transponder (PIT) tags for tracking fish movement) and techniques (remote sensing and GIS) can aid understanding ecohydrological linkages to meet the increasingly tight challenges facing sustainable water resources management.

The ecohydrological approach to water resources management utilizes the functions of various ecosystems present in a catchment to maintain natural flow patterns, year-round water availability, flood protection, and water quality (Hunt and Wilcox 2003). Ecosystems confer resiliency to a watershed from extremes of high and low rainfall years, thereby buffering against uncertainty associated with climate change. Hence, the understanding of the hydrology ecosystem links in a catchment and the maintenance and utilization of ecosystem services confer sustainability to water resources management (McClain et al. 2012).

Even though forests and wetlands have been utilized in the past for their ecosystem services, the challenges today of vastly increased water demand, rising costs, and degrading ecosystems impose the need for a detailed understanding of the basic ecohydrological processes that affect water quality, dynamics, and ecosystems. Both seasonal and interannual patterns in precipitation are changing and getting more uncertain as a consequence of climate change. This uncertainty in precipitation inputs is transferred into a streamflow and the water balance of a catchment. For instance, Setegn et al. (2014) examined the impact of downscaled precipitation predictions by an ensemble of general circulation models for the Blue Nile river basin upon streamflow, with the finding of a high likelihood of agricultural drought on account of the water balance being very sensitive and tightly coupled to rainfall. The preservation of natural water storage on the landscape can partially buffer a catchment against the vagaries of climate change.

8.2 Ecohydrology of Watershed Ecosystems

The essence of the ecohydrological understanding of a catchment is knowing how much water enters and leaves the catchment, followed by how natural ecosystems influence quality, quantity, and flux of water, and finally how to maintain these natural ecosystems, so as to avail ecosystem services for water management. Developing this understanding typically involves the following steps:

1. Characterizing the water cycle in a catchment by monitoring water inputs and outputs, by analyzing long-term meteorological and hydrological data (if that exists), and by calculating a water budget for the catchment
2. Noting the climatic, edaphic, biotic, and anthropogenic factors that affect water availability and quality
3. Investigating the links between hydrology/quality and aquatic and terrestrial plant and animal communities present in the region

Ecohydrology thus not only seeks to use ecosystem services to ensure the availability and quality of water in a practical and economical manner, it also aims to understand how to preserve ecosystem structure and function, so as to maintain ecosystem services. For instance, Saha et al. (2009) used stable isotope analysis to detect the specific water and nutrient sources of different plant communities in the Everglades; maintenance of community diversity requires maintaining seasonal water levels to avoid undue flood/drought stress to the communities. While animal communities do influence water quality, most ecohydrological investigations concern plant communities; plants are not only the primary producers, they also exert important feedbacks on the hydrological cycle, such as soil water uptake/transpiration (e.g., Eamus et al. 2006), creating microclimates that affect local precipitation and evaporation, and in wetlands, plant communities influence water flow and biogeochemical cycles (Rodriguez-Iturbe 2000; Rodriguez-Iturbe et al. 2001). It is this interlinked set of communities, ecosystem, and hydrological processes that provide ecosystem services, the most crucial of which is year-round water availability and quality.

To briefly illustrate the range of topics that come under the umbrella of ecohydrology, some examples of specific ecohydrological questions pertinent for water resources management are: (1) How does the flow, depth, and seasonal availability of water in wetlands/savannas determine nutrient cycling and vegetation zonation? (2) What are minimum environmental flows required in rivers to maintain aquatic ecosystems and fish populations and thereby self-purification processes? (3) How does land cover change in watersheds alter the rainfall-runoff infiltration relationship and thereby affect the flow regime in streams? (4) How can one accurately estimate evapotranspiration of different plant communities, such as evergreen forests and monoculture plantations? This section lays out some of the fundamental areas of ecohydrological understanding that govern the quality, quantity, and flux of water in watersheds.

A river basin or catchment typically includes different ecosystem types that each have their individual sets of hydrological processes and behavior affecting water availability and quality in the catchment. Figure 8.1 illustrates some typical ecosystem types present in a catchment such as headwater forests, wetlands, lowland forests, farmland, and urban areas. It is helpful to visualize the ecosystems that lie along the hydrological path of water overland in the catchment, i.e., from the headwaters to the outputs, keeping in mind that a large fraction of total water in a catchment short-circuits this overland path by entering the atmosphere via evapotranspiration. Some of the water infiltrates underground in parts of the catchment to reemerge above the surface as springs; inputs into streams, rivers, and wetlands; as well as submarine groundwater discharge. Widely occurring ecosystem types are categorized at the broadest level into forest, grassland/cropland, and wetlands. Each of these broad categories is further classified; for instance, the hydrological processes in an old growth primary forest differ considerably from a single-species plantation forest, or a regenerating secondary forest. Sections 17.3–17.5 examine some of the widely occurring ecosystem types from a hydrological perspective.

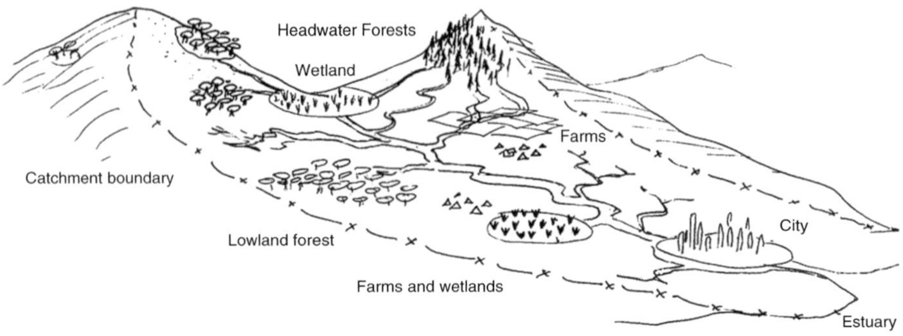

Fig. 8.1 Ecosystems within a typical river catchment

8.2.1 Catchment-Level Water Balance

Determining water availability for various human uses and ecosystem needs requires the computation of a water balance for the catchment. Computing a water balance or water budget involves quantifying how much water enters and leaves the catchment over a period of time and what is the change of the water stored in the catchment. The primary input into most catchments is precipitation in the form of rain and/or snow, although a few catchments like the Okavango Delta in Botswana have river inflow as the main input. A major output or flux of water out of a catchment is evapotranspiration (ET) that is often 50–100 % of incoming precipitation in the tropics and subtropics. River discharge can be another important output. In some parts of a catchment, surface water percolates down to recharge groundwater, while in the dry season, groundwater discharges to the surface as springs and seeps. While rain and river discharge can be measured and ET estimated, it is far more difficult to estimate net groundwater recharge accurately, which is often obtained from the residual term of a water balance. There are several methods and modeling tools to estimate the water balance of a watershed system.

Many of the water balance equations defined based on the law of conservation of mass and are expressed in a form as given below:

$$\mathbf{P - ET - Q \pm \Delta S \pm Residual = 0}$$

where **P** = precipitation, **ET** = evapotranspiration, **Q** = water yield (streamflow), **S** = storage (**Δ** signifies “change”), and **Residual** = error in all terms plus seepage or leakage in or out of the watershed.

If there is significant human activity in the catchment, water abstractions and return flows are included in the term Q. A water balance can be calculated at daily, monthly, and annual time scales, depending on the frequency of data availability.

The hydrological model SWAT (soil and water assessment tool) simulates the hydrological cycle based on the water balance equation:

$$SW_t = SW_0 + \sum_{i=1}^{t} \left(R_{\mathrm{day}} - Q_{\mathrm{surf}} - E_a - w_{\mathrm{seep}} - Q_{\mathrm{gw}}\right)_i$$

where SW_t is the final soil water content (mm), SW_o is the initial soil water content on day i (mm), t is the time (days), R_{day} is the amount of precipitation on day i (mm), Q_{surf} is the amount of surface runoff on day i (mm), E_a is the amount of evapotranspiration on day i (mm), W_{seep} is the amount of water entering the vadose zone from the soil profile on day i (mm), and Q_{gw} is the amount of return flow on day i (mm).

Water balance is deceptively simple and is also remarkably powerful as a conceptual model and analytical tool. It serves as a constant reminder of the compensatory changes in water movement and storage that are continually occurring in forest ecosystems. Natural or anthropogenic disturbance alters the water balance and initiates compensatory changes. Apart from determining the water available in a catchment (or at any scale over which a water balance is computed), this exercise can also yield quantitative estimates of water inputs and outputs that are difficult to measure directly. For instance, Saha et al. (2012) performed a water balance for the Everglades National Park with daily data over 2002–2008 that indicated net groundwater discharge in the summer months and groundwater recharge during the rainy season along with quantitative estimates.

Evapotranspiration is a very large component in the water cycle of almost all regions on our planet. But there is no single method to measure ET due to a wide variety of vegetation all over the world. In particular, the estimation of ET in woody vegetation is very difficult. Plant species and communities vary widely in their water uptake and transpiration (Douglass 1966) that also change seasonally. Hence, evapotranspiration over woody vegetation is an active area of research. It is necessary to be aware that there is considerable uncertainty in ET estimates obtained in commonly used hydrological models as well as in global datasets. For better accuracy, it is advisable to compare ET estimates from various approaches such as vapor transport models (Saha et al. 2012), eddy flux (Schedlbauer et al. 2011), diurnal water table levels and stand-level sap flow (Villalobos-Vega 2010), and remote sensing (Nouri et al. 2013). However, all these methods require location-specific studies. In the absence of such studies and/or detailed meteorological data, global ET datasets can be used to get an idea, as discussed below.

8.2.2 *Data Needs and Limitations*

The availability of data is vital for performing water balances, for monitoring water availability, and as inputs into ecohydrological research. Unlike the Everglades which has the benefit of having very spatiotemporally detailed meteorological and hydrological data, much of the world has very sparse or no data available. In such data-limited areas, the use of global and remote sensing data provides a first approach. Precipitation estimates for tropical regions are available in tropical rainfall measurement mission (TRMM) datasets (NASA) while evapotranspiration datasets moderate-resolution imaging spectroradiometer (MODIS) are available for most of the global terrestrial surfaces at a 1 km resolution. While there is active research on estimating river level/discharge as well as soil moisture from remote sensing measurements (Brocca et al. 2013), there still is the need for establishing monitoring stations and networks on the ground, for calibration/validation of these products. The cost of installing, operating, and maintaining such monitoring programs is worthwhile given that reliable data enables better assessment of water dynamics and availability in regions subject to increasing human demands as well as uncertainties under climate change conditions.

8.3 Forest Ecosystems and Hydrology

8.3.1 Headwater Catchment Forests

Rivers typically originate as streams in elevated parts of a catchment. These hilly or mountainous regions have high-altitude grasslands and/or forests and occasional wetlands as natural ecosystems; forests exist today either from protection on account of their water-harvesting functions or because they occur in the steepest areas unsuitable for agriculture or large-scale human settlement. In the subtropics and tropics, depending upon the location (latitude/longitude) and altitude, such forests can be broadly classified as deciduous forests (typically lowland to 1,500 m), tropical montane evergreen cloud forests occurring at altitudes between 1,000 and 2,000 m, and evergreen coniferous forests (2,000–3,500 m). In addition, there are single- and mixed-species plantations, often with fast-growing exotic species such as *Eucalyptus* in Asia, Africa, and Latin America. The differences in canopy structure, plant species water uptake, and soil type in these different forests lead to differences in the partitioning of precipitation into canopy interception, throughflow and stemflow, percolation and infiltration, evaporation, water uptake and transpiration, and surface runoff. Figure 8.2 illustrates the typical hydrological processes in the forest canopy and stand.

8.3.1.1 Hydrological Processes

Forests essentially function as sponges on the landscape, by intercepting rainfall and allowing time for rain to percolate into underground and recharge groundwater (Fig. 8.2). In the absence of forest cover, rainwater immediately flows off overland as surface runoff. This is why streams in forested catchments flow longer in the dry season than streams in deforested watersheds. For instance, the removal of shola evergreen tropical montane cloud forests in the Western Ghats ranges in India has resulted in the inability of the land to retain water, thus resulting in a destructive

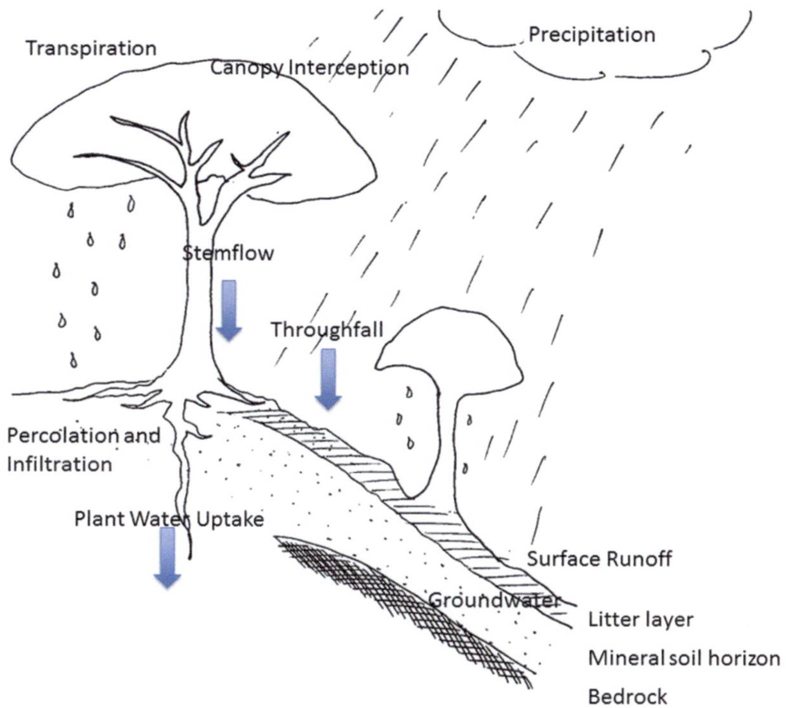

Fig. 8.2 Hydrological processes in a forest canopy

cycle of monsoon flooding and subsequent drought (Kadur and Bawa 2005). The author has also noticed this in streams of thickly forested Bori National Park in central India that has streams flowing all through the dry summer, while streams outside the National Park in catchments converted completely to agriculture run dry months earlier.

Interception of rain and snow by forest vegetation is an important hydrologic process. It may account for a substantial proportion (as much as 20–30 %) of annual precipitation, even though many factors are involved in accurately measuring the fraction of precipitation that is intercepted, making it difficult to characterize interception by forest type (Crockford and Richardson 2000). Some of the intercepted precipitation evaporates back into the atmosphere, while another fraction percolates down the leaves, stems, branches, and trunk into the soil and is termed stemflow.

8.3.1.2 Forest Soil and Litter Layer

While the canopy forms part of the sponge, the other part is the forest floor. Forest vegetation and forest soils develop together, one influencing the other, over the course of centuries; once washed away upon deforestation, forest soil can take centuries to form again. The combination of (a) annual additions of leaf litter and woody debris; (b) root growth; (c) the actions of microbes, insects and other invertebrates, and small mammals; and (d) biogeochemical cycling leads to the development of unique soil properties in forests relative to most other land covers and land uses. Forest soils typically have high organic matter with water-holding capacity and permeability. As a result, incident rainfall or snowmelt rarely exceeds the infiltration capacity of forest soils, and overland flow along with surface erosion is rare, except under very high rainfall events that saturate the soil. The soil is the nexus for many ecological processes (energy exchange, water storage and movement, nutrient cycling, plant growth, and carbon cycling at the base of the food web).

The forest canopy intercepts the kinetic energy of rain and along with the protective influence of the litter layer ensures that the porosity and permeability of forest soils remains intact – and that soil particles are not detached and converted to sediment. One centimeter of rain on 1 ha has a total mass of 100,000 kg (110 t), which exerts a considerable erosive force on soil that is stripped of its protective canopy and litter layers.

8.3.1.3 Forest-Climate Linkages

Forest vegetation has an obvious influence on microclimate (air temperature, humidity, and wind speed) under the canopy; for instance, the average nighttime temperature inside a shola cloud forest in the Western Ghats is around 10° higher than the exposed grassland (Meher-Homji 1991), with forest edge effects of lower humidity penetrating about 15–20 m inside the forest studied (Jose et al. 1996). Primary forests thus shield the soil from high evaporative demand.

Given the complexity of wind circulation patterns and other factors that define climate, the influence of forests on regional, continental, and global climate is not as straightforward. Despite the complexity, there is a large body of experimental and modeling work investigating the role of forests in influencing rainfall that has found evidence of considerable effects that forests have on local precipitation. For instance, Moreira et al. (1997) used stable isotopes of oxygen and hydrogen to find that almost half of the rainfall in the eastern Amazon resulted from transpiration from local forests. Shukla et al. (1990) used numerical models to arrive at a similar conclusion that deforestation would lead to changes in rainfall patterns, with negative implications for regeneration of many Amazonian forest tree species. Similarly, forests along the fog-enshrined Pacific coastline of the Americas intercept moisture on their leaves (Cavelier and Goldstein 1989). Dawson (1998) estimated that 34 % of the ecosystem water input in a California redwood forest resulted in dripping from trees, and only 17 % in a deforested catchment bereft of tree condensation. Tropical cloud forests have from fog a similar interception of moisture from clouds that enshroud these forests much of the year, thus constituting an important moisture source even in the absence of direct rainfall (Bruijnzeel 2001).

8.3.2 Single- or Mixed-Species Plantations

Plantation forestry is common and widespread and is managed for timber and pulp. Being much younger than old growth forest and having typically uniform stands of even-aged trees with similar architecture, their canopy is more open as compared to native forests that have an interwoven closed canopy. It follows that hydrological partitioning in such forests is more skewed toward runoff with lesser infiltration than native primary forests. Krishnaswamy et al. (2012) report that exotic *Acacia* plantations in the Western Ghats had higher fraction of rainfall leave as runoff as compared to primary evergreen forest in the same region, while degraded heavily used forest had the highest runoff fraction. Furthermore, many exotic fast-growing species have water uptake rates higher than native vegetation which results in higher evapotranspiration and consequently lower streamflow (Putuhena and Cordery 2000).

8.3.3 Lowland Forests

Lowland forests span a range of types associated with rainfall, from evergreen forests in high rainfall areas to increasing deciduousness and finally scrub forests at the arid end of the rainfall gradient. The processes of moisture interception, percolation, and transpiration occurring in headwater catchment forests also occur in lowland forests. Streams passing through lowland forests flow for longer durations than streams in adjacent watersheds; in addition, streams running through forests are clear and cool. Deforestation in lowland forests thus affects water quality (soil erosion) as well as reduced infiltration and baseflow.

8.3.4 Savannas: Grassland and Woodland Matrix

Savannas occur in areas where potential evapotranspiration exceeds rainfall and occupy large areas throughout the tropics and subtropics. Fires both natural and anthropogenic occur frequently, almost on an annual basis. There is usually considerable heterogeneity in topography and moisture, which results in a mosaic of vegetation (D'Odorico and Porporato 2006). Gallery forests occur alongside river courses in savannas, on account of the availability of year-round moisture, and usually have very different plant species from the surrounding savanna.

There is considerable pressure on savannas; for instance, deforestation in Tanzania has been the highest in the savanna woodlands (2000–2012) on account of ease of access, felling of woodland trees for charcoal, and agricultural expansion following increases in irrigation. Similarly, efforts to control Amazon deforestation have increased the pressure on the Cerrado savanna ecosystem in central Brazil for conversion to agriculture for soybean and biofuel demands. The effects of deforestation in savanna woodlands are similar to those in lowland forests.

8.3.5 *Forests and Water Management*

Paired watershed studies that are comparative studies in adjacent forested and deforested catchments (e.g., Brown et al. 2005) support traditional evidence that watersheds with forest cover have a more regulated river flow than watersheds within the same climatic zone that are deforested. Forests dampen high flows immediately following heavy rainfall events while prolonging baseflow in streams in the dry season (e.g., Bruijnzeel 2001; Krishnaswamy et al. 2013). Deforested watersheds exhibit high runoff following heavy rainfall events that lead to soil erosion, landslides, and floods.

Now there is considerable confusion over the role of forests in catchments when the goal of water management is to maximize water yield from a catchment. It is important not to confuse annual water yield with the duration of flow. A deforested watershed typically sees a higher water yield as a result of much lower evapotranspirative losses, which also varies by forest type (Brown et al. 2005). It is important to note that while the annual water yield can increase following deforestation and decrease following reforestation, looking at the flow duration curves gives an idea of how long the rivers flow in the dry season. For instance, Lele et al. (2008) report on a case of paddy farming in South India that was irrigated from a reservoir, which in turn was filled by a river arising in the Western Ghats hills. There were concerns that reforestation in the hills would lower streamflow in the wet season months, thereby decreasing water stored in the reservoir during the paddy season, which in turn would reduce the irrigated area under paddy cultivation. This is an example of contrasting watershed uses of paddy farmers not favoring reforestation to ensure high water yields, even though reforestation would lead to longer baseflows in the river (Krishnaswamy et al. 2013) and other watershed benefits such as water quality and reduced soil erosion.

8.3.6 *Riparian/Gallery Forests: The Last Defense Against Nonpoint Pollution*

Riparian areas or riverbanks include the transition or ecotone between terrestrial and aquatic ecosystems that have special implications for biogeochemical reactions (McClain et al. 2003). Forests growing along these riverbanks are called riparian or gallery forests (Fig. 8.3) and provide a host of essential functions, including (a) shade that cools water temperature and thereby increases dissolved oxygen concentration; (b) leaf litter inputs to microbes and invertebrates at the base of

Fig. 8.3 Natural vegetation on the *right bank* of the stream (riparian buffer) protects stream ecosystems and water quality; in contrast, farming on the *left bank* results in soil eroding into the stream

the food web; (c) structural support of stream banks; (d) large woody debris that stabilizes channels, diversifies stream habitat, and provides essential cover; and (e) hydraulic resistance to flood flows and sediment transport. Riparian buffers also intercept sediment in runoff arriving from adjoining slopes, especially if these slopes contain farmland and roads. Riparian forests, often being the last patches of forest left on the landscape, provide the last remaining habitat for regional biodiversity (Naiman et al. 1993).

The importance of riparian trees in maintaining water quality and the aquatic ecosystem has been known for a long time. *Terminalia arjuna* trees that grow along stream courses in peninsular India have been protected as sacred trees since many centuries. Many countries have riparian buffer guidelines under best management practices for watersheds, whereby no cultivation or settlement is allowed within 30–50 m on either side of a stream bank. In essence, riparian buffers provide the last defense in multiple use watersheds with large tracts of farmland, pastureland, and settlements, hence need to be critically enforced. Most watersheds in the world today do not have the benefit of large tracts of pristine old growth forests to stabilize the soil.

Investigations over the past couple of decades have focused on biogeochemical cycling in riparian zones where periodic inundation creates a fluctuating aerobic/anaerobic environment in the soil that in turn accelerates the processes of nitrification and denitrification (Pinay et al. 1993; Orr et al. 2007). Results indicate that the soil and leaf litter in riparian zones are able to entrap chemical fertilizers present in various forms of nitrogen and phosphorus; nitrogen forms then undergo various biogeochemical transformations depending upon the residence time of the groundwater in the soil and the degree of anoxia (Hill 1996; Reed and Carpenter 2002). Phosphorus is the limiting nutrient in many aquatic ecosystems (Schindler 1977; Elser et al. 2007).

The width of a riparian buffer necessary for a certain desired level of sediment entrapment and possibly nutrient retention depends upon many factors: rainfall, the slope of the watershed, the soil type, presence of floodplains, and watershed land

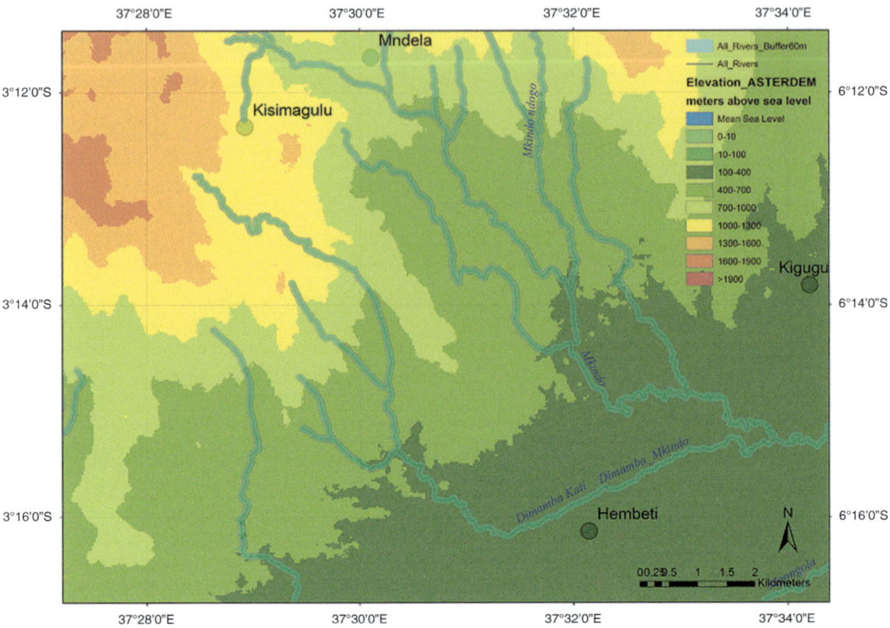

Fig. 8.4 Illustration of a 50 m riparian buffer on either side of rivers (Map shown for the Mkindo catchment, Wami river basin, Tanzania)

use. GIS models using the universal soil loss equation predict a required width that can be compared against the recommended uniform regulation to examine whether the recommended width is adequate (Xiang 1993; Baker et al. 2001). Steep channels can also short-circuit a riparian buffer adjoining a stream; for instance, Wenger (1999) suggests three approaches to determine buffer widths, and a slope greater than 25 % calls for a wider buffer. However, the use of models requires trained personnel and data such as land cover, topographic and meteorological data, and resources that are unavailable for much of the developing world. Hence, fixed width buffer (Fig. 8.4), such as the recommended 30 m buffer width in the flat agricultural plains of Ohio, USA, or the 50 m buffer law in Tanzania, is a first step toward protection of stream banks that achieve a level of sediment entrapment that is vastly preferable to no protection at all. Location-specific studies are necessary to analyze the sources of sediment and the efficiency of varying buffer widths. For instance, Isabirye et al. (2014) indicated that sediment trapping efficiencies of up to 80 % were obtained from just 10 m buffer widths and that the current 100–200 m buffer width may need to be rethought. However, as a cautionary note, buffer efficiencies depend not only upon the particular watershed topographic, soil, climate, and land use, it also depends upon the vegetation stage and type in the buffer that can change from area to area. Hence, it is better to recommend a buffer width as wide as is feasible in a given area.

8.4 Wetlands: Kidneys of the Landscape

8.4.1 Freshwater Wetlands

Wetlands occupy the land-water interface and are some of the most biologically productive environments on earth. Wetlands occur in depressions in every river basin throughout the world and can be as small as an acre. Even though wetlands occupy less than 9 % of the planet's area, they contribute greatly to biodiversity support, water quality improvement, flood abatement, and carbon sequestration (Mitsch and Gosselink 1993; Zedler and Kircher 2005). A large section of the world's population depends upon wetlands for growing rice, palm, sago, and fisheries. Wetland drainage and loss has been significant worldwide, and hence 144 countries signed the Ramsar Convention in 1971 to identify and promote the protection of major wetlands which are deemed as Ramsar sites. Large freshwater wetlands such as the Pantanal, the Everglades, the Okavango Delta, and the Tonle Sap in the Mekong Basin are well known.

The US Fish and Wildlife Service developed a non-regulatory, technical definition of wetlands that emphasizes these concepts via the following three points: (1) hydrology – the degree of flooding or soil saturation is such that at some time during the growing season, the substrate is saturated or covered by shallow water; (2) vegetation – plants adapted to grow in water or in a soil or substrate that is occasionally oxygen deficient due to water saturation (hydrophytes) are found; and (3) soils, those saturated long enough during the growing season to produce oxygen-deficient conditions in the upper part of the soil, which commonly includes the majority of the root zone of plants, predominate (i.e., hydric soils).

8.4.2 Benefits of Wetlands

8.4.2.1 Flood Control, Water Storage, and Groundwater Recharge

The most significant social and economic benefit that wetlands provide is flood control. Wet grasslands alongside river basins and marshes with centuries of peat buildup in the soil act like sponges, absorbing rainfall, controlling its flow into streams and rivers, and, at the same time, recharging groundwater in the dry season when the water table falls. When peat becomes completely saturated and unable to absorb any more water, surface pools and peatland vegetation – including sedge meadows and some types of forest – help to slow and reduce runoff. Similarly, floodplains alongside the lower reaches of major rivers, such as the Nile, Parana, Yangtze, Ganges-Brahmaputra, and Danube, allow heavy rainfall or spring snowmelt to spread out slowly. When the peat bogs are drained or the floodplains reduced, the risk of flash floods is increased.

8.4.2.2 Water Quality

Wetlands act as filters on the landscape, cleaning up water in a number of ways. Excess nutrients (chemical fertilizers in runoff) are entrapped in the soil, transformed by microbial processes to less ecologically harmful forms, or taken up by wetland plants. Similarly, heavy metals and toxins in runoff are trapped in wetland sediments.

Perhaps the most important water quality ameliorative function of wetlands is denitrification, or the transformation of nitrate to nitrogen gas by soil microbes (Forshay and Stanley 2005; Craig et al. 2008). Because of extensive wetland and riparian forest loss, nitrification of waterways increased drastically during the twentieth century (Malakoff 1998; Walter and Merritts 2008). Excessive nitrate in the water can contribute to eutrophication. Eutrophication creates extensive algal blooms; upon death, the algal mats are decomposed by microbial activity that lowers dissolved oxygen in bottom waters. This leads to dead zones/hypoxia with attendant extermination of marine and estuarine life and an abrupt change in ecosystem structure. Dead zones have spread exponentially in coastal oceans since the 1960s and have now been reported from more than 400 systems worldwide, affecting a total area of more than 245,000 km^2 (Diaz and Rosenberg 2008). To avoid collapse of marine ecosystems along with their fisheries, it is imperative to reduce nutrient loading into rivers, for which the only feasible solution from a water management perspective involves the use of natural wetlands and riparian buffers to entrap and prevent some fraction of nutrients from reaching rivers.

A growing area of ecohydrological application is the design and use of artificial or constructed wetlands for the treatment of municipal wastewater as well as certain types of industrial effluents. The next chapter gets into that in bit more detail.

8.4.3 Estuarine Wetlands: Freshwater Is the Lifeline

8.4.3.1 Ecosystem Services

Coastal wetlands such as mangroves and salt marshes act as frontline defenses against devastation from periodic storms and wave surges. The roots of wetland plants bind the shoreline together, resisting erosion by wind and waves and providing a physical barrier that slows down storm surges and tidal waves, thereby reducing their height and destructive power. In the Caribbean, the shoreline protection services provided by coral reefs are valued at up to US$2.2 billion annually. Worldwide, an estimated 200 million people who live in low-lying coastal regions are at potential risk from catastrophic flooding.

Mangroves and seagrass beds also constitute nurseries for marine fish and are the basis of coastal fisheries. The ever-changing environment of fresh and saline water along with the nutrients brought by both water pools provides one of the world's most productive ecosystems – the estuarine and coastal ecosystems. In the tropics, seagrass beds cover the estuary and coastal offshore muddy/sandy bottom, where marine fish come to breed, the seagrass providing both shelter for juvenile fish from larger marine predators of the open sea and food in the form of submerged aquatic vegetation and marine invertebrates.

8.4.3.2 Estuaries and Freshwater Management: An Optimal Range of Freshwater Inflows

From a water management standpoint, the challenge in maintaining estuarine forests and wetlands is to ensure adequate freshwater inflow via rivers that follows the natural seasonal cycle of wet and dry season flows. The inflow of freshwater has been long recognized as a crucial factor affecting the biological productivity of estuarine areas worldwide (Powell et al. 2002), since freshwater affects bays at physical, biochemical, and ecological levels. Estuaries have a unique environment with a constantly varying mix of freshwater and seawater. This mix varies seasonally from a pulse of freshwater flowing far out to the sea during the rainy season to very saline conditions in the estuary during the dry season when the freshwater flow in the river has decreased. The mix of freshwater and saline seawater also varies diurnally with tides; at high tide, the seawater opposes the freshwater and moves into the river as a wedge of denser water flowing in underneath the freshwater that is flowing seaward in the opposite direction.

Estuarine communities are adapted to this natural seasonal fluctuation in freshwater inflows. A decrease in freshwater inflow to levels lower than the natural seasonal flow regime results in increased seawater intrusion into the estuary (Nguyen and Savenije 2006). Prolonged exposure to high salinity reduces water uptake in mangroves by stressing the salt-exclusion mechanisms in roots and leaves (Parida and Das 2005). Even though mangrove species differ in their tolerances to salinity, high levels of flooding with saline water can stress even the most salinity-resistant species, resulting in eventual mangrove dieback. Coastal forests typically have a range of tree species that vary in their salinity tolerance, including very intolerant species existing on slightly higher elevations that depend upon a rain-derived freshwater lens that floats above groundwater and occupies the vadose zone. Higher salinity in groundwater arising from decreased freshwater inflows decreases the freshwater lens (e.g., Saha et al. 2011).

Similarly, hypersaline conditions in bays stress seagrasses, as well as the various organisms that reside in these habitats. Decreased river inflows into estuaries also lead to decreased nutrient inputs. At the same time, very high freshwater flows can also disrupt lifecycle process of estuarine ecosystems (Powell et al. 2002; Tolley et al. 2013). Keeping all this in mind, there is an optimal range of freshwater inflows into estuaries necessary to maintain estuarine ecosystems.

Freshwater flows to the estuary thus balance seawater coming in with the tide. Hence, any large decrease in freshwater inflows leads to seawater intrusion into the estuary and possibly into coastal aquifers near the estuary in areas where the estuary and underlying aquifers are hydrologically connected, or in low elevation flat areas along the riverbanks where seawater floods in overland during low tide. Once shallow well water gets saline, wells often have to be abandoned. Sotthewes (2008) notes increasing saltwater intrusion occurring in the Pangani estuary over the past several decades and attributes it to two major factors: decreasing freshwater discharge on account of irrigation and hydropower reservoir abstractions and increasing erosion at the marine end on the account of less deposition of river sediment. Similarly, the drainage of the Everglades in the early twentieth century has led to decreases in freshwater discharges along the east coast of South Florida that has progressively brought the freshwater-seawater interface inland into the Biscayne Aquifer (USGS 2013), leading to the salinization and the eventual abandonment of six out of eight water supply wells in the city of Hallandale in 2011 (Reid 2011).

Policymakers thus are faced with the difficult task of developing water resource management programs that allocate freshwater between changing human and ecosystem needs in a sustainable manner. In their review of the prevailing understanding and experiences of the restoration and recovery of estuarine, coastal, and marine ecosystems, Elliott et al. (2007) caution that even though restoration is worthwhile, rarely can it replace lost habitat or ecosystem diversity. Theoretical ecological concepts related to restoration are well understood, such as ecosystem structure and function; however, other factors such as assimilative capacity, resilience, and ecosystem services are specific to the particular region or ecosystem and typically are poorly quantified in much of the world. The linking between these ecological concepts and the management framework is required to impart a holistic approach to understanding and managing these ecosystems and the services they provide to mankind.

8.5 Aquatic Ecosystems and Water Quality

8.5.1 Assimilative Capacity and Self-Purification

Aquatic ecosystems have an inherent capacity to maintain water quality (e.g., McClain 2008) that is referred to as the overall assimilative capacity of the particular stream, river, or wetland. Ostroumov (2005, 2006) has reviewed the array of physical, chemical, and biological processes that contribute to maintaining water quality; physical processes include filtering, deposition, and dilution, chemical processes include sorption/release of substance from sediments and organic matter and transformation via biogeochemical reactions, while biological processes include sequestration, microbial transformation, uptake by plants and animals, and nutrient spiraling. These processes are interconnected and depend upon the existence of different habitat types and zones such as stream, floodplain, and riparian vegetated zone.

The inherent capacity of a particular water body is assessed by ecohydrologists which then along with a factor of safety is used by water managers to determine the total daily maximum loads (TMDLs) of pollutants in discharges by different point sources along with prevailing nonpoint sources. Seasonal variation in hydrology and ecology influences the TMDLs to a water body. By recognizing and supplying the self-purifying functions that a natural stream or river provides, water quality can be maintained at source which vastly decreases the expense of treatment at the user's end. In most developing countries, maintaining good water quality in streams, rivers, and wetlands is the only way to ensure water quality, given the infeasibility and unsustainability of large treatment plants.

8.5.2 Maintaining Aquatic Ecosystems: Threats and Opportunities

Water flow and quality in streams and rivers are influenced by the landscapes they drain (Hynes 1970; Allan 2004). A major threat to stream ecosystems worldwide is sedimentation (Malmqvist and Rundle 2002) arising from soil erosion due to deforestation and/or inadequate soil conservation measures in hillslope farms and road building. Siltation of stream bottoms covers up the spaces underneath and in between streambed stones, thereby removing habitat for the aquatic larvae of many insects. These aquatic macroinvertebrates form a prey base for fish, and in addition, many of them facilitate the breaking down and subsequent decomposition of leaf litter and organic matter in streams. Adequate soil conservation methods such as terracing, strip mulching, dykes, and bunds are necessary through the cooperation of farmers, agriculture extension, and NGOs along with riparian buffers as a last defense against nonpoint pollution.

Another threat to aquatic ecosystems arises from flow alterations resulting from straightening stream courses or channelization. Straightening removes heterogeneity of in-stream habitats such as depositional gravel bars, riffle zones, and deep pools that arise from the effect of meandering channels upon water flow and deposition. The past decade has seen an increase in river channel restoration activities across North America and Europe, with manuals available (such as Soar and Thorne 2001) that detail the hydraulic engineering and ecohydrological inputs necessary for restoring channel meander and stabilized stream banks.

A third threat to ecosystems arises from the introduction of exotic fish, mollusk, and plant species, which alter the relationships between different species and communities and thereby affect ecosystem function as well as ecosystem services. A global example is water hyacinth, *Eichhornia crassipes*, of South American origin that has been spreading over inland waterbodies in all other continents and if not controlled covers lakes and wetlands completely, cutting off sunlight for native plants and lowering oxygen that results in large fish and turtle kills. Exotic species can also affect infrastructure, such as the small mollusk *Corbicula fluminea* of Asian origin that has been seen to clog up water intake pipes in cooling systems in the USA. However, on the other hand, the European zebra mussel that was accidentally introduced to the Great Lakes in the USA and Canada has been found to improve water quality by filtration. The control of exotic species is especially difficult and labor intensive. Biological control solutions abound, usually based on introducing species that prey upon the target exotic species from its native area; however they need to be studied very carefully as this intentional introduction can in turn cause other problems.

8.5.3 Minimum Seasonal Environmental Freshwater Flows

Tropical rivers exhibit enormous seasonal variations in flow and depth; for instance, the tributaries of the Wami and Ruvu Rivers in Tanzania vary from 4 to 5 m depth in the wet season to less than a meter in the dry season. Over time, riverine species and communities have evolved to adapt to a natural seasonal flow cycle. Changing that cycle by changing flow magnitude, removing seasonal flow variation, or sudden releases of water, typically from dam and reservoir operations or large water abstractions, can completely change the aquatic environment with changes in community structure. Migratory fish in particular are seriously affected. For example, McClain et al. (2014) describe flow-ecology relationships in the Mara River of Kenya/Tanzania by comparing the seasonal flow regime, channel hydraulics, and biological communities; such relationships constitute valuable inputs to river management.

An environmental flow assessment (EFA) in a river aims to determine the quality, quantity, and timing of freshwater flow required to maintain the aquatic ecosystem (Poff et al. 1997). Having determined the minimum flow requirements for a river and incorporated them into policies, the far bigger challenge involves actual implementation. Part of the challenge is technical in that monitoring river flows is often absent or patchy. But the greater challenge rests in allocating a limited quantity of water among multiple stakeholders. To accept this, the stakeholders need to be aware of the importance of maintaining aquatic ecosystems. Dickens (2011) reviews EFAs recently carried out in four major rivers in Tanzania, looking at both the methodologies and direct relevance for water management with the heartening conclusion that the initiative, progress, and implementation of EFAs in Tanzania have been exemplary for any nation. The group International Rivers has developed guidelines based upon the experience in implementing programs to restore natural flow regimes in rivers in many countries (Kendy and LeQuesne 2014). Foremost is the suggestion to take deliberate, incremental steps in a multi-stakeholder process that do not exceed the technical, financial, and logistical capacity in place. It is a hopeful sign that projects in many countries have succeeded in restoring some degree of natural channel geomorphology and flow, as described in the case studies by International Rivers (Kendy and LeQuesne 2014). This agrees with the assessment by McClain (2008) that large river basins in the humid tropics still retain a high degree of ecosystem function.

8.6 Conclusion

This chapter has described the general hydrological behavior of various forest and wetland types commonly encountered in watersheds. Forests and wetlands act as water storage units on the landscape, thereby both regulating high flows and extending the period of low flows in the dry season. They also ensure good water quality. It has always been known that forests are beneficial for regulating water and natural resources. However, in the present era, only a fraction of watershed area is covered with natural forests, while most wetlands have been drained. Hence, the ecohydrological features of the forests and wetlands present in a catchment have to be studied and understood in order to maintain these remaining ecosystems as well as employ their beneficial hydrological services to mankind. An understanding of the ecosystem-hydrology linkages is also necessary for the restoration of ecosystems, thereby increasing ecosystem services and enabling sustainability in water availability, quality, and management. Watershed ecosystems also buffer hydrological processes and water availability from the adverse effects of climate change, namely, uncertainty associated with precipitation and increasing evaporative demand.

References

Allan JD (2004) Landscapes and riverscapes: the influence of land use on stream ecosystems. Annu Rev Ecol Evol Syst 35:257–284

Bradshaw C, Sodhi NS, Peh KSH, Brook BW (2007) Global evidence that deforestation amplifies flood risk and severity in the developing world. Glob Chang Biol 13(11):2379–2395

Baker ME, Wiley MJ, Seelbach PW (2001) Gis-based hyirologic modeling of riparian areas: implications for stream water quality. JAWRA J Am Water Resour Assoc 37:1615–1628. doi: 10.1111/j.1752-1688.2001.tb03664.x

Breshears DD (2005) An ecologist's perspective of ecohydrology. Bull Ecol Soc Am 86:296–300

Brocca L, Melone F, Moramarco T, Wagner W (2013) A new method for rainfall estimation through soil moisture observations. Geophys Res Lett 40(5):853–858. doi:10.1002/grl.50173

Brown A, Zhang L, McMahon TA, Western AW, Vertessy RA (2005) A review of paired catchment studies for determining changes in water yield resulting from alterations in vegetation. J Hydrol 310(1–4):28–61

Bruijnzeel LA (2001) Hydrology of tropical montane cloud forests: a reassessment. Land Use Water Resour Res 1:1.1–1.18

Cavelier J, Goldstein G (1989) Mist and fog interception in elfin cloud forests in Colombia and Venezuela. J Trop Ecol 5(03):309–322

Craig LS, Palmer MA, Richardson DC, Filoso S, Bernhardt ES, Bledsoe BP, Doyle MW, Groffman PM, Hassett BA, Kaushal SS, Mayer PM, Smith SM, Wilcock PR (2008) Stream restoration strategies for reducing river nitrogen loads. Front Ecol Environ 6:529–538

Crockford RH, Richardson DP (2000) Partitioning of rainfall into throughfall, stemflow and interception: effect of forest type, ground cover and climate. Hydrol Process Spec Issue Link Hydrol Ecol 14(16–17):2903–2920

D'Odorico P, Porporato A (2006) Dryland ecohydrology, eco-hydrology defined, William Nuttle. ISBN 1-4020-4261-2

Diaz RJ, Rosenberg R (2008) Spreading dead zones and consequences for marine ecosystems. Science 321:926–928

Dawson T (1998) Fog in the California Redwood Forest: ecosystem inputs and use by plants. Oecologia 117:476–485

Dickens C (2011) Critical analysis of environmental flow assessments of selected rivers in Tanzania and Kenya. IUCN ESARO Office/INR, Nairobi/Scottsville, viii+104 pp

Douglass JE (1966) Effects of species and arrangement of forests on evapotranspiration. In: Proceedings of a National Science Foundation advanced science seminar international symposium on forest hydrology held at The Pennsylvania State University, Pennsylvania Aug 29-Sept 10, 1965. Pergamon Press, Oxford/New York

Eamus D, Hatton T, Cook P, Colvin C (2006) Ecohydrology: vegetation function, water and resource management. Csiro Publishing, Melbourne, 360 pp

Elliott M, Burdon D, Hemingway KL, Apitz SE (2007) Estuarine, coastal and marine ecosystem restoration: confusing management and science: a revision of concepts. Estuar Coast Shelf Sci 74:349–366

Elser JJ, Bracken MES, Cleland EE, Gruner DS, Harpole WS, Hillebrand H, Ngai JT, Seabloom EW, Shurin JB, Smith JE (2007) Global analysis of nitrogen and phosphorus limitation of primary producers in freshwater, marine and terrestrial ecosystems. Ecol Lett 10(12): 1135–1142. doi:10.1111/j.1461-0248.2007.01113.x

Forshay KJ, Stanley EH (2005) Rapid nitrate loss and denitrification in a temperate river floodplain. Biogeochemistry 75:43–64

Giambelluca TW, Gerold G (2011) Chapter 11: Hydrology and biogeochemistry of tropical montane cloud forests. In: Levia DF, Carlyle-Moses DE, Tanaka T (eds) Forest hydrology and biogeochemistry: synthesis of research and future directions, Ecological studies series, no. 216. Springer, Heidelberg

Hill A (1996) Nitrate removal in stream riparian zones. J Environ Qual 25(4):743–755

Hunt RJ, Wilcox DA (2003) Ecohydrology – why hydrologists should care. Ground Water 41(3):289

Hynes HBN (1970) The ecology of running waters. University Toronto Press, Toronto, xxiv+555 p

Isabirye M, Kimaro D, Semalulu O, De Meyer A, Magunda M, Poesen J, Deckers J (2014) Sediment generation and evaluation of vegetation buffer strips filters in the riparian zone of Lake Victoria. National Agricultural Research Laboratories, Uganda. http://www.narl.go.ug/index.php/publications/41-sediment-generation-and-evaluation-of-vegetation-buffer-strips-filters-in-the-riparian-zone-of-lake-victoria

Iyango L, Kiwazi F, Tindamanyire T, Kaganzi E, Busulwa H, Mafabi P (2012) Traditional wetland practices in rural communities in Uganda. Lake Victoria Environmental Management Project Phase I (LVEMP I) [120]. Available from the online repository of Lake Victoria Commission http://repository.lvbcom.org/handle/123456789/65

Jackson RB, Jobbagy E, Nosetto MD (2009) Ecohydrology bearings: invited commentary – ecohydrology in a human-dominated landscape. Ecohydrology 2:383–389

Jose S, Gillespie AR, George SJ, Kumar BM (1996) Vegetation responses along edge-to-interior gradients in a high altitude tropical forest in peninsular India. For Ecol Manag 87(1–3):51–62

Kadur S, Bawa K (2005) Sahyadris: India's Western Ghats – a vanishing heritage. Ashoka Trust for Ecology and Environment, Bangalore. ISBN 9 780977 021109

Kendy E, le Quesne T (2014) Environmental flow policies: moving beyond good intentions. Report for International Rivers available at http://www.internationalrivers.org/resources/environmental-flow-policies-moving-beyond-good-intentions-1671. Accessed 28 Mar 2014

Khan ML, Khumbongmayum ADV, Tripathi RS (2008) The sacred groves and their significance in conserving biodiversity: an overview. Int J Ecol Environ Sci 34(3):277–291

Krishnaswamy J, Bonell M, Ventatesh B, Purandara BK, Lele S, Kiran MC, Reddy V, Badiger S, Rakesh KN (2012) The rain-runoff response of tropical humid forest ecosystems to use and reforestation in the Western Ghats of India. J Hydrol 472–473:216–237

Krishnaswamy J, Bonell M, Ventatesh B, Purandara BK, Lele S, Kiran MC, Reddy V, Badiger S (2013) The groundwater recharge response and hydrologic services of tropical humid forest ecosystems to use and reforestation: support for the "inflitration-evapotranspiration trade- off hypothesis". J Hydrol 498:191–209

Lele S, Patil I, Badiger S, Menon A, Kumar R (2008) The economic impact of forest hydrological services on local communities: a case study from the Western Ghats of India. No 45, Working papers from The South Asian Network for Development and Environmental Economics, Kathmandu, Nepal

Malakoff D (1998) Restored wetlands flunk real-world test. Science 280(5362):371–372. doi:10.1126/science.280.5362.371

Malhotra KC, Gokhale Y, Das K (2001) Sacred groves of India: an annotated bibliography. Indian National Science Academy and Development Alliance, New Delhi. http://www.sacredland.org/media/Malhotra_Sacred-Groves-of-India.pdf

Malmqvist B, Rundle S (2002) Threats to the running water ecosystems of the world. Environ Conserv 2002:134–153. doi:10.1017/S0376892902000097

McClain ME (2008) Ecohydrology as a tool in the sustainable development of large tropical rivers. In: Harper D, Zalewski M, Pacini N (eds) Ecohydrology: processes, models and case studies. CABI, Oxfordshire

McClain ME, Boyer EW, Dent CL, Gergel SE, Grimm NB, Groffman PM, Hart SC, Harvey JW, Johnston CA, Mayorga E, McDowell WH, Pinay G (2003) Biogeochemical hot spots and hot moments at the interface of terrestrial and aquatic ecosystems. Ecosystems 6:301–312

McClain ME, Chicharo L, Fohrer M, Gavino N, Windhorst W, Zalewski M (2012) Training hydrologists to be ecohydrologists and play a leading role in environmental problem solving. Hydrol Earth Syst Sci 16:1685–1696

McClain ME, Subaluski AL, Anderson EP, Dessu SB, Melesse AM, Ndomba PM, Mtamba JOD, Tamatamah RA, Mligo C (2014) Comparing flow regime, channel hydraulics and biological communities to infer flow-ecology relationships in the Mara river of Kenya and Tanzania. Hydrol Sci J 59(3–4):1–19

Meher-Homji VM (1991) Probable impact of deforestation on hydrological processes. Clim Chang 19:163–173

Mitsch WJ, Gosselink JG (1993) Wetlands, 2nd edn. Van Nostrand Reinhold, New York, xii +722 pp. ISBN 0 442 00805 8

Moreira M, Sternberg L, Martinelli L, Victoria R, Barbosa E, Bonates L, Nepstad D (1997) Contribution of transpiration to forest ambient vapor based on isotopic measurements. Glob Chang Biol 3(5):439–450. doi:10.1046/j.1365-2486.1997.00082.x

Munishi PKT, Shear TH (2005) Rainfall interception and partitioning in afromontane rainforests of the Eastern Arc Mountains, Tanzania: implications for water conservation. J Trop For Sci 17(3):355–365

Naiman RJ, Dé camps H, Pollock M (1993) The role of riparian corridors in maintaining regional biodiversity. Ecol Appl 3(2):209–212

Nguyen AD, Savenije HHG (2006) Salt intrusion in multi-channel estuaries. Hydrol Earth Syst Sci 10:743–754

Nouri H, Beecham S, Kazemi F, Hassanli AM, Anderson S (2013) Remote sensing techniques for predicting evapotranspiration from mixed vegetated surfaces. Hydrol Earth Syst Sci Discuss 10:3897–3925. www.hydrol-earth-syst-sci-discuss.net/10/3897/2013/. doi:10.5194/hessd-10-3897-2013

Nuttle WK (2002) Eco-hydrology's past and future in focus. Eos 83:205

Orr CH et al (2007) Effects of restoration and reflooding on soil denitrification in a leveled Midwestern floodplain. Ecol Appl 17(8):2365–2376

Ostroumov SA (2005) On some issues of maintaining water quality and self-purification. Water Resour 32(3):305–313

Ostroumov SA (2006) Biomachinery for maintaining water quality and natural water self-purification in marine and estuarine systems: elements of a qualitative theory. Int J Ocean Oceanogr 1(1):111–118. ISSN 0973-2667

Parida AK, Das AB (2005) Salt tolerance and salinity effects on plants: a review. Ecotoxicol Environ Saf 60(3):324–349

Pinay G, Roques L, Favre A (1993) Spatial and temporal profiles of denitrification in a riparian forest. J Appl Ecol 30:581–591

Poff NL, Allan JD, Bain MB, Karr JR, Prestegaard KL, Richter BD, Sparks RE, Stromberg JC (1997) The natural flow regime: a paradigm for river conservation and restoration. Bioscience 47(11):769–784

Powell GL, Matsumoto J, Brock DA (2002) Methods for determining minimum freshwater inflow needs of Texas bays and estuaries. Estuaries 25(6B):1262–1274

Putuhena WH, Cordery I (2000) Some hydrological effects of changing forest cover from eucalypts to Pinus radiate. Agric For Meteorol 100(1):59–72

Reed T, Carpenter SR (2002) Comparisons of P-yield, riparian buffer strips, and land cover in six agricultural watersheds. Ecosystems 5:568–577. doi:10.1007/s10021-002-0159-8

Reid A (2011) South Florida drinking water faces salt water threat. Sun Sentinel http://articles.sun-sentinel.com/2011-09-12/health/fl-saltwater-intrusion-20110912_1_saltwater-intrusion-saltwater-threat-drinking-water. Accessed 27 Mar 2014

Rodríguez-Iturbe I (2000) Ecohydrology: a hydrologic perspective of climate-soil-vegetation dynamics. Water Resour Res 36(1):3–9

Rodriguez-Iturbe I, Porporato A, Laio F, Ridolfi L (2001) Plants in water-controlled ecosystems: active role in hydrologic processes and response to water stress. I. Scope and general outline. Adv Water Resour 24:695–705

Saha AK, Sternberg LSLO, Miralles-Wilhelm F (2009) Linking water sources with foliar nutrient status in upland plant communities in the Everglades National Park, USA. Ecohydrology 2:42–54

Saha AK, Saha S, Sadle J, Jiang J, Ross MS, Price RM, Sternberg LS, Wendelberger KS (2011) Sea level rise and South Florida's coastal forests. J Clim Change. doi:10.1007/s10584-011-0082-0

Saha AK, Moses C, Price R, Engel V, Smith TJ, Anderson G (2012) A hydrological budget (2002–2008) for a large subtropical wetland indicates seawater intrusion accompanies diminished freshwater flow. J Estuar Coast 35(2):459. doi:10.1007/s12237-011-9454-y

Schedlbauer JL, Oberbauer SF, Starr G, Jimenez KL (2011) Controls on sensible heat and latent energy fluxes from a short-hydroperiod Florida Everglades marsh. J Hydrol 411:331–341

Schindler DW (1977) Evolution of phosphorus limitation in lakes. Science 195:260–262

Setegn SG, Rayner D, Melesse AM, Dargahi B (2014) Climate Change Impact on Water Resources and adaptation strategies in the Blue Nile River Basin. Ch. 20. In: Melesse A, Abtew W, Setegn SG (eds) Nile River Basin: ecohydrological challenges, climate change and hydropolitics. Springer, Cham

Sheridan MJ, Nyamweru C (2007) African sacred groves: ecological dynamics and social change. Ohio University Press, Athens, 240 pp. ISBN 978-0-8214-1789-8

Shukla J, Nobre C, Sellers P (1990) Amazon deforestation and climate change. Science, New Series 247(4948):1322–1325

Soar P, Thorne CR (2001) Channel restoration design for Meandering Rivers. Report for US Army Corps of Engineers, Washington, DC

Sotthewes W (2008) Forcing on the salinity distribution in the Pangani estuary. Thesis report. University of Delft, Delft, The Netherlands, 82p

Tolley SG, Brosious BB, Peebles EB (2013) Recruitment of the Crabs *Eurypanopeus depressus, Rhithropanopeus harrisii, and Petrolisthes armatus* to Oyster Reefs: the influence of freshwater inflow. Estuar Coasts 36:820–833. doi:10.1007/s12237-013-9590-7

UNEP (2004) Integrated watershed management – ecohydrology & phytotechnology: a manual. United Nations Environment Program, Osaka, Japan. Available online at http://www.unep.org/ietc/Portals/136/Publications/Water&Sanitation/Integrated%20Watershed%20Management%20-%20Ecohydrology%20&%20Phytotechnology%20-%20Part%201%20&%202.pdf

UNICEF (2013) Water, sanitation and hygiene report. http://www.unicef.org/wash/. Accessed 31 Mar 2014

USGS (2013) Development of water-management system and impact on the hydrology of Southeastern Florida: assessment of saltwater intrusion. Circular 1275. United States Geological Survey. http://sofia.usgs.gov/publications/circular/1275/saltintrusion.html. Accessed 27 Mar 2014

Villalobos-Vega R (2010) Water table and nutrient dynamics in neotropical savannas and wetland ecosystems. Open access dissertations, Paper 389. http://scholarlyrepository.miami.edu/oa_dissertations/389

Walter RC, Merritts DJ (2008) Natural streams and the legacy of water-powered mills. Science 319:299–304

Wenger S (1999) A review of the scientific literature on riparian buffer width, extent and vegetation. Office of Public Service & Outreach, Institute of Ecology, University of Georgia, Athens

Xiang W (1993) A GIS method for riparian water quality buffer generation. Int J Geogr Inf Syst 7(1):57–70. doi:10.1080/02693799308901939

Zalewski M (2000) Ecohydrology – the scientific background to use ecosystem properties as management tools toward sustainability of water resources. Ecol Eng 16:1–8

Zedler P, Kircher J (2005) Wetland resources: status, trends, ecosystem services, and restorability. Annu Rev Environ Resour 30:39–74

Chapter 9
Assessment of Agricultural Water Management in Punjab, India, Using Bayesian Methods

Tess A. Russo, Naresh Devineni, and Upmanu Lall

Abstract The success of the Green Revolution in Punjab, India, is threatened by a significant decline in water resources. Punjab, a major agricultural supplier for the rest of India, supports irrigation with a canal system and groundwater, which is vastly overexploited. The detailed data required to estimate future impacts on water supplies or develop sustainable water management practices is not readily available for this region. Therefore, we use Bayesian methods to estimate hydrologic properties and irrigation requirements for an under-constrained mass balance model. Using the known values of precipitation, total canal water delivery, crop yield, and water table elevation, we present a method using a Markov chain Monte Carlo (MCMC) algorithm to solve for a distribution of values for each unknown parameter in a conceptual mass balance model. Model results are used to test three water management strategies, which show that replacement of rice with pulses may be sufficient to stop water table decline. This computational method can be applied in data-scarce regions across the world, where integrated water resource management is required to resolve competition between food security and available resources.

Keywords Agricultural water management • Punjab • India • Markov chain Monte Carlo • Groundwater overdraft • Mass balance model

---

T.A. Russo (✉)
Department of Geosciences, The Pennsylvania State University, 310 Deike Building, University Park, PA 16802, USA

Columbia Water Center, Columbia University, 500 W 120th St., New York, NY 10026, USA
e-mail: russo@psu.edu

N. Devineni
Columbia Water Center, Columbia University, 500 W 120th St., New York, NY 10026, USA

Department of Civil Engineering, The City College of New York, 160 Convent Ave, New York, NY 10031, USA

U. Lall
Columbia Water Center, Columbia University, 500 W 120th St., New York, NY 10026, USA

S.G. Setegn, M.C. Donoso (eds.), *Sustainability of Integrated Water Resources Management*, DOI 10.1007/978-3-319-12194-9_9

9.1 Introduction

9.1.1 The Green Revolution and Groundwater

India made tremendous strides toward self-sufficiency in food grains in the last half century. The Green Revolution, marked by higher-yielding varieties, chemical fertilizers and pesticides, and the provision of irrigation, are some of the key factors that contributed to a successful food grain economy and a transition to food self-sufficiency since the 1960s. The national food security goals have led to targeted regions for procurement of major food grains such as rice and wheat, with minimum guaranteed prices for the farmers, and a variety of subsidies for fertilizer, energy, and seeds. The success of the Green Revolution in India is now threatened by a significant decline in water resources. Punjab (Fig. 9.1), a major agricultural supplier for the rest of India, supports irrigation demand by overexploiting local groundwater resources. Competition between food security and water resources will only become more persistent with increasing populations and changing climate (Brahmanand et al. 2013; Sidhu et al. 2011). A solution is required that meets food security needs, uses water resources sustainably, and maintains farmer income.

Since the onset of the Green Revolution, the total area of irrigated agriculture as well as the intensity of agriculture as measured by the number of cropping seasons

Fig. 9.1 Map showing the location of Punjab in northwestern India and the districts of Punjab. The three districts analyzed in this study: Gurdaspur, Jalandhar, and Sangrur are shown in *gray*

per year has increased drastically. Most notably, the total area of high-water-demand crops has increased. Rice, a crop with a high water demand, typically requires ~1,800 mm in Punjab, while average annual rainfall is only ~650 mm. The area under rice increased from 7 % of the total Punjab land area in 1970 to 50 % in 2001, while gross cropped area increased from 95 % to 152 %, respectively (Takshi and Chopra 2004). Groundwater extraction rates exceed natural groundwater recharge rates, and as a result, water tables are decreasing on average at a rate of 0.4 m/year and up to 1.7 m/year locally (Kahn et al. 2007). In Punjab, the net annual groundwater availability is estimated to be 21.44×10^9 m^3, and the annual amount of groundwater extracted is 31.16×10^9 m^3 (Chatterjee and Purohit 2009). According to the Central Groundwater Board of India, the number of overexploited Blocks (subdistrict divisions) has increased from 53 in 1984 to 110 in 2009 of the total 138 Blocks in Punjab (CGWB 2012). Declining water tables can have negative consequences including increasing pumping costs and require a shift from centrifugal to submersible pump technology due to greater lifts (Samanpreet Kaur et al. 2011).

This study focuses on districts in Punjab that are experiencing groundwater overdraft and subsequent water table decline. Conversely, several districts in southwestern Punjab are experiencing rising water tables and water logging at the surface. Due to farming practices in these regions, soil water is highly saline and 50 % of the region reports saline groundwater deeper than 35 m (CGWB 2012), so irrigation water must come primarily from the canal system. Though not addressed in this study, water management in Punjab must ultimately accommodate both rising and falling groundwater, in addition to degradation of water resource quality from agricultural activity.

9.1.2 Institutional and Economic Obstacles

The two major incentives for growing rice and wheat in Punjab are the guaranteed purchase prices set by the central government and a fixed connection charge for electricity with no per unit charge. The policy of electricity subsidies adopted by the state government permits farmers to freely pump groundwater required to irrigate their crops without a concern for efficiency or conservation (Sidhu et al. 2011). As a result, almost 97 % of the total agricultural land in Punjab is now irrigated, and agricultural pumping accounts for 40–60 % of the total electricity consumption and leads to unreliable electricity for all uses. Electricity supply to agricultural areas is intermittent and unpredictable, often on for only a few hours per day. To overcome the unreliability, farmers commonly leave their pumps on all the time leading to over-irrigation.

Driven by a national food security concern, the central government established procurement and price support systems for selected grains to protect producers against sudden price declines in the unregulated market. Government procurement policies have shifted the major crops in Punjab from vegetables, pulses, and oil seeds to rice and wheat. Punjab, which accounts for only 1.5 % of India's land area, now supplies 40–50 % of the rice and 60–65 % of the wheat that India consumes (Aggarwal et al. 2009). Farmers have little incentive to grow low-water-demand crops because of the subsidized electricity and the guaranteed market price provided by government procurement programs. This scenario is typical in much of the country. A recent study shows that the major regions that contribute to the food security of the country are chronically falling short of water and are under severe stress (Devineni et al. 2013). Theoretically, replacing rice with a cash crop like vegetables or flowers, which also require less irrigation, benefits the hydrologic balance as well as farmer income. However, most farmers are hesitant to switch to cash crops due to the high market price variability.

Many authors have suggested a variety of adaptation strategies to mitigate groundwater overdraft in Punjab (Aggarwal et al. 2009; Ambast et al. 2006; Hira 2004; Humphreys et al. 2010; Shah 2009), though most lack rigorous quantitative analysis or simulation of proposed management strategies due to an overall lack of access to hydrogeologic and water consumption data. In lieu of a dynamic numerical model, a hydrologic mass balance approach can provide initial results to inform water management decisions. In this study, we developed a Bayesian approach to constrain the mass balance while estimating the probability density for unknown parameters. Bayesian methods are becoming increasingly common for hydrologic system modeling (e.g., Engeland and Gottschalk 2002; Winslow et al. 2013).

The objectives of this study are to address the following questions: (1) How do we estimate the unknown hydrogeologic and agricultural parameters in the hydrologic mass balance equation? (2) How and why might these parameters vary within the state of Punjab? (3) Which types of irrigation management reduce or reverse groundwater elevation decline? While we focus on Indian Punjab, our methodology is transferrable to similar data-scarce regions around the world.

9.2 Study Location and Data

9.2.1 Study Area

The state of Punjab located in northwestern India has an area of 50,362 km^2 (Fig. 9.1). The state is underlain by Quaternary alluvium eroded from the Himalayas and contains clay, loam, and silt. Secondary porosity in the alluvium comes from buried channels within the greater alluvial deposits. The alluvium is bounded in the northeast by the Sivalik formation, primarily the Tertiary period sandstone and conglomerate. The region is known for particularly fertile soils and is a critical agricultural region within India.

Average annual precipitation in Punjab ranges between 360 and 1,120 mm/year, with wetter areas in the northeast and dryer areas in the southwest. Even in the wettest areas, the monsoon rainfall is not sufficient to support current year-round rice-wheat agriculture; therefore, irrigation is required. Irrigation water can be sourced from a network of surface canals or groundwater wells. Canals are supplied by the Bhakra reservoir, the second largest in India, which holds up to 9.34 billion cubic meters (bcm), and diversions from the Ravi River which supplies water to the districts of Gurdaspur and Amritsar. Despite the large canal network, use of surface water has been surpassed by groundwater due to unreliable surface flows (Tyagi et al. 2005) and because many farmers are not directly adjacent to a canal.

We focus on three districts for which agricultural and hydrologic data were available: Gurdaspur, Jalandhar, and Sangrur (Fig. 9.1). The three districts are all intensively farmed and have differing water table depths and average precipitation (Table 9.1). In 1973, the average depth to groundwater was between 5 and 7 m for all three districts (Fig. 9.2). By 2010, the depth ranged from 8 to 23 m. The regional variability in water table elevation change over time is due to differences in irrigation requirements, climate, aquifer properties, and recharge parameters.

9.2.2 Data Sources

Groundwater elevation data were obtained from the Central Groundwater Board. Crop area and yield records were obtained from colleagues at the Punjab Agricultural University. Daily release values from the Bhakra Dam were obtained from Bhakra Beas Management Board. We assume the magnitude of flow from the Bhakra reservoir to correlate with flows in the Bari Doab canal system in Gurdaspur. Gridded daily rainfall data from 1971 to 2005, available at $1^\circ \times 1^\circ$ spatial resolution from Indian Meteorological Department (IMD) (Rajeevan and Bhate 2008), were spatially averaged over each district using the geographic information system (GIS). Gridded daily temperature data (at 6 hourly time step) from 1948 to 2000, available at the spatial resolution $1^\circ \times 1^\circ$ from National Centers for Environmental Prediction/National Center for Atmospheric Research (Ngo-Duc et al. 2005), is used in this study to estimate the potential evapotranspiration. Crop water requirements were determined from crop data provided by colleagues at the Punjab Agricultural University and annual potential evapotranspiration. Specific yield (S_y) of the aquifer was assumed to be 0.15 for all districts (MWR 1997). The intersection of all available data sets spans from 1973 to 2002, which were used as model input for the Bayesian model.

Table 9.1 Geographic, agricultural, and hydrologic data for each study district

	Gurdaspur	Jalandhar	Sangrur
Area (km^2)[a]	3,513	2,662	3,737
Gross irrigated area (km^2)[a]	4,270	4,137	5,754
Average annual rainfall (mm)[b]	831	654	522
Average depth to GW, 2011 (m)	8.2	17.8	22.4

[a]Values from Central Groundwater Board district reports (2007)
[b]Average calculated from IMD gridded precipitation data, 1979–2012 (Rajeevan and Bhate 2008)

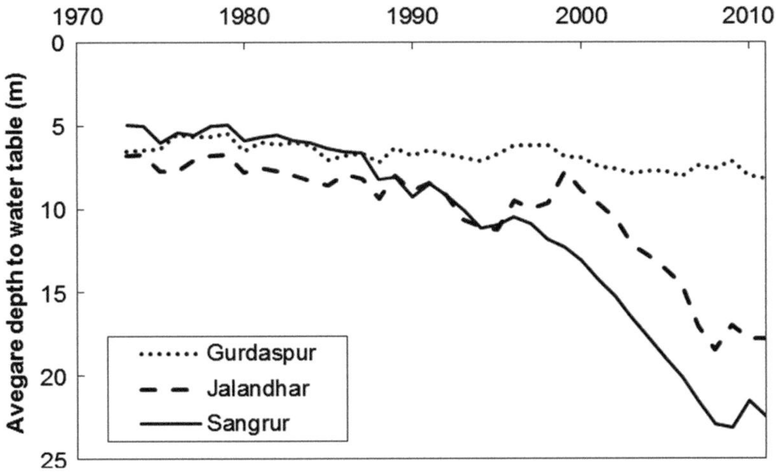

Fig. 9.2 Annual average water table elevation between 1973 and 2011 for Gurdaspur (*dotted line*), Jalandhar (*dashed line*), and Sangrur (*solid line*) districts

9.3 Methods

9.3.1 *Parameter Estimation: Markov Chain Monte Carlo Method*

We use a Bayesian method to estimate parameters in the under-constrained hydrologic mass balance for each study district in Punjab. Groundwater inputs to Punjab include recharge from precipitation and leakage from the canal network. Groundwater outputs include evapotranspiration and extraction for irrigation. We use the following groundwater mass balance equation:

$$R + L - ET - X - \Delta S = 0 \tag{9.1}$$

where R is recharge from precipitation, L is canal leakage, ET is evapotranspiration, X is total groundwater pumped for irrigation, and ΔS is change in groundwater storage which includes lateral groundwater flow in and out of the district. For this study, we account for ET in irrigation requirements (X) and assume additional

Table 9.2 List of known and unknown agricultural, hydrologic, and climate variables

Known variable	Unknown variable
Precipitation, $P(t)$	Precipitation recharge, $\beta P(t) = R$
Canal discharge from Bhakra reservoir, $B(t)$	Canal leakage, $\varepsilon B(t) = L$
Total area, A_T	Pumped irrigation, $x_i(t)$
Crop area, $A_i(t)$	Net groundwater influx, γ
Crop water requirements, $\mathrm{CWR}_i(t)$	
Maximum yield, $Y_i^{\max}$	
Crop yield, $Y_i(t)$	
Groundwater elevation change, ΔGW	
Specific yield, S_y	

groundwater loss is equal to zero, because the average groundwater depth is greater than the expected evapotranspiration extinction depth (Shah et al. 2007). Both annual precipitation and canal discharge are known, and we assume a constant coefficient to represent the percentage of each source that reaches the groundwater. Net groundwater flow into the district is unknown but is assumed to originate as recharge in the Sivalik foothills in and northeast of Punjab.

The coefficients for precipitation recharge (R), canal leakage (L), and net groundwater influx (γ), and volumes of pumped irrigation (X), were estimated using a Markov chain Monte Carlo (MCMC) method. For each individual district (Gurdaspur, Jalandhar, and Sangrur), we use the MCMC method to simulate the complete posterior distribution of values for each unknown model parameter (Table 9.2). Uniform prior distributions based on physical properties constrain acceptable values for each unknown parameter.

The two governing equations of the MCMC algorithm are shown in Eqs. (9.2) and (9.3). Every iteration proposes new values for the unknown parameters, which are accepted or rejected, based on the prior constraints and the assumption that the known values are normally distributed about the mean, and have a variance with a prescribed distribution. Equations (9.2) and (9.3) describe the relationship between the parameters of the mass balance equation and two known values, change in water table elevation and observed annual yield, respectively:

$$\Delta \mathrm{GW}(t) \sim N\left(\gamma\left(\frac{\beta p(t) A_T(t) + \varepsilon B(t) - \sum_{i=1}^{n} \left(x_i(t) A_i(t)\right)}{S_y A_T}\right), \tau_{gw}^2\right) \tag{9.2}$$

$$Y_i(t) \sim N\left(\frac{x_i(t)}{\mathrm{CWR}_i(t)} Y_i^{\max}(t), \tau_y^2\right) \tag{9.3}$$

where t is simulation year, i refers to each of the eight most common crops in Punjab (rice, wheat, groundnut (peanut), maize, sugarcane, potato, and two varieties of cotton), and all other variables are defined in Table 9.2. We assume non-informative prior distributions for each unknown variable are as follows:

$$\beta \sim U(0, 0.7) \tag{9.4}$$

$$\varepsilon \sim U(0, 0.7) \tag{9.5}$$

$$\gamma \sim U(0, 5) \tag{9.6}$$

$$x \sim U(0, \mathrm{CWR}_i(t)) \tag{9.7}$$

$$\tau_{gw} \sim U(0, 1000) \tag{9.8}$$

$$\tau_y \sim U(0, 400) \tag{9.9}$$

The joint posterior distribution $p(\boldsymbol{\theta}/y)$ of the complete parameter vector $\boldsymbol{\theta}$ (that includes β,ε, γ, X, and τ) is derived by defining the posterior distribution function as follows:

$$\begin{aligned} &p(\Delta\mathrm{GW}, Y) \ \propto \\ &\prod\nolimits_{t=1}^{T} N\left(\Delta\mathrm{GW}(t)|\gamma\left(\frac{\beta p(t)A_T(t) + \varepsilon_B(t) - \sum\nolimits_{i=1}^{n}(x_i(t)A_i(t))}{S_y A_T}\right), \tau_{gw}^2\right) \cdot \\ &N\left(\mathrm{Y}_i(t)|_{\frac{x_i(t)}{\mathrm{CWR}_i(t)}} \mathrm{Y}_i^{\max(t)}, \tau_y^2\right) \cdot U(\beta|0, 0.7) \cdot \\ &U(\varepsilon|0, 0.7) \cdot U(\gamma|0, 5) \cdot U(x|0, \mathrm{CWR}_{i(t)}) \cdot \\ &U(\tau_{gw}|0, 1000) \cdot U(\tau_y|0, 400) \end{aligned} \tag{9.10}$$

The unknown parameters are estimated using WinBUGS (Lunn et al. 2000; Spiegelhalter et al. 1996). WinBUGS employs the Gibbs sampler, an MCMC method for simulating the posterior probability distribution of the parameters conditional on the current choice of parameters and the data. The Gibbs sampler sequentially samples one parameter from the conditional distribution of that parameter relative to the others and provides an effective sampling-based numerical solution for parameter estimation (Gilks et al. 1996). We simulated three chains starting from random initial values for the parameters to verify the convergence of the posterior distribution based on the shrink factor (Gelman and Rubin 1992). The shrink factor compares the variance in the sampled parameters within and across the chains to describe the improvement in the estimates for an increasing number of iterations. For this application, each chain was run for a 1,000 cycle burn-in to discard the initial state, followed by 2,000 simulations of model parameters until the shrink factor was <1.1.

9.3.2 *Model Projections: Evaluation of Water Management Scenarios*

To make projections about future water management scenarios, we use the resulting parameter distributions from the MCMC algorithm to solve for the distribution of water table change. We simulate three management scenarios for 30 years to estimate the cumulative influence on water table elevation of each strategy: (1)

continue current irrigation practices, (2) reduce irrigation by 30 %, and (3) replace all rice crops with pulses. The first scenario assumes that crop areas and irrigation practices remain constant from 2002 through the end of the 30-year simulation. The second scenario represents expected irrigation savings if farmers use soil moisture sensors to inform irrigation timing and duration. Previous experiments carried out by the Punjab Agricultural University demonstrated savings as high as 30 % using a low-cost soil tensiometer (Perveen et al. 2012). The third management scenario represents irrigation requirements if each district replaced high-water-demand rice crops with lower-water-demand pulses. Precipitation and evapotranspiration values from the original model (1973–2002) are repeated in order for the management scenario simulations.

9.4 Results and Discussion

9.4.1 Model Results: Variability Across Districts

The model parameters were fit using observed groundwater and yield data. Annual groundwater elevation changes were summed to give each the cumulative water table elevation change for each simulation year (Fig. 9.3). The modeled and observed water table elevation levels are given for each study district. As expected, 50–56 % of the observations are within the 50 % confidence interval of the modeled values for each district.

The three study districts vary in climate and cropping patterns, in addition to geologic properties and density of surface canals. Theoretically, the latter two should control β and ε, while γ depends on both aquifer transmissivity and hydraulic gradient between the district and neighboring regions. The medians of each distribution of values for Gurdaspur, Jalandhar, and Sangrur are as follows: $\beta = 0.25$, 0.39, and 0.23; $\varepsilon = 0.039$, 0.017, and 0.12; and $\gamma = 0.060$, 0.086, and 0.058, respectively (Fig. 9.4). The distribution and median vary across districts for each unknown parameter.

Groundwater recharge comprises a percentage of annual precipitation and a percentage of total water release from the irrigation canals. Recharge from precipitation ranges between 20 and 40 % of total annual rainfall, which is reasonable for subhumid to humid regions. The drier climate in Sangrur theoretically should have a lower recharge coefficient, and though it has the lowest median recharge value, the posterior distribution of values does not differ significantly from those of Gurdaspur or Jalandhar. Model results suggest that seepage to groundwater from canals accounts for 20–50 % of total average surface recharge. Gurdaspur has the greatest total recharge from canal leakage, which may come from the major branch of the Upper Bari Doab canal, the oldest canal network in Punjab. Sangrur is relatively far from the Bhakra headwaters and may have a smaller amount of recharge from the canals because less water reaches the district.

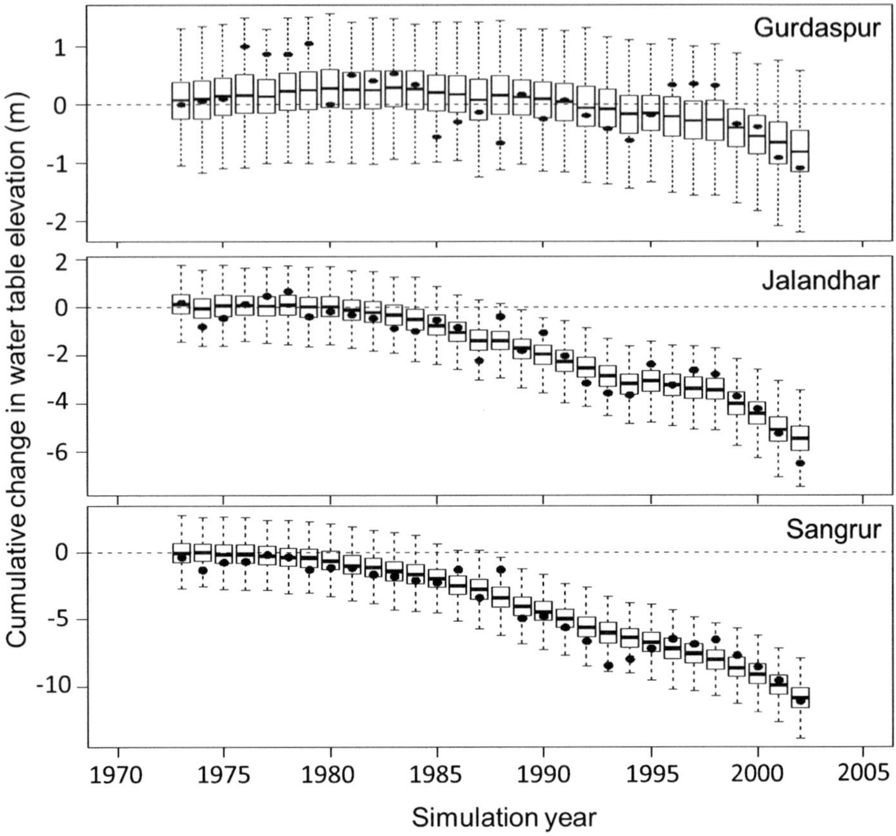

Fig. 9.3 Observed and modeled water table elevations for Jalandhar, Sangrur, and Gurdaspur districts. The *boxplots* show the calculated distribution of water table elevations for each simulated year, and the *filled circles* represent observations

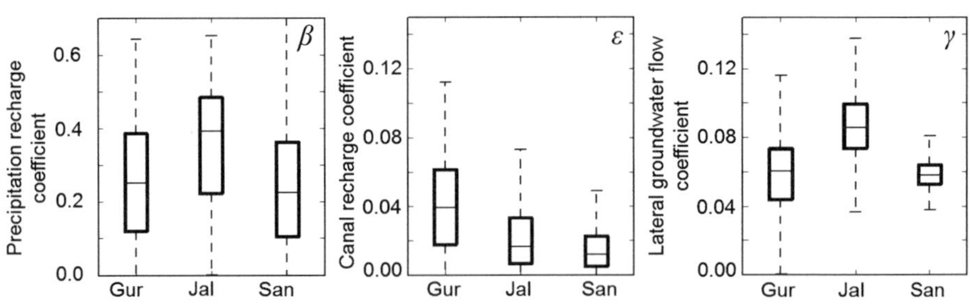

Fig. 9.4 *Boxplots* show the first and third quartiles of each distribution of coefficients for precipitation recharge (β), canal seepage (ε), and lateral groundwater flow (γ) for Gurdaspur, Jalandhar, and Sangrur districts

Gamma, representing the net contribution of lateral groundwater flow to the district, has similar medians for Gurdaspur and Sangrur and a higher value for Jalandhar. Values less than one indicate a net groundwater inflow, where a value of 1 means groundwater inflow equals outflow. Theoretically, gamma represents an external source for the net volume change of groundwater and should be quantified as an additive term in the numerator of Eq. (9.2). However, due to a lack of physical constraints on the value, the model was not able to converge with γ in the numerator. Therefore, γ is represented as a multiplier of the estimated change in water table elevation, which implies that net lateral flow to a district will be a function of current groundwater overdraft. For all districts, the mass balance cannot be completed without net groundwater inflow. The median simulated volume of groundwater inflow is similar to recharge from precipitation. Gurdaspur may have high inflow due to its proximity to the high surface recharge region in the Sivalik foothills, while Sangrur is experiencing the greatest groundwater overdraft; therefore, the gradient drawing groundwater in is high.

9.4.2 Model Assumptions and Limitations

Limitations of the study stem from uncertainty in the data, assumptions made about individual parameters, and assumptions in the underlying model. Assumptions made about variables in the mass balance equation include simplification and lumping of parameters that may not accurately represent the heterogeneity of the system. However, these assumptions and the use of Bayesian methods are opportune given the lack of hydrogeologic data and observations. We assume that the aquifer is continuous and has uniform storage properties. We also assume that irrigation application (X) is 100 % efficient, meaning there is no return flow to the aquifer of excess irrigation water. Actual pumping volumes may be greater in practice, with some of the surplus water consumed by ET and some of it recharging the aquifer. Our model does not simulate extraction of groundwater for irrigation that immediately returns to the aquifer, because it is not required for the mass balance to agree. However, our estimates must be conservative because we do not account for extra losses due to ET on the flooded fields. The use of constant coefficients to represent recharge from precipitation and the canals is another simplification of the system. Actual coefficient values will change seasonally and with long-term changes in climate and land use.

Our assumptions combined with the macro-level aggregated model structure will not capture each interannual water table fluctuation (Fig. 9.3). This is primarily because the model time step is 1 year; therefore the resolution at that time step is coarse. However, the model capacity for representing the decadal-scale trend in groundwater depletion implies that we have accounted for the major components of the mass balance correctly. This model should not be used to estimate groundwater response to water management strategies on short time scales, such as 1–3 years, unless more frequent water table elevation data can be found to reduce the model time step length. The range of each parameter distribution can be used to inform future field data collection efforts to constrain the most uncertain parameters.

9.4.3 Hydrologic Response to Management Scenarios

Water management strategies often involve widespread political change and economic investments and may take several years to evaluate the benefit. Modeling offers a quick and low-cost platform for systematic evaluation of the solutions for estimating the influence of agricultural water management strategies on water table elevation and should be used as a preliminary step to inform management decisions. We calculate the cumulative groundwater response to three water management scenarios over 30 years for each of the three study districts (Fig. 9.5).

The three strategies are not intended to represent easily implementable management strategies, but rather to illustrate the type and extent of change required in order to reduce the groundwater deficit in each district. The first strategy is essentially what the farmers are doing now, though may underestimate current (2013) irrigation requirements due to increases in rice area since 2002. The second strategy will require extensive outreach to farmers, likely a combined effort by the Punjab Agricultural University, extension programs, farmer cooperatives, and government programs (Humphreys et al. 2010; Kaur 2009; Sidhu et al. 2011). Funding for such a program could come from electricity cost savings due to reduced groundwater pumping; for example, the government would save over 150 million USD per year with a 30 % decrease in groundwater pumping. The third strategy will require a change in the government procurement program, which we show to be the most effective method for mitigating groundwater overdraft in Punjab (Devineni and Perveen 2012).

In all cases, the first management scenario (no change) results in further water table decline, with the median rate of decline varying between 0.16 and 0.43 m/year across the three districts (Fig. 9.5). Groundwater extraction is unsustainable in scenario 1 because more water is pumped for irrigation than enters the system via surface recharge (from precipitation and canal leakage) and lateral groundwater inflow. The second management scenario (30 % irrigation savings) results in a slower water table elevation decline (0.07 to 0.28 m/year) compared to the first management strategy. In Jalandhar and Gurdaspur, the third management scenario (replacing rice crops with pulses) has a median result of sustainable water consumption, with greater uncertainty in Jalandhar. In Sangrur, scenario 3 is not sufficient to balance water supply and consumption; however, the median water table decline is ~60 % less than for scenario 1. Scenario 3 represents the most effective of the three simulated management strategies for mitigating damage and potentially restoring depleted aquifers.

Though we show that irrigation efficiency measures and crop replacement are not sufficient to balance and restore aquifer levels in all districts, they can be significant components of a multifaceted management plan. Other options for

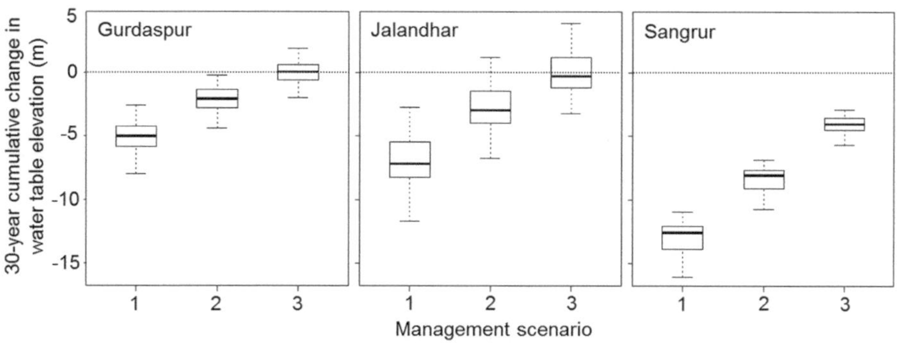

Fig. 9.5 Simulated 30-year-cumulative water table response to three management strategies in Gurdaspur, Jalandhar, and Sangrur. *Strategy 1* represents continuing current practices; *Strategy 2* represents 30 % reduction in irrigation consumption; and *Strategy 3* represents replacement of rice with pulse crops

management include more aggressive crop shifting, such as replacement of rice, sugarcane, and cotton, and a combination of irrigation efficiency with crop replacement. Alternatives to pulses include other low-water-demand crops such as maize, millet, gram (chick pea), and barley (Jalota and Arora 2002). In addition to water conservation methods like scenarios 2 and 3, balance can be attained by enhancing local water supply. Strategies that enhance water supply include managed aquifer recharge and water harvesting.

Model projections do not account for changes in rainfall patterns and temperature in Punjab. Changes in rainfall will impact irrigation requirements and groundwater recharge rates, while temperature will influence *ET* and yields. Intensity and frequency of extreme precipitation events decreased in eastern Punjab between 1951 and 2003, while much of eastern and central India experienced increases (Goswami et al. 2006; Krishnamurthy et al. 2009). Assessment of rainfall changes should be conducted at the district scale, rather than state scale to capture local heterogeneities (Ranade and Singh 2013; Russo et al. 2013). Future work should include projected climate changes when evaluating future agricultural management strategies, especially crop shifting scenarios that utilize local precipitation in lieu of irrigation.

9.5 Conclusion

We present a method using an MCMC algorithm to solve for a distribution of values for each unknown parameter in an under-constrained hydrologic mass balance. The model accounts for variation in observed groundwater overexploitation, climate, crop patterns, and surface and lateral groundwater recharge parameters. The MCMC method quantifies unknown hydrogeologic and agricultural parameters, which vary across districts. Parameter distributions provide estimates of hydrologic fluxes through the ground surface and laterally into each district. We found that the observed water table elevation changes could not be explained by pumping extraction and surface recharge alone, and therefore the system must include a net lateral groundwater inflow approximately equivalent to the flux received from precipitation recharge. We attribute this source to be lateral flow that recharges in the Sivalik foothill range of the Himalayas and flows to the southeast through Punjab.

The model provides quantitative estimates of projected groundwater pumping and water table change under specified water management scenarios. Calculated posterior distributions are used to evaluate three water management scenarios: continuing current practices, reducing irrigation by 30 %, and replacing high-water-consuming rice with low-water-consuming pulses. Projected water table responses to the three scenarios show that replacement of rice crops with pulses may be sufficient to balance or restore the depleted groundwater in Gurdaspur and Jalandhar.

This paper illustrates the hydrologic benefit associated with each strategy, but the results could also be used as input to a cost-benefit analysis, further optimizing management decisions. The feasibility and cost to implement each scenario must be considered in future work. There are economic, political, and cultural elements embedded in agricultural management actions. Changing decisions will require consent and effort from local and regional politicians, farming cooperatives, and agricultural extension services. Results of this study can be used to inform and motivate the management changes required to ensure sustainable agricultural production in Punjab.

Acknowledgments We thank colleagues at the Punjab Agricultural University (Ludhiana, India), the Central Groundwater Board, and the Bhakra Beas Management Board (both from Chandigarh, India) for sharing valuable data. This work was completed with funding from the Columbia University Earth Institute Postdoctoral Fellowship program and the International Development Research Centre (Canada).

References

Aggarwal R, Kaushal M, Kaur S, Farmaha B (2009) Water resource management for sustainable agriculture in Punjab, India. Water Sci TechnolA J Int Assoc Water Pollut Res 60(11):2905–2911. doi:10.2166/wst.2009.348

Ambast SK, Tyagi NK, Raul SK (2006) Management of declining groundwater in the Trans Indo-Gangetic Plain (India): some options. Agric Water Manag 82(3):279–296. doi:10.1016/j.agwat.2005.06.005

Brahmanand PS, Kumar A, Ghosh S, Chowdhury SR, Singandhupe RB, Singh R, Nanda P, Chakraborthy H, Srivastava SK, Behera MS (2013) Challenges to food security in India. Curr Sci 104(7):841–846

CGWB (2012) Dynamic ground water resources of Punjab state. Water Resources and Environment Directorate, and Central Ground Water Board, Chandigarh

Chatterjee R, Purohit RR (2009) Estimation of replenishable groundwater resources of India and their status of utilization. Curr Sci 96(12):1581–1591

Devineni N, Perveen S (2012) Securing the future of India's "Water, energy and food." GWF discussion paper 1240 (p. GWF discussion paper 1240). GWF discussion paper 1240, Global Water Forum, Canberra

Devineni N, Perveen S, Lall U (2013) Assessing chronic and climate-induced water risk through spatially distributed cumulative deficit measures: a new picture of water sustainability in India. Water Resour Res 49(4):2135–2145. doi:10.1002/wrcr.20184

Engeland K, Gottschalk L (2002) Bayesian estimation of parameters in a regional hydrological model. Hydrol Earth Syst Sci 6(5):883–898. doi:10.5194/hess-6-883-2002

Gelman A, Rubin D (1992) Inference from iterative simulation using multiple sequences. Stat Sci 7(4):457–511

Gilks W, Richardson S, Spiegelhalter D (1996) Markov chain Monte Carlo in practice. Chapman and Hall, London, p 486

Goswami BN, Venugopal V, Sengupta D, Madhusoodanan MS, Xavier PK (2006) Increasing trend of extreme rain events over India in a warming environment. Science 314(5804):1442–1445. doi:10.1126/science.1132027

Hira GS (2004) Status of water resources in Punjab and management strategies. In: Abrol I, Sharma B, Sekhon G (eds) Groundwater use in North-West India – workshop papers. Centre for Advancement of Sustainable Agriculture, New Delhi, pp 65–71

Humphreys E, Kukal SS, Christen EW, Hira GS, Sharma RK (2010) Halting the groundwater decline in North-West India — which crop technologies will be winners? Adv Agron 109: 155–217. Elsevier Ltd. doi:10.1016/B978-0-12-385040-9.00005-0

Jalota S, Arora V (2002) Model-based assessment of water balance components under different cropping systems in north-west India. Agric Water Manag 57(1):75–87. doi:10.1016/S0378-3774(02)00049-5

Kahn S, Kumar G, Kumar S, Marwaha S, Pandey S, Rani V, Saigal SK, Sharma A, Singh AK, Singh GP, Singh S, Singh T (2007) In: Gupta S (ed) Punjab district reports. Central Ground Water Board, Government of India, Chandigarh

Kaur S (2009) On-farm water management practices in Punjab. Curr Sci 97(3):307–309

Kaur S, Aggarwal R, Soni A (2011) Study of water-table behaviour for the Indian Punjab using GIS. Water Sci Technol 63(8):1574. doi:10.2166/wst.2011.212

Krishnamurthy CKB, Lall U, Kwon H-H (2009) Changing frequency and intensity of rainfall extremes over India from 1951 to 2003. J Clim 22(18):4737–4746. doi:10.1175/2009JCLI2896.1

Lunn DJ, Thomas A, Best N, Spiegelhalter D (2000) WinBUGS – a Bayesian modelling framework: concepts, structure, and extensibility. Stat Comput 10:325–337

MWR (1997) Ground water estimation methodology. Ministry of Water Resources, Government of India, New Delhi

Ngo-Duc T, Polcher J, Laval K (2005) A 53-year forcing data set for land surface models. J Geophys Res 110:D06116. doi:10.1029/2004JD005434

Perveen S, Krishnamurthy CKB, Sindu RS, Vatta B, Kaur B, Modi V, Fishman RM, Polycarpou L, Lall U (2012) Restoring groundwater in Punjab, India's breadbasket: finding agricultural solutions for water sustainability. Columbia Water Center – White Paper

Rajeevan M, Bhate J (2008) A high resolution daily gridded rainfall data set (1971–2005) for Mesoscale meteorological studies. National Climate Center, IMD, Pune, p 14

Ranade A, Singh N (2013) Large-scale and spatio-temporal extreme rain events over India: a hydrometeorological study. Theor Appl Climatol (in press). doi:10.1007/s00704-013-0905-1

Russo T, Fisher A, Winslow D (2013) Regional and local increases in storm intensity in the San Francisco Bay Area, USA, between 1890 and 2010. J Geophys Res Atmos 118(8):3392–3401. doi:10.1002/jgrd.50225

Shah T (2009) Climate change and groundwater: India's opportunities for mitigation and adaptation. Environ Res Lett 4(3):035005. doi:10.1088/1748-9326/4/3/035005

Shah N, Nachabe M, Ross M (2007) Extinction depth and evapotranspiration from ground water under selected land covers. Groundwater 45(3):329–338. doi:10.1111/j.1745-6584.2007.00302.x

Sidhu R, Vatta K, Lall U (2011) Climate change impact and management strategies for sustainable water-energy-agriculture outcomes in Punjab. Indian J Agric Econ 66(3):328–339
Spiegelhalter D, Thomas A, Best N, Gilks W (1996) BUGS 0.5: Bayesian inference using gibbs sampling (version ii). Medical Research Council Biostatistics Unit, Cambridge
Takshi K, Chopra R (2004) Monitoring and assessment of groundwater resources in Punjab state. In: Abrol I, Sharma B, Sekhon G (eds) Groundwater use in North-West India – workshop papers. Centre for Advancement of Sustainable Agriculture, New Delhi, pp 8–15
Tyagi NK, Agrawal A, Sakthivadivel R, Ambast SK (2005) Water management decisions on small farms under scarce canal water supply: a case study from NW India. Agric Water Manag 77(1–3):180–195. doi:10.1016/j.agwat.2004.09.031
Winslow DM, Fisher AT, Becker K (2013) Characterizing borehole fluid flow and formation permeability in the ocean crust using linked analytic models and Markov Chain Monte Carlo analysis. Geochem Geophys Geosyst 14(9):3857–3874

Chapter 10
Ecohydrology for Sustainability of IWRM: A Tropical/Subtropical Perspective

Amartya K. Saha and Shimelis Gebriye Setegn

Abstract Water resources in the tropics and subtropics are under severe pressure from burgeoning populations, ad hoc development, and the degrading environment. Uncertainty in precipitation due to climate change adds to the pressure to equitably provide adequate and safe water to all of humanity. Hence, utilizing the beneficial roles played by forests and wetlands upon water availability and quality is the only way to enable the equitable provision of water to all sections of society as well as for buffering water resources against climate change. Understanding the links between different ecosystems in a catchment and local/regional hydrology enables restoration and maintenance of the ecosystems along with the services they provide. This chapter describes the ecohydrological approach to water resources management along with some examples of the application to a range of areas, from river basin management to wastewater treatment and reuse.

Keywords Ecohydrology • IWRM • Sustainability • Ecosystem services • Watershed ecosystems • Water quality and ecosystems

10.1 Introduction

Ecosystem services or the beneficial roles that forests and wetlands have on water availability and quality are being increasingly recognized worldwide. Forests and wetlands store water during the rains, promote groundwater recharge, and feed streams and springs in the dry season. Forests also control soil erosion, thereby maintaining water quality and aquatic ecosystems. Wetlands remove excess nutrients and are thus termed kidneys of the landscape. This awareness is not recent;

A.K. Saha (✉)
Global Water for Sustainability (GLOWS), Department of Earth and Environment, Florida International University, North Miami, FL 33181, USA
e-mail: asaha@fiu.edu

S.G. Setegn
Department of Environmental and Occupational Health and Global Water for Sustainability Program (GLOWS), Florida International University, 11200 SW 8th Street, Miami, FL 33199, USA
e-mail: ssetegn@fiu.edu

S.G. Setegn, M.C. Donoso (eds.), *Sustainability of Integrated Water Resources Management*, DOI 10.1007/978-3-319-12194-9_10

communities have historically protected forests for various reasons, an important one being that forests assure clear running streams.

Over the past half a century, however, growing population and unplanned development have led to large-scale deforestation, drainage of wetlands, an alarming increase in wastelands, and advancing desertification. This has been especially acute in many parts of Africa, Asia, and South America. Such degradation has greatly reduced the hydrological and water quality regulatory services provided by these ecosystems. The consequences have been spiraling water quality declines in rivers, estuaries, and oceans as well as disruptions of human and aquatic ecosystems on account of flooding and flow changes in rivers. At the same time, water management over the past 70 years has been a patchwork of regional strategies with an added focus on purely technological solutions for wastewater treatment, which are prohibitively expensive to build, operate, and maintain.

The limited freshwater supplies on Earth, along with the increasing uncertainty in water availability due to climate change and escalating demand, make sustainable water management imperative. Incorporating ecosystem services, using nature as an ally, is the only way to achieve this. Ecosystems maintain water flow and quality and also increase the buffering capacity or the resilience of a catchment to climate change. Given the declining state of ecosystems worldwide, restoring ecosystem function and reversing environmental degradation is the need of the hour. Such a task requires understanding the link between hydrology and the ecosystem structure/function, which has been illustrated in the previous chapter. This chapter describes a few selected case studies of ecohydrological approaches used in different aspects of water resources management in the tropics, from river basin water management to wastewater treatment and reuse.

10.1.1 Incorporating Ecohydrology into Management: Enhancing Water Availability and Wastewater Treatment

Ecosystem services can be utilized for both water supply and in water treatment and reuse. Some major areas of ecohydrological application are:

- Increasing catchment water retention and flow duration through maintenance of existing forest cover, reforestation, and wetland protection
- Decreasing the loading of nonpoint pollution by soil conservation and maintaining riparian vegetation along stream courses
- Maintaining in-stream habitat by maintaining/restoring natural river channels and floodplains, ensuring a natural seasonal flow regime
- Employing biogeochemical processes in natural and constructed wetlands to treat organic matter and nutrient-laden sewage

10.2 Developing an Ecohydrological Approach

Figure 10.1 illustrates some major benefits of ecosystem services together with typical steps to incorporate an ecohydrological approach in IWRM. The first step in integrating an ecohydrological approach in catchment-level water resources management is to identify the ecosystems, followed by research on their characteristics, processes, and services they provide. In their review of the ecohydrological processes in estuaries, Wolanski et al. (2004) emphasize the necessity of a basin-wide approach for estuary and coastal management. They call for the broadening of the vision of water management to unify the narrowly focused approaches of constituent departments, such as forestry and water or urban water supply or industrial pollution control. As a precursor to forming management plans, they stress developing an understanding of the linkages between the ecosystem communities and hydrology, sediment, and nutrient flux. This understanding is then used to develop science-based remediation measures at the basin scale to maintain the ecosystem.

The second step thus involves the analysis of how to incorporate ecosystem services into local/regional water resources management planning and policies. Approaches include protecting existing primary forests, managing secondary and plantation forests to minimize sediment-generating operations, maintaining riparian forests along all stream banks, protecting wetlands, and restoring wetlands, river channels, and forests. Integrated Water Resources Management necessarily involves coordination and cooperation between all stakeholders; hence, a common understanding of water availability and use between different sectors is necessary amongst all stakeholders.

Finally, the changing nature of both the environment and human water demand necessitates the continued monitoring and evaluation of ecosystem services, ecosystem health, and the success of management efforts to adapt and adjust IWRM strategies and plans.

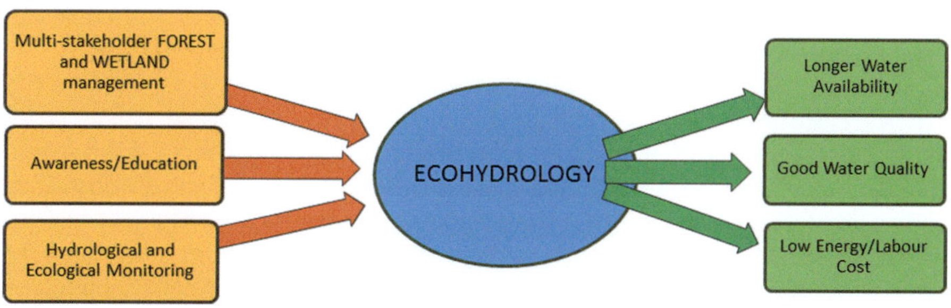

Fig. 10.1 Benefits of ecohydrology (*boxes on the right*) with requirements for applying the ecohydrological approach (*boxes on the left*)

10.2.1 Ecohydrology Research on Local Ecosystems

The previous chapter outlined the general ecosystem-hydrology-water quality linkages of different ecosystem types along with the respective ecosystem services. While these linkages form a framework for understanding the ecohydrological functioning of an ecosystem, it is necessary to understand the ecohydrology of the specific ecosystem at hand in order to maximize ecosystem services as well as maintain the ecosystem. This is because ecosystems in different regions vary widely in terms of climate, soil and geology, topography, biodiversity, and human activities. Hence, in order to understand the ecohydrology of a particular catchment, it is necessary to study and investigate the hydrology-ecosystem links in different natural areas present in that catchment.

Developing an ecohydrological approach in a region creates a large opportunity for institutions and universities to carry out research on local ecosystem-water linkages, train a new generation of professionals, and involve local communities resident around the ecosystems. Specific areas of research depend upon the specific ecosystems present, the types of linkages they have upon water availability/quality, and issues regarding their management, resource use, and conservation.

Linkages between plant communities and hydrology are central to the understanding of the ecohydrologic function of a specific ecosystem, given the central role plants play in governing the water cycle via soil water uptake and transpiration, which varies between 50 and 100 % of precipitation. Apart from water uptake, plants also play a role in nutrient cycling, organic matter decomposition, and soil stabilization – factors that affect water quality in the watershed.

Just as plant communities affect water availability and quality, plant communities are also dependent on seasonal hydrology patterns. This is especially the case in wetlands that can have a variety of plant communities arranged along a hydrological (and nutrient) gradient. Each plant community is closely linked with the seasonal hydrological regime, and thus maintaining plant diversity and ecosystem function in wetlands requires understanding plant community water and nutrient sources and how these vary with hydrology (e.g., Saha et al. 2009, 2010). For instance, tree islands in the Everglades are the only areas that are not flooded and thus constitute critical wildlife habitat as well as biogeochemical hotspots that accelerate nutrient cycling. Conserving tree islands requires basin water management to maintain the natural seasonal hydrology of the wetlands.

Similarly, deciphering the links between hydrology and fish communities enables basin water management to maintain the ecosystem and fisheries; for example, Boucek and Rehage (2013) have examined the amplitude (water depth) and duration of the flood season in connection with fish surveys (both electrofishing and angler catches) for two focal fish species in the Everglades. The fact that the two focal species differed greatly in their responses to hydrology indicates the necessity of maintaining seasonal hydrology that is as close to the natural regime as possible.

While it is necessary to understand the ecohydrological functioning of individual communities and ecosystems, basin-scale water management also requires understanding ecohydrology at the landscape scale. A fast-advancing area of ecohydrological research involves the calibration of remotely sensed data for a wide range of applications, from estimating evapotranspiration to river discharge. This approach can yield valuable information when extrapolated across large areas where monitoring networks are nonexistent and ground data is scarce or absent. For example, linking hydrology to land cover change by using a combination of hydrological modeling, remote sensing, and field approaches has that done by Nobert and Jeremiah (2012) for the Wami River basin in Tanzania.

10.2.1.1 Monitoring and Data Analysis

Understanding ecohydrological processes in a catchment requires hydrological data, which is typically obtained by a regional monitoring program. Such programs or networks exist in some regions of developing countries; however, oftentimes, the monitoring instruments get dysfunctional, outdated, or vandalized, or data collection is not done regularly owing to logistical issues or budgetary constraints. Chances of vandalism are minimized by actively involving local communities in the monitoring program. Gomani et al. (2010) described the setting up of a hydrological monitoring network in the Ngerengere River catchment in Tanzania and strongly urge the inclusion of local communities in the decision-making process as to the locations of the monitoring stations as well as to the purpose of monitoring. This ensures the assimilation of local and expert knowledge of hydrology.

The greater the accuracy and spatiotemporal resolution of hydrological data, the better is the understanding of ecohydrological processes in a catchment. However, a single institution may not have the resources to maintain numerous monitoring stations and either the staff to read data or the finances required for automated monitoring. Community water monitoring programs can significantly contribute to data collection in a catchment, such as daily rainfall measurements in primary and secondary schools or water-level measurements if a stream or river is situated near a school. Community water monitoring programs also can get a community more involved in the active management of local water resources as well as in monitoring/reporting illegal deforestation, mining, and fish kills and facilitate the cooperation between villages in a catchment by forming early flood warning systems. Examples of such programs are Waterwatch Victoria (Australia) and Save Our Streams in the USA. Community monitoring programs are all the more important in developing countries where in most cases there exists a disconnect between local communities and the agencies and institutions charged with maintaining monitoring networks.

After data collection, the next steps are data storage and analysis. Open source database systems have the advantage of being able to be used effectively in institutions that do not have the financial and technical capacity to obtain commercial software. There should be resources allocated to maintaining and enhancing the capacity of an institution to analyze hydrological data through periodic training of employees.

10.2.2 Awareness Among Multiple Stakeholders: Frameworks for Cooperation

Everyone depends on water. Therefore, sustainability requires universal awareness of the finite nature of freshwater resources along with the issues imperiling water and ecosystems today, the very lifelines we have. A multi-stakeholder approach is therefore essential in sustainable water management. For example, the conservation of a wetland requires coordination between ministries of water, agriculture, land zoning development, fisheries, and human resources as well as local communities. Awareness generation, education (primary, secondary, college, nonformal), and community monitoring programs are forms of outreach.

The handbook of IWRM at the basin level (2009) documents numerous examples of approaches necessary to facilitate multi-stakeholder dialogue, cooperation, and the institutional aspects of IWRM that ultimately influence policy and legislation. As an example, the Joint Danube Survey (JDS) was set up to investigate pollution in the Danube basin. Collaborating countries collect hydrological, biochemical, and ecological data; the data serves to identify pollution sources and pathways and is ultimately used to develop the Joint Danube Action Programme and Integrated River Basin Management Plans. In addition, the participation of all countries in the Danube basin facilitates opportunities to share experiences, coordinate sampling and monitoring procedures, and in some cases standardize analytical methodologies. Similar initiatives occur in many other river basins, such as the Niger River Initiative, the Lake Victoria Commission, and the Mekong River Commission, where physicochemical, ecological, hydrological, and socioeconomic data are shared to better manage the basin from a shared vision approach of sharing not just the water but also the benefits arising from sound management.

10.3 Ecohydrology and Integrated Water Resources Management

10.3.1 River Basin Water Management

Ecohydrology is the subdiscipline shared by ecological and hydrological sciences that is concerned with the effects of hydrological processes on the distribution, structure, and function of ecosystems and on the effect of biological processes on the elements of the water cycle (Nuttle 2002). A guiding hypothesis of the UNESCO Ecohydrology Programme is that "the ecohydrological approach can be a tool towards the sustainable use of aquatic resources by enhancement of the resistance, resilience and buffering capacity of fluvial corridors" (Zalewski et al. 1997 as cited by El-Sadek et al. 2008). Many rivers, lakes, and reservoirs are continuously affected by human activities causing enormous environmental problems related to biodiversity, ecosystem functioning, and preservation of the water cycle. In most parts of the world, urbanization has caused progressive occupation and development of open land and land reclamation from water basins, causing changes in ecology and hydrology. In developing countries, providing enough safe water to meet basic human needs is a serious problem.

A watershed planning and management strategy within a hydrologically defined area provides a coordinating framework for water supply protection, pollution prevention, and ecosystem preservation. Although watershed strategies vary, they should be based on an integrated study of ecosystems and hydrological characteristics and processes and their combined potential to influence water dynamics and quality. Ecohydrology requires an understanding of the temporal and spatial patterns of catchment-scale water dynamics which are determined by four fundamental components: climate, geomorphology, plant cover/biota dynamics, and anthropogenic modifications (UNEP 2003).

10.3.2 Water Quality and Ecosystems

Water quality is a term used to describe the chemical, physical, and biological characteristics of water, usually in respect to its suitability for a particular purpose. Excessive usage of agrochemicals (fertilizers, herbicides, and pesticides) and change in land use endanger and deteriorate the quality of soil as well as the quality of the available surface and groundwater resources (El-Sadek et al. 2003).

A more efficient approach to water quality and ecosystem integrity requires not only the reduction or elimination of pollution but, in parallel, an augmentation of the effectiveness of potential tools to manage the dynamics of excess nutrients, pollutants, mineral, and organic matter in the landscape. This can be done by reducing human impacts and by regulating the aquatic and terrestrial biota in the catchment. One of the most efficient ways to control biota dynamics is through the regulation of hydrological processes by increasing watershed water retention through reforestation and restoration of land/water ecotones, enhancing in-stream retention of water sediments and nutrients through river re-naturization and wetland restoration, and amplifying biogeochemical cycles such as denitrification through wetland inundation.

The application of ecohydrological concepts in watershed management relies on the existence and manipulation of plants understood in terms of species distribution and interactions.

Whigham et al. (1988) have suggested that sediment and nutrient retention in headwater areas should be proportional to the area covered by wetlands. Riparian zones and wetlands are the most important natural components of riverine ecosystems regulating the quantity and quality of water entering from uplands. Together they function as effective buffers against extreme flooding and excess sediment loads linked to runoff from the landscape. In developed areas, they are also effective buffers against excess erosion, nutrient inputs, and contaminant runoff from agricultural fields, pastures, and residential areas (Haycock et al. 1997). Riparian zones and wetlands are most effective in the headwater portions of watersheds, where flow is distributed among a large number of smaller streams. In downstream portions of rivers and streams like the Nile Delta, water quality and fluxes are regulated by riverine wetlands and floodplains.

At the level of implementation, the conservation of ecohydrological processes clearly should not be viewed as an independent and stand-alone solution to water quality management problems. Natural ecohydrological processes should be viewed as a fundamental component of more multifaceted solutions, including additional systems to treat water for domestic supply and to treat wastes prior to disposal.

In urban and industrial areas, ecohydrological processes should be maintained to fortify nearby aquatic systems and increase the likelihood that other engineering-based controls will achieve water quality goals. In rural areas, natural ecohydrological processes may be recognized as primary tools for controlling surface water quality in conjunction with simple drinking water treatment and waste treatment systems (Zalewski et al. 1997).

10.3.3 Ecohydrological Analysis in the Lake Tana Basin, Blue Nile

The Blue Nile River originates from the Lake Tana in Ethiopia. The Blue Nile eventually joins the White Nile at Khartoum, Sudan, and the Nile continues through Egypt to the Mediterranean Sea. The Blue Nile has a drainage area of 199,812 km^2 and supplies nearly 84 % of the water of the Nile River. The flow of the Blue Nile reaches its maximum discharge in the main rainy season (from June to September), when it supplies more than two-thirds of the water of the Nile River. The basin has an average annual runoff of about 50 billion m^3.

MoWR 1999 (the Abbay River Basin Master Plan Study) stated that the range of soil erosion for the Blue Nile basin on an average is between 2.3 to 212.9 tons per hectare per year.

Lake Tana is located in Ethiopia's northwest highlands (Lat 12° 0′ North, Lon 37° 15′ East). The lake is a natural type which covers 3,000–3,600 km^2 area at an elevation of 1,800 masl and with a maximum depth of 15 m. Lake Tana basin comprises a total area of 15,096 km^2 including the lake area. It is the largest lake in Ethiopia and the third largest in the Nile basin. The mean annual rainfall of the catchment area is about 1,280 mm. The annual mean actual evapotranspiration and water yield of the catchment area are estimated to be 773 mm and 392 mm, respectively (Setegn et al. 2008). The Lake and its watershed constitute Ethiopia's largest water body. With its large area of wetlands, it has high biodiversity and many endemic species and cattle breeds. It is home to many endemic birds and cultural and archaeological sites. This basin is of critical national significance as it has great potentials for irrigation, hydroelectric power, cash crops and livestock production, ecotourism, and others.

Gilgel Abbay, Ribb, Gumera, and Megech rivers are the main tributary rivers feeding the lake which contributes more than 93 % of the inflow. It is the main source of the Blue Nile River that is the only surface outflow for the Lake. The climate of the region is "tropical highland monsoon" with the main rainy season between June and September. The air temperature shows large diurnal but small seasonal changes with an annual average of temperature of 20 °c. The Lake Tana basin can be divided into four distinct physiographic units comprising: the lake water body (including the islands), lower catchments (shorelines and surrounding wetlands), middle catchments (floodplains and gentle hill slopes), and upper catchments (low plateau plains, ridges, and mountains).

Different studies have been conducted in the Lake Tana basin to determine the ecohydrological status of the Lake as a tool for integrated water resource management in improving the sustainability of the quantity and quality of freshwater resources.

Studies indicated that reduced lake water level with its annual fluctuations and seasonal floods associated with high flows are becoming amplified and frequent, and the total average annual sediment load of the four major tributaries shows an increasing trend. Source of pollution from urban waste and rural agriculture and degradation of biota in the catchment are the two main environmental threats for the lake ecosystem.

The land and water resources of the basin and the Lake Tana ecosystem are in danger due to the rapid growth of population, deforestation and overgrazing, soil erosion, sediment deposition, storage capacity reduction, drainage and water logging, flooding, pollutant transport, and overexploitation of specific fish species. Sediments and organic and inorganic fertilizers from the agricultural fields that enter the Lake by runoff may result in eutrophication.

The Lake Tana basin is one of the most affected area by soil erosion, sediment transport, and land degradation. Using a physically based watershed SWAT (Soil and Water Assessment Tool) model and GIS-based decision support system (MCE analysis), to identify the most erosion-prone areas in the Lake Tana basin, Setegn et al. (2008) reported that in average about 22 % of the watershed area has high potential for soil erosion.

The lack of decision support tools and limitation of data concerning weather, hydrology, topography, and soil and land use are factors that significantly hinder research and development in the area. Lake Tana basin is one of the major basins that significantly contribute to the livelihoods of people living in the basin.

The hydrological water balance analysis showed that baseflow is an important component of the total discharge within the study area that contributes more than the surface runoff. More than 60 % of losses in the watershed are through evapotranspiration. The correlation between the rainfall, discharge, and sediment yield has shown that the amount and intensity of rainfall play an important role for the sediment yield (Setegn et al. 2010).

The Blue Nile (locally called Abbay basin) exhibits alarming degradation of biota (riparian vegetation, grassland, and wetlands) crucial for maintaining the resilience of the ecosystem and improving livelihoods. Many studies show that the wetlands surrounding the lake and the flora and fauna they sustain are increasingly degrading (WRDA 1990; Lee and Zewdie 1997; Tesfahun and Demissie 2004; Ligdi 2008). Wetlands around the lake and along the river banks are increasingly degrading and quickly disappearing and it brings dramatic reduction of their buffering role in the ecosystem as well as the economic benefits and the livelihoods they sustain.

Many studies, including (Setegn et al. 2008), point out that the land and water resources of the basin and the Lake Tana ecosystem are endangered by soil erosion, sediment transport, land degradation, and other natural and man-made factors. The sediment load outgoing from the lake at Bahir Dar outlet is almost negligible, so the lake acts as continuous sediment sink that ultimately affects the storage capacity of the lake and reduces the quantity and quality of soil and water resources. There must be an immediate action and an integrated approach to reduce erosion and sedimentation problem in the upstream subbasins. Appropriate soil and water conservation strategies should be designed involving communities and other stakeholders in the area.

According to Ligdi et al. 2010, the identified ecosystem functions, which could be integrated into the basin-scale resource management through engineering and ecohydrological phytotechnologies, were acknowledged as: lake water-level regulation, habitat regulation, species regulation, control of eutrophication and cyanobacterial toxicity, restraining of erosion, sediment transport and sedimentation, improvement of water quality, and reducing pollutant load, besides the support of livelihoods.

10.3.4 Urban Regions

The combination of high population density and commercial and industrial activities in urban regions necessitates year-round planned water supply and the daily treatment of enormous quantities of municipal and industrial wastewater. This creates special challenges in both the supply and safe disposal aspects of urban water management.

Zalewski and Wagner (2005) explain the advantages of considering urban regions as ecological systems in themselves, where the fundamental processes such as water circulation and matter and energy flow are extremely condensed. Understanding the flow paths of these components can help to regulate them and enhance the effectiveness of Integrated Urban Water Management (IUWM). Urban regions have the highest domestic and industrial water consumption in catchments, as well as return flows in both treated and untreated forms. The high degree of impervious surfaces also leads to high instantaneous runoff. While permeable concrete has yet to receive widespread acceptance, especially in developing countries, vegetated zones in the form of parks and grassy traffic islands for improving percolation as well as green roofs and community gardens on vacant plots of land can absorb rainwater and thereby reduce stormwater runoff which otherwise significantly contributes oil, grease, and other pollutants to watercourses in a city.

Since the 1950s, the trend in water pollution control has been to build large centralized concrete and steel structures. However, with the increasing energy prices and labor costs, operating, maintaining, and upgrading these systems have become more and more unsustainable for communities. For small communities in particular, this cost represents a higher percentage of the budget than the previously allocated to water pollution control. Therefore, processes that use relatively more land and are lower in energy use, and labor costs are therefore becoming attractive alternatives. This is especially applicable in developing countries, where very few large urban areas have wastewater treatment plants that can handle most of the city's wastewater. Even cities in the developed world are discovering significant cost savings as well as the ability to treat wastewater as a resource by using wetland-based treatment (Mitsch and Gosselink 2000; Nichols 1983). New York City found that it could avoid spending USD$3–8 billion on new wastewater treatment plants by investing USD$1.5 billion in the purchase of land around the reservoirs upstate. This land purifies the water supply for free. Other notable examples of the use of natural wetlands for domestic wastewater treatment are the East Kolkata Wetlands and the city of Arcata, California, which created the Arcata Marsh and Wildlife Sanctuary in 1986, a model for small towns globally (Suutari 2006).

10.3.5 The East Kolkata Wetlands (EKWs)

The East Kolkata Wetlands (India) achieve the natural treatment of domestic wastewater (~250 million gallons/day) from the city of Kolkata (Calcutta) along with the production of fish (150 metric tons/day) and vegetables (15,000 metric tons/year), thereby providing livelihoods for almost 0.2 million people and forming the backbone of food security for the Kolkata region (EKWMA 2010). The East Kolkata Wetlands (EKWs) is a hydrologically linked network of 264 shallow fish ponds and paddy fields spread over 12,500 ha amidst natural wetlands and floodplains in the Gangetic delta system. The tremendous ecological and socioeconomic significance of the East Kolkata Wetlands has led to the declaration in 2003 by the Government of India as a Ramsar wetland site and its inclusion among 17 case studies worldwide on wise wetland use by the Ramsar Convention. The East Kolkata Wetlands (Conservation and Management) Act was passed in 2006 which laid the foundation of the East Kolkata Wetland Management Authority and systematic implementation of wise use principles for the management of the Wetlands. The Authority includes the representation of all the major stakeholders, from various government ministries and institutions to local fishing cooperatives and NGOs.

Treatment Process Municipal wastewater is pumped into the wetlands through a network of drainage channels that flow into the canals and fish ponds. The fish ponds, locally known as *bheris*, occupy 40–50 ha each. The shallow depth (0.5–1.5 m) permits sunlight penetration to the bottom as well as full vertical circulation of the wastewater to the surface where algal blooms occur. Exposure to sunlight deactivates pathogens and facilitates algal photosynthesis; the resulting oxygenation supports bacterial populations carrying out a range of biochemical reactions leading to decomposition of organic matter. Each hectare of a shallow water body can remove about 235–240 kg of Biochemical Oxygen Demand (BOD) per day. Photosynthesis is one of the major processes promoting waste matter reduction in these ponds. Water hyacinths present in these ponds are associated with the uptake of heavy metals as well as buffering canal banks from the erosive forces of wastewater flow; these plants are periodically harvested for buffalo fodder. Many fish species are algae eaters and thereby maintain algal populations as well as convert nutrients in sewage to a form useful for consumption via biomass. About one-third of the fish demand of Kolkata is met from these wetlands, very significant given that fish is the main protein source for most of the population.

Aquaculture The flow of sewage is controlled for maintaining successful pisciculture, as the main requirement for fish production is the proper supply and quality of wastewater. Three types of ponds are needed according to the stage of production: the nursery pond, the rearing pond, and the stocking pond. The flow of water is mostly directed by gravity but in some areas diesel-powered pumps are used. Fish are raised in five major phases: pond preparation (done in the coolest months when ponds are drained and maintenance or repairs of dikes are carried out), primary fertilization (initial introductions of wastewater into the pond and allowed to undergo natural purification and stirring of the pond in order to reduce anaerobic conditions in the sediments), fish stocking (where farmers initially stock a small number of fish to test for water quality, subsequently stocking up to four times), secondary fertilization (periodic introductions of wastewater into the ponds throughout the growth cycle), and finally fish harvesting (taken at different times according to species).

Soil erosion control and maintenance: Pits (3 m wide, 30–40 cm deep) are located at the edges of the ponds to trap silt in runoff. These are periodically dredged and used to strengthen pond dikes. Ponds themselves are periodically drained before the monsoon season (which will naturally help to refill them) for various reasons: Draining them helps to free up some of the nutrients in the sediments if they are exposed to air, and it also kills some of the parasites that affect fish production.

Agriculture The effluent from the fish ponds then drains further southeast into paddy fields located strategically, as well as vegetable fields. Vegetable farming is a household activity, with people renting small plots or subletting smaller plots for household sustenance and income. Jobs include farming, garbage sorting, fish farming, trading, auctioneering, selling, raising fish seed, making nets, maintaining drainage, and reinforcing banks. Many people depend on the wetlands not only for livelihood but also for the fish and vegetables produced in the area. Thus, the EKWs not only treat sewage in a very cost-effective manner, thereby decreasing malarial outbreaks and waterborne disease epidemics, but also provide livelihoods to a large section of the population that would else be below the poverty line, as well as locally produced food in large quantities for the urban region of Kolkata.

Challenges to the Resilience of the EKW Rapid urban development in the immediate vicinity of the East Kolkata Wetlands is posing several challenges to the protection and continued existence of these wetlands. Pressure for land reclamation and drainage for real estate development, siltation of the canals and fish ponds, land subsidence from excessive groundwater extraction, and heavy metal and industrial effluents constitute threats that decrease the efficiency of the wetland system. Internally, policies are skewed toward larger landholding farmers, while exotic fish pose threats to the native aquatic ecosystem.

Bunting et al. (2010) examined the resiliency of the East Kolkata Wetlands system to change. They identified aspects of wastewater-fed aquaculture in the EKW that contribute to its sustained operation using a The Driving Forces, Pressures, State, Impact, Response (DPSIR) framework. Resilience within the EKW was attributed to the scale of operation, adaptive production strategies optimizing resource utilization while minimizing risks, stakeholder organization, timely legislation, and institutional interventions to preserve the ecological functioning of the wetlands. The introduction of externalizing technologies, erosion of social capital, and loss of traditional ecological knowledge threaten to undermine this resilience.

Steps for Sustainable Management Bhattacharjee et al. 2012 detail a list of steps necessary for sustainable management of the EKW. These include developing a comprehensive management plan for all canals and ponds to coordinate de-siltation, further research on biodiversity and conservation actions especially for water birds, prevention of direct release of industrial (tannery) effluents, and stakeholder awareness generation. According to a report by the World Bank and the International Development Research Center in Canada, the reuse of sewage in fish culture, algal and aquatic plant production, and energy production have been identified as new, promising technologies that radically change the context of urban sanitation. The EKWs are an opportunity for research and study that may influence the development of low-cost sanitation technology, especially for developing countries.

10.3.6 Constructed Wetlands

Cities that do not have extensive natural wetlands in their vicinity can use artificial or constructed wetlands for the treatment of domestic wastewater. Constructed wetlands are treatment systems engineered to use natural processes involving wetland vegetation, soils, and their associated microbial assemblages to improve water quality (Kadlec and Knight 2004; US EPA 2004) and are of two types, surface flow wetlands and subsurface flow wetlands (Vyzamal 2005), based upon whether the wastewater is directly exposed to the atmosphere or is underground. They are relatively inexpensive to build and maintain, require little or no energy to operate, can accomplish tertiary treatment, generate wildlife habitat, and are viewed as a green technology by the public.

The major constraint in using wetlands to treat domestic wastewater is the space required to accommodate the large volumes generated daily in today's cities; wetlands discharging treated water to surface water bodies typically require 4–10 times the surface area of conventional wastewater treatment plants. In addition, if not properly designed and maintained, surface flow wetlands can attract mosquitoes and other pests. Furthermore, the treatment of wastewater and the breakdown of organic matter are carried out by living organisms (microbes and plants) which depend in turn upon the prevailing temperature and climate and biotic factors such as nutrition and disease. Operating a constructed wetland therefore requires some initial time of trial and error while the plants are in the growing phase, especially if the operators do not have prior experience. Design considerations also take into account wastewater loading rates, the time taken for treatment, and problems such as clogging of substrate (Blazejewski and Murat-Blazejewska 1997).

Apart from treating wastewater in an affordable manner, wastewater can be used as a resource as illustrated in the East Kolkata Wetlands case. Treated water can be reused for irrigation, aquaculture, and other uses in the local region, while solid wastes resulting from wastewater treatment can be turned into manure or digested to create methane gas that can be used as fuel. Thus, constructed wetlands are being used all over the world in various configurations, from large city wards to smaller towns, communities, and even specific industries, which require prior wetland design based upon the pollutant type, loading, end-water use, and pollution discharge limits.

Lastly, it needs to be emphasized that different water management strategies in a watershed with wetlands and urban areas in close proximity can lead to very different outcomes upon wetland hydrology with effects on the wetland ecosystem (e.g., Sullivan et al. 2013).

10.4 Conclusion

The planet is still alive. There still are functioning ecosystems in every part of the world with a high degree of resiliency. Ecohydrology gives us the tools and the approach to engage natural processes and ecosystem services that render water resources management sustainable. At the same time, in the same process, the ecohydrological approach helps conserve and maintain ecosystem biodiversity, habitats, and integrity. Ecohydrology is now in the process of being mainstreamed; the Ecohydrological Programme of the United Nations Educational, Scientific, and

Cultural Organization (UNESCO) works with other organizations to promote a broader understanding of the combined benefits of bioremediation and ecohydrological applications worldwide.

In this era of water availability and quality that is being squeezed between escalating human demand and increasing uncertainty from climate and land cover change, incorporating an ecohydrological approach allows water resources management to be more economical and thus available to all, without jeopardizing natural processes that help sustain life on this planet.

References

Blazejewski R, Murat-Blazejewska S (1997) Soil clogging phenomena in constructed wetlands with subsurface flow. Water Sci Technol 35(5):183–188

Boucek RE, Rehage JS (2013) A tale of two fishes: using recreational angler records to examine the link between fish catches and floodplain connections in a subtropical coastal river. Estuar Coasts 38:124–135. doi:10.1007/s12237-013-9710-4

Bhattacharjee S, Ganguli A, Bose S, Mukhopadhyay A (2012) Biodiversity, traditional practices and sustainability issues of East Kolkata Wetlands: a significant Ramsar site of West Bengal (India). BioSciences Rev 6(11):340–347

Bunting SW, Pretty JN, Edwards P (2010) Wastewater-fed aquaculture in the East Kolkata Wetlands, India: anachronism or archetype for resilient ecocultures? Rev Aquacult 2(3): 138–153, ISSN 1753–5123

East Kolkata Wetlands Newsletter (2010) Vol 1. Published jointly by the East Kolkata Wetlands Management Authority and Wetlands International: South Asia, November. Available at http://www.ekwma.com

El-Sadek A, Oorts K, Sammels L, Timmerman A, Radwan M, Feyen J (2003) Comparative study of two nitrogen models. Irri Drain Eng 129(1):44–52

El-Sadek A, Kahloun ME, Meire P (2008) Ecohydrology for integrated water resources management in the Nile Basin. Ecohydrol Hydrobiol 8(2–4):237–244

Gomani MC, Dietrich O, Lischeid G, Mahoo H, Mahay F, Mbilinyi B, Sarmett J (2010) Establishment of a hydrological monitoring network in a tropical African catchment: an integrated participatory approach. Phys Chem Earth 35(2010):648–656

Handbook of Integrated Water Resources Management in Basins (2009) Published by the Global Water Partnership (GWP) and the International Network of Basin Organizations (INBO). Available online at http://www.inbo-news.org, http://www.gwpforum.org

Haycock NE, Burt TP, Goulding KWT, Pinay G (1997) Buffer zones: their processes and potential in water protection. Quest Environmental, Harpenden

Kadlec RH, Knight RL (2004) Treatment wetlands. Lewis Publishers, Boca Raton

Lee R, Zewdie G (1997) Population growth, environmental stress and innovation in Ethiopian peasants. Agric Sci Technol Dev 15(1):104–260

Ligdi EE (2008) Ecohydrology as an important tool for integrated water resources management in the Nile basin portion in Ethiopia: The case of the Lake Tana sub-basin: a case study report. Paper presented to the 2nd eco-hydrology component workshop held in, Nairobi, Kenya from 26th–30th January 2009. UNESCO-IHP FRIEND/Nile project (Phase II), Eco-hydrology component, Nairobi, Kenya

Ligdi EE, Kahloun ME, Meire P (2010) Ecohydrological status of Lake Tana — a shallow highland lake in the Blue Nile (Abbay) basin in Ethiopia: review. Ecohydrol Hydrobiol 10(2–4):109–122

Mitsch WJ, Gosselink JG (2000) The value of wetlands: importance of scale and landscape setting. Ecol Econ 35:25–33 (Special issue: the values of wetlands: landscapes and institutional perspectives)

Nichols D (1983) Capacity of natural wetlands to remove nutrients from wastewater. J Water Pollut Control Fed 55(5):495

Nobert J, Jeremiah J (2012) Hydrological response of watershed systems to land use/cover change: a case of Wami River basin. Open Hydrol J 2012(6):78–87

Nuttle WK (2002) Is ecohydrology one idea or many? Hydrol Sci J 47:805–807

Saha AK, Sternberg LSL, Miralles-Wilhelm F (2009) Linking water sources with foliar nutrient status in upland plant communities in the Everglades National Park, USA. Ecohydrology 2:42–54

Saha AK, Sternberg LSL, Ross MS, Miralles-Wilhelm F (2010) Water source utilization and foliar nutrient status differs between upland and flooded plant communities in wetland tree islands. Wetl Ecol Manag 18(3):343–355

Setegn SG, Srinivasan R, Dargahi B (2008) Hydrological modeling in the Lake Tana Basin, Ethiopia using SWAT model. Open Hydrol J 2(2008):49–62

Setegn SG, Dargahi B, Srinivasan R, Melesse AM (2010) Modeling of sediment yield from Anjeni-Gauged Watershed, Ethiopia using SWAT model. J Am Water Resour Assoc (JAWRA) 46(3):514–526. doi:10.1111/j.1752-1688.2010.00431.x

Sullivan P, Price R, Schedlbauer J, Saha A, Gaiser E (2013) The influence of hydrologic restoration on groundwater-surface water interactions in a karst wetland, the Everglades. Wetlands 34:S23. doi:10.1007/s13157-013-0451-8

Suutari A (2006) Constructed Wetland – a cost effective alternative for wastewater treatment. Arcata, California, USA. http://www.ecotippingpoints.org/our-stories/indepth/usa-california-arcata-constructed-wetland-wastewater.html. Accessed 27 Mar 2014

Tesfahun G, Demissie S (2004) Lake Tana: a socioeconomic synopsis. Amhara Regional Agricultural Research Institute, Bahir Dar. Paper presented to The Lake Tana Resource Management Workshop Bahirdar, Ethiopia

UNEP (2003) IETC freshwater management series no. 7, phytotechnologies, a technical approach in environmental management. ISBN 92-807-2253-0. http://www.unep.or.jp/ietc/Publications/Freshwater/FMS7/8.asp

US EPA (2004) http://www.water.epa.gov/type/wetlands/upload/design.pdf. Accessed 27 Mar 2014

Vyzamal J (2005) Horizontal sub-surface flow and hybrid constructed wetlands systems for wastewater treatment. Ecol Eng 25(5):478–490

Whigham DF, Chitterling C, Palmer B (1988) Impacts of freshwater wetlands on water quality: a landscape perspective. Environ Manag 12(5):663–671

Wolanski E, Boorman LA, Chícharo L, Langlois-Saliou E, Lara R, Plater AJ, Uncles RJ, Zalewski M (2004) Ecohydrology as a new tool for sustainable management of estuaries and coastal waters. Wetl Ecol Manag 12:235–276

WRDA (1990) Environmental impact assessment, Tana Beles project study. Water Resources Development Authority, Addis Ababa, Ethiopia

Zalewski M, Wagner I (2005) Ecohydrology – the use of water and ecosystem processes for healthy urban environments. Ecohydrol Hydrobiol 4:263–268 (Special issue: Aquatic Habitats in Integrated Urban Water Management)

Zalewski M, Janauer GA, Jolankaj G (1997) Ecohydrology: a new paradigm for the sustainable use of aquatic resources, Technical documents in hydrology No. 7. UNESCO, Paris

Part III
Climate Change and Integrated Water Resources Management (IWRM)

Chapter 11
Sustainability of Water Resources in Tropical Regions in the Face of Climate Change

Fernando González-Villarreal, Malinali Domínguez-Mares, and Jorge Arriaga-Medina

Abstract The most important impacts of climate change will be exerted on water resources. In Mexico, due to environmental conditions, human factors, and water uses, different degrees of social vulnerability will take place because of the increased temperature, decreased precipitation, and higher recurrence of extreme events projected to take place in the second half of the twenty-first century. Based on these considerations, this chapter presents the case study of the Grijalva River basin, the largest dam system used for hydropower generation in Mexico. Using temperature and precipitation scenarios, the impacts of climate change on sea level rise, the production of electricity, and water quality and availability are analyzed. Finally, it presents some adaptation measures that contribute to reduce the vulnerability of societies living in the Grijalva River basin.

Keywords Climate change • Mexico • Grijalva basin • Hydropower • Adaptation • Water resources • Water quality • Water availability • Vulnerability • Extreme events

11.1 Climate Change Impacts on Water Resources in Mexico

The most important impacts of climate change will be exerted on water resources and water management systems, reflecting the importance of water resources to social development. These impacts will affect strongly populations with lack of financial resources to implement adaptation plans, and there will be different vulnerability degrees, even within the same territory.

F. González-Villarreal (✉)
Universidad Nacional Autónoma de México, Torre de Ingeniería, Piso 5, Ala Norte, Cubículo 10, Circuito Escolar, Ciudad Universitaria, Mexico, CP 04510, Mexico
e-mail: contacto@agua.unam.mx

M. Domínguez-Mares • J. Arriaga-Medina
Red del Agua UNAM, Universidad Nacional Autónoma de México, Torre de Ingeniería, Piso 5, Ala Norte Cubículo 5, Circuito Escolar, Ciudad Universitaria, Mexico, CP 04510, Mexico
e-mail: mdominguezm@iingen.unam.mx; jarriagam@iingen.unam.mx

S.G. Setegn, M.C. Donoso (eds.), *Sustainability of Integrated Water Resources Management*, DOI 10.1007/978-3-319-12194-9_11

According to Bates et al. (2008), *globally, the negative impacts of future climate change on freshwater systems are expected to outweigh the benefits (high confidence). By the 2050s, the area of land subject to increasing water stress due to climate change is projected to be more than double that with decreasing water stress. Areas in which runoff is projected to decline face a clear reduction in the value of the services provided by water resources. Increased annual runoff in some areas is projected to lead to increased total water supply. However, in many regions this benefit is likely to be counterbalanced by the negative effects of increased precipitation variability and seasonal runoff shifts on water supply, water quality and flood risks (high confidence).*

For the Mexican case, based on regionalized simulations of coupled general circulation models (CGCM) made in recent years by the Mexican Institute of Water Technology (IMTA), it is estimated that the hydrological variables behavior under climate change conditions is as follows (Martínez et al. 2010):

(a) Precipitation. In general, widespread declines in rainfall are expected with greater effects in regions that already suffer shortages, especially in the central and northern areas of the country.
(b) Evapotranspiration. The temperature rise in the coming decades increases evapotranspiration, resulting in a stronger irrigation needed.
(c) Lakes. As a result of temperature, precipitation, and evaporation variations, an increase in eutrophication is expected and also a biodiversity decrease associated to water resources.
(d) Floods. Although models used in climate change studies provide limited information about the increase rainfall intensity, precipitation pattern changes are expected – brief rain but with high intensity brings, together with flood generation, increased soil erosion and recurrence of extremes events, resulting in a high number of flooding.
(e) Drought. Temperature increases and reduced rainfall will combine, resulting in an increased frequency and severity of hydrological drought, and these are most important in northern basins.
(f) Water Quality. Atmospheric temperature rise will also increase water temperatures and decrease the oxygen amount dissolved; thus, water self-purification capacity will be reduced, adversely affecting the quality, if it is not treated with a higher level of treatment, in treatment plants.
(g) Saline Intrusion. As a result of sea level rise, the coastal aquifers will face saline intrusion and/or soils and groundwater salinization.

Mexico will experience strong pressures on water availability and quality as a result of climate change and other social factors as population growth and sprawl. In order to make planning actions consistent with local realities, climate models are needed to be applied at the basin level. Based on these considerations, this chapter presents the results of the research conducted by the Engineering Institute of the National Autonomous University of Mexico (UNAM) and funded by the National Institute of Ecology to assess vulnerability at the Grijalva River basin to climate change impacts, with scenarios projected to 2050.

11.2 The Grijalva River Basin: Geographical, Social, and Environmental Characteristics

The Grijalva River basin is located between meridians 91° 30′ and 94° 30′ west longitudes and parallel 14° 30′ and 19° north latitude, in southeastern Mexico. This river is born in the Cuchumatanes, Guatemalan Highlands, and empties into the Gulf of Mexico after crossing the Central Depression of Chiapas and joins the Usumacinta River in Tabasco. It results from the union of two major streams, San Miguel River and San Gregorio River and it has an extension of 58,000 km^2.

The basin climate conditions are not uniform. In the upper watershed, there are tropical and winter systems, with some of the highest rainfall areas in the world, about 4,200 mm annually. The other areas in the basin have an average rainfall between 1,200 and 1,700 mm per year. The decrease in precipitation is the result of its location between weather barriers in Chiapas, the northern mountainous areas, and the coast. The estimated annual runoff in this basin is 46,796 Mm^3, one of the largest in Mexico. The Grijalva River basin is impacted by extreme hydrometeorological events, primarily tropical storms and cyclones. These hydrometeorological events mixed up with cloudiness, moisture influx, tropical maritime air, and cold air masses cause heavy rains, leading to major flooding in low-lying areas (CONAGUA 1999), as experienced in the city of Villahermosa and Chiapas in 2007.

Since 1958, the Federal Electricity Commission (CFE), considering the volumes of rainfall and runoff, decided to set up the largest hydroelectric system of Mexico in the Grijalva River. In 2012, the Grijalva system represented 49 % of the national hydro energy production (Hernández de la Torre 2009). The system has four dams: La Angostura, Chicoasen, Malpaso, and Peñitas. These dam systems operate in cascade, as Fig. 11.1 shows. Table 11.1 presents the dams main features, namely, ordinary high water level (NAMO), which indicates the maximum water level of the dams operation in order to meet demands; the extraordinary maximum water level (NAME) which corresponds to the highest level achieved in a reservoir under any circumstances; and the minimum operation water level (NAMINO) which is the lowest level of a hydroelectric dam operation. According to the Public Registry of Water Duties, 99 % of the total volume of the water resources allocated is used for electricity generation, while the remaining 1 % is divided between urban (29 %), industrial (4 %), and agricultural (67 %) use (SEMARNAT 2010).

Early in 2010, the Grijalva River basin had a population of 4.9 million people, mainly working in the primary sector (64 %), and dropout rates located between high and very high. The population is distributed in localities of less than 2,500 inhabitants, while in the four largest cities (San Cristobal de las Casas and Tuxtla Gutierrez in Chiapas and Villahermosa and Cárdenas in Tabasco) which only represent 16 % of the population, the average population density is 54 in hab./km^2 (INECC 2013). This structure hinders public services provision. Drinking water coverage in the region, in cities, is 78 % and sewerage 76 %. In rural areas the coverage is 68.8 % and 64.4 %, respectively (CONAGUA 2011), below the

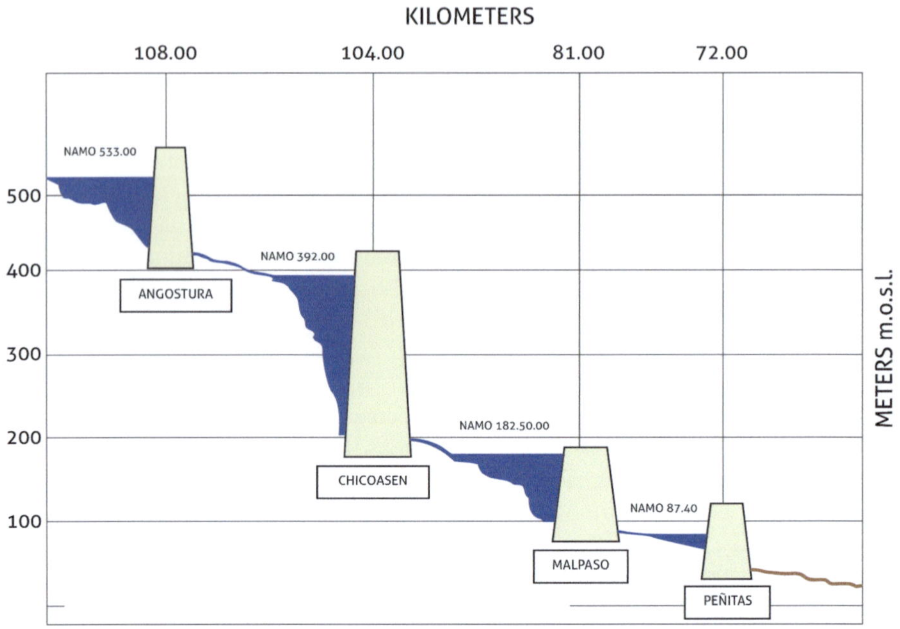

Fig. 11.1 Grijalva River basin dam system

national average of 91.6 % for water supply and 90.2 % for sewer (CONAGUA 2012). The delay in the development of water services is added to the dam operation system policy. During the dry season, May–December, it is required to meet the power generation commitments, damaging communities located in the upper and middle hydrological subregions of the Grijalva River basin.

Although the region is not a major national industrial area, there are some plants of Petroleos Mexicanos (PEMEX), which is the largest industrial water consumption in the region with an annual average expenditure of 1,078 liters per second; 86 % is groundwater and 14 % the surface water remaining (SEMARNAT 2010). PEMEX is also the largest supplier of wastewater discharges to water bodies. Between 1990 and 1994, it paid over 1,200 million pesos to compensate about 3,15,000 claims for spills, leaks, and corrosion of fishing wire (INECC 2013). The presence of oil spills, agricultural activities, and certain services has deteriorated the water quality in the region, not in a meaningful way. According to the National Water Commission (CONAGUA) which uses biological oxygen demand, total suspended solids, and chemical oxygen demand as criteria to determinate water quality, no water body is classified as heavily polluted in this area (CONAGUA 2011).

The Grijalva River basin territory is one of highest ecological biodiversity areas in Mexico; even though it represents less than 4 % of the total national area, it includes 64% of the known biodiversity (INECC 2013). Coniferous forests in upland areas and numerous plant associations in coastal plains and shorelines, in

Table 11.1 Characteristics of the Grijalva River basin dams

Data	Unit	La Angostura	Chicoasén	Malpaso	Peñitas
Basin area	Km^2	18,099	7,940	9,952	1,275
Maximum flow recorded	m^3/s	3,820	6,214	7,200	5,650
Average annual precipitation	mm				1,923
Evaporation	10^6 m^3	10	0	1	0
Dam					
Elevation to NAMINO	mosl	500	380	144	85
Elevation to NAMO	mosl	533	394	183	87
Elevation to NAME	mosl	540	395	188	94
Volume to NAMINO	10^6 m^3	2,380	1,169	3,056	961
Volume to NAMO	10^6 m^3	13,170	251	9,317	1,091
Volume to NAME	10^6 m^3	17,357	274	11,001	1,485
Total capacity	10^6 m^3	19,736	1,443	14,058	1,485
Elevation to the average discharge level	mosl	422	203	85	53
Power generation plant					
Turbines		5	5	6	2
Design load	m	94	191	95	34
Maximum monthly volume for turbines	10^6 m^3	3,075	2,452	3,784	3,784
Design flow	m^3/s	1,170	933	1,440	1,440
Maximum height	m	147	245	138	45
Elevation of the crown	mosl	543	405	192	98
Crown width	m	10	25	10	8
Crown length	m	324	584	478	750
Freeboard	m	4	10	4	5
Curtain total volume	10^6 m^3	4	15	5	2
Spillway					
Crest length	m	50	76		116
Crest elevation	mosl	520	373		77
Design flood	m^3/s	23,000	17,400	20,000	22,877
Maximum capacity of discharge	m^3/s	6,900	15,000		18,700

addition to pastures, provide environmental services of ample importance such as flood control, freshwater supply, erosion control, sediment trapping, and shelter to a wide variety of wildlife. Also, coastal wetlands bordering the Gulf of Mexico provide important environmental services. In fact, they are considered among the most important wetlands of Mesoamerica (Lazcano Barrero et al. 1992) The vast biodiversity of the basin is directly competing with agricultural and livestock activities, which occupy about 43,000 ha (INECC 2013). The increasing food demand of the local population and the link of producers with national and international chains led to the advance of forest clearing to increase food cultivation. After the market sophistication, agriculture took a turn toward commercial plantations and ranching, reducing more than half the coverage of cloud forests and conifers, making predominant secondary vegetation (Challenger and Dirzo 2009).

11.3 Climate Change Scenarios for the Grijalva River Basin

Considering the scenarios given by the Intergovernmental Panel on Climate Change scenarios, that use regional scales of continental dimensions with a spatial resolution which hampers adaptation action planning at a basin level, it was decided to use regional scenarios generated by assembling models regionalized by the principal component regression technique.

First, a network of 109 base stations was built, under the National Weather Service unit (CLICOM), to collect precipitation and temperature data. This network allowed performing an historical analysis of variable patterns, as well as runoff and extreme events recurrence. From this analysis, the following information resulted:

Temperature. The higher annual average temperatures were recorded in coastal areas and the minimum in the high mountains, with an average annual temperature of 22 °C in the basin. The hottest month is May, followed by June, April, and July, but November and December are the only months with temperatures below 21 °C. The monthly average maximum temperature is 38.9 °C at an elevation range from 1,000 to 1.500 m above sea level, while the minimum is 8.1 °C at an altitude higher than 2,000 m above sea level.

Precipitation. The combination of cold air masses from the north and Atlantic and tropical humid ones from the Pacific causes most of the annual precipitation in the region, resulting in a value of average annual rainfall of 1,289 mm. Chiapas presents an average annual rainfall of 1,969 mm, while in Tabasco it is slightly higher, with 2,093 mm. From June to October, larger precipitation volumes occur, e.g., 82.3 % of the average annual rainfall in Angostura, 80 % in Chicoasén, 74 % in Malpaso, and 71 % in Peñitas. From September to November, rivers and lakes reach their highest levels.

Runoff. Runoff analysis was performed only in the upper half of the basin, where the dam's system for electricity generation is located. The observations show that maximum monthly average volumes are presented in September, with the exception of the Peñitas basin, where the maxima occur in October. The average annual runoff at Angostura basin has a value of 10,061 Mm^3, Chicoasén of 2,205 Mm^3, Malpaso of 5,515 Mm^3, and Peñitas of 1,576 Mm^3, with variation coefficients of 0.24, 0.29, 0.38, and 0.31, respectively.

Extreme Events. There are few studies developed in Mexico and the Caribbean analyzing trends in extreme events, so it was decided to use three of the 27 climate change indices proposed by the World Climate Change Research Programme (WCRP) for this purpose: cold period duration index, warm period duration index, and daily maximum rainfall index. From the weather station networks, the stations located at Tuxtla Gutierrez in Chiapas and Pueblo Nuevo in Tabasco were selected. In conclusion, it was found that there is a decreasing trend of cold periods in Tuxtla Gutierrez and an increasing trend in the number of hot days at the Pueblo Nuevo station.

Subsequently, historical behavior analysis of selected variables, cumulative monthly precipitation departures, and regionalized average monthly temperature were used under the scenarios A2, A1B, and 20c3m.[1] For an analysis of observed climate, an accumulated monthly temperature and precipitation database developed at the Climate Research Unit of the University of East Anglia was used, which is provided in a grid with dimensions of 50 km × 50 km and contains information from 1901 to 2002. The validation of this ensemble of model methodologies considered the 1970–1999 period as a baseline.[2]

11.3.1 Temperature Scenarios

All climate change literature, regardless of the emission scenarios, converges in showing an increase of temperature in Mexico. However, the uncertainty is presented in its magnitude and its exact location at finer scales, at the basin level, and in this case the emission scenario is significant. According to estimations, changes in temperature for the southern region of Mexico in the coming decades are shown in Table 11.2.

As can be seen in Table 11.2, using as baseline the 1970–1999 period, the highest temperatures are presented for the second half of the century, reaching 1.75 °C in the higher GHG emission scenario. For the A1B scenario, the period of 2010–2030 shows the lowest values of anomaly, which are found in the southern and southeastern border of the Chiapas State. The same conditions are found during the 2030–2050 period, but this time for the A1B scenario. These results can be identified in Figs. 11.2 and 11.3.

11.3.2 Precipitation Scenarios

In contrast to the high degree of uncertainty existing in the projection of increase of temperature for Mexico in the coming years, the decrease in annual precipitation accumulated and assumed by the GCM for the country is certainly less projected as

[1] 20c3m scenario refers to experiments of increased greenhouse gases as recorded in the twentieth century and serves as baseline scenario or control.

[2] The detailed description of the validation of the ensemble of model methodologies can be found at González-Villarreal et al. (2009). Evaluación de la vulnerabilidad del sistema de presas del río Grijalva ante los impactos del cambio climático. Informe Final. INE-IINGEN, Mexico.

Table 11.2 Median anomaly temperature (°C) for the ensemble of models. A2 and A1B scenarios for 2010–2030 and 2030–2050 periods

Scenario	Temperature anomaly for 2010–2030 (°C)	Temperature anomaly for 2030–2050 (°C)
A2	0.75–1.0	1.5–1.75
A1B	0.75–1.25	1.75

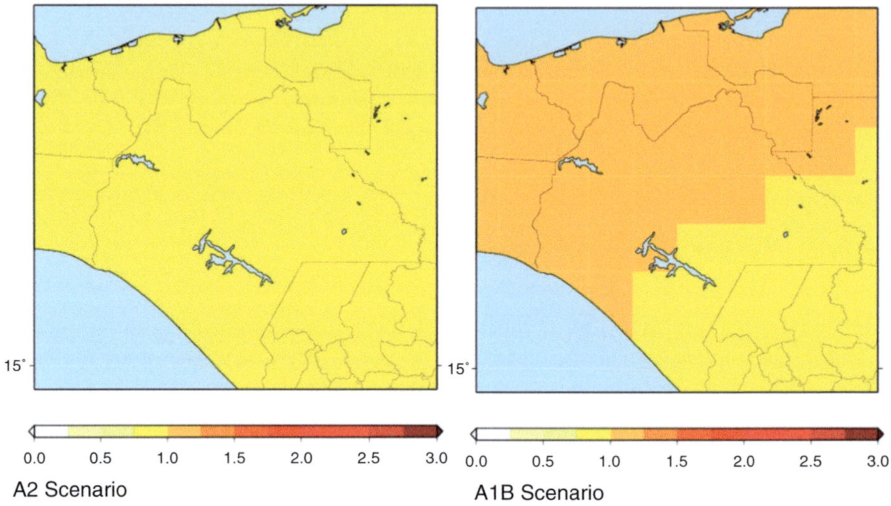

Fig. 11.2 Median temperature anomaly (°C) for the ensemble of models. A2 and A1B scenarios for 2020

they approach the end of the century; this is because the rainfall pattern is defined largely by hurricane activity and these systems are not yet sufficiently modeled by the GCM (Meehl et al. 2007). Considering these limitations and assuming the projected annual precipitation anomaly is higher in the regions where it rains, projections were made for the change of annual precipitation in the southern region.

Broadly, a decreased precipitation is observed in the South. For the 2010–2030 period, under the A2 scenario, results show that the most important changes are in the northern part of eastern Tabasco and Chiapas, where a reduction of 5–10% is expected, while a ± 5 % anomaly is observed for the rest of the basin. In the same period, but under the A1B scenario, a change of +5 % to −10 % is expected. The largest negative anomaly is registered in the northwest of Tabasco and for Chiapas in the eastern region, although the upper Usumacinta, in the Guatemalan territory, has an anomaly of −35 %.

Meanwhile, for the 2030–2050 period under the A2 scenario, eastern Chiapas presents the most negative anomaly of precipitation, with a reduction of up to 15 %. For the rest of the region, changes are estimated to be ±5 %. When A1B scenario is used, the most significant negative changes are also registered in the same

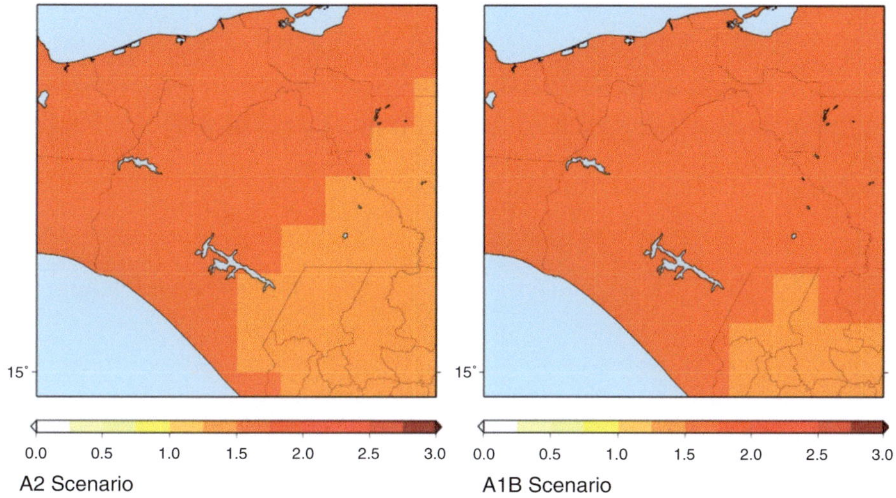

Fig. 11.3 Median temperature anomaly (°C) for assembly models. A2 and A1B scenarios for 2050

geographic area. The positive anomalies amount to 5 %. These results can be seen in Figs. 11.4 and 11.5.

11.3.3 Extreme Events Projection

Using the *time stochastic weather generator* (MTSG) technique, which used the observed data for the period 1970–2000 CLICOM-specific stations, Oxolotán and El Triunfo in Tabasco and Abasolo in Chiapas, scenarios were analyzed and generated for Mexico. Changes in extreme events in watershed were estimated. Analysis of the data provided the following information:

Temperature. Under the climate change for 2010–2030 and 2030–2050 periods, the maximum increases projected in both are the average of the distribution in variability range. In the case of severely warm events as heat waves, the magnitude of temperature could increase by 3–4 °C for such periods, respectively. Increasing higher minimum temperatures, meanwhile, will be around 2 °C, which coincides with the trend of minimum and maximum temperatures in the area since the second half of the twentieth century.

Precipitation. It is expected a decrease of the maximum rainfall associated with a return period (Tr) of 100 years for the A1B scenario in both periods. The largest decrease is 8 % and is presented in the station of El Triunfo, Tabasco.

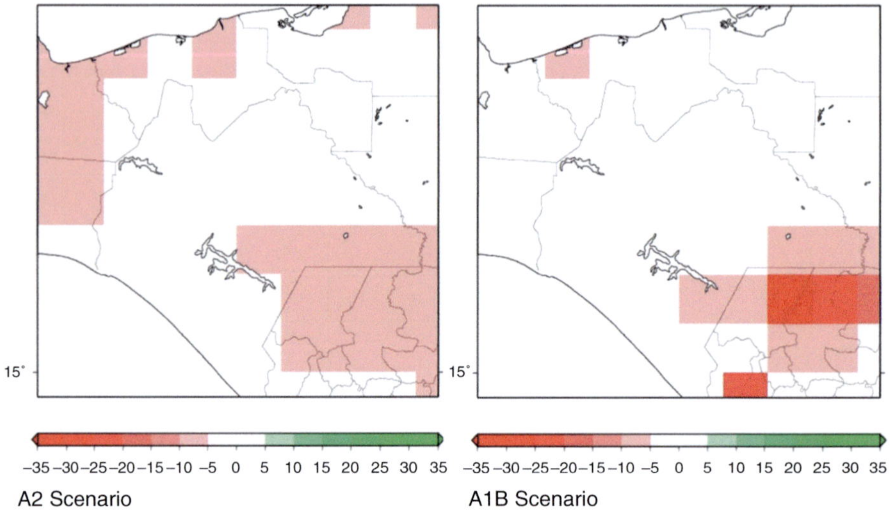

Fig. 11.4 Median precipitation anomaly (%) for the ensemble of models. A2 and A1B scenarios for 2020

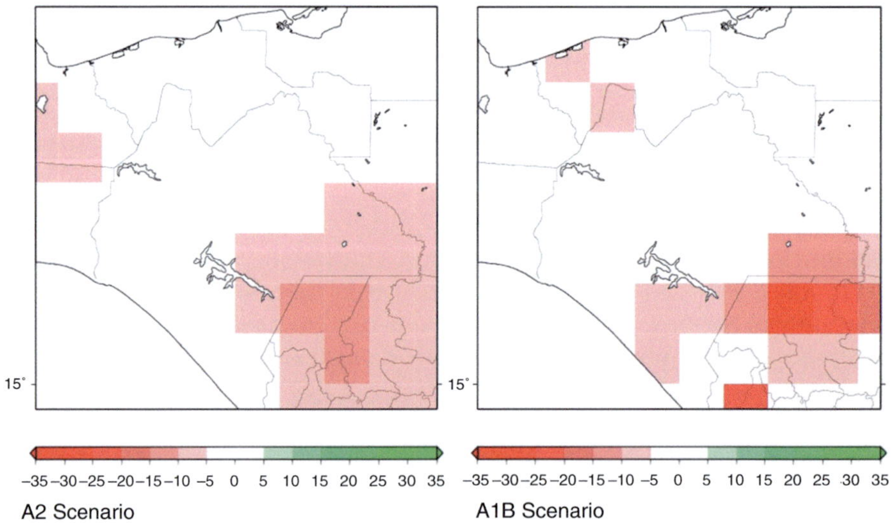

Fig. 11.5 Median precipitation anomaly (%) for the ensemble of models. A2 and A1B scenarios for 2050

11.4 Impacts of Climate Change on the Grijalva River Basin

The temperature increases, the precipitation decreases, and the occurrence of extreme events projected under climate change scenarios will impact the Grijalva River basin in different ways, particularly affecting the dam system. The effects analyzed are identifiable in water availability affecting the ability to maintain the optimal levels for electric-power generation, the rising sea level which may affect infrastructure and increase the vulnerability of coastal populations, and the water quality which necessarily affects people's health.

11.4.1 Water Availability in the Diversion Dam System

To identify the impacts of climate change on water availability in the diversion dam system, a simulation was made, based on the monthly historical volumes for each of the subbasins – Angostura, Chicoasén, Malpaso, and Peñitas – with an analysis period of 1952–2008. The simulations allowed to know the average annual runoff volume in billion cubic meters and annual hydropower generation in thousands of GWh. For this purpose, the values of temperature increase and precipitation decrease obtained from the A1B scenario for the periods 2030 and 2050 by basin were used. These values are shown in Table 11.3.

The increased temperature and decreased precipitation necessarily turn into a reduction of system productivity. To know the extent of the decline, we fixed critical thresholds for both climatic variables, and it was related with the maximum loss that can be accepted per year. The critical thresholds were defined considering an average cost of 1.7 pesos per kWh. Thus, for a maximum acceptable loss of 500 million pesos per year, the critical climate thresholds are a temperature increase of 0.1 °C and a decrease in precipitation of −1.0 %. If the maximum loss amounts to one billion pesos per year, the critical thresholds are 0.29 °C and −1.9 %, respectively.

According to the analysis, it was estimated for the entire system that, without action by 2030, there will be an annual loss due to the decreasing electricity generation of about 2,000 million pesos per year, while in 2050 it will reach 3,500 million pesos. High costs for 2050 are the result of a decrease in the benefits of electricity generation estimated at 16 % for the entire basin, −20 % for Angostura, 19 % for Malpaso, and 13 % for Chicoasén and Peñitas.

The study also considered the impacts of extreme events. As noted above, there is a tendency for a decrease in the number of days with rain and the presence of extreme events, despite an increase in temperature. The estimates made using a Gumbel distribution for maximum rainfall associated with a return period of 100 years indicate a reduction of 9 %, and thus the maximum rates would experience a decrease in the same proportion, so there is no need to worry about the

Table 11.3 Values of temperature (°C) and precipitation (%) by the subbasin of Grijalva River. A1B scenario for 2030 and 2050

	2030	2050
Upper part of the watershed	0.9 °C	1.4 °C
	−3.8 %	−6.7 %
Medium part of the watershed	1.1 °C	1.6 °C
	−3 %	−5.3 %
Lower part of the watershed	1.1 °C	1.6 °C
	−3 %	−3.8 %

damages in infrastructure caused by rain; however, the hypothesis requires further study for confirmation. In terms of tropical cyclone activity under climate change scenarios, it is expected to see a decrease in number but an increase in intensity, although there remains a high degree of uncertainty. Finally, an increase in both the mean of distribution and the range of variability is projected for 2030 and 2050. For severe warmer temperature events, such as heat waves, the magnitude could increase between 3 and 4 °C. An increase in minimum warm temperatures, meanwhile, is estimated at 2 °C by the end of the century, according to the trend that has been observed as a continuous process in the area since the second half of the twentieth century.

11.4.2 Effects of Sea Level Rise

So far there is no final conclusion on the overall average of sea level rise; however, the IPCC has projected a rise of approximately 60 cm by 2100. It is also essential to know the possible effects of this rise at the regional level. In this context, the sea level rise has been analyzed in the Grijalva River basin. Given the geographical irregularities on the area, it was decided to focus on the Samaria River in Tabasco, which has a length of 52 km and hosts locations and infrastructure with considerable importance to the state.

From the information obtained from the cross sections of the Samaria river bed and interpolating sections with information from the National Institute of Statistics and Geography of Mexico (INEGI), sections were generated to apply a one-dimensional model of steady flow calculation developed by the Institute of Engineering of the National Autonomous University of Mexico (UNAM) for the flow of 500, 1,000, 4,000, and 6,500 m^3/s under the mean sea level at 0, 1, and 2 m.

According to the information obtained, it was identified that 50 kilometers upstream from the sea, there is no trace of sea level rise, even when it is calculated at 2 m. Thus, Villahermosa City, the capital and most important economic center of Tabasco, does not experience any kind of impact. However, when the sea level rise is 2 m with a flow of 6,500 m^3/s, an overflow is observed in the lagoon area that goes from the coast to kilometer 44. In this territory the presence of flood damages is expected. See Fig. 11.6. These occurrences are expected by mid-century.

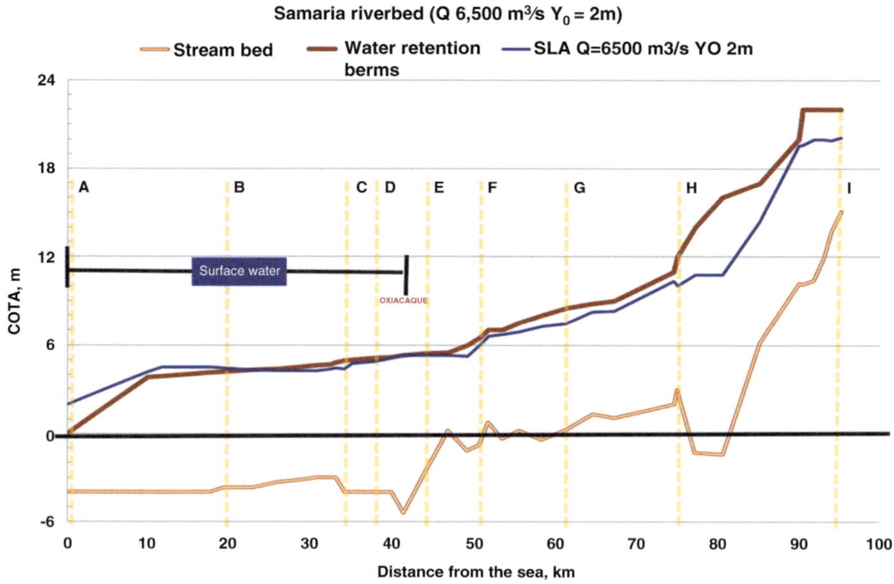

Fig. 11.6 Elevations on indicated sections by the effect of sea level rise and river flow of 6,500 m^3/s, for a sea level rise of 2.0 m. In *brown* are shown water retention berms and in blue surface water

11.4.3 Water Quality

The effects of climate change on water resources have been mostly analyzed from changes in availability, but few studies can be found about the impact on quality. In addition to the generation and implementation of regionalized methodologies, whose difficulty has been already highlighted in this text, the analysis of the impact of climate on water quality requires the existence of databases with consistent and reliable information. Despite the challenges of this type of studies, Antonio García et al. (2010) analyzed water quality in the Grijalva River basin using deterministic-conservative methodology and scenarios A1B and A2 regionalized to Mexico for the years 2020, 2050, and 2080. To estimate the impact, they used five-day biochemical oxygen demand (BOD_5) and chemical oxygen demand (COD) indicators, which determine the amount of biodegradable organic matter and total organic matter, respectively (García et al. 2010).

As a result of this research, it was found that, in the case of BOD5 indicator, water quality will be acceptable throughout the basin in 2020, implying that surface water will maintain even their self-purification capacity. For 2050 and 2080, the water quality of the overall basin will change to polluted – mainly due to local untreated water discharges – or heavily polluted, with local and nonmunicipal untreated water discharges, regardless of the scenario considered. The COD indicator shows a different behavior. For 2020, the basin quality is good and acceptable

– with a low presence of organic matter – but it will change for 2050 and 2080 to polluted, regardless of the scenario considered (García et al. 2010).

11.5 Conclusions

Climate change will increase the vulnerability of socio-ecosystems located in the Grijalva River basin as long as adaptation measures to the effects on water resources are not executed in the short, medium, and long terms. The geographic, social, and economic characteristics of the area involving a number of different measures and broad-scale actions, especially designed to prevent flood damages, either by the increase in mean sea levels and river flow or the possibility of dam overflow, and to address the decline in energy production.

For Mexico, which is not counted within the major generators of greenhouse gases and its development priorities focusing in poverty alleviation of more than half of its population, it has been widely discussed that it is more advisable to invest in adaptation measures than in mitigation. This doesn't mean to stop taking decisive actions to reduce the number of emissions but define actions with the greatest impact based on a limited budget. Thus, it is more efficient to develop adaptation works for the Grijalva River basin, as established in the Tabasco Comprehensive Water Plan (PHIT), formulated by the Engineering Institute of UNAM at the request of the National Water Commission, in response to the worst flooding it has experienced in the state of Tabasco, and has as its main objective to reduce the risk conditions and vulnerability of people in the region and their economic activities and ecosystems.

For the Tabasco Comprehensive Water Plan, adaptation to climate change must consider building infrastructure at scales appropriate to the changes that will be experienced. Overall, the measures are focused on increasing the water storage capacity in the basin and improving the efficiency of protective dikes. In connection with the construction of infrastructure, it's necessary to update the hydropower project's portfolio of the Grijalva watershed. In addition to the construction of infrastructure, which is further inserted into a traditional view for the treatment of water problems, we recommend improving soil conservation practices and integrated watershed management, specifically through protection plans for mangroves and wetlands, and making a review of land management mechanisms to prevent the proliferation of informal settlements on river and coastal areas identified as highly vulnerable to experience flooding due to the increase of sea level.

Studies on water quality in the context of climate change show that it is necessary to prevent impacts in other sectors, particularly in the health sector. It is recommended a more precise analysis of the subject, focusing specifically on the behavior of vector-borne diseases, e.g., dengue. It will be necessary to also promote more restrictive legislation in terms of discharge of contaminants.

According to this study, there is an urgent need to scale climate change scenarios at lower levels, in which decision making on adaptation is simpler. The publication in Mexico of a new Climate Change Action on June 2012 provides an important opportunity, as required by the states to develop and implement their own climate change programs. In this context, it becomes necessary to provide technical support to the centers involved in research in this field. Major efforts should be made also to disseminate the results of this research among stakeholders to facilitate and attain concrete commitments to adaptation and mitigation objectives.

References

Bates BC et al (2008) Climate change and water. UNEP, Geneva

CONAGUA (1999) Consejo de Cuenca de los ríos Grijalva y Usumacinta. SEMARNAP, Mexico City

CONAGUA (2011) Estadísticas del agua en México. SEMARNAT, Mexico City

CONAGUA (2012) Situación del Subsector Agua Potable, Alcantarillado y Saneamiento. SEMARNAT, Mexico City

Challenger, Dirzo (2009) Factores de cambio y estado de la biodiversidad. In: José Sarukhán (Coord.) Capital Natural de México. CONABIO, Mexico

García et al (2010) Calidad del agua. In: Martínez, Patiño (eds) Atlas de vulnerabilidad hídrica en México ante el cambio climático. SEMARNAT, Mexico City

González-Villarreal et al (2009) Evaluación de la vulnerabilidad del sistema de presas del río Grijalva ante los impactos del cambio climático. Informe Final. INE-IINGEN, Mexico

Hernández de la Torre J (2009) Plan Hidráulico de Tabasco. Segundo Seminario de Potamología, Villahermosa

INECC (2013) La cuenca de los ríos Grijalva y Usumacinta. INECC. http://www2.inecc.gob.mx/publicaciones/libros/402/cuencas.html. Accessed 24 July 2013

Lazcano Barrero MA et al (1992) Características socioeconómicas de la Selva Lacandona. In: Ramos MA, Vázquez MA (eds) Reserva de la Biosfera Montes Azules, Selva Lacandona: Investigación para su conservación. Centro de Estudios para la Conservación de los Recursos Naturales, Mexico City

Martínez et al (2010) Efectos del cambio climático en los recursos hídricos. In: Jiménez et al (eds) El agua en México: cauces y encauces. Academia Mexicana de Ciencias-CONAGUA, Mexico City

Meehl G et al (2007) Global climate projection. In: Solomon S et al (eds) Climate change 2007: the physical science basis. UNEP, New York

SEMARNAT (2010) Acuerdo por el que se dan a conocer los estudios técnicos de aguas nacionales superficiales de las subregiones hidrológicas: Alto Grijalva, Medio Grijalva y Bajo Grijalva de la Región Hidrológica no. 30 Grijalva Usumacinta. Diario Oficial de la Federación, México

Chapter 12
Sustainable Development and Integrated Water Resources Management

José Alberto Tejada-Guibert, Shimelis Gebriye Setegn, and Ryan B. Stoa

Abstract Sustainable development integrates economic development, social development, and environmental protection, with three overarching objectives and essential requirements: (1) poverty reduction, (2) changing unsustainable patterns of production and consumption, and (3) protecting and managing the natural resource base of economic and social development. Integrated water resources management (IWRM) is capable of promoting all three objectives by providing stakeholders with a framework for integrating and coordinating the various aspects of water management in a sustainable and holistic manner. This chapter relates the concept of IWRM to development in the context of the international community's sustainable development paradigm.

Though there would seem to be a universal acceptance of the cross-cutting importance of water in development, relative marginalization of water at different levels has taken place because of competing interests. In the ongoing Millennium Development Goals (MDGs), there is only one target dealing specifically with water (drinking water and sanitation) within one of the eight goals (goal 7—ensure environmental sustainability). There is now a call from the water community for the adoption of a dedicated and comprehensive sustainable development goal on water in the post-MDG follow-up.

Some current topics that need to be encompassed by IWRM are succinctly addressed with reference to the recent international developments, including water security, human rights to water, gender equity, implications of climate change, governance, and the water–energy nexus.

J.A. Tejada-Guibert (✉)
Global Waters for Sustainability Program – GLOWS, Florida International University, Miami, FL 33199, USA
e-mail: jatejada@globalwaters.net

S.G. Setegn
Department of Environmental and Occupational Health and Global Water for Sustainability Program (GLOWS), Florida International University, 11200 SW 8th Street, Miami, FL 33199, USA
e-mail: ssetegn@fiu.edu

R.B. Stoa
Florida International University, College of Law and Global Water for Sustainability Program – GLOWS, Miami, FL 33199, USA
e-mail: rstoa@fiu.edu

S.G. Setegn, M.C. Donoso (eds.), *Sustainability of Integrated Water Resources Management*, DOI 10.1007/978-3-319-12194-9_12

The ability of making operational the nexus of water with energy, food, health, and ecosystems by thinking "out of the water box" has become critically important.

Keywords Water resources management • IWRM • Sustainability • Sustainable development • Water security • Climate change adaptation • Human right to water • Gender equity • Governance • Water–energy–food nexus

12.1 Sustainable Development

12.1.1 Foundations of Sustainability

The World Commission on Environment and Development (WCED), or Brundtland Commission, appointed in 1983 by the UN Secretary-General whose final report—"Our Common Future"—delivered in 1987 attempts to overcome the problem of the multiplicity of the notions of sustainability, even in terms of water conservation. The report's definition based sustainability on three pillars—economic growth, environmental protection, and social equality—thus going beyond the original notions of sustainability as a goal exclusively connected with the environment. The Commission chose a simple definition—sustainability as the ideal scenario where each generation refrains to limit society's ability to meet its needs in the future (United Nations 1987). It also described sustainable development as "development that meets the needs of the present without compromising the ability of future generations to meet their own needs."

Sustainable development integrates economic development, social development, and environmental protection. In the current conception of the United Nations, sustainable development has three overarching objectives and essential requirements: (1) poverty reduction, (2) changing unsustainable patterns of production and consumption, and (3) protecting and managing the natural resource base of economic and social development (UNDESA 2008). It is not surprising that the pillars of SD and the overarching objectives are fully compatible with the IWRM principles and objectives cited before, as IWRM is one of the key instruments to support SD.

12.1.2 Measuring Sustainability

There are multiple differing definitions of sustainability; these determine the characteristics that would need to be monitored or measured to ascertain if sustainability is being achieved. The traits that different organizations ascribe to sustainability reflect their institutional functions and objectives. For instance, the European Union has adopted the principles of general sustainability (EU, Med Programme 2007–2013 2010). These principles are:

1. Integration: the effective integration of environmental, social, and economic considerations in decision-making.
2. Community involvement: recognition that sustainability cannot be achieved, nor significant progress made toward it, without the support and involvement of the whole community.
3. Precautionary behavior: where there are threats of serious or irreversible environmental damage, a lack of full scientific certainty should not be used as a reason for postponing measures to prevent environmental degradation.
4. Equity within and between generations: fairness and equal access to opportunities both in our lifetimes, as well as future generations.
5. Continual improvement: the declining environmental situation means there is an imperative to take immediate action to become more sustainable and to make continual improvement.
6. Ecological integrity: to protect biological diversity and maintain essential ecological processes and life support systems.

Most of the principles listed above flow naturally from the discussion on sustainable development. Two concepts not usually cited in an explicit fashion in IWRM literature are (1) the "precautionary behavior," which reflects the precautionary principle (which had been already incorporated into the Rio Declaration of 1992 on Environment and Development), and (2) "continual improvement," which is a concept that is espoused by, among others, the Global Compact of the United Nations (UN Global Compact Office 2011). Launched in 2000, the Global Compact is both a policy platform and a practical framework for private sector companies committed to sustainability and responsible business practices.

The desirability of achieving sustainability is evident in the application of IWRM. However, the assessment of sustainability presents a serious challenge. For instance, a report prepared for the US Environmental Protection Agency stated: "While much discussion and effort has gone into sustainability indicators, none of the resulting systems clearly tells us whether our society is sustainable" (Hecht 2007). The inherent difficulty of assessing sustainability is likewise captured in the book *Sustainability Indicators: Measuring the Immeasurable?* (Bell and Morse 2008).

Nonetheless, serious efforts are being currently expended to develop effective methodologies to handle this elusive matter in connection to water resources management. For example, a sustainability check tool is at an advanced stage of development for the specific aspect of water, sanitation, and hygiene (WASH) (Rotary International and USAID 2012); also the State of California has likewise launched a major effort in this direction (Shilling et al. 2012).

12.1.3 The Millennium Development Goals and Recent International Developments

12.1.3.1 The Rio+20 Conference

The Millennium Development Goals (MDGs) (http://www.un.org/millenniumgoals/) have been a major instrument during the past decade for lending thrust and direction to the pursuit of sustainable development. The MDGs, adopted through the intergovernmental mechanism of the United Nations, set eight major goals to be achieved by 2015; among them, goal 7—ensure environmental sustainability—had a specific target asking the international community to “halve, by 2015, the proportion of the population without sustainable access to safe drinking water and basic sanitation.” To date (United Nations 2013) the MDG drinking water target was met 5 years ahead of the target date, despite significant population growth; however, gains in sanitation are impressive—but not good enough. . . more rapid progress is needed to meet the MDG target. Sub-Saharan Africa shows an especially critical situation with regard to this goal.

More recently, the United Nations Conference on Sustainable Development (UNCSD), also known as Rio+20, adopted the document “The Future We Want” (United Nations 2012) supporting the development of Sustainable Development Goals (SDGs), a set of measurable targets aimed at promoting sustainable development globally for the post-2015 period. The SDGs would pick up where the Millennium Development Goals leave off. The 192 governments present declared “We recognize that water is at the core of sustainable development as it is closely linked to a number of key global challenges. We therefore reiterate the importance of integrating water in sustainable development and underline the critical importance of water and sanitation within the three dimensions of sustainable development.” This is quite a significant statement as normally within the global scenario of intergovernmental sustainable development debate, water resources have normally been assigned a supporting role, as happened with the current MDGs that had only one target on drinking water and sanitation appended to one of the eight major goals.

The Rio+20 conference has likewise chosen Green Economy in the context of sustainable development and poverty eradication as an overarching theme for realizing transformational change. One commonly used definition of Green Economy is “one that results in improved human well-being and social equity, while significantly reducing environmental risks and ecological scarcities. It is low carbon, resource efficient and socially inclusive” (UNEP 2011). It incorporates the direct valuation of natural capital and ecological services unlike the prior economic regimes.

12.1.3.2 Post-Rio+20 Developments

After Rio+20 some significant international follow-up events regarding water resources management have been held to provide a richer backdrop to the process of formulating and implementing the SDGs. In early 2013 an ample all-inclusive stakeholder consultation was carried out conducted by several UN bodies and hosted by several governments. As a result the document "The World We Want" (UN-Water et al. 2013) was produced. Among the points made are as follows:

- Improving the management of water resources is of capital importance, but given differing governance situations are different, and no blueprints apply, pointing out that improved coordination is required across the different water-using sectors.
- Shortage of capacities on water governance and integrated management remains an important constraint.
- Concerning the coming SDGs, it was suggested that water resources goals and targets could be based on two topics: (1) water for socioeconomic development and environmental protection and (2) implementing an integrated water resources management approach.

The recently held Budapest Water Summit (October 2013), an international but not intergovernmental meeting, called for the adoption of a dedicated and comprehensive SDG on water, a "Water-Secure World" (Government of Hungary 2013). The meeting made a strong case to adopt a water–energy–food nexus approach to water resources management planning so as to achieve cross-sectoral integration which has chronically been missing or coming up short (see Sect. 12.2.5). It also championed achieving universal access to clean water and sanitation, completing the achievement of the MDGs, but pointedly indicated "Drinking water supply and sanitation should be fully integrated into water resources management policies with the recognition that water use and sanitation not only consume the resource, but also pollute water and, therefore, negatively influence the water cycle as a whole, if countermeasures are not applied"—this statement may seem puzzling, even contradictory, since access to clean water and sanitation was and is indeed supported as major immediate targets. It constitutes a cautionary message as the emphasis at the international scale (e.g., targets of the MDGs and focus of bilateral aid agencies, such as USAID) and at the national scale in the past decade has been on water and sanitation, also known as water, sanitation, and hygiene (WASH). The intense focus on WASH during this past decade may be perceived as not having kept an overall balance with all other pressing needs of the countries related to WRM.

12.2 Focus on Some Current Topics in IWRM

12.2.1 Climate Change Impact and Adaptation

12.2.1.1 Impact of Climate Change

Water resources are crucial to both ecosystem and well-being of the society. A reliable and clean supply of drinking water is necessary to sustain human health.

Water is also needed for agriculture, energy production, navigation, recreation, and manufacturing. These demands place pressures on water resources that are likely to be exacerbated by climate change. In many areas, climate change is likely to reduce surface and groundwater resources, accompanied by an increasing water demand. A major effect of climate change is likely to be alterations in hydrologic cycles and changes in water availability. Increased evaporation, combined with changes in precipitation, has the potential to affect runoff, the frequency and intensity of floods and droughts, soil moisture, and the availability of water for irrigation and hydro-electric generation.

The Intergovernmental Panel on Climate Change's (IPCC 2007) findings suggest that developing countries will be more vulnerable to climate change due to their economic, climatic, and geographic settings. According to the IPCC (2007) report, the population at risk of increased water stress in Africa is projected to be between 75–250 and 350–600 million people by the 2020s and 2050s, respectively. Moreover, yields from rain-fed agriculture could be decrease by up to 50 %, in countries that depend mainly on rain-fed agriculture.

In some areas, climate change increases runoff, flooding, or sea level rise. Changes in the amount of rain falling during storms provide evidence that the water cycle is already changing. Warming winter temperatures cause more precipitation to fall as rain rather than snow. Furthermore, rising temperatures cause snow to begin melting earlier in the year altering the timing of streamflow in rivers that have their sources in mountainous areas (USGCRP 2009). In Western United States, the future projections of reduced total annual rainfall, less snowpack in the mountains, and earlier snowmelt mean that less water will likely be available during the summer months when demand is highest. This will make it more difficult for water managers to satisfy water demands throughout the course of the year (CCSP 2008). Setegn et al. (2011) investigated how changes in temperature and precipitation might translate into changes in streamflows and other hydrological components using downscaled outputs from four climate models.

These effects can reduce the quality of water and can damage the infrastructure that we use to transport and deliver water. The quality of water supply in coastal and island regions is at risk from rising sea level and changes in precipitation. Rising sea level and the occurrence of drought can increase the salinity of both surface water and groundwater through saltwater intrusion. For example, the freshwater Everglades currently recharge Florida's Biscayne Aquifer, a natural underground area that collects water and is the primary water supply to the Florida Keys. If rising sea levels submerge the low-lying areas of the Everglades, portions of the aquifer would become saline. Sea level rise can also push salty water upstream in coastal areas, threatening surface water supplies (CCSP 2009).

Assessing the impact of climate change on streamflows, soil moisture, groundwater, and other hydrological parameters essentially involves taking projections of climatic variables (e.g., precipitation, temperature, humidity, mean sea level pressure, etc.) at a global scale, downscaling these global-scale climatic variables to local-scale hydrologic variables, and computing the hydrological components for water resources variability and risks of hydrologic extremes in the future.

12.2.1.2 Adaptation to Climate Change

Adaptation to the global climate change and variability is considered a cornerstone for the application of IWRM for it to be truly effective and sustainable. The adaptive nature of IWRM is considered a good platform to incorporate climate change adaptation. Currently, there are available valuable references on the topic—a sampler would include the manual on IWRM as a tool for adaptation to climate change (Cap-Net and UNDP 2009), the set of 16 perspective papers on water and climate change adaptation produced by a consortium comprising the World Water Council, IUCN, and others, presented on the occasion of the 5th World Water Forum (WWC and IUCN 2009). The European Union, through the Sustainable Water and Integrated Management (SWIM) project, has issued a guideline for mainstreaming adaptation options in IWRM plans (SWIM-EU 2013).

The major driving instrument for the international efforts in climate change adaptation is the United Nations Framework Convention on Climate Change (UNFCCC), an international environmental treaty negotiated at the Rio 1992 UN Conference on the Environment and Development that entered into force in 1994. Since 1995 there has been an annual Conference of the Parties (COP). Though adaptation to climate change has been part of the UNFCCC mandate from the beginning, it was only in 2005 at the 10th Conference of the Parties (COP 10) in Buenos Aires, Argentina, that it decided to systematically address the vulnerability and climate change adaptation aspects. This was the origin of the program currently known as the Nairobi Work Programme (NWP) (http://www.siwi.org/knowledge-services/climate-change-water/siwis-messages-at-cop19/). It has been designed to assist the states party to the convention, particularly the developing countries to:

- Improve their understanding and assessment of impacts, vulnerability, and adaptation to climate change
- Make informed decisions on practical adaptation actions and measures to respond to climate change on a sound scientific, technical, and socioeconomic basis, taking into account the current and future climate change and variability

In another significant complementary decision, COP 17 (2011, Durban, South Africa) established the national adaptation plan (NAP) process as a way to facilitate effective adaptation planning in least developed countries and other developing countries; the NAP guidelines were finalized in 2012 (UNFCC 2012).

However, the issue of water resources still remained not formally addressed by the Conference of the Parties for an extended period. On the occasion of the 16th COP, held in Cancun, Mexico, in December 2010, the Mexican Government held a significant international pre-event with the specific objective of promoting the inclusion of water resources as an important issue to be considered at the COP; however, it was not taken up then by the conference. In COP 19 held in Warsaw, Poland, in November 2013, the issue of water resources was addressed, particularly in regard to the Nairobi Work Programme, with respect to the areas of impacts, vulnerability, and adaptation to climate change, and to the National Adaptation Plans (NAPs) and for whose implementation the COP requested wide support and funding.

This augurs well for the international and national water resources management community, and it is a powerful reminder that major decisions have to be obtained outside of the water "box." No matter how cogent and real a water-related argument is, decision makers and their scientific advisors in other sectors have to understand the issue before leaning to a favorable pronouncement.

12.2.2 Water Security

Water security is a topic that has recently been propelled to the forefront as an issue of international priority and as an objective of IWRM. According to UN-Water, "Water security is defined as the capacity of a population to safeguard sustainable access to adequate quantities of acceptable quality water for sustaining livelihoods, human well-being, and socio-economic development, for ensuring protection against water-borne pollution and water-related disasters, and for preserving eco-systems in a climate of peace and political stability" (UN-Water 2013).

Another definition of water security states "the availability of an acceptable quantity and quality of water for health, livelihoods, ecosystems and production, coupled with an acceptable level of water-related risks to people, environments and economies" (Grey and Sadoff 2007, cited in GWP 2012). Both definitions embrace the concept of sustainable development, with its aim of ensuring a triple bottom line of social, environmental, and economic development outcomes. The first definition, in addition, makes water security hinge on "the capacity of the population" to safeguard the provision of the resource and the societal and environmental well-being. The concept of water security as used in this chapter and generally in the water resources community does not extend to operational risk and security issues linked to deliberate and wanton external interventions to render nonoperational or unsafe the water service. It refers rather to the process of improving water resources management practices aimed to a greater sustainability and security of water services as a contribution to development.

Addressing this complex goal of achieving water security requires interdisciplinary collaboration across sectors, communities, and political borders, so that the competition or potential conflicts are adequately managed to yield the desired degree of security. Furthermore, the recognition of the human right to safe drinking water and sanitation is considered an important step toward ensuring water security at the individual and community levels.

12.2.3 Human Rights to Water and Sanitation

Since the adoption in 2010 of the UN resolutions on the human right to water and sanitation by the UN General Assembly and the UN Human Rights Council, an increasing number of countries have integrated this right into the national policy and/or legislation through new strategies, laws, and constitutional amendments.

This instrument allows setting up a human rights approach for access to water and sanitation addressing the concerns of those who have traditionally been vulnerable and marginalized and makes possible the full realization of the rights of individuals and communities (UN-Water 2013).

A recent report of the UN Human Rights Council (2013), prepared by the Special Rapporteur to the UN on human rights, considers sustainability to be a fundamental human rights principle essential for realizing the human rights to water and sanitation. It decries retrogressive measures that lessen the enjoyment of human rights. Thus, this understanding of sustainability directly counters retrogression; fulfilling human right to water further "requires that services be available and accessible to everyone on an almost permanent basis, without discrimination, while ensuring beneficial change through quality services and sustained behavior change. Water and sanitation must be available for present and future generations, and the provision of services today should not compromise the future ability to realize these human rights." This is a natural extension of the concept of sustainable development embracing the human right to water that strengthens the framework of IWRM and its sustainability considerations.

12.2.4 The Role of Gender in Sustainable Water Resources Management

As seen in Chap. 6, the importance of the role of women has already been prominently characterized in the third Dublin principle on gender mainstreaming; as a globally accepted approach to achieving gender equality, it has been recognized as a general principle for international action for close to 20 years and nearly as long as one of the cornerstones of IWRM. However, up to now implementation of this principle, despite many gains, can be termed at best as very partially successful. The means of rating the sustainability of these measures in the IWRM context are virtually nonexistent in the currently proposed analytical frameworks for water resources initiatives sustainability inclusion and evaluation.

Regarding the inclusion of gender aspects in IWRM, particularly those linked to sustainability, there are a number of multisectoral issues such as ensuring equal educational opportunities (including suitable sanitary facilities for female students) and greater financial independence and access to the work market and countering strict traditional gender roles, which are deeply culturally ingrained. If these are tackled successfully, it will certainly assist in the attainment of sustainability in the projects and in the relevant women aspects. There are aspects directly linked to the implementation of IWRM that can be targeted in the design of the projects and in the assessment tools. Women constitute more than half of humanity and should have equal say and participation in the management and conservation of the planet's water resources. In line with the scheme suggested by the Global Water Partnership (GWP 2006), the following issues should be of especial interest with regard to the IWRM sustainability issues:

- Is the enabling environment being created and/or improved? Assess if relevant policies, legislation, and financing include:
  - Political commitment
  - Focus on gender in policy-related documents
  - Legal status for water management institutions stipulating the share of women
  - Budgetary allocation for proactive measures for women interest
- Are the institutional roles and organizational frameworks in water resources management evolving toward enhancing the participation of women? This includes the promotion of gender parity and capacity building for gender analysis and the creation or strengthening of gender units—the concept of empowerment of women at all levels is contained here.
- Are the proper management instruments in place: IWRM strategies and plans? These are not static instruments so the evolution considering the gender role would need to be monitored.

The deep social and cultural significance of women's issues was graphically illustrated in a recent event on the topic (FIU/GLOWS 2013). The meeting recognized that even nowadays in developing nations, women are overwhelmingly responsible for performing household tasks that involve water, resulting from strict gender roles that are deeply rooted, often for centuries, and that almost always have gone unquestioned and the lack of access of women to the decision-making process. The meeting emphasized that gender inequalities persist related to the allied fields of water, sanitation, and hygiene (WASH) and water resources management throughout the world. Among others, they highlighted aspects such as:

- Ingrained attitudes regarding the perceived obligation of women (and children) of being water carriers that must be constructively dispelled
- The curtailment of educational opportunities for girls that still persists in many places due to the lack of provision of appropriate sanitary installations, which particularly once they reach puberty denies privacy and exposes them to other risks
- The need to undertake more vigorous processes of empowerment of women in the water management process

12.2.5 *Water–Energy Nexus*

12.2.5.1 Overview: The Economy-Wide Water Nexus

The central concept and origin of the water–energy–food–health nexus arise naturally when viewing the economy as a whole. Throughout the discourse of integrated water resources management (IWRM), the accent is placed on the term "integrated"—except that for a long time it has been circumscribed to the water resources segment of the economy as seen by stakeholders, experts, practitioners, and officials of the water community. The shortcomings of such narrow base have been since clearly recognized, to the point that "thinking out of the water box" and "intersectoral coordination" are an accepted part of the water resources professionals' lore.

However, the needs go much farther than just communicating and coordinating. The complexities of the modern world reflecting growing uncertainties, clashing trends, and the disparate effects of many of the drivers are leading to attempts to identify, characterize, and analyze the concerned processes. The very powerful interconnectivities and interdependences of these processes must be represented competently in order to produce results that approximate reality. This has led to the use of the term "nexus" when referring to multisectoral relationships, that is, more than one-way streets, but a richer connection and interdependence. As indicated in WWAP (2014), "While it is true that all aspects of social and economic development – often referred to as the *food–energy–health–environment 'nexus'* – depend on water, that is only half of the truth; the relationship is one of interdependency." The Budapest Water Summit (Government of Hungary 2013) acknowledged "A cross-sectoral or 'Nexus' perspective integrating water, energy, agriculture and other sectors, as well as ecosystems, should be applied. In this case trade-offs will be identified, synergies seized and resources used more efficiently."

One fact that makes this approach especially relevant to our focus on water resources management is that water is increasingly accepted as a major all-permeating element. For instance, even though the MDGs did not have a single dedicated goal on water, it is evident that water is the common thread that runs through all the goals, whether they deal with the environment, health, education, poverty eradication, or gender equity. At the same time water is impacted by water management and use in the different sectors, individually and collectively.

A key accompanying concept that enriches and gives greater texture to the meaning and implications of the nexus is that of water security—addressed here in Sect. 12.2.2. The World Economic Forum (WEF 2011) articulates this idea expressively when it points out in its introduction: "Water security is the gossamer that links together the web of food, energy, climate, economic growth, and human security challenges that the world economy faces over the next two decades." There is a powerful double-fold message in this statement: (1) it identifies water as the shared element running through these fundamental economic and developmental activities, and (2) the primacy of water is implicit in this context. Beck and Villarroel Walker (2013) argue that attaining water security amounts to achieving basic, single-sector water development as a precursor of the more general self-sustaining multisectoral development. Water is key in achieving security around the nexus of water, food, energy, and climate; hence, it can be treated as "first among equals" (primus inter pares). WEF (2011) stresses that unless water is managed competently across the web of global economy, it may affect adversely water security, tearing into the global economic system.

12.2.5.2 Focus on the Water–Energy Nexus

The World Bank (2013) provides the following definition: "As almost all energy generation processes require significant amounts of water, and water requires energy for treatment and transport, these two resources are inextricably linked. This relationship is the energy-water nexus." The UN system has been keenly aware of this crucial link for a number of years and of the need of incorporating it explicitly in our water management analyses. For instance, the International Hydrological Programme of UNESCO included in its 6-year 7th Phase that started in 2008 a focal area "Addressing the water-energy nexus in basin-wide water resources" (IHP 2007). More recently, the Water–Energy Nexus was the widely visible theme adopted for the UN-organized World Water Day 2014.

WEF explains that currently already 70 % of the world's current freshwater withdrawals are used for agriculture, posing serious limitations for further sustainable increments in food production. Moreover, energy production is the largest industrial user of water, and its expansion requires more access to freshwater. The mining and utility industries withdraw about half of all freshwater withdrawals in the USA today.

By 2030, hydropower will constitute the world's dominant renewable energy source, providing more than twice the amount of onshore wind power. But hydropower consumes water in a nontrivial amount—for instance, it is estimated that to produce one megawatt hour, a hydropower plant evaporates 17 m^3 of water from its reservoir, while the water consumption of a thermal electric plant in its cooling process is but a fraction—of course, thermal plants have other serious environmental implications. Energy production in general impacts water quality. Thermal, chemical, radioactive, or biological pollution can have direct impacts on downstream ecosystems (WWAP 2014). Coping with climate change adds to the complexity, as it affects both water and energy; mitigation measures normally call for the reduction of energy consumption and of carbon emission, while adaptation translates into foreseeing increased hydrological variability and extreme events.

For a more precise understanding of what water use implies, the concepts of water withdrawal and of consumptive use must be clearly grasped. As cited by Brian Richter (2014), water withdrawal is defined as "water diverted or withdrawn from a surface water or groundwater source." Consumptive water use, on the other hand, is defined as "water use that permanently withdraws water from its source; water that is no longer available because it has evaporated, been transpired by plants, incorporated into products or crops, consumed by people or livestock, or otherwise removed from the immediate water environment." Thus, not all water withdrawn is used; as a matter of fact, in many cases, just a fraction of the water extracted is consumed. If this is not kept firmly in mind, many apparently contradictory situations will arise in which totals of water availability, use, and consumption will not add up. This is common where water is recycled/reutilized in a man-made or natural environment.

In the past, the US Geological Survey has periodically estimated the US consumptive water use, but has not done so in the last few years due to lack of resources. The Electric Power Institute (EPRI) has just completed an evaluation of the water consumption of agriculture, municipal supply, and thermoelectric plants in the USA at the national scale (EPRI 2014). Among its findings, as summarized by Richter (2014):

- Only 23 % of water withdrawals is consumed—the rest remains basically available for further use.
- Though withdrawals to electricity-generating plant are 41 % of the total withdrawals in the USA, it consumes less than 5 % of the total consumptive use.
- Irrigated agriculture is the largest consumptive user, nearly two thirds of all water consumed.

From the foregoing discussion, it is clear that the amount of water withdrawn for the various uses compared to availability is enormous and the quantity of water consumed though much smaller is still large. Consequently, for an accurate assessment of actual demand and consumption of water, the water circulation (and recirculation) in natural and man-made systems must be clearly and carefully characterized.

12.2.5.3 Quo Vadis Nexus?

The consideration of the water and energy nexus has added a complex dimension to the already difficult exercise of IWRM. Nevertheless, there is no way to circumvent the issue—either we handle this capably or the consequences can be grave or even catastrophic for the economy of a region, a country, or even the community of nations. Thus, this calls for various measures that will challenge this generation and undoubtedly the next ones. These include:

1. Opening further the "water box" so that the thinking and action behind the water resources management decision-making process embrace the consideration of the nexus with energy (and the other sectors) with an explicit link to the sustainable development process
2. Working with the national policymaking mechanisms (and the corresponding ones at the international and intergovernmental spheres) to set up the proper governance framework, including the integration of the Sustainable Development Goals that will hopefully incorporate water as the bonding element of the various goals
3. Updating the formal framework of IWRM with these considerations

12.2.6 Water Governance and the Enabling Environment

The sustainable implementation of IWRM is not possible without effective and transparent governance structures in place that translate theoretical objectives into practical policies. There are many elements of the enabling environment that are critical to securing a sustainable IWRM framework, including legislation and policymaking, institutions, enforcement mechanisms, and the broader social, political, and economic developments that indirectly affect water management. Because water resources are a key input for so many sectors—including agriculture, industrial and commercial development, municipal supply and sanitation, transportation, recreation, and environmental conservation—a water governance strategy should be coupled with rigorous enabling mechanisms capable of meeting the demands of water users in the present while incorporating the future generations. The IWRM framework does so by promoting three interrelated governance concepts: stakeholder participation and decentralized decision-making, equity of use, and cross-sectoral coordination.

The 1992 Dublin Statement on Water and Sustainable Development established stakeholder participation as one of the four pillars of IWRM. In its explanation of the principle, the agreement claimed that a participatory approach "means that decisions are taken at the lowest appropriate level, with full public consultation and involvement of users in the planning and implementation of water projects" (ICWE 1992). Because IWRM embraces a hydrological approach to water governance, the lowest appropriate level typically means that decision-making should take place at the basin level. Decentralization is attractive for several reasons, primarily because: (1) the governance level can be reduced to reflect environmental characteristics, such as the hydrological borders of a watershed that would otherwise cross administrative boundaries; (2) decentralization promotes community and stakeholder engagement when decision-making is localized; (3) inefficiencies are reduced by eliminating reliance on central government bureaucracies and budgetary constraints; and (4) laws and institutions can be adapted to reflect localized conditions at a scale where integrated natural resources management and climate change adaptation is more focused. However, it is important to note that decentralized water management must take place at an appropriate pace, with mechanisms and tools to support basin-level institutions with the complex and daunting challenges of integrated water management. Otherwise, decentralized water management can have adverse and unsustainable consequences (Stoa 2014).

In many cases, basin-level management requires international or transboundary cooperation. Around 563 rivers, lakes, and aquifers are located in two or more countries (UN Water 2008). Accordingly, it is critical that mechanisms are in place for countries to cooperatively manage shared water resources. In 1997, the United Nations General Assembly approved the Convention on the Law of the Non-Navigational Uses of International Watercourses. Unfortunately, however, it has not received sufficient state ratifications to enter into force as a binding treaty. The United Nations Economic Commission for Europe has its own Water Convention, and while it was born as a regional treaty, it has since expanded to allow ratifications from countries outside the UNECE region. Finally, the International Law Association has developed a set of customary international laws governing water resources, the latest in 2004. Ultimately, the corpus of the international water law is still nascent, and it remains to be seen if the various legal instruments can promote IWRM and sustainable development. Nonetheless, progress is being made toward establishing framework principles capable of providing co-riparian countries with the tools and legal rights and duties necessary to cooperatively manage transboundary water resources.

Next, IWRM promotes equity of use. In legal terms, equity implies a balancing of competing interests that will typically be determined on a case-by-case basis. This is true in the IWRM context as well. Whether legislation proscribes water quantity and quality schemes, administrative agencies devise rules and regulations, or courts settle disputes after the fact, the enabling environment will play a large role in ensuring a fair distribution of water resources between sectors and between humans and the environment. Although a particular use is generally not given priority over other uses, customary water law may suggest that domestic needs—such as water and sanitation—that are necessary to fulfill the basic human needs may take priority over competing uses in line with the Berlin Rules on Water Resources (International Law Association 2004). No other uses, however, enjoy an inherent preference, and the particular characteristics of the water resources in question, as well as the needs of the water users, will dictate judgments of equity.

Finally, IWRM promotes sustainable development by encouraging coordinated lawmaking and regulation across sectors and stakeholders. On the conceptual level, this requires an understanding that water resources management includes surface water, groundwater, riparian land, and coastal and marine environments. Effective governance requires vertical coordination incorporating all governance levels, with mechanisms in place to ensure that decentralized institutions (e.g., river basin commissions) are making allocation decisions that are in line with broader national and regional social, economic, and environmental objectives. Horizontal water management, similarly, requires coordinated interministerial or interinstitutional strategy development in order to spur sustainable economic growth.

Ultimately, the principles of integrated management and sustainable development may themselves already represent customary water laws. While IWRM articulates relatively uncontroversial principles for water management, carrying out those principles in reality requires a robust enabling environment with institutions, laws, and policies that are capable of ensuring stakeholder participation, equity of use, and coordinated water management.

12.3 Final Remarks

The water-related interdependencies and interlinkages between many of the key development aspects are growing more evident as illustrated by the water nexus; thus, water management issues cannot be resolved unilaterally by the water community. The implications for the effective application of IWRM are clear: the IWRM implementation framework must be equipped with the ability to handle the complexities arising from the rapidly shifting conditions of the countries, regions, and the world and the requisite intersectorality.

The intergovernmental processes, particularly through the UN system with the support of a network of collaborating organizations, have resulted from the fundamental importance to thrust forward the IWRM approach. Nonetheless, a considerable effort is still needed in positioning water in the intergovernmental arena so that its importance is properly reflected in the overall sustainable development debate. Two clear examples of its relative marginalization are (1) "underrepresentation" of water in the current MDGs, reduced to a supporting role in the attainment of the drinking water and sanitation targets (in response the freshwater community of the world has launched a strong initiative so that the intergovernmental mechanisms incorporate a dedicated water goal among the forthcoming SDGs that will be adopted for the post-2015 period), and (2) the absence of in-depth consideration by the UNFCCC (United Nations Framework Convention on Climate Change) mechanism of water resources climate change adaptation despite the 20 years of operational history (only in the latest conference of the parties, Warsaw, December 2013), there is a decision to include water resources as one of four focus areas in the climate adaptation implementation program NWP (Nairobi Work Programme).

The two shortcomings in the intergovernmental arena mentioned above, and some others, have to be overcome. The action of the countries ("member states") is called for now; for far too long the communities of water professionals have kept to themselves with little interaction with other sectors limiting its transformational impact. This is a case for the freshwater community to go out of the water box at different levels and get the message out so that it reaches the broader international and intergovernmental venues enabling the changes.

References

Beck B, Villarroel Walker R (2013) On water security, sustainability, and the water-food-energy-climate nexus. Front Environ Sci Eng 7(5):626–639. Higher Education Press/Springer, Berlin/Heidelberg

Bell S, Morse S (2008) Sustainability indicators: measuring the immeasurable? 2nd edn. Earthscan, London

Cap-Net, UNDP (2009) IWRM as a tool for adaptation to climate change training manual and facilitator's guide, Delft, New York

CCSP (2008) The effects of climate change on agriculture, land resources, water resources, and biodiversity in the United States. A report by the U.S. Climate Change Science Program and the Subcommittee on Global Change Research. Backlund P, Janetos A, Schimel D, Hatfield J, Boote K, Fay P, Hahn L, Izaurralde C, Kimball BA, Mader T, Morgan J, Ort D, Polley W, Thomson A, Wolfe D, Ryan MG, Archer MG, Birdsey R, Dahm C, Heath L, Hicke J, Hollinger D, Huxman T, Okin G, Oren R, Randerson J, Schlesinger W, Lettenmaier D, Major D, Poff L, Running S, Hansen L, Inouye D, Kelly, Meyerson L, Peterson B, Shaw R. U.S. Department of Agriculture, Washington, DC, 362 pp

CCSP (2009) Coastal sensitivity to sea-level rise: a focus on the mid-Atlantic region. A report by the U.S. Climate Change Science Program and the Subcommittee on Global Change Research. [Titus JG (Coordinating Lead Author), Anderson KE, Cahoon DR, Gesch DB, Gill SK, Gutierrez BT, Thieler ER, Williams SE (Lead Authors)]. U.S. Environmental Protection Agency, Washington, DC, 320 pp

EPRI (Electric Power Research Institute) (2014) Evaluating thermoelectric, agricultural, and municipal water consumption in a national water resources framework – final report, Palo Alto, CA
EU, Med Programme 2007-2013 (2010) Guideline for developing an indicators system for monitoring sustainability of water management, Project 1G-MED08-515 Water-in-Core
FIU/GLOWS (2013) Report on Panel - women and water: the role of gender equality in defining sustainability of water resources management, Miami (internal)
Government of Hungary (2013) Budapest water summit statement: a sustainable world is a water-secure world, Budapest. http://www.budapestwatersummit.hu/data/images/Budapest_Water_Summit_Statement___Final___11_October_2013.pdf
Grey D, Sadoff C (2007) Sink or swim? Water security for growth and development. Water Policy 9:545–557
GWP (2006) Gender mainstreaming – an essential component of sustainable water management, Policy Brief 3, Stockholm
GWP (2012) Increasing water security – a development imperative, Perspectives paper 2, Stockholm. www.gwptoolbox.org
Hecht JE (2007) Can indicators and accounts really measure sustainability? Considerations for the U.S. Environmental Protection Agency – environmental documents. Paper prepared in conjunction with the USEPA workshop on sustainability. http://www.epa.gov/sustainability/pdfs/hecht-epa-ord-paper.pdf
ICWE (1992) Proceedings, international conference on water and the environment, Dublin, Ireland, 26–31 January
IHP (International Hydrological Programme) (2007) IHP/Bur-XL/11, updated draft strategic plan for the 7th phase of the IHP (2008–2013). UNESCO, Paris
IPCC (United Nations Intergovernmental Panel on Climate Change) (2007) Climate Change 2007: The Fourth Assessment Report (AR4), Geneva, Switzerland
International Law Association (2004) Berlin rules on water resources, London
Richter B (2014) To understand water, learn the math, Water Currents, National Geographic. http://newswatch.nationalgeographic.com/2014/03/28/to-understand-water-learn-the-math/
Rotary International, USAID International H2O Collaboration (2012) Sustainability check tool methodological guide
Setegn SG, Rayner D, Melesse AM, Dargahi B, Srinivasan R, Wörman A (2011) Climate change impact on agricultural water resources variability in the Northern Highlands of Ethiopia. In: Melesse AM (ed) Nile river basin: hydrology, climate and water use, 1st edn. Springer, X, 480 p. 200 illus; Part 4, 241-265. doi:10.1007/978-94-007-0689-7_12
Shilling F, Khan A, Juricich R, Fong V, Hodge D (2012) The California water sustainability indicators framework, California Water Plan, UC Davis, Department of Water Resources – CA, USEPA Region 9
Stoa RB (2014) Subsidiarity in principle: decentralization of water resources management. Utrecht Law Rev 10(2):31–45. http://www.utrechtlawreview.org/index.php/ulr/article/view/267
SWIM-EU (Sustainable Water Integrated Management (SWIM)-European Union) (2013) Guideline for mainstreaming adaptation options in IWRM plans
UNDESA (2008) Climate change and indicators of sustainable development. Power Point presentation by Bruckner M to Conference on Climate Change, Development and Official Statistics, Seoul (Dec 2008)
UNEP (2011) Towards a green economy: pathways to sustainable development and poverty eradication. www.unep.org/greeneconomy, ISBN:978-92-807-3143-9
UNFCCC (UN Framework Convention on Climate) Change (2012) National Adaptation Plans-Technical guidelines for the national adaptation plan process. UN Climate Change Secretariat, Bonn
UN Global Compact (2011) United Nations Global Compact Annual Review 2010, United Nations Global Compact Office, New York

United Nations (1987) Our common future, Report of the world commission on environment and development. Published as Annex to General Assembly document A/42/427, Development and International Co-operation: Environment, New York

United Nations (2012) The future we want, General Assembly Sixty-sixth session. Resolution 66/288, New York

United Nations (2013) The millennium development goals report, New York

United Nations Human Rights Council (2013) Report of the Special Rapporteur on the human right to safe drinking water and sanitation, Catarina de Albuquerque

UN-Water (2013) Water security & the global water agenda. A UN-Water analytical brief, UNU-INWEH, UNESCAP, Hamilton, Canada

UN-Water, UN-DESA, UNICEF (2013) The post 2015 water thematic consultation – Water resources management report, the world we want

UN Waters (2008) Transboundary waters: sharing benefits, sharing responsibilities. http://www.unwater.org/downloads/UNW_TRANSBOUNDARY.pdf

UN WWAP (World Water Assessment Programme) (2014) The United Nations world water development report 2014: water and energy. UNESCO, Paris

USGCRP (2009) Global climate change impacts in the United States. Karl TR, Melillo JM, Peterson TC (eds) United States Global Change Research Program. Cambridge University Press, New York

WEF (World Economic Forum) (2011) Water security – the water-food-energy-climate. Nexus, Washington/Covelo/London

World Bank (2013) Thirsty energy. Water paper, Rodriguez D, Delgado A, DeLaquil P, Sohns A (authors), Water Unit, Transport, Water and ICT Department, Sustainable Development Vice Presidency, Washington, DC

World Water Council, IUCN (2009) Perspectives on water and climate change adaptation. Series of 16 papers

Chapter 13
Water-Resource Management in Mexico Under Climate Change

R.T. Montes-Rojas, J.E. Ospina-Noreña, C. Gay-García, C. Rueda-Abad, and I. Navarro-González

Abstract In Mexico, pressure on water resources exists in all sectors. This, coupled with population growth and the consequent increase in demand for water, could drive communities and cities whose water supply is limited to become more vulnerable in the future. Also, the effects of climate change have been felt through a series of hydrometeorological disasters, such as floods and droughts, that have produced impacts on the economic system and the safety of the population.

The country studies that have addressed the impact of climate change on supply systems and flood management are reduced. It is therefore urgent to give attention to areas of the country that are most vulnerable. But it is also important to address other aspects of water resources that will be affected by the effects of climate change.

This chapter presents some of the studies that have addressed the effect of climate change on water resources. It includes studies examining the effects at the country level and at the local level such as a city. An analysis to determine the future demand and vulnerability of a city due to local scale effects and the feasibility of capturing rainwater as a potential adaptation are presented. The studies focus on the quality and quantity of water that could be expected under future climate scenarios for Mexico. Other studies consider urban, rural, and coastal areas.

Some adaptation measures that could be implemented in the country are suggested in this document.

R.T. Montes-Rojas
General Coordination of Adaptation to Climate Change, National Institute of Ecology and Climate Change, Mexico City, Mexico

J.E. Ospina-Noreña (✉)
Professor Assistant, Faculty of Agricultural Sciences, Department of Agronomy, National University of Colombia, Campus Bogota, Bogotá, Colombia
e-mail: jeospina@atmosfera.unam.mx; jeospinan@unal.edu.co

C. Gay-García • C. Rueda-Abad
Research Program in Climate Change (PINCC), National Autonomous University of Mexico (UNAM), Mexico City, Mexico

I. Navarro-González
Institute of Engineering, National Autonomous University of Mexico (UNAM), Mexico City, Mexico

S.G. Setegn, M.C. Donoso (eds.), *Sustainability of Integrated Water Resources Management*, DOI 10.1007/978-3-319-12194-9_13

Keywords Water-resource management • Mexico • Climate change • Adaptation • Water quality • Water use

13.1 Introduction

According to the Intergovernmental Panel on Climate Change (IPCC) (2007), the continuous increase in greenhouse gases in the atmosphere can cause an increase in the average global temperature of between approximately 1.8 and 4.0 °C. As a result, the hydrological cycle and the water-supply reliability will be affected by changes in precipitation levels, modifications of the temporal and spatial distribution of rainfall, and the increase or acceleration of evaporation due to higher temperatures (IPCC 2007; Bates et al. 2008).

During the twentieth century, there was an increase in rainfall in the regions located in northern latitudes and a decrease in regions located between 30°N and 10°S. Moreover, precipitation projections for the twenty-first century in climate-change models predict increased precipitation in northern latitudes and decreased precipitation in the tropics and subtropics. Mexico is located in the tropics and subtropics of the Northern Hemisphere, where rainfall is expected to decrease during the twenty-first century (Bates et al. 2008).

In addition, the availability of drinking water may be affected by the decline in water quality due to severe runoff or decreased dilution capacity, increased saline intrusion due to sea-level rise, modification of groundwater reserves, and damage to hydraulic infrastructures by more intense and frequent extreme weather events (Kundzewicz et al. 2008). This situation particularly affects countries and regions that are already suffering water shortages. The impacts on water resources will affect all sectors of society and the environment itself.

According to the 3rd chapter of the IPCC Working Group II (WG II) (2007), there are both climatic and non-climatic drivers of water quantity and quality, in addition to other factors acting on water resources.

The main climatic factors that determine water quantity and quality are precipitation, temperature, and the demand for water resources involved in the processes of evaporation. The latter, in turn, is determined by the net radiation at ground level, humidity, wind speed, and temperature. Temperature is especially important in snow-affected watersheds and in coastal areas (IPCC 2007).

Some of the main non-climatic drivers of water quality and quantity are changing land use, the construction and management of reservoirs, pollutant emissions, and wastewater treatment. Additionally, water demand is driven by demographic changes, food consumption, politics, economics, technology, lifestyle, and priorities in water consumption (IPCC 2007).

There is an expected increase in both the use of wastewater, with the inherent risk of being unable to remove all of the pathogenic microorganisms, and the use of desalination techniques, with adverse effects such as the presence of other chemicals and high energy consumption. The use of these water-supply techniques will most likely increase in semiarid and arid regions (IPCC 2007).

An important contribution to the understanding of the interconnection of water resources to climate variations was made by Kundzewicz et al. (2008). Various key points brought up in his work are presented in this chapter.

The sensitivity of water resources to climate change impacts is high. At higher temperatures, the atmospheric retention of water increases because evaporation is higher, leading to higher climate variability. Temperature increases are expected throughout the world, together with changes in precipitation, introducing uncertainty in sensitivity analyses of water resources. The semiarid and arid areas of non-developed countries are particularly vulnerable because rainfall and river flow are concentrated in a few months and register large annual variations (Lenton 2004).

River flow and water levels in lakes and wetlands will depend on the changes in the volume, duration, and intensity of precipitation. Changes in temperature, radiation, humidity, and wind speed affect the potential evapotranspiration. The latter change can offset small increases in precipitation and further aggravate the effect of decreasing surface-water runoff. Consequently, the river flow and volume of water in lakes and wetlands have been reduced mainly due to water demand and reduced precipitation (Kundzewicz et al. 2007).

Groundwater responds to climate change more slowly than surface water. Groundwater levels have been reported to be strongly correlated with precipitation. Additionally, brackish-water intrusion has been registered as a consequence of reductions in water recharge and the infiltration of seawater. However, there is a lack of data to model groundwater resources in relation to climate change, and the available databases are poor (Kundzewicz et al. 2007).

Other important aspects are climate change-driven droughts and floods. The latter depend on the intensity, volume, and timing of precipitation as well as on the biotic and physical characteristics of the watersheds. Climate can have a direct impact on increased flooding. The number of major disasters for the period 1996–2005 was twice the number that occurred between 1950 and 1980, and the affected population is estimated at 140 million people/year on average (WDR 2003, 2004).

Meanwhile, droughts affect agricultural production and crop yield as well as water supply for domestic, commercial, industrial, and other productive sectors. The semiarid regions and sub-humid areas of the world, such as Australia, the western USA, southern Canada, and the Sahel, have registered more intense multiyear droughts, and these areas are highly vulnerable to projected climate change. It is expected that extreme drought conditions will increase by 30 % by the end of the century (Kundzewicz et al. 2007).

13.2 Water Quality

In many lakes and reservoirs of the world, climate-change effects are mainly due to variations in water temperature affecting oxygen regimes, oxidation/reduction (redox) reactions, stratification, mixing rates, and the development of biota.

For example, increasing the temperature decreases the depuration capacity of rivers by reducing the amount of dissolved oxygen, which is used for biodegradation. An increase in heavy precipitation leads to increased nutrients, pathogens, and toxins in water bodies (Kundzewicz et al. 2007). The presence of pathogens in the water supply has been associated with extreme rainfall events (Yarze and Chase 2000; Curriero et al. 2001; Fayer et al. 2002; Cox et al. 2003; Hunter 2003).

The effects of dry periods on water quality have not been properly studied yet. This topic is a priority research area due to the predicted increase in aridity in several regions of the world.

Globally, health problems due to arsenic and fluoride in groundwater are more important than those related to other chemicals. Examples of affected regions are India, Bangladesh, China, North Africa, Mexico, and Argentina, with over 100 million people suffering from arsenic poisoning and fluorosis (United Nations 2003; Clarke and King 2004).

A quarter of the world's population lives in coastal regions, which are characterized by water shortage. Rapid population growth, increased salinization of water sources, and reduced freshwater availability are predicted for these regions (Small and Nicholls 2003; Millennium Ecosystem Assessment 2005). Studies conducted in different parts of the world have found that a third of the urban water-supply systems in Africa, Latin America, and the Caribbean and more than half of such systems in Asia are intermittently available during dry periods. This pattern has a negative effect on the water quality of these systems (WHO/UNICEF 2000).

13.3 Water Uses, Availability, and Stress

There are many regions in the world suffering from water stress. The process has been documented and studied in watersheds of Africa, the Mediterranean region, the Near East, the south of Asia, northern China, Australia, the USA, Mexico, and northwestern Brazil (Kundzewicz et al. 2007).

For many of the processes related to climate change (e.g., increases in temperature), there is documented evidence that the changes have already begun. However, the information on the derived impacts of such changes is still uncertain and has been poorly documented (Cromwell et al. 2007). The direct impacts on water systems and supplies due to climate change are as follows (adapted from Cromwell et al. 2007):

(a) Warmer and drier seasons:

Changes in watershed vegetation and in the aquifer-recharge areas

Aquifer-recharge alterations
Changes in the quantity and quality (e.g., total organic carbon and alkalinity) of runoff entering surface waters

Higher water temperature

Evaporation and eutrophication increases for surface water
Lower reservoir levels

Increased temperature in shallow water bodies
Increases in evaporation and eutrophication

Challenges for water purification and distribution (e.g., disinfection, chlorination by-products, and regrowth)

Increasing water demand

Increased water demand for irrigation
Increased urban water demand, heat waves, and droughts
Increased loss of groundwater level needed to meet the demand

(b) More intense precipitation events:

Increased turbidity and sedimentation, challenging water purification
Challenges to water-filtration treatments
Increased risk of direct flood damage to the water-distribution facilities

(c) Less precipitation:

Reduced availability of surface water
Reduced availability of water for groundwater recharge
Increased difficulty in meeting environmental flows in surface waters

(d) Sea-level rise:

Increased saline intrusion into aquifers

Challenges to water purification, increased bromide concentrations, and demand for water desalination

Increased salinity in brackish coastal water bodies

Challenges to water purification, increased bromide concentrations, and demand for water desalination

One of the most important findings of Bates et al. (2008) is that the global warming observed in recent decades is linked to large-scale changes in the hydrological cycle, such as changes in vapor content in the atmosphere, precipitation patterns, rainfall intensity and frequency of extraordinary storms, snowpack depth, glacier cover, soil moisture, and runoff processes.

13.4 The Case of Mexico

In Mexico, the agricultural, industrial, urban, and tourism sectors all exert great pressure on water resources. Continued population growth causes increased water demand, making those communities and cities whose water supply is limited more vulnerable. In addition, the effects of climate change have emerged as a sequence of hydrometeorological disasters (e.g., floods and droughts) with serious social, ecological, and economic consequences and impacts on the economic system and the safety of the population. For example, the droughts of 1997, 1998, and 2005 led to forest fires and crop loss. Additionally, extreme rainfall events have increased since 1999, causing great human and economic losses, especially in the southeast (Soto and Herrera 2009).

There are a limited number of studies of the impacts of climate change on water-supply systems and flood management in Mexico. Moreover, these types of studies are especially needed for the most vulnerable areas.

In Mexico, the average water availability per capita has fallen by 66 % in 50 years, from 11,500 m^3 in 1955 to 3,822 m^3 in 2005. In only 5 years, the water availability was reduced by 22 %, from 4,900 m^3 in 2000 to 3,822 m^3 in 2005. The mean water availability is expected to be 3,610 m^3 in 2012, 3,285 m^3 in 2030, and 3,260 m^3 in 2050, assuming no substantial changes in precipitation and constant population growth (SEMARNAT 2009). However, climate projections indicate that as global warming increases, leading to less frequent and more concentrated precipitation, the average annual water availability per capita will decrease rapidly and at an increasing rate, especially in the arid and semiarid regions of the country (González et al. 2011).

Furthermore, the overexploitation of aquifers, the lack or inadequate treatment of municipal and industrial wastewaters, and the lack of efficient technologies in the agricultural sector, among other factors that impair water quality and quantity, increase the future vulnerability of water resources. Water conservation will largely depend on how water resources are managed as well as on the implementation of mitigation and adaptation programs in every economic activity generated in the river basins (González et al. 2011).

The risk corresponding to the management of water resources in Mexico has been addressed by institutions such as the Mexican National Center for Disaster Prevention (CENAPRED, based on its initials in Spanish) and the Mexican National Civil Protection System (SINAPROC, based on its initials in Spanish). These institutions have studied and characterized flood hazard and risk zones in Mexico, developing state and local maps for areas where floods occur frequently (CENAPRED 2001). These maps include physical and social vulnerability factors, which determine the risk levels among sites. However, as mentioned by Herzer (1998), it is often difficult to select the social, economic, and environmental factors that will define risk (González 2011).

Below are some examples of studies on the effects of climate change in Mexico.

13.4.1 Water Demand in Mexico City

In many cities of Mexico, water availability and pollution problems are becoming increasingly severe, putting the future provision of water at risk and making the population vulnerable to problems with the future water supply. The government authorities have acknowledged that the Metropolitan Area of the Valley of Mexico (MAVM) is the region of the country with the most serious problems of sustainability regarding water-resource management. Due to the deterioration of the water supply, it is of great importance to know the additional risk imposed by changes in rainfall patterns and temperature. Therefore, it is necessary to generate future climate-change scenarios to assess vulnerability to extreme events and to propose strategies to address these adverse situations (Soto and Herrera 2009).

Mexico City is the second largest city in the world, with 19.2 million inhabitants (SEDESOL-CONAPO-INEGI 2007). As a whole, Mexico City consumes 61 m^3/s of water. More than 80 % of the water volume distributed in Mexico City is intended for domestic purposes. To meet that demand, 67 % of the water consumed comes from the local aquifer, and the remaining is mainly supplied from the state of Mexico. According to estimates, the consumed water volume exceeds the natural recharge capacity of the aquifer by between 40 and 80 %, making groundwater a nonrenewable resource. The scarcity of sources to meet the demand of the current population and its future needs, coupled with the potential pressures derived from climate change, might cause a further depletion of water resources, driving MAVM to a crisis. Acknowledging this problem, Soto and Herrera (2009) evaluated the potential effects of climate change on water resources to generate information that could help in the design of adaptation strategies for efficient water management.

The aforementioned study used the GFDLCM2[1] model, which is one of the models that best reproduces the historic climate conditions of Mexico City. The model was run with available daily data on maximum and minimum temperature and precipitation. The analysis was conducted for the 2046–2085 period for the A1B[2] and B1 scenarios. The data set was retrieved from the World Climate Research Program (WCRP) Coupled Model Intercomparison Project Phase 3 (CMIP3) multi-model dataset (http://www-pcmdi.llnl.gov/ipcc/about_ipcc.php). The methodology proposed by Arnell et al. (1997) and Yusoff et al. (2002) was followed to develop scaling factors. These factors were applied to the series of future scenarios. The adjusted series were used to calculate the potential

[1] The Centro de Ciencias de la Atmósfera from the Universidad Nacional Autónoma de México (CCA-UNAM) analyzed 22 general circulation models for evaluating the performance of the different models to reproduce the observed climate at global scale and for the region of México (Conde et al. 2011); the statistics used were pattern correlation (*r*), root mean square error (RMSE), and bias and root mean square error corrected by bias (RMSE-corr) (Wigley 2008), and taking into account the representativeness criterion suggested by IPCC-TGICA (2007), three models were chosen to reasonably represent the uncertainty range. These models provide a broad range of possible temperature increases and, more importantly, they provided increases as well as reductions in precipitation. These models are ECHAM5, HADGEM1, and GFDLCM2.

[2] *A1*: Very rapid economic growth; population peaks mid-century; social, cultural, and economic convergence among regions; market mechanisms dominate. Subdivisions: *A1FI*, reliance on fossil fuels; *A1T*, reliance on non-fossil fuels; *A1B*, a balance across all fuel sources.

A2: Self-reliance, preservation of local identities, continuously increasing population, economic growth on regional scales.

B1: Clean and efficient technologies; reduction in material use; global solutions to economic, social, and environmental sustainability; improved equity; population peaks mid-century.

B2: Local solutions to sustainability, continuously increasing population at a lower rate than in *A2*, less rapid technological change than in *B1* and *A1* (Hulme et al. 2002).

evapotranspiration following the Hargreaves method. Finally, the actual evapotranspiration and effective precipitation were estimated using the improved soil moisture balance method.

The effects of climate change simulated by the GFDLCM2[3] model for the period 2046–2085 for the A1B scenario show a decrease in mean annual precipitation and an increase in potential evaporation, resulting in a decrease in effective precipitation. The simulation for the B1 scenario shows increased precipitation and decreased evapotranspiration, resulting in an increase in the mean effective annual precipitation. This increase would be expected during the rainy season, when the mean temperature could decrease. The results of these simulations would result in a volumetric flow rate of 15.54 m^3/s for the A1B scenario and 20.17 m^3/s for the B1 scenario in the MAVM urban area and of 11 m^3/s in the conservation area for B1 scenarios.

Two consequences can be predicted based on these results, which can have a significant impact on the water-supply system and the sewer service:

1. Increased water demand due to higher mean temperatures in the Valley of Mexico.
2. Changes in rainfall that would differentially affect the uses of urban land and conservation land. In the conservation areas, the major impact would be on the amount of recharge water in the aquifer, while in urban areas, the main impact would be on the quantity of water received by the rainwater-drainage system.

Water-consumption data in the domestic sector in Mexico for the period 2004–2008 were analyzed to study the first consequence. The aggregated data did not show the expected trends, as several municipalities exhibited reductions in the water volume consumed during the dry season. Additionally, the reduction in the flow levels in the Cutzamala system also had an impact on water supply during the dry season in 2009. The data confirm that demand is limited by reductions in supply. This phenomenon was particularly evident for the Benito Juárez (2004 and 2006), Coyoacán (2004 and 2006), and Cuauhtémoc (2007 and 2008) municipalities.

The information from those municipalities that did show the expected trend during the dry season was analyzed to explore the relationship of temperature with increasing water demand. The most representative and consistent cases were Gustavo A. Madero (2006 and 2008), Miguel Hidalgo (2004, 2006, 2007, and 2008), Azcapotzalco (2005 and 2006), and Álvaro Obregón (2006, 2007, and 2008).

When analyzing the mean water consumption of the selected municipalities, a mean increase of 2.3 % was registered for the cold years, and an 8.3 % increase was observed during the warm year. However, it is worth mentioning that in the Gustavo A. Madero municipality, water demand increased by 19 % in 2006. In the B1 scenario, an annual increase of 30×10^6 m^3 was predicted (an additional 9.83 %),

[3] For further information about this model, see Stoufferr et al. (2006), Wittenberg et al. (2006), Gnanadesikan et al. (2006), and Delworth et al. (2006).

with the greatest impact in the first half of the year. The A1B scenario predicted an increase of 93×10^6 m^3 (an additional 30.3 %), with a constant increase in water demand throughout the year.

The water-supply reduction registered in some municipalities could be extended to other regions in the future, leading to continuous and severe water scarcity during the dry season in the B1 scenario and during most of the year in the A1B scenario.

Regarding the second consequence, rainwater has two main destination points: aquifer recharge in conservation soil and drainage in urban land. This pattern means that for the A1B scenario, which presents conditions more similar to the present ones, the rainwater would be reduced by 5 % in the urban drainage system, moderately reducing the problems associated with flooding. In soil conservation reservoirs, the volume of natural recharge would be reduced by 6 %. In the B1 scenario, in which precipitation patterns are substantially modified, the amount of water captured in the conservation soil would increase. The estimated actual recharge value is 6.68 m^3/s, and under this scenario, it would be possible to increase the recharge to 11 m^3/s, which represents a 66 % increase. From the increase in infiltration, the authorities may decide not to remove the surplus to ease the situation of overexploitation of the aquifer or to have an additional 4.4 m^3/s input to the water-supply system. Regarding rainwater in urban areas, the actual flow in drainage systems is 16.33 m^3/s. Under this scenario, the flow would increase to 20.17 m^3/s. This change represents an increase of 23 % and could lead to increased flood risk and associated problems.

In spite of the uncertainties, the study provides evidence of the low resilience of MAVM. This low resilience means that there is a risk of water-supply reduction that can threaten the access of the population to water resources, even without taking climate-change impacts into account. A change in precipitation pattern would differently affect conservation and urban soils. In the conservation soil, the amount of recharge water in the aquifer would be impacted; in the urban soil, the rainwater received by the drainage system would be impacted. One of the scenarios shows a potential risk of inadequate drainage infrastructure, with associated flooding and civil protection risks (Soto and Herrera 2009).

13.4.2 *Vulnerability of Mexico City*

Ospina et al. (2014a) described the most important variables regarding water availability in Mexico City, creating vulnerability indices using *historical data* from climate and hydrometric stations. The indices were derived from *actual and retrospective analyses*. The future water availability under climate change was estimated by drawing baselines and taking into account uncertainty regarding hydrological processes and modeling.

Mexico City has two clearly different climatic zones: dry and wet. For each zone, weather stations providing consistent data with existing records and periods from hydrometric stations were selected. In the case of the dry zone, the 1976–1988 period was selected as having consistent data from nine weather stations and three hydrometric stations. For wet areas, the 1973–1988 period was selected as having consistent data for nine climatological and hydrometric stations. Records of daily maximum temperature (T_{max}), average temperature (T_{avg}), minimum temperature (T_{min}), evaporation (Evp), and precipitation (PP) were retrieved from the weather stations, while river surface flow records were obtained from hydrometric stations; these gauging stations are located in the south and west of the city of Mexico. Results are provided mainly for the wet area, with some comparisons to the dry area.

The analysis of historical data showed that T_{max}, T_{avg}, and T_{min} were much higher in the dry areas than in the wet, while precipitation was lower in the dry area; therefore, the dry area shows a more deficient and critical condition in terms of natural water availability.

The results show that as PP decreases, Evp increases, a behavior that could be related to the increase in temperature, changes in land use, and loss of systems that regulate the water cycle, such as forests.

A *multiple-variable analysis* was conducted to look for correlations between the variables. The pairs of variables showing strong correlations were T_{max} vs. T_{avg}, T_{max} vs. Evp, T_{min} vs. T_{avg}, T_{min} vs. PP, T_{med} vs. Evp, and PP vs. flow in the wet zone (Q_Z_W). Additionally, the different temperatures were correlated to each other and with evaporation and temperature. As expected, flow registered a strong linear relationship with flow. A multiple correlation analysis was performed between flow (Q) (as the dependent variable) and T_{max}, T_{avg}, T_{min}, PP, and Evp (as independent variables) to find the best-fitting model. The resulting model confirmed the strong relationship between PP and Q in that zone based on the 95 % confidence level.

Once the relationship between the variables was established, the sensitivity of water-resource availability was analyzed in each of the zones. The vulnerability index was built using the variables retained by the model (Ospina et al. 2014b).

When applying the models to the data obtained from each of the selected stations as representatives of the climate conditions of the area, it is assumed that the obtained results represent the real sensitivity to water availability. In the dry area, the percentage of reduction in flow (Q) was equal to the percentage of reduction in PP, while in the wet area, the flow experienced a 60 % reduction compared to PP. The runoff in the dry areas is 0.11 m^3/s and represents 22 % of the runoff in the wet area (0.50 m^3/s).

The vulnerability index was built taking into account the relationship between variables and the sensitivity analysis.

Therefore, the vulnerability to water availability was estimated for each year or studied period, allowing inference of the past, actual, and future impact (high, moderate, or low) on the system.

The index can be generated from the sensitivity analyses, taking into account some additional criteria:

In the wet area, the reduction in Q is directly proportional to the reduction in PP, representing 60 % of the PP reduction. Additionally, the effect on the flow rate caused by the reduction in PP is less severe than that reported for the dry zone, where changes in Q are directly proportional and of equal magnitude to the percentage of change in PP.

This pattern allows for certain ranges of influence of rainfall on runoff and water availability to be set for both of the zones: low, moderate, high, and very high for the dry zone and slightly low, low, moderate, high, and very high for the wet zone. In the wet zone, the index ranged from 0.07 to 0.33. Within this range, the vulnerability to water availability was established as follows: low vulnerability, values lower or equal to 0.15; moderate vulnerability, values ranging between 0.16 and 0.24; and high vulnerability, values equal or higher than 0.25.

Given the water deficit registered in Mexico City, any water reduction is critical in both of the zones. Although the modeled reductions in precipitation and flow might account for some errors, the estimates can be adjusted to be more precise and objective. The estimates can also be adjusted to the experts' judgment, to policy strategies, or to specific water-resource management plans.

According to the results, the runoff in Mexico City is directly dependent on rainfall, in both the dry and the wet zones. The most adverse weather conditions occurred in the dry zone, due to higher maximum, mean and minimum temperatures, increased evaporation, and lower precipitation. All of these factors directly affect runoff and water availability in the dry zone, where the volumes are lower than those recorded in the wet zone.

13.4.3 Vulnerability of the Tamaulipas Coastal Zone

González et al. (2011) developed a methodology for assessing the vulnerability to climate change of water availability in coastal watersheds following the method proposed by Rivas et al. (2010). This methodology was applied to the coastal zone of the state of Tamaulipas. The study area corresponded to the hydrological basins flowing into the coastal zone located within the hydrological regions RH 24 Río Bravo-Conchos, RH 25 Río San Fernando-Soto La Marina, and RH 26 Río Pánuco. The water availability was assessed prior to the vulnerability assessment, considering future changes in precipitation under climate-change scenarios A1B and A2 and considering a time horizon of 10 years for the period 2010–2060. The variables selected to assess the water vulnerability in the watershed were grouped in three categories: *the exposure degree*, *the susceptibility or sensitivity degree*, and *the adaptation capacity*.

Three indices were used to measure the *exposure degree*: (1) the precipitation index under climate change (PICC), which seeks to identify those watersheds impacted by changes in precipitation; (2) the index of surface and groundwater runoff under climate change (ISGRCC), which captures the reduction in the average annual runoff volume relative to the historical runoff; and (3) the index of water availability under climate change (IWACC), which estimates the mean annual decrease in available water.

The *susceptibility or sensitivity degree* comprises two indicators: (1) the exposed population, which represents the number of people likely to be affected by the lack of water, and (2) the water consumption index (WCI), which identifies the water uses; identifies and evaluates water scarcity, prioritizing water-resource needs; and allows the monitoring of the situation of societies facing water scarcity to design public policies for water management.

The *adaptation capacity* identifies the elements exposed to modifications in water availability and seeks to identify the resilience of the watershed to decreased water availability due to climate change. This capacity comprises three indicators: (1) the aquifer exploitation index (AEI) represents the degree of exploitation of the aquifer; (2) the social exclusion index (SEI) represents the degree of urban infrastructure that facilitates coping with the adverse effects of climate change; and (3) the water reuse index represents the sustainability effort and captures the percentage of reclaimed water.

The methodology was applied following four sequential steps: (1) identification of the variables described above, (2) calculation and normalization of the indices, (3) weighting of the indices, and (4) calculation of global vulnerability.

The index weighting is performed following an analytic hierarchy process (AHP) developed by Saaty (2001).

This proposal represents a first approach to the assessment of vulnerability to changes in water availability due to climate change.

The authors continue the research and are still working on finding other indicators and the weights corresponding to each variable and group of variables. Currently, they are performing water availability modeling and identification of urban, agricultural, livestock, and industrial zones, with the purpose of initiating a full assessment.

13.4.4 Rainwater Harvesting

Sanchéz et al. (2011) assessed the required infrastructure for water supply based on a system for rainwater harvesting from the main building of the Earth and Lime Institute, A.C. (ITyC), in San Miguel Allende, Guanajuato.

The aim of this study was to determine the dimensions of the required infrastructure for drinking water supply based on a system of rainwater harvesting located in the main building of the ITyC. An analysis of climate records and climate modeling was conducted in the study area. Then, the size of the water-storage tank was estimated together with the percentage of reuse of treated wastewater needed to meet the daily demand of water users in the main building of the institute.

Thirteen stations were selected within the study area to analyze the mean monthly precipitation. The historical records of these stations were obtained using the CLICOM Quick Finder on Climate Information (ERIC) III, the CONAGUA Climatological Station Network, and National Weather Service records. The quadrant method was then used on these historical records to obtain the mean monthly precipitation in the ITyC. The referred study followed Mihelcic et al. (2009), who, after studying rainwater-harvesting facilities around the world, reported that this type of system is feasible in regions with an average annual precipitation equal to or greater than 400 mm. The ITyC matches these criteria because the annual average rainfall in the site is 505.63 mm.

To observe the variation in the amount of rainwater that could be harvested under climate-change scenarios, the annual outputs (every 5 years) of the UKHADGEM (Martin et al. 2006; Ringer et al. 2006), UKHADCM3 (Gordon et al. 2000 and Pope et al. 2000), MPIECH-5 (Jungclaus et al. 2006), GFDLCM21 (Delworth et al. 2006) y Gnanadesikan et al. 2006), MIROCMED (Abe-Ouchia et al. 2004), CSIRO-30 (Gordon et al. 2002), and CCCMA-31 models were used, as were the A2 and B2 scenarios using the MAGICC-SCENGEN. V.5.3 at coordinates 21.25 north and 101.25 west, which are the closest to the institute's coordinates. This analysis was performed to estimate the increase or reduction in water resources. Because these scenarios do not provide information related to infrastructure-design programs, the integrated and efficient use of water-resource programs, or appropriate irrigation schedules regarding plant crops, the authors propose to generate a weighted average scenario taking into account all the models and scenarios because all may have the same probability of occurrence. In this case, the most adverse effects would be related to decreased precipitation. For this purpose, the four models showing such behavior under scenarios A2 and B2 (CSIRO-30, CCCMA-31, UKHADGEM, and MPIECH-5) were analyzed. The results of each model were analyzed by season—winter, spring, summer, and autumn—and by decade. Finally, the monthly outputs of the most adverse scenarios (corresponding to the CCCMA-31 and CSIRO-30 models for the A2 scenario) were analyzed. Regarding the B2 scenario, the CSIRO-30, CCCMA-31, UKHADGEM, and MPIECH-5 models were the most adverse for rainwater harvesting in this case study.

According to the analyses of the weighted average, the greatest reduction (19.6 %) was expected during summer, and the annual reduction could be approximately 14.2 %.

The authors recommend implementing infrastructure design and integrated and efficient use of water-resource programs together with irrigation schedules according to the species that will be planted, taking into account the minimum and maximum available water and the maximum decrease expected under climate-change scenarios. That is, the results for winter and summer seasons are used as reference values, while accounting for the weighted average scenario.

The water-harvesting system was sized according to the most adverse climate-change scenario (A2) and the CCCMA-31 model. The available area in the main ITyC building was 1,045.60 m^2. The runoff coefficient was 0.85. The system was designed to supply 70 people with 50 L/day of water. The estimated volume for washing the roof was 40.0 L/100 m^2. The monthly water demand in the main ITyC building for January, March, May, July, August, October, and December was 108.50 m^3; for April, June, September, and November, the demand was 105.00 m^3; and for February, the demand was 98.00 m^3. The authors estimated the hydraulic balance of the availability of rainwater harvested using the roof of the

Table 13.1 Results from the analysis for the sizing of the hydraulic system of drinking water for the Earth and Lime Institute, San Miguel Allende, Guanajuato

	Models not taking into account climate change	CCCMA-31 model, accounting for the A2 climate-change scenario
Optimum percent of treated WWR (%)	69.0	71.0
Maximum monthly WWR volume (m^3)	74.87	77.04
Mean monthly WWR volume (m^3)	73.46	75.59
Mean monthly influent into the WWTS (L/s)	0.34	0.35
Mean monthly volume in the storage tank (m^3)	346.41	355.28
DWSS efficiency (%)	97.50	95.0
Maximum monthly volume of excess water (m^3)	40.16	112.84
Proposed volume for the storage tank (m^3)	450.0	450.0

WWR wastewater reuse, *WWTS* wastewater-treatment system, *DWSS* drinking water-supply system

main ITyC building during the 2001–2030 period. A monthly analysis of the historical precipitation records projected into the future (not accounting for climate change) was conducted based on the climatic station 11051 (Peñuelitas, Dolores Hidalgo, Guanajuato) located 10.9 km away from the study area.

The following parameters were calculated: the optimal reuse percentage of treated wastewater, the maximum monthly volume of wastewater reuse, the average monthly influent entering the wastewater-treatment system of the institute, the mean monthly volume stored at the end of each period, the efficiency of the drinking water supply, and the maximum monthly volume of excess water. Table 13.1 shows the results of the analysis performed with and without climate change.

The approach to estimate the weighted average of climate-change projections from different general circulation models was suitable for preliminary studies that require an estimation of an average value (weighted) of the temperature and precipitation projections. This method can also be applied to guide decisions in feasibility studies for hydraulic infrastructure sizing.

The methodology proposed and applied in this study and the modeling process for capturing rainwater are simple and easy to apply. Additionally, the method allows the establishment of the hydraulic design parameters required for the sizing and design of water infrastructure. An advantage of this modeling process is that it can be applied to any type of installation that includes the capture of rainwater as a source of drinking water.

The results of climate-change modeling for the Guanajuato region show reductions in precipitation. In the most extreme scenario (CCCMA-31 model, A2 scenario), the reductions are −5.6 % for the year 2020, −7.0 % for the year 2025, −8.7 % for the year 2030, and up to −17.34 % for the year 2060. Unfortunately, projections of future precipitation for this region are not optimistic, nearing the minimum value of annual rainfall that can be considered to implement rainwater-harvesting systems. The authors believe that eventually, rainwater-harvesting and wastewater reuse and treatment systems should be implemented for the entire Mexican plateau region because such systems will be an essential element for the supply of drinking water.

13.4.5 Water Footprint of Ethanol Production

Biofuels offer advantages but also have negative impacts on the environment and food security. Haro (2012) estimated the water footprint for the production of ethanol from sugarcane under various climate-change scenarios. Biofuel production has an impact on water resources because large amounts of water are consumed for biofuel production. The author quantified the water used for sugarcane irrigation in Tamazula (Jalisco). This site has one of the best national yields in sugarcane production, with modern and efficient facilities. A water-balance model was developed for the specific conditions of the site. In addition to assessing the impact of the water footprint of ethanol production, the water availability and demand for other uses, such as the production of molasses for ethanol, were estimated.

The water footprint was estimated following Hoekstra (2007). This approach requires the calculation of sugarcane evapotranspiration. This parameter was calculated using the Penman-Monteith method (FAO 2006). The water availability within the region was estimated as the difference between the precipitation and evapotranspiration volumes. The latter was estimated using the Turc model.

Estimates of this investigation indicate that the water footprint for ethanol production in the region of Tamazula was 99.8 m^3/t in 2010. This value could be increased on average up to 100 and 102 m^3/t by 2020 and 2050, respectively, according to the general circulation GFDLCM21 (Delworth et al. 2006) and ECHAM5 (Jungclaus et al. 2006) models under climate-change scenarios A2 and B1. This low increase can be explained by the estimations of precipitation increase during the wet season and precipitation reduction during the dry season according to the GFDL and ECHAM5 models. Therefore, there will be more water available during the last 4 months of the irrigation period.

It was also estimated that the total water demand for the production of this biofuel (137 million m^3 for each agricultural cycle) has a serious impact on the availability of water, placing the region in extreme water stress (stress index of 37 %) in the year 2010. This situation corresponds to a high-pressure level and confirms the estimates of the basin organization. A decrease in water availability of 20 % by 2020 and 30 % by 2050 is expected. The future water stress will also be affected, with a 39 % reduction by 2020 and 58 % by the year 2050.

The author emphasizes the lack of historical data for both crop production and climate because these data could reduce the uncertainty of the water-footprint estimates. One of the variables that should be better estimated is evapotranspiration because it depends on precipitation and temperature. Another important variable for estimating the water footprint is the type of irrigation. Therefore, the author suggests including this parameter in the estimation when possible to guarantee more accurate results.

A conclusion of this study refers to the use of water for sugarcane production as much greater than that required to obtain sugar ethanol. Therefore, agricultural water demand is the most important category in terms of water footprint and has the highest impact on water availability.

13.4.6 Coastal Municipalities

Given that the flood risk is related not only to natural causes but also to anthropogenic activities, efforts should be concentrated on the analysis of the latter, requiring in turn the development of methodologies to address, mitigate, or prevent anthropogenic impacts. Clearly, this action is a priority to confront climate change. Regarding the assessment and management of flood risk in coastal areas, González (2011) developed a methodology that addresses flood assessment and risk management from a holistic perspective. The methodology includes physical risk, economic and social variables, and the responsiveness of the authorities and the affected population. This approach also identifies circumstances that favor the transformation of an intense phenomenon into a disaster to anticipate such circumstances and be able to intervene, reducing the impact of future hazard events. In this sense, the proposed methodology adds additional environmental complexity to the study of flood risk under climate-change scenarios. The method addresses both the conservation and use of exposed areas and the need to establish adaptation, mitigation, and risk-reduction strategies.

This study defines the *flood-risk index* (FRI) as a function of physical risk and existing or prevailing vulnerability. The FRI includes two indices: the *physical risk index* (PRI) and the *prevalent vulnerability index* (PVI). The former index is estimated from the physical damage to the territory and society. The second index is estimated from three categories of vulnerability: exposure and physical susceptibility, socioeconomic fragility, and lack of resilience.

The FRI is based on the assessment of the effects or economic impacts generated by a natural disaster according to the experience of the Economic Commission for Latin America and the Caribbean (CEPAL 2003). The FRI results from the product of the physical risk-associated indicators and their weight. The authors propose to calculate FRI as the product of each risk descriptor multiplied by its associated weight. The PVI is the weighted average of three vulnerability categories: the physical exposure and susceptibility, the socioeconomic fragility, and the lack of resilience. Each of these categories is weighted according to their importance to global vulnerability, and the descriptors are the components of each category. The weights must capture the importance of each descriptor in the determination of each index. The weights are intended to reflect the expert knowledge. For this weighting, the AHP is advised.

The methodology for the assessment and management of flood risk was applied to the municipality of Pueblo Viejo, which is located in the state of Veracruz, on the Gulf of Mexico coast. This municipality is constituted as an urban center, consisting of a set of suburban areas. The municipality has 73 localities, 5 suburban and 68 rural, and it has a population of 52,593 inhabitants. This municipality has a lagoon surface area of 74.49 km^2, divided into four water bodies. The most important water body in terms of surface area and fish production is the Pueblo Viejo lagoon, with a surface area of 41.61 km^2. The average altitude is between 10 and 50 m above sea level. These two conditions, together with the municipality's geographic location at the mouth of the Panuco River, make the municipality vulnerable to flooding.

To evaluate the FRI, a flood-risk scenario exceeding 50 cm of flooding was considered. A total of 35 descriptors were evaluated and grouped into four groups: the *flood physical risk index*, the *prevailing vulnerability associated with exposure and physical susceptibility index*, the *prevailing vulnerability associated with socioeconomic fragility index*, and the *prevailing vulnerability due to lack of resilience index*.

Regarding the *flood physical risk index*, assuming that 50 cm of flooding occurs during an extreme rainfall event, 12.3 % of the municipal territory could be affected. The affected population would exceed 20 %; the potential damage to homes would exceed 46 %; damage to agriculture and livestock would exceed 50 % of the occupied territory for this activity; and half of the territory would suffer environmental damage.

Regarding the *prevailing vulnerability associated with exposure and physical susceptibility index*, the descriptors registering the highest weight were the population density (0.750), the agricultural and livestock density (0.858), the physically dependent population (0.623), and the number of poor or overcrowded homes (0.715). These values were higher than 0.50, indicating that the exposure and physical susceptibility in the municipality is medium to high. It can be concluded that there is a large number of people under exposure to threat.

Regarding the *prevailing vulnerability associated with socioeconomic fragility index*, the descriptors registering the highest weight were social backwardness (0.750), social security (0.632), and socioeconomic dependence (0.913). These values correspond to a socioeconomic fragility ranging from medium to high or even very high. The results show that a large number of individuals are dependent on the economically active population.

Regarding the *prevailing vulnerability due to lack of resilience index*, the descriptors registering the highest weight were the capacity of communicating information (0.920), the municipal development index (0.518), the index flood-risk perception (0.750), and the rate of emergency operation (0.750). These values indicate that this area has a moderate to high capacity to respond to emergencies.

The estimation of the FRI, ranging from 0 to 1 (high risk), showed that the site is at moderate risk (0.450). According to the PRI (0.305) and PVI (0.475), the site is at moderate risk. The risk level depended on the hazard and the vulnerability.

The vulnerability/capacity of the population regarding its socioeconomic fragility contributed to lower risk. Both exposure and physical susceptibility showed a high risk, involving a large population and many homes.

Regarding the lack of resilience, the population is aware of the flood risk in the regions where they live and is able to perceive the recurrence of the events and identify the risk areas. However, they do not know to whom to turn in case of disaster.

The authorities lack the operational infrastructure to manage risk, and they have failed to raise public awareness of risk. Civil protection authorities act mainly reactively, and they have no mitigation or risk-reduction plans. This pattern, coupled with the apparent lack of financial and material resources, the small number of staff resources, and the continuous movement of personnel, prevents continuity in mitigation and risk-reduction projects.

An effective measure would be to develop plans and programs for participatory integrated flood-risk management. These programs present an alternative to indirectly increase the adaptive capacity of the population to climate change. The application of this methodology allowed the identification of future research areas, such as the identification of new indicators, and highlighted the importance of conducting multidisciplinary work involving human, economic, and social losses as well as organizational and institutional issues related to the level of development of the community.

13.4.7 Considering Sea-Level Rise

Posada and Vega (2010) evaluated the flood-prone areas in the city of Campeche due to a north wind effect for actual conditions (no change in mean sea level) and considering a rise of +0.59 m in sea level, which corresponds to the worst-case scenario provided by the IPCC (2007).

The coastal flooding was evaluated using a numerical model proposed by Posada et al. (2008). The model solves shallow water equations on a quadtree-type hierarchical mesh vertically averaged on a finite volume using a second-order numerical approximation of the Godunov type. The Riemann solver was used to solve the approximation of Roe (1981), used to determine non-viscous flows. The time integration was performed using the first-order method of Adams-Bashforth. This method has the ability to simulate the drying and wetting of cells during the evaluation of hydrodynamic cases. This model was applied to the bay of the city of San Francisco de Campeche to model a wind of constant magnitude and direction of 120 km/h and 315° north. According to Gregow et al. (2008), this wind is associated with a return period of 100 years. The modeling interval was 2 days, at which numerical stability was achieved in the solution.

The modeled region is located between the coordinates 19°47′4.65″, 90°37′14.63″ and 19°52′27.79″, 90°28′46.36″. The bathymetry was calculated from information collected during a bathymetric campaign conducted by the Institute of Engineering at the National Autonomous University of Mexico (UNAM) in 2004 and based on the 28265 letter published by the US Navy. The topography was obtained from a digital elevation model with a resolution of 3 m per pixel. The maximum resolution of the numerical mesh cell for the scenarios modeled was 30.4 m for the X-direction and 18.8 m for the Y-direction. This resolution was applied to all the points located between +2.0 m and −7.0 m when modeling the actual conditions and to those between 2.5 and −7.0 m when modeling climate-change scenarios. There was no displacement of the coast inland when the sea level rose to +0.59 m. This pattern is mainly due to the protection provided by the levee built in the city of Campeche. The time step used was 1.0 s. The wind force gradually increased from 0.0 to its peak following a cubic function. Once it reached its maximum, the wind force remained constant until the numerical solution was stable, i.e., registering no change in the flood area at an interval of 30 min.

In actual conditions, the flooding of the open surface of the coast begins approximately 4 h after the wind has reached its maximum. The maximum open surface is 0.38 m, and the maximum wind speed is approximately 0.72 m/s. The stable flood level is reached when the maximum wind has acted for 15 h in the city of Campeche. The actual flooding in coastal areas is between 30 and 35 cm.

For climate-change models, the wind conditions were the same as those of the actual case, but an increase in mean sea level of 0.59 m was considered. For this case, before the flood in the city of Campeche, the wind speeds were similar to those modeled for actual conditions. The free surface of the sea on the beachfront was also similar, considering that the data here referred to the mean modeled sea level. When flooding was stabilized, the maximum values were close to 0.60 m and 2.0 m/s for the flood and the wind speed, respectively.

When analyzing the configuration of the coastline with and without climate change for the city of San Francisco de Campeche under an increase of +0.59 m of mean sea level (without considering the effect of wind), only the northern part of the modeled area (Los Petenes) was flooded, and the city was not affected. That pattern is related to the defense provided by the waterfront along the coastline, with an average height of 1.5 m. It is noteworthy that this rise of +0.59 m caused an increase in the retention of seawater due to the astronomical tide in the northern mangroves.

For present conditions in the city of Campeche, flooding begins 5 h after the wind begins acting with full intensity. For the climate-change scenario, this time interval decreased to 3 h. The area susceptible to flooding was three times greater than that modeled for the present conditions. The main areas that could be flooded were the Petenes, the Akim shopping area, and the central area (walled). The latter two areas were reclaimed during the 1950s and 1960s after the construction of the first jetty.

The modeled wind corresponds to that expected for a return period of 100 years. Winds with longer return periods will produce lower flood levels. An extreme wind analysis and modeling was obtained for 1948–2007 records. The authors suggest performing continuous updates to assess the wind variation according to climate change.

In addition to their research, the authors suggest analyzing the flooding caused by the overflow of the drainage channels. A storm surge level of +0.60 m acts as a plug, which forces flow to increase, thus causing flooding in areas far from the coastline.

13.4.8 Water-Resource Management

Currently, the analysis of water management needs to consider the effect of climate change, that is, how the increase or decrease in precipitation and temperature affects the demand and supply of water. Acknowledging this need, Hernández et al. (2011) evaluated the effect that climate change could have on the water supply in the Rio Grande de Morelia basin. This evaluation was performed using the decision support tool AQUATOOL.

The Rio Grande de Morelia basin is located in the north-central portion of the state of Michoacán, between parallels 19°35′ and 20°05′ north latitude and the meridian 100°45′ and 101°25′ west longitude. The basin comprises an area of approximately 1,487 km^2. The topographic elevations observed in the area range around 1,920 masl in the north-central portion and around 2,500 masl in the western part of the region. The average altitude of the valley is approximately 2,200 masl.

The simulation was conducted in two modules comprising reservoirs, pipelines, and the most representative water demands in the Rio Grande basin. The data were calibrated to evaluate the difference between the historical and the simulated water storage.

The A2 scenario for the state of Michoacán in 2050 predicts a decrease in total precipitation of 8 % and a variation of the average annual temperature between +1 and +2.5 %. The authors proposed two scenarios for the year 2050. Scenario 1 considers minimal effects of climate change and predicts an increase in water demand of 15 %. Scenario 2 considers the maximum effects of climate change and predicts an increase in water demand of 30 %. The present conditions are represented by scenario 0.

The authors conducted a flow-assurance assessment following the UTHA criteria,[4] which establishes that the deficit in 1 year cannot exceed 50 % of the

---

[4] The UTAH criterion states:

The deficit in 1 year should not exceed 50 % of current demand.
The deficit in 2 consecutive years may not exceed 75 % of annual demand.
The deficit in 10 years in a row should not exceed 100 % of the annual demand.

present demand, the deficit in 2 consecutive years cannot exceed 75 % of the annual demand, and the 10-year deficit cannot exceed 100 % of the annual demand.

The application of the UTAH criteria showed that the 1-year deficit is acceptable for the current situation, although it is close to 50 % for module 1, while the maximum expected deficit is 66.90 %. The 2-year deficit for the scenarios 0, 1, and 2 shows critical problems for module 1, demanding risk evaluation in that area. Moreover, the maximum deficit could reach 103.4 %. This deficit is caused by increasing water demand and reductions in incoming flows. This problem could occur for the three scenarios considering a 10-year deficit. In module 1, a deficit problem could arise in the three scenarios. The urban water demand will experience problems in scenarios 1 and 2. In module 2, problems will be experienced in scenario 2. The annual flow assurances, which should be above 70 % for scenarios 1 and 2, will be compromised if urban water demand increases. For module 1, these flow assurances range between 51 and 68 %. Flow assurance indicates the amount of water being delivered relative to that being requested. For the case study, the minimum requirement (lower than 60 %) is not covered. Therefore, serious problems associated with water supply could arise. The authors conclude that flow assurance decreased once climate change was included in the Río Grande basin management model.

13.4.9 Water Quality

There are regions in Mexico where, due to the geology, arsenic and fluoride are naturally dissolved into groundwater (Vega-Gleason 2001). Concentrations of arsenic above the permissible limits according to the groundwater quality criteria in Mexico have been recorded and documented mainly in four states: Coahuila, Chihuahua, Durango, and Hidalgo. Arsenic has also been sporadically detected in wells located in Guanajuato, Zacatecas, Oaxaca, Morelos, and Puebla. Arsenic and fluoride have been detected in 26 municipalities of the abovementioned states, in which the exposed population could be more than two million inhabitants (Vega-Gleason 2001; Martínez 2002). Arsenic and fluoride concentrations in groundwater are likely to increase in the future because there is less water infiltration and an increased demand for water with increasing temperature. This change will require greater monitoring efforts of affected areas, as well as areas at risk, and the implementation of removal systems that improve the quality of water for human consumption (Leal et al. 2008).

Given the lack of studies of the water quality in surface-water bodies of the country, especially regarding the self-purification capacity of rivers, Montes (2013) proposes a methodology that includes the process of self-purification and the temperature and precipitation projections according to climate change. The method relies on a modification of the dissolved oxygen (DO) model initially proposed by Streeter and Phelps (1925) to evaluate the modification of the self-purification capacity of rivers and on the outputs of the general circulation models ECHAM5,

HADGEM1 (Martin et al. 2006; Ringer et al. 2006; Stott et al. 2006; McLaren et al. 2006), and GFDLCM2.0 for the emission scenarios A1B, A2, and B1 in the 2020 and 2050 time horizons. The analyses were performed for the rainy and the dry seasons. The proposed methodology consists of four stages: selecting the DO model that best describes the water body under study, building the baseline self-purification scenario, generating climate-change scenarios for the DO, and comparing the baseline and climate-change scenarios to establish critical situations and identify possible adaptation measures.

This methodology was applied to the Magdalena River, located to the west of Mexico City. The river supplies water to 5,600 people. The modeled section is located in the conservation area and has a length of 7 km, within which the intake of the water-treatment plant "Rio Magdalena" is located. A discharge event was simulated approximately 2 km from the beginning of the modeled section. This area was selected because it is impacted by tourism activity, discharging wastewater into the river.

The A1B, A2, and B1 scenarios were modeled for both of the time horizons, and the general circulation models ECHAM5 and HADCM3 were used. The latter was chosen because it provides precipitation and temperature outputs for the three scenarios.

The modeling showed that the removal of organic matter from the discharge modeled for 2020 in the dry season could reach 95.8 % but decrease by 28.9 % compared to the baseline scenario. The recovery of DO could reach 76.7 % and decrease by 43.9 %. These percentages suggest that the removal of the organic load is virtually unchanged compared to the baseline scenario. However, in the case of DO recovery, an improvement could be expected. During the rainy season, the removal of the organic load may decrease between 60.99 % and 33.36 %, and the recovery of DO would be between 3.79 % and 21.71 %. This change suggests that an improvement in both of the parameters compared to the baseline scenario could be expected for this season. For this horizon and both seasons, the organic load fed to the treatment plant could be expected to decrease compared to that of the baseline scenario. This decrease can be attributed to the dilution effect associated with flow increase.

In general terms, the climate-change scenarios for the 2050 time horizon register a lower organic load removal during the dry season, but they show an increase in DO recovery compared to the 2020 time horizon. Both of these parameters could be expected to improve during the rainy season.

According to the obtained results, it can be concluded that the effects of climate change will not affect the self-purification capacity of the Magdalena River. In fact, the actual conditions might even be improved. This pattern allows monitoring the temperature of this water body, as has been performed in other countries with pristine rivers (Pekárová et al. 2011), because the analysis of this river has minimal anthropogenic impacts (compared to other water bodies in the country). These data may provide insight into how climate change would affect the hydrological cycle components and other hydrological processes, such as the concentration of DO and the degradation of organic matter. However, this study highlights the poor quantity and quality of water quality and hydrometric data available. Even with this drawback, the proposed methodology is considered a good first approximation for the evaluation of the self-purification capacity of rivers under climate-change scenarios.

13.5 Adaptation in the Water Sector

Regarding the issue of adaptation in the water sector, Leal et al. (2008) propose that the adaptation to climate change should start in those regions where overexploitation, pollution, and other pressures are exerted on the water resources. Regarding the urban use of water, Mexico City has developed the Climate Action Plan to establish measures, actions, and budgets for saving water in homes, enhancing the pumping infrastructure, reducing leakage, conducting pipeline rehabilitation and distribution system sectorization, harvesting rainwater, expanding the capacity of wastewater treatment, recharging groundwater with rainwater, improving the water-collection system, and expanding the water reuse network (SMAGDF 2008). The most important water source should be identified in each administrative region, taking actions for its protection, such as water treatment and reuse, and adapting to reductions in water availability. Locations such as Nuevo Leon, Aguascalientes, Sonora, and Queretaro have advanced in the reinjection of water into aquifers, in environmental education, in the reduction of leaks, and in the sectorization of drinking water networks. In regions where agriculture is responsible for the increased use of water, action must be taken to increase the efficiency of the agricultural sector to attain the same performance with a reduced availability of water. If necessary, the cultivation should be considered of crop species that use less water and are resistant to higher temperatures, saline conditions, and levels of dissolved solids greater than 2,000 mg/L.

13.6 Adaptation in Water Quality

Leal et al. (2008) suggest that a greater monitoring effort of water-supply systems will be required to adapt to climate change. In those regions using surface runoff water, such as Mexico and Guadalajara valleys, the surveillance of eutrophication, chemical and biological pollution, persistent organic compounds, metals, and cyanotoxins will need to be intensified. Particular attention should be given to the water in agricultural areas, subject to metal and nitrate pollution, if eutrophication and pollution are to be reduced in the water-supply system.

One critical adaptation measure is the restoration of water quality during the next 10 years, improving the health of aquatic ecosystems. Some localities, such as Nuevo León, Aguascalientes, Sonora, and Querétaro, have already made some advances in this direction, improving their water quality.

Measures to prevent the resolubilization of metals from the sediments into the water column of reservoirs could already be implemented, especially in those sites where anoxic conditions are registered during large periods. Additionally, preventive actions could be taken in polluted water bodies to reduce the risk of exposure of the human population and the local organisms to potentially hazardous substances.

In those locations where groundwater is the main water supply, the concentrations of nitrates, arsenic, fluorine, phenols, benzene, iron and manganese, aromatic petroleum hydrocarbons and other aromatic compounds, trichloroethylene and other chlorinated compounds, persistent organic compounds, and fecal coliforms, as indicators of bacteriological quality, should be monitored. This monitoring would reduce the cost of water treatment.

13.7 Conclusions

The problem when determining the impacts of climate change on water-supply systems in developing countries, and particularly in Mexico, is the lack of high-resolution long data series required for the analyses (Soto and Herrera 2009). Hence, it is important that the methodologies proposed by the studies reported in this chapter are reproduced elsewhere, to improve and gain knowledge on the effects of climate change on water resources and to guide decision-making.

In Mexico, there are many pressures on water resources by the agricultural, industrial, urban, and tourism sectors. Continued population growth leads to increasing water demand, increasing the vulnerability of those communities and cities whose water supply is limited.

In many cities of Mexico, the problems of water availability and pollution are becoming more critical. This challenge threatens the water supply and makes the public vulnerable to future water shortages. Due to the potential degradation of the resource, it is important to know the risks that the water-supply systems are subjected to in these cities. A reduction in the water volume of overexploited basins due to changes in precipitation patterns and evapotranspiration can result in a catastrophe. Therefore, it is necessary to generate future scenarios of climate change for Mexico to explore the vulnerability to extreme climate conditions and to develop strategies to cope with these adverse situations (Soto and Herrera 2009).

Regarding water quality, it is necessary to monitor the actual changes associated to climate change.

Due to the lack of consistent public data on water quality, there are few studies providing information on the actual and future state of water resource. It is necessary to generate reliable public data to at least detect changes in the temperature and the concentrations of DO, biological oxygen demand (BOD), chemical oxygen demand (COD), and nitrates.

According to the material discussed in this chapter, it can be concluded that there are several studies, methodologies, and tools regarding water resources and climate change that have been developed in Mexico. However, these studies have been developed independently, dispersedly, and even disjointedly. The studies address a variety of issues, such as the supply-demand relationship, the natural availability of the resource and its projections according to climate change, pressures on the resource, the water footprint of some crops, rainwater harvesting, water infrastructure, water quality, and the social and cultural aspects related to water use. We suggest testing all these tools and applications in a pilot basin to have a more comprehensive view and advance toward the comprehensive integrated management of water resources. The studies should be conducted at the basin or sub-basin level because the basin is a basic unit that is easy to define and that includes all of the physical and biotic elements and other environmental aspects of ecosystem complexity.

Acknowledgments The authors thank the Virtual Center for Climate Change (CVCCCM, its initials in Spanish), the Research Program on Climate Change (PINCC, its initials in Spanish), and the Institute of Science and Technology of Mexico City (ICyTDF, its initials in Spanish) for the financial support enabling the development of this chapter.

References

Abe-Ouchia A, Emoris S, Hasegawa A, Hasumi H, Inoue T, Kimoto M, Matsumura S et al (2004) K-1 coupled model (MIROC) description, K-1 technical report, 1, In: Hasumi H, Emori S (eds) Center for Climate System Research, University of Tokyo, 34 pp

Arnell NW, Reynard NS, King R, Prudhomme C, Branson J (1997) Effects of climate change on river flows and groundwater recharge: guidelines for resource assessment, U.K. Water Industry Research Environment Agency, report 97/CL/04/1. UKWIR, London, p 32 + appendix

Bates BC, Kundzewicz ZW, Wu S, Palutikof JP (eds) (2008) Climate change and water. Technical paper of the Intergovernmental Panel on Climate Change, IPCC Secretariat, Geneva, 210 pp

CENAPRED (2001) Diagnóstico de peligros e identificación de riesgos de desastres en México. Secretaría de Gobernación. Centro Nacional de Prevención de Desastres, México, 225 p

CEPAL (2003) Manual para la evaluación del impacto socioeconómico y ambiental de los Desastres. Comisión Económica para la América Latina y el Caribe (CEPAL) y el Banco Internacional de Reconstrucción y Fomento (El Banco Mundial), 95 p

Clarke R, King J (2004) The atlas of water. Earthscan, London, 128 pp

Conde C, Estrada F, Martínez B, Sánchez O, Gay C (2011) Regional climate change scenarios for México. Atmosfera 24(1):125–140

Cox P, Fisher I, Kastl G, Jegatheesan V, Warnecke M, Angles M, Bustamante H, Chiffings T, Hawkins PR (2003) Sydney 1998 – lessons from a drinking water crisis. J Am Water Work Assoc 95:147–161

Cromwell JE, Smith JB y Raucher RS (2007) Implications of climate change for urban water utilities. Association of Metropolitan Water Agencies, Washington, DC, 20 pp

Curriero F, Patz J, Rose J, Lele S (2001) The association between extreme precipitation and waterborne disease outbreaks in the United States, 1948–1994. Am J Public Health 91:1194–1199

Delworth T, Broccoli A, Rosati A, Stouffer R, Balaji V, Beesley J et al (2006) GFDL's CM2 global coupled climate models. Part I: Formulation and simulation characteristics. J Clim 19:643–674

FAO (Food and Agriculture Organization of the United Nations) (2006) Evapotranspiración del cultivo, guía para la determinación de los requerimientos de agua de los cultivos. Organización de las Naciones Unidas para la Agricultura y la Alimentación, Roma

Fayer R, Trout J, Lewis E, Xiao E, Lal A, Jenkins M, Graczyk T (2002) Temporal variability of *Cryptosporidium* in the Chesapeake Bay. Parasitol Res 88:998–1003

Gnanadesikan A, Dixon K, Griffies S, Balaji V, Barreiro M, Beesley A et al (2006) GFDL's CM2 global coupled climate models. Part II: The baseline ocean simulation. J Clim 19:675–697

González DME, Haces MA y Rangel L (2011) Metodología para valorar índices de vulnerabilidad ante el cambio climático y acciones de compensación en las costas de Tamaulipas. 1er Congreso Nacional de Investigación en Cambio Climático. Ciudad de México, del 17 al 21 de octubre del 2011

González DM (2011) Valoración y gestión del riesgo por inundaciones en municipios costeros (Una aproximación con base en indicadores), pp 591–606. En: Rivera-Arriaga E, Azuz-Adeath I, Alpuche Gual L y Villalobos-Zapata GJ (eds) Cambio Climático en México un Enfoque Costero-Marino. Universidad Autónoma de Campeche, Cetys-Universidad, Gobierno del Estado de Campeche, 944 p

Gordon HB, Rotstayn LD, McGregor JL, Dix MR, Kowalczyk EA, O'Farrell SP, Waterman LJ, Hirst AC, Wilson SG, Collier MA, Watterson IG, Elliott TI (2002) The CSIRO Mk3 climate system model. CSIRO atmospheric research technical paper no. 60, Commonwealth Scientific and Industrial Research Organisation Atmospheric Research, Aspendale, Victoria, Australia, 130 pp. http://www.cmar.csiro.au/e-print/open/gordon_2002a.pdf

Gordon C, Cooper C, Senior CA, Banks HT, Gregory JM, Johns TC, Mitchell JFB, Wood RA (2000) The simulation of SST, sea ice extents and ocean heat transports in a version of the Hadley Centre coupled model without flux adjustments. Clim Dyn 16:147–168

Gregow H, Venäläinen A, Peltola H, Kellomäki S, Schultz D (2008) Temporal and spatial occurrence of strong winds and large snow load amounts in Finland during 1961–2000. Silva Fennica 42(4):515–534

Haro ME (2012) Evaluación de la huella hídrica por la producción de bioetanol en Tamazula, Jal. Tesis de Maestría, Posgrado en Ingeniería, Universidad Nacional Autónoma de México

Hernández J, Sánchez S, Domínguez C y Lara B (2011) Evaluación del efecto de cambio climático en la gestión de sistemas de recursos hídricos. En Décimo Seminario Iberoamericano de Planificación, Proyecto y Operación de Sistemas de Abastecimiento de Agua SEREA. Morelia, Michoacán, México, 10 al 14 de Enero

Herzer H (1998) Construccion del riesgo, desastre y gestion ambiental urbana (Perspectivas en debate). International conference. Research Community for the Habitat Agenda. Linking research and policy for the sustainability of human settlements. Forum of Researchers on Human Settlements, Geneva, July 6–8

Hoekstra AY (2007) Human appropriation of natural capital: comparing ecological footprint and water footprint analysis. Institute for water education. Ecol Econ 68:1963–1974

Hulme M, Jenkins GJ, Lu X, Turnpenny JR, Mitchell TD, Jones RG, Lowe J, Murphy JM, Hassell D, Boorman P, McDonald R, Hill S (2002) Climate change scenarios for the United Kingdom: the UKCIP02 scientific report, Tyndall Centre for Climate Change Research, School of Environmental Sciences, University of East Anglia, Norwich, UK, 120 pp. Consultado el 07 mayo de 2013 de. http://www.ukcip.org.uk/wordpress/wp-content/PDFs/UKCIP02_tech.pdf

Hunter P (2003) Climate change and waterborne and vector borne disease. J Appl Microbiol 94:37–46

IPCC (2007) Cambio climático 2007: Informe de síntesis. Contribución de los Grupos de trabajo I, II y III al Cuarto Informe de evaluación del Grupo Intergubernamental de Expertos sobre el Cambio Climático ipcc, Ginebra, Suiza, 104 p

IPCC-TGICA (2007) General guidelines on the use of scenario data for climate impact and adaptation assessment. Version 2. Prepared by T. R. Carter on behalf of the Intergovernmental Panel on Climate Change, Task Group on Data and Scenario Support for Impact and Climate Assessment, 66 p. Available at: http://www.ipcc-data.org/guidelines/TGICA_guidance_sdciaa_v2_final.pdf

Jungclaus JH, Keenlyside N, Botzet M, Haak H, Luo JJ, Latif M, Marotzke J, Mikolajewicz U, Roeckner E (2006) Ocean circulation and tropical variability in the coupled model ECHAM5/MPI-OM. J Clim Spec Sect 19:3952–3972

Kundzewicz ZW, Mata LJ, Arnell NW, Döll P, Kabat P, Jiménez B, Miller KA, Oki T, Sen Z, Shiklomanov IA (2007) Freshwater resources and their management. In: Parry ML, Canziani OF, Palutikof JP, van der Linden PJ, Hanson CE (eds) Climate change 2007: impacts, adaptation and vulnerability. Contribution of Working Group II to the fourth assessment report of the Intergovernmental Panel on Climate Change. Cambridge University Press, Cambridge, pp 173–210

Kundzewicz ZW, Mata LJ, Arnell N, Döll P, Jimenez B, Miller K, Oki T, Şen Z, Shiklomanov I (2008) The implications of projected climate change for freshwater resources and their management. Hydrol Sci J 53(1):3–15

Leal MT, Millán DV, Mendez C-G, Servín CA (2008) Evaluación de la afectación de la calidad del agua en cuerpos de agua superficiales y subterráneos por efecto de la variabilidad y el cambio climático y su impacto en la biodiversidad, agricultura, salud, turismo e industria. Disponible en: http://www.inecc.gob.mx/descargas/cclimatico/ev_calidad_agua_cc.pdf

Lenton R (2004) Water and climate variability: development impacts and coping strategies. Water Sci Technol 49(7):17–24

Martin GM, Ringer MA, Pope VD, Jones A, Dearden C, Hinton TJ (2006) The physical properties of the atmosphere in the new Hadley Centre Global Environmental Model, HadGEM1. Part 1: Model description and global climatology. J Clim 19(7):1274–1301

Martínez A (2002) Calidad de agua para consumo humano en México. En: Memorias del Encuentro sobre uso y resultados de la aplicación de tecnologías económicas para la purificación de aguas en América Latina. Buenos Aires, 8 y 9 de noviembre

McLaren AJ, Banks HT, Durman CF, Gregory JM, Johns TC, Keen AB, Ridley JK, Roberts MJ, Lipscomb WH, Connolley WM, Laxon SW (2006) Evaluation of the sea ice simulation in a new coupled atmosphere-ocean climate model (HadGEM1). J Geophys Res 111:1–17

Mihelcic JR, Fry LM, Myre EA, Phillips LD, Barkdoll BD (2009) Field guide to environmental engineering for development workers. Water, sanitation, and indoor air. ASCE Press, Reston

Millennium Ecosystem Assessment (2005) Ecosystems and human well-being: synthesis. Island Press, Washington, DC, 155 pp

Montes RT (2013) Metodología para evaluar la modificación de la capacidad de autodepuración de los ríos por efecto del cambio climático. Tesis Doctoral. Programa de Maestría y Doctorado en Ingeniería, Universidad Nacional Autónoma de México

Ospina J, Conde A, y García C (2014a) Capítulo V: Relación de Variables Climáticas como Líneas Base e Índice de Vulnerabilidad para la Disponibilidad de los Recursos Hídricos. En Ospina JE, García C, Conde A (eds) Historia del Clima de la Ciudad de México: Efectos Observados y Perspectivas. Centro Virtual de Cambio Climático de la Ciudad de México en colaboración con el Instituto de Ciencias y Tecnología del Gobierno Federal, ahora Secretaría de Ciencias, Tecnología e Innovación, 149 pp

Ospina JE, García C, Conde A (eds) (2014b) Historia del Clima de la Ciudad de México: Efectos Observados y Perspectivas. Centro Virtual de Cambio Climático de la Ciudad de México en colaboración con el Instituto de Ciencias y Tecnología del Gobierno Federal, ahora Secretaría de Ciencias, Tecnología e Innovación, 149 pp

Pekárová P, Mikláneka P, Halmová D, Onderka M, Pekár J, Kucarovac K, Liová S, Škoda P (2011) Long-term trend and multi-annual variability of water temperature in the pristine Bela River basin (Slovakia). J Hydrol 400:333–340

Pope V, Gallani ML, Rowntree PR, Stratton RA (2000) The impact of new physical parameterizations in the Hadley Centre climate model: HadCM3. Clim Dyn 16:123–146

Posada G, y Vega BE (2010) Evaluación de zonas inundables para la ciudad de San Francisco de Campeche, pp 607–622. En: Rivera-Arriaga E, Azuz-Adeath I, Alpuche Gual L, y Villalobos-Zapata GJ (eds) Cambio Climático en México un Enfoque Costero-Marino. Universidad Autónoma de Campeche, Cetys-Universidad, Gobierno del Estado de Campeche, 944 pp

Posada G, Silva R, y Medina R (2008) Modelo numérico tridimensional para transporte de un contaminante conservativo. *Ingeniería Hidráulica en México segunda época*, imta xxiii(1): 5–19, II época

Ringer MA, Martin GM, Greeves CZ, Hinton TJ, James PM, Pope VD, Scaife AA, Stratton RA, Inness PM, Slingo JM, Yang G-Y (2006) The physical properties of the atmosphere in the new Hadley Centre Global Environmental Model, HadGEM1. Part 2: Aspects of variability and regional climate. J Clim 19(7):1302–1326

Rivas I, Güitron A, y Ballinas H (2010) Vulnerabilidad hídrica global: aguas superficiales. En Efectos del cambio climático en los recursos hídricos de México. Atlas de Vulnerabilidad hídrica en México ante el Cambio Climático. Volumen III. Secretaria de Medio Ambiente y Recurso Naturales- Instituto Mexicano de Tecnología del Agua

Roe PL (1981) Approximate Riemann solvers, parameter vectors, and difference schemes. J Comput Phys 43:357–372

Saaty T (2001) The analytic network process: decision making with dependence and feedback. RWS Publications, Pittsburg

Sánchez G, Arcos G, y Barragán R (2011) Modelación Hidrológica y Disponibilidad de Agua ante el Cambio Climático en la Zona Costera de Tamaulipas. Primer Congreso Nacional de Investigación en Cambio Climático. Ciudad de México del 17 al 21 de octubre del 2011

SEDESOL-CONAPO-INEGI (2007) Delimitación de las zonas metropolitanas de México 2005. Primera edición. Secretaría de Desarrollo Social, Consejo Nacional de Población y Instituto Nacional de Estadística, Geografía e Informática Impreso en México

SEMARNAT (2009) Atlas del Agua en México. Secretaría del Medio Ambiente y Recursos Naturales. Disponible en. http://futurocostaensenada.files.wordpress.com/2010/02/sgp-25a-atlas.pdf

SMAGDF (2008) Programa de Acción Climática de la Ciudad de México 2008–2012. Secretaría del Medio Ambiente del Distrito Federal, 172 pp

Small C, Nicholls RJ (2003) A global analysis of human settlement in coastal zones. J Coast Res 19:584–599

Soto G, y Herrera M (2009) Estudio sobre el impacto del cambio climático en el servicio de abasto de agua de la Zona Metropolitana de la Ciudad de México

Stott PA, Jones GS, Lowe JA, Thorne PW, Durman CF, Johns TC, Thelen JC (2006) Transient climate simulations with the HadGEM1 climate model: causes of past warming and future climate change. J Clim 19(12):2763–2782

Stoufferr J, Broccoli AJ, Delworth TL, Dixon KW, Gudgel R, Held I, Hemler R, Knutson T, Lee H-C, Schwarzkopf MD, Soden B, Spelman MJ, Winton M, Zeng F (2006) GFDL's CM2 global coupled climate models. Part IV: Idealized climate response. J Clim Spec Sect 19:723–740

Streeter H, Phelps E (1925) A study of the pollution and natural purification of the Ohio River, III, Factor Concerned in the Phenomena of Oxidation and Reaeration. U.S. Public Health Bulletin (146) pp 75

United Nations (2003) World water development report: water for life, water for people. UNESCO/Berghahn Books, Paris/Barcelona, 544 pp

Vega-Gleason S (2001) Riesgo sanitario ambiental por la presencia de arsénico y fluoruros en los acuíferos de México. Comisión Nacional del Agua, Gerencia de Saneamiento y Calidad del Agua, México D.F

WDR (2003) World disaster report: focus on ethics in aid. International Federation of Red Cross and Red Crescent Societies, Geneva, 240 pp

WDR (2004) World disaster report: focus on community resilience. International Federation of Red Cross and Red Crescent Societies, Geneva, 240 pp

WHO/UNICEF (2000) Global water supply and sanitation assessment 2000 report. World Health Organization with UNICEF, Geneva, 79 pp

Wigley TML (2008) MAGICC/SCENGEN 5.3: user manual (versión 2). NCAR, Boulder, 80 pp

Wittenberg AT, Rosati A, Lau N-C, Ploshay JJ (2006) GFDL's CM2 global coupled climate models. Part III: Tropical Pacific climate and ENSO. J Clim 19:698–722

Yarze JC, Chase MP (2000) E. coli O157:H7 – another waterborne outbreak! Am J Gastroenterol 95(4):1096

Yusoff I, Hiscock KM, Conway D (2002) Simulation of the impacts of climate change on groundwater resources in eastern England. In: Hiscock KM, Rivett MO, Davison RM (eds) Sustainable groundwater development, Geological Society special publication 193. Geological Society, London

Chapter 14
Prediction of Hydrological Risk for Sustainable Use of Water in Northern Mexico

Alfonso Gutiérrez-López, Thierry Lebel, Israel Ruiz-González, Luc Descroix, and Marcela Duhne-Ramírez

Abstract Among surface hydrologic phenomena, it is common to find series of events of random occurrence in time. Poisson processes lead to probabilistic models that are appropriate to explain the number of events produced by certain phenomena. For instance, in surface hydrology, it is quite frequent to relate the Poisson distribution to the occurrence of rainfall events. The so-called leak distribution consists of the simultaneous use of a Poisson law to represent the probability of occurrence of an event and an exponential distribution applied to the mean magnitude of such event. Originally introduced to simulate gas leaks in distribution networks in France, from where it takes its name, the leak distribution has important applications in hydrology. In this paper, the theoretical basis of the law and the method for the estimation of its parameters are introduced. Some applications are included, such as further knowledge of the precipitation regime of hydrologic region No. 10 in Mexico. In this case, through the knowledge of the two parameters of this law, which can be associated to physical variables, it is possible to determine the temporal and spatial distribution of precipitation in detail. As an additional application, the use of this law in drought analysis is shown. Here, the distribution parameters are related to the Standardized Precipitation Index, SPI, allowing the construction of a modified SPI that much better represents the spatial variability of drought periods in the watershed. According to the

---

A. Gutiérrez-López
Centro de Investigaciones del Agua, CIAQ, Universidad Autónoma de Querétaro, Cerro de las Campanas, s/n Qro., Col. Las Campanas 76010, México
e-mail: alfonso.gutierrez@uaq.mx

T. Lebel • L. Descroix
LTHE, Bâtiment OSUG-B, Domaine universitaire, BP 53, 38041 Grenoble cedex 09, France

I. Ruiz-González (✉)
Centro de Investigaciones del Agua, CIAQ, Universidad Autónoma de Querétaro, Santiago de Querétaro, Mexico
e-mail: israel.ruiz@uaq.mx

M. Duhne-Ramírez
Laboratoire d'Etude des Transferts en Hydrologie et Environnement, LTHE, Grenoble, France

S.G. Setegn, M.C. Donoso (eds.), *Sustainability of Integrated Water Resources Management*, DOI 10.1007/978-3-319-12194-9_14

results presented in this chapter, the use of the leak distribution in surface hydrology processes allows deeper knowledge of regional climatology.

Keywords Leak distribution • Poisson distribution • Mexican precipitation regime • Standardized precipitation index • Hydrological risk

14.1 Introduction

In daily activity, it is common to observe series of events of random occurrence in time. Poisson processes provide probabilistic models that are appropriate to explain the number of events that occur or that are produced by a certain phenomenon (López-Segovia et al. 2002). In hydrology, the Poisson distribution is frequently related to the occurrence of precipitation events to explain, for example, the interval of occurrence of precipitations in which the intensity and duration are presented as two random independent variables (Tapsoba 1997; Bacchi et al. 1994; Le Barbé et al. 2002; and Gutiérrez-López 2003). Likewise, the use of this distribution can be extended to construct stochastic models of hourly precipitation that will preserve the pattern of occurrence of rain events throughout the year, as well as to vary the characteristics of duration, magnitude, and intensity of precipitation among events (Istok and Boersma 1989; Abi-Zeid et al. 2004). Even studies as precise as quantifying the number of drops are based on this distribution. Calder (1986) proposes a stochastic model for precipitation modeling which associates, through a Poisson distribution, the average number of raindrops retained in a given surface with the average number of total raindrops. This type of Poissonian model provides a rational explanation of why soil surfaces dampen gradually and asymptotically. Likewise, this distribution allows considering intensity and duration as random bivariate variables, each with an exponential marginal distribution (Bacchi et al. 1994). This scheme of using jointly a Poisson distribution and an exponential one is very useful to characterize the movement of storms with different structures and positions. This fact establishes the affirmation that the Poisson process is an adequate tool to represent the behavior of hydrometeorological phenomena.

When the mixed Poisson-exponential model is used, it is also possible to evaluate the hydrological risk associated with floods and droughts in a simultaneous manner (Gutiérrez-López 2003). In this mixed distribution, it is shown that the parameters of both distributions are related through a variance reduction factor, which is a function of the spatial correlation structure of precipitation. Supposing that the mean depth per precipitation event follows an exponential distribution, the scale parameter adjusted from the daily rainfall series acquires the same value as the scale parameter of the Gumbel-type extreme value distribution, adjusting the maximum values of monthly rainfall (Sivapalan and Blöschl 1998; Gutiérrez-López 2003). Similar works replacing the exponential distribution with the gamma distribution have been developed in watersheds with hourly precipitation records in the center of the United Kingdom and the United States (Onof and Wheater 1994; Wilks 1998).

Among the main applications of exponential and Gumbel distributions in a joint manner, it is worth to mention studies that relate precipitation with relief (regional topographic pattern). If we consider that precipitation follows an exponential tendency to decrease with altitude and, thus, with the topographic characteristics of the relief, a principal component analysis (of a digital model in elevation) provides a set of variables that describe topography, linked to the parameters of Gumbel distribution and precipitation intensity (Singh and Kumar 1997; Wotling et al. 2000). Likewise, the Gumbel distribution is used as an approximation to the probability distribution of the extremes of the bivariate exponential model (Bacchi et al. 1994).

Following the ideas raised so far, the first objective of this work is to obtain a probabilistic model whose parameters have a well-defined physical meaning. For example, one of them should be able to represent the number of events (rainy days) and the other, the mean magnitude of such event. It would even be desirable for a probability distribution to allow a temporal disaggregation of estimated events, with a simple multiplication of the parameters by the value of the time interval. This possibility of disaggregation in time is raised as the second objective of this work.

In the same fashion, the Poisson distribution has been used to model periods of drought (Sharma 1996; Cameron et al. 2000). The periods of drought and humidity of a rainy season are usually conserved and can be modeled using a Markov process of order one. A traditional way of analyzing these extreme phenomena consists of mixing two or more distributions. In this way, a Poisson probability function describes the occurrence of drought periods, the geometric distribution describes the duration of the periods, and the Weibull distribution models the total precipitation during a wet period. A complete analysis, which requires the joint estimation of the parameters of these distributions, gives results related to the probability of drought in any given day, the probability of a dry day following another dry day, or the probability of a humid day following another humid day. Although this type of results describes adequately the time interval between periods of drought, it is desirable to correlate the parameters of a probability distribution with, for example, an aridity index. This is precisely the third objective of this work. For the case of river flows, a similar scheme of detection of shortage or excess of flow in a time interval can be modeled by a Poisson distribution (Coles et al. 2003; Hughes 2003).

Because in most countries civil engineers estimate design flows for the design of hydraulic structures using the extrapolation of extreme values, hydrological forecast is conducted starting with an interpretation of the frequencies of occurrence probabilities of maximum events (Pandey and Nguyen 1999). Since the adjective "extreme value" is assigned to these distributions, because they can be obtained as limit distributions of the large or small values between random independent variables, each having a continuous distribution (Pandey and Nguyen 1999; Escalante and Reyes 2002), it is possible to hypothesize that the mixed Poisson-exponential distribution represents extreme phenomena of drought or flood because it contains a distribution of extremes. That is, raising a fourth objective where it is possible to show that one of the parameters of the mixed Poisson-exponential distribution is the same as the Gumbel distribution parameter, when the latter is adjusted to a series of maxima. This will allow associating the mixed Poisson-exponential distribution with an extreme value law.

Finally, it is worth mentioning that although the Poisson distribution is adequate for modeling processes of the hydrological cycle, for the case of the occurrence of maximum flows in a river (indicated by their peaks modeled with a Poisson process), it has been shown that the variance of a series of annual excesses, in some series, is significantly lower (or higher) than the mean of the original series. In these cases, the binomial distribution (or negative binomial) shows a better fit than a Poisson distribution. It has also been shown that the estimation of design events associated with return periods adjusts adequately to the binomial model (or negative binomial) when combined with the exponential distribution (magnitudes of peaks). However, one of the great advantages of the Poisson model, over binomial distribution, is that it has simpler expressions, generally obtaining nearly identical results (Önöz and Bayazit 2002).

14.2 Presentation of the Leak Distribution

This law has its origin in the study of time series of events separated by random durations. The two basic hypotheses are stationarity, which means that the statistical properties of the time series do not change with time, and independence, which means that every one of these events occurs independently from the others. The name "leak distribution," originally coined in French as *Loi des Fuites,* comes from the study conducted by M. Morlat (Babusiaux 1969) on the distribution of gas flows that leaked from a distribution network in France. It consists of simultaneously using a Poisson law to represent the probability of occurrence of an event and an exponential distribution to express the mean magnitude of this event. This law has been used to describe precipitation regimes in Africa, in Benin (Lebel and Le Barbé 1997), in Togo (Seguis 1989), in Nigeria (Le Barbé and Lebel 1997), and in Burkina Faso (Tapsoba 1997).

If a random variable is considered, $y \geq 0$, which follows a simple exponential distribution with scale parameter β, its density function can be expressed as:

$$f(y) = \frac{1}{\beta} e^{-y/\beta} \tag{14.1}$$

Let ℓ be a random whole number that follows a Poisson law with shape parameter $\lambda > 0$. Its probability will be given by:

$$p(\ell) = \frac{e^{-\lambda}\,\lambda^{\ell}}{\ell\,!} \tag{14.2}$$

In this way, a random variable $x = y_1 + y_2 + \cdots + y_\ell$ can be represented as the sum of $\ell > 0$ independent random variables that follow a leak distribution with shape parameter λ and scale parameter β. In regard to duration, it is equivalent to say that a number of events $n(t)$ will follow a Poisson law for a duration t, or that the interval T between two consecutive events follows an exponential law of scale parameter μ_T, over the same time t.

$$f(T) = \frac{1}{\mu_T} e^{-T/\mu_T} \tag{14.3}$$

where:

μ_T is shape parameter of the exponential law.
T is interval between two events.

It is worth reminding that the exponential distribution is a particular case of the so-called gamma distribution family (Saporta 1990).

Let $t_r = T_1 + T_2 + \cdots + T_r$ be the sum of r variables that follow an exponential distribution (expression 14.3). It is known that the density function of the sum of r independent random variables is the convolution product of the densities of each one of these variables (Tapsoba 1997). This way, the conditional law of t_r will be an incomplete gamma distribution with a probability density function equal to:

$$f(t_r) = \frac{\left(\frac{t_r}{\mu_T}\right)^{r-1} e^{\frac{-t_r}{\mu_T}}}{\mu_T(r-1)!} \tag{14.4}$$

We have that $\mathrm{Prob}(t_r \leq t) = \int\limits_0^t f(t_r)\mathrm{d}t_r = F_{t_r}(t)$; therefore, $\mathrm{Prob}[n(t) = r] = \mathrm{Prob}$ $(t_r > t) = 1 - F_{t_r}(t)$ which can also be expressed as: $\mathrm{Prob}[n(t) = r] = \mathrm{Prob}[n(t) < r + 1] - \mathrm{Prob}[n(t) < r] = F_{t_r}(t) - F_{t_{r+1}}(t)$.

Then, we get:

$$\mathrm{Prob}[n(t) = r] = \int\limits_0^t \frac{\left(\frac{x}{\mu_T}\right)^{r-1} e^{\frac{-x}{\mu_T}}}{\mu_T(r-1)!}\mathrm{d}x - \int\limits_0^t \frac{\left(\frac{x}{\mu_T}\right)^{r} e^{\frac{-x}{\mu_T}}}{\mu_T r\,!}\mathrm{d}x \tag{14.5}$$

Integrating by parts:

$$\text{Prob}[n(t) = r] = \frac{e^{\frac{-t}{\mu_T}} \left(\frac{t}{\mu_T}\right)^r}{r\,!} \tag{14.6}$$

Starting from an exponential distribution as the duration between two events, with shape parameter μ_T, it has been shown that the number of events $n(t)$, over a period t, follows a Poisson law with parameter (t/μ_T).

14.2.1 Density Function

Let y be a positive random variable that follows a simple exponential distribution with scale parameter β and that has a probability density function given by the expression (14.1). ℓ is a random whole number that follows a Poisson law of shape parameter $\lambda > 0$, of probability:

$$p(\ell) = e^{-\lambda} \frac{\lambda^\ell}{\ell\,!}$$

Being $x = y_1 + y_2 + \cdots + y_\ell, \ell > 0$, the sum of a number of independent variables that follows an incomplete gamma distribution, of probability density equal to:

$$f(x, \ell) = \frac{1}{(\ell - 1)!} \left(\frac{x}{\beta}\right)^{\ell - 1} \frac{e^{-x/\beta}}{\beta} \tag{14.7}$$

If the reduced variable $U = x/\beta$ is introduced, we get:

$$f(U, \ell) = \frac{1}{(\ell - 1)!} U^{\ell - 1} e^{-U} \tag{14.8}$$

From these, we obtain the marginal distribution of U for ℓ taking the values from 1 to infinite, resulting in:

$$f(U) = \sum_{\ell=1}^{\infty} \frac{e^{-\lambda} \lambda^\ell}{\ell!} \frac{1}{(\ell - 1)!} U^{\ell - 1} e^{-U} \tag{14.9}$$

or

$$f(U) = e^{-\lambda} \lambda e^{U} \left[1 + \frac{\lambda U}{2} + \cdots + \frac{(\lambda U)^i}{i!(i + 1)!} + \cdots\right] \tag{14.10}$$

With the help of the modified Bessel function of the first kind, the terms in brackets can be expressed as:

$$I_1\left(2\sqrt{\lambda U}\right) = \sum_{n=0}^{\infty} \frac{\left(\sqrt{\lambda U}\right)^{2n+1}}{n!(n+1)!}$$

$$= \sqrt{\lambda U} + \frac{\left(\sqrt{\lambda U}\right)^{3}}{1!\,2!} + \cdots + \frac{\left(\sqrt{\lambda U}\right)^{2i+1}}{i!\,(\mathrm{i}+1)!} + \cdots \tag{14.11}$$

This way, the probability density function of the leak distribution for the reduced variable $U = x/\beta$ will be given by the expression:

$$f(U) = \mathrm{e}^{-\lambda}\lambda \mathrm{e}^{U} \frac{I_1(2\sqrt{\lambda U})}{\sqrt{\lambda U}} \tag{14.12}$$

$$f(0) = f_0 = \mathrm{e}^{-\lambda}$$

where:

λ is the shape parameter, $\lambda > 0$ (number of events within a period of time T).
β is the scale parameter, $\beta > 0$ (mean depth of precipitation per event).
I_1 is the Bessel function of the first order.

This law is a mixed distribution that has a discontinuity for the zero value (except when $\lambda = 1$), that is, the ordinate to the origin has a value of $\lambda \mathrm{e}^{-\lambda}$. Figure 14.1 shows the functions $f(U = x/\beta)$ and $F(U = x/\beta)$.

The main hypotheses to consider for using this law in the analysis of the regime and distribution of precipitation are: rain follows a memory-less occurrence process; that is, the probability distribution of the interval between the start of any given day and the first event observed (day with rain) is the same. The depth of rain of each event follows an exponential distribution, and the number of events (days with rain) recorded at a station in an interval t follows a Poisson distribution. If we consider that in the first hypothesis, the process is also stationary, the distribution of the duration between each event will follow an exponential distribution law and the time interval T of the number of events λ will follow a Poisson law. This way, it can be said that the proposed distribution is a Poisson process with an exponential distribution, like the so-called white noise. Also, the parameter β has an important property, which represents an additional advantage making possible the derivation of the distribution of rainfall in any time interval T^*, according to the distribution calculated for the period T. This property must be used as a validation of the law in a given region (Gutiérrez-López et al. 2002).

14.3 Parameter Estimation

Considering the hypotheses of this law and the meaning of the parameters, it is inferred that the most effective procedure depends on the number of null events n_0 that exist in the sample; that is, the number of dry days in the historical series of

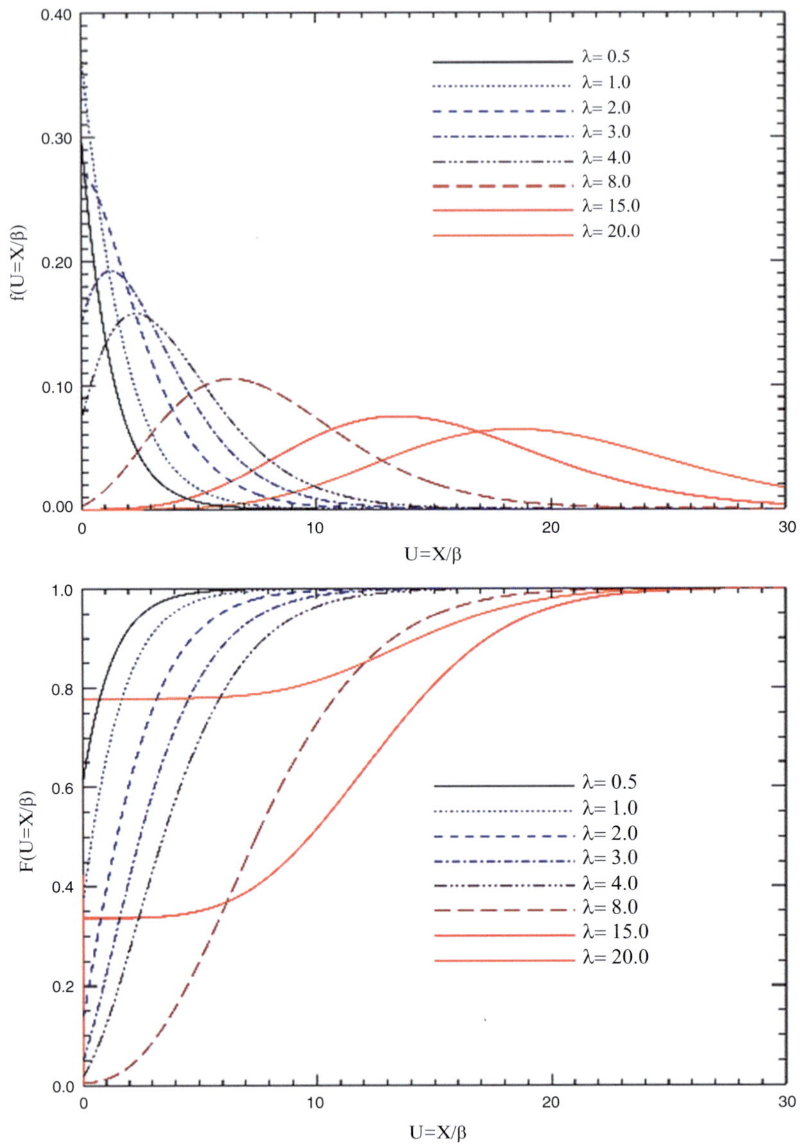

Fig. 14.1 Density and cumulative functions for different values of λ

data. If $n_0 \neq 0$, the estimation procedure requires, for a correct solution, considering the sample size n (Tapsoba 1997). This way, the parameter estimators can be obtained as:

$$\hat{\lambda} = \left[1 - \left(2 - \sqrt{2}\right)\sqrt{\frac{n_0}{n}}\right]\left[\frac{2\bar{x}^2}{s_x^2} - \mathrm{Log_e}\frac{n}{n_0}\right] + \mathrm{Log_e}\frac{n}{n_0} \quad (14.13)$$

$$\hat{\beta} = \left[1 - \left(2 - \sqrt{2}\right)\sqrt{\frac{n_0}{n}}\right]\left[\frac{s_x^2}{2\bar{x}} - \frac{\bar{x}}{\mathrm{Log_e}\frac{n}{n_0}}\right] + \frac{\bar{x}}{\mathrm{Log_e}\frac{n}{n_0}} \quad (14.14)$$

For the case where $n_0 = 0$, substituting in the previous expressions, the parameters of the leak distribution (method of moments) can be estimated as:

$$\hat{\lambda} = \frac{2\bar{x}^2}{s_x^2} \qquad \hat{\beta} = \frac{s_x^2}{2\bar{x}} \quad (14.15)$$

where:

$\bar{x}$ is the estimator of the first moment of the probability distribution.
S_x is the sample's standard deviation.

14.4 Method of Maximum Likelihood

The logarithm of the likelihood function, from which a maximum is sought, is the following:

$$L = -\lambda n + \frac{(n - n_0)}{2}\mathrm{Log_e}\frac{\lambda}{\beta} - \sum_{i=1}^{n-n_0}\frac{x_i}{\beta} - \frac{1}{2}\sum_{i=1}^{n-n_0}\mathrm{Log_e}x_i + \sum_{i=1}^{n-n_0}\mathrm{Log_e}I_1\left(2\sqrt{\frac{\lambda x}{\beta}}\right) \quad (14.16)$$

The cancellation of the partial derivatives of L, with respect to λ and μ, leads to a system of equations of the form:

$$-n + \frac{n - n_0}{2\lambda} + \sum_{i=1}^{n-n_0}\frac{I_1'(z_i)}{I_1(z_i)}\frac{z_i}{2\lambda} = 0 \quad (14.17)$$

$$-\frac{n - n_0}{2\beta} + \sum_{i=1}^{n-n_0}\frac{x_i}{\beta^2} - \sum_{i=1}^{n-n_0}\frac{I_1'(z_i)}{I_1(z_i)}\frac{z_i}{2\beta} = 0 \quad (14.18)$$

where:

$$z_i = 2\sqrt{\frac{\lambda x_i}{\beta}} \tag{14.19}$$

$I_1'(z)$ is the first derivative of $I_1(z)$. Solving the equation system (14.17) and (14.18), member to member, it can be obtained that:

$$\overline{x} = \hat{\lambda}\hat{\beta} \tag{14.20}$$

The properties of the Bessel function allow writing the derivative of I_1 as:

$$I_1'(z) = I_0(z) - \frac{1}{z}I_1(z)$$

where I_0 is the Bessel function of zero order. Using this expression in Eq. (14.17) and substituting $\overline{x}$ in the expression (14.19) $\frac{z_i}{2\lambda} = \sqrt{\frac{x_i}{\overline{x}}}$, we get:

$$g(\lambda) = \sum_{i=1}^{n-n_0} \frac{I_0(z_i)}{I_1(z_i)}\sqrt{x_i} - n\sqrt{\overline{x}} = 0$$

whose solution is a unique and positive value of $\hat{\lambda}$. The function $g(\lambda)$ is monotonous and decreasing; its solution lies within the interval $\left[10^{-4}, 5\hat{\lambda}_{\text{mom}}\right]$, where $\hat{\lambda}_{\text{mom}}$ corresponds to the value of the parameter estimated by the method of moments. The value of β is obtained in expression (14.20). Le Barbé and Lebel (1997) proposed an alternative procedure for the estimation of parameters of the leak distribution. This method proposes using the mean number of dry days n_0 in a time period T of the total of n days. This procedure is of singular importance because the data of dry days is not used directly in the estimation of the parameters of the methods of moments. This way, the parameters will be given by:

$$\hat{\lambda} = -\text{Log}_e\left(\frac{n_0}{n}\right) \qquad \hat{\beta} = \frac{\overline{\mathrm{X}}}{\hat{\lambda}} \tag{14.21}$$

14.5 Application of the Leak Distribution

To exemplify the application of this law in the modeling and analysis of the spatial variability of precipitation, hydrologic region No. 10, located in northwest Mexico, is chosen as shown in Fig. 14.2. This is one of the most important regions in the country because it constitutes a productive area from an agricultural point of view, with the Huites dam as one of the main sources of water supply. With an approximate surface area of 80,000 km^2, the region is divided into nine basins: Fuerte, Sinaloa, Mocorito, Laguna de Caimanero, Culiacán, San Lorenzo, Elota, Piaxtla, and El Quelite. There are 93 weather stations in the region with historical records going back 35 years in average.

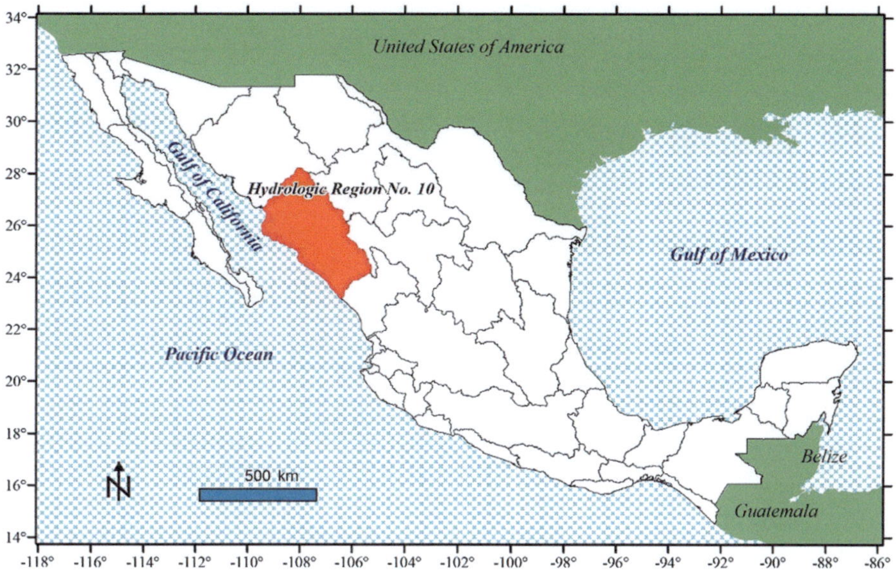

Fig. 14.2 Map with the location of hydrologic region No. 10

The mean annual precipitation in this region is 800 mm, close to the national average which reaches 770 mm. Rain in this region presents a well-defined seasonality, being the most abundant period takes place between June and October. However, it is also a sensitive region, exposed to climatic changes. Long periods of drought and long periods of floods have been recorded, induced mainly by extreme meteorological event. Also, because this region is near the Sierra Madre, it is exposed to what is known as "Mexican Tropical Front."

14.6 Model Validation

The historical series of the 93 weather stations within hydrologic region 10 were used, and the parameters of the leak distribution were estimated using the three procedures described earlier. Based on the χ^2 statistic, the best fits were selected and the method of moments showed, in most cases, the best fit to the data samples (Gutiérrez-López et al. 2002). Figure 14.3 shows the fit of the leak distribution to the historical series of daily rainfall at the Tamazula station, for the 1947–1985 period, using parameters estimated by the method of moments. As mentioned earlier, parameter β offers an additional advantage, because with it the distribution of rain can be inferred considering different time intervals. This way, parameters λ^* and β^*, inferred in a simplified manner, can be estimated directly by multiplying by the desired time interval. That is, the parameters have a specific physical meaning and can be used for different values of T. This way, parameter β remains

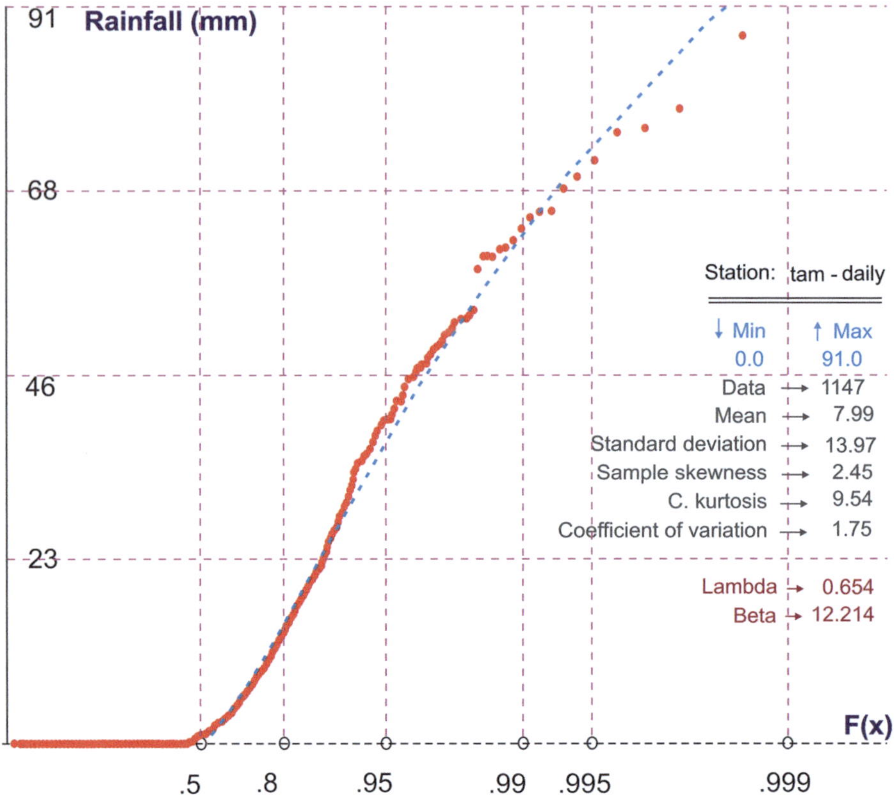

Fig. 14.3 Daily rainfall distribution for the month of August in the Tamazula station (1947–1985)

constant through time and scale parameter λ can be calculated for any duration T, that is:

$$\beta^* = \hat{\beta}; \quad \lambda^*{}_T = k\hat{\lambda}_T \tag{14.22}$$

It will be accepted that there is no stationarity in cases where the value of λ has a variation higher than 20 % between the maximum and the minimum value, within the considered period. That is to say that it will be considered a stationary process while remaining constant for 80 % of the period; otherwise, the scale parameter must be estimated as (Gutiérrez-López et al. 2002):

$$\lambda^*{}_T = \frac{1}{T}\int_0^T \hat{\lambda}(t)\mathrm{d}t \tag{14.23}$$

When daily rainfall data is available, it is recommended to choose T equal to one day (Lebel and Le Barbé 1997; Le Barbé and Lebel 1997). Figures 14.3, 14.4, and

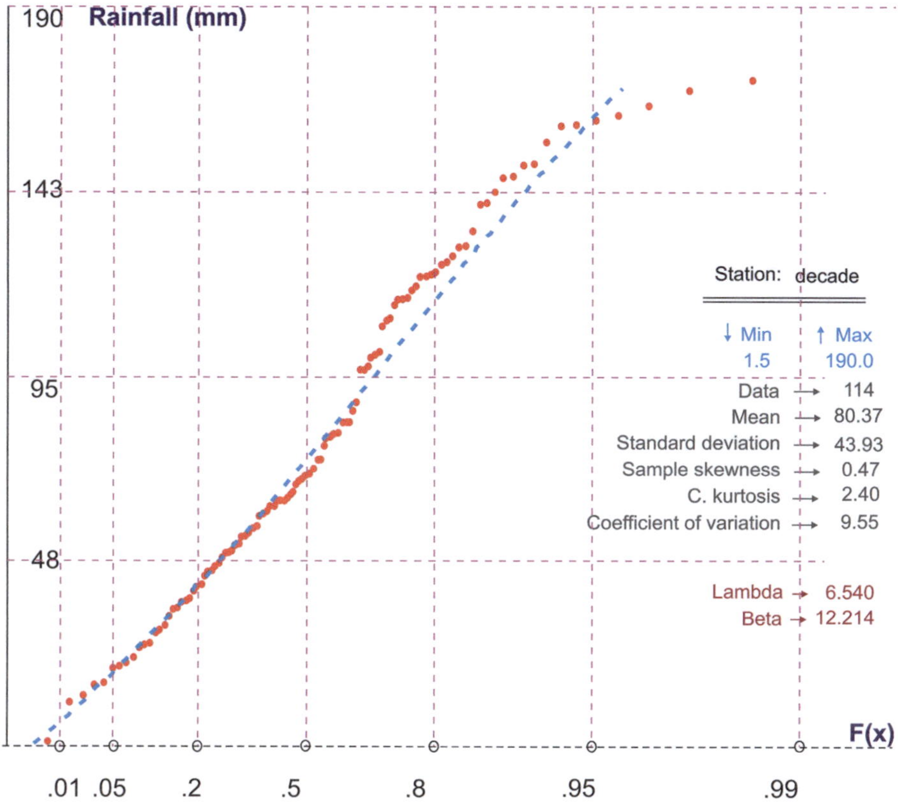

Fig. 14.4 Ten-day cumulative rainfall distribution for the month of August in the Tamazula station (1947–1985)

14.5 and Table 14.1 show the model's validation for the Tamazula station for the 1947–1985 period. The estimated parameters from the daily rainfall analysis, using the method of moments, were $\hat{\lambda} = 0.654$ and $\hat{\beta} = 12.214$. From them, it can be deduced parameters, for example, for a 10-day interval, using Eq. (14.22), that is, $\lambda^* = 0.654 \times \ 10 = 6.54$ and $\beta^* = 12.214$, whereas the real parameter values, estimated with the original data series, for the 10-day interval, are $\hat{\lambda} = 7.071$; $\hat{\beta} = 11.465$. Likewise, parameters for any duration T can be estimated.

Likewise, for validation purposes, a calculation was conducted for the 93 stations analyzed in hydrologic region 10. Figure 14.6 shows the distribution of the relative differences of that validation $\Delta\lambda = \left(\lambda_T^* - \hat{\lambda}_T\right)/\hat{\lambda}_T$ and $\Delta\beta = \left(\beta_T^* - \hat{\beta}_T\right)/\hat{\beta}_T$. It is observed that 48 % of the stations show differences of less than a 10 % absolute value for parameter λ 40 % for parameter β. Only the Huahuapan station has a difference of more than 90 %. The distribution of parameter $\Delta\lambda$ is positive asymmetric, whereas the distribution of β is negative (Gutiérrez-López et al. 2002).

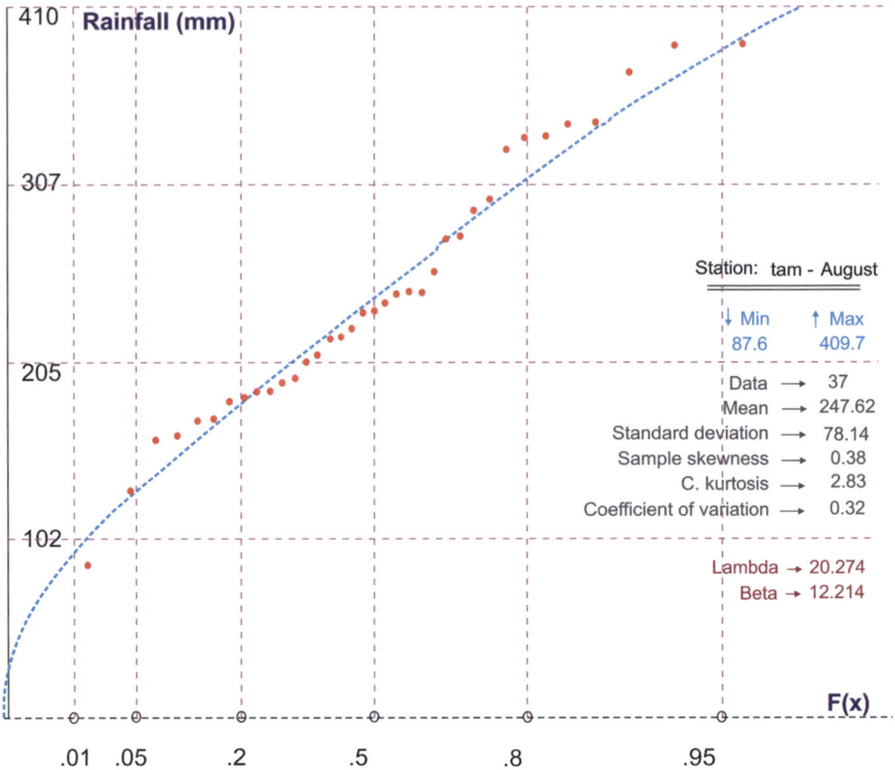

Fig. 14.5 Monthly rainfall distribution for the month of August in the Tamazula station (1947–1985)

Table 14.1 Validation of the values of scale and form parameters of the leak distribution

	Estimated from the original data series	Estimation with Eq. (14.23)
$t = 1$ day	$\hat{\lambda} = 0.654$ and $\hat{\beta} = 12.214$	----------------------
$t = 10$ days	$\hat{\lambda} = 7.071$ and $\hat{\beta} = 11.465$	$\lambda^* = 6.54$ and $\beta^* = 12.214$
$t = 30$ days	$\hat{\lambda} = 20.083$ and $\hat{\beta} = 12.33$	$\lambda^* = 20.274$ and $\beta^* = 12.214$

When observing a map of region 10, with the spatial evolution of these two parameters ($\hat{\lambda}$ and $\hat{\beta}_T$), it is evident that there is a relationship between the forms of the contour lines and the region's topographic environment. Figures 14.7 and 14.8 show, respectively, the number of events with rain and mean depth per event. It is observed that the spatiotemporal distribution of parameter $\hat{\lambda}$ is strongly influenced by the topography and remains similar throughout the rainy season. This is because the complicated topography of the Sierra Madre changes quickly with elevation and can cause strong fluctuations between two consecutive seasons. This effect is illustrated when comparing the results obtained at Las Truchas station at an

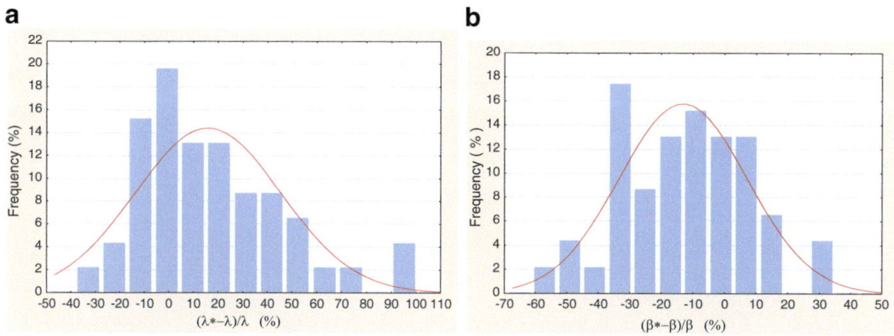

Fig. 14.6 Histograms of $\Delta\lambda = \left(\lambda_T^* - \hat{\lambda}_T\right)/\hat{\lambda}_T$ (**a**) and $\Delta\beta = \left(\beta_T^* - \hat{\beta}_T\right)/\hat{\beta}_T$ (**b**) for the 93 stations of hydrologic region 10

elevation of 2,700 m and Huahuapan station at 1,150 m (Fig. 14.9). These stations are separated by only 7.3 km but with a difference in elevation of 1,550 m, which causes a singularity in the spatial variation of parameter $\hat{\lambda}$ (a depression). See, for example, the map from July 5 and August 10 (Fig. 14.7). The relationship between topography and $\hat{\lambda}$ is more uniform in the north, a region that experiences a dry subtropical climate typical of a coastal desert (310-mm annual precipitation in Topolobampo); this is because the most humid areas are located on the southern portion of the region, at the slopes of the Pacific Ocean.

It is also observed that the behavior is different for parameter $\hat{\beta}$. Maximum values are located near the coast, indicating high probability of rain in the coastal hills, compared with little rainfall in the plateau of the Sierra Madre. In general, it can be said that, altogether, region 10 presents a strong correlation between the number of rain events, average height of rain, and topography. The average amount of rain is between 13 and 17 mm per event, which corresponds to similar values for deserts at the same latitude, for example, in West and Central Africa (Sahel desert), where the average rain per event is between 12 and 13 mm (Tapsoba 1997). These results can also be expressed according to a spatiotemporal variation of the rainfall regime, that is, according to latitude or distance to the ocean. This type of analysis can be illustrated with the help of Hovmöller diagrams (Gutiérrez-López et al. 2005).

Finally, another application of the physical meaning that the parameters of the leak distribution represent lies in the possibility of associating them with concepts of hydrological risk management, because parameter $\hat{\beta}$ can be associated with a measure of probability of extreme rainfall and $\hat{\lambda}$ can be associated with the rainfall occurrence probability (drought), as proceeded in the next section.

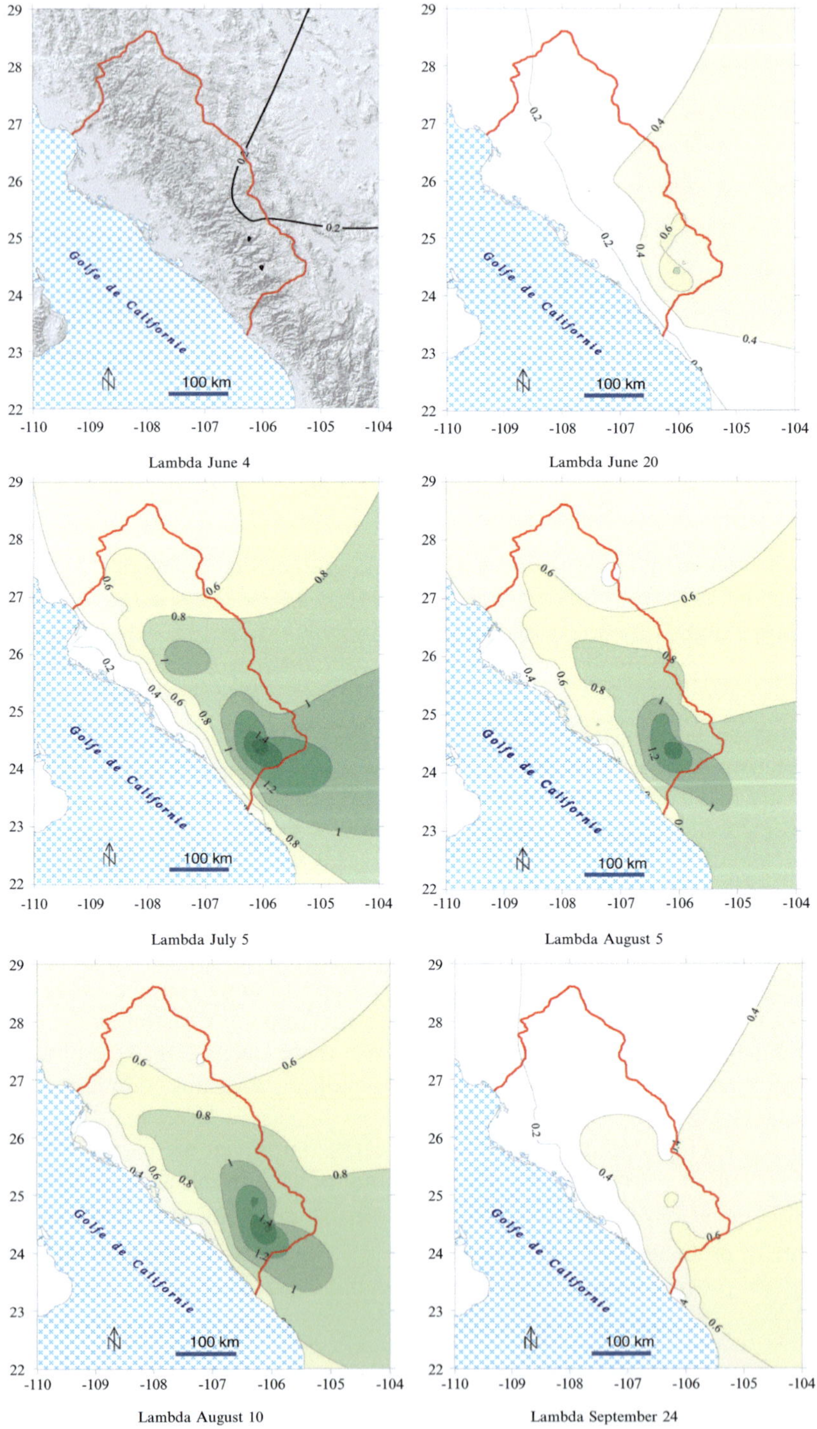

Fig. 14.7 Cartography of spatial evolution of the number of rainy days (λ)

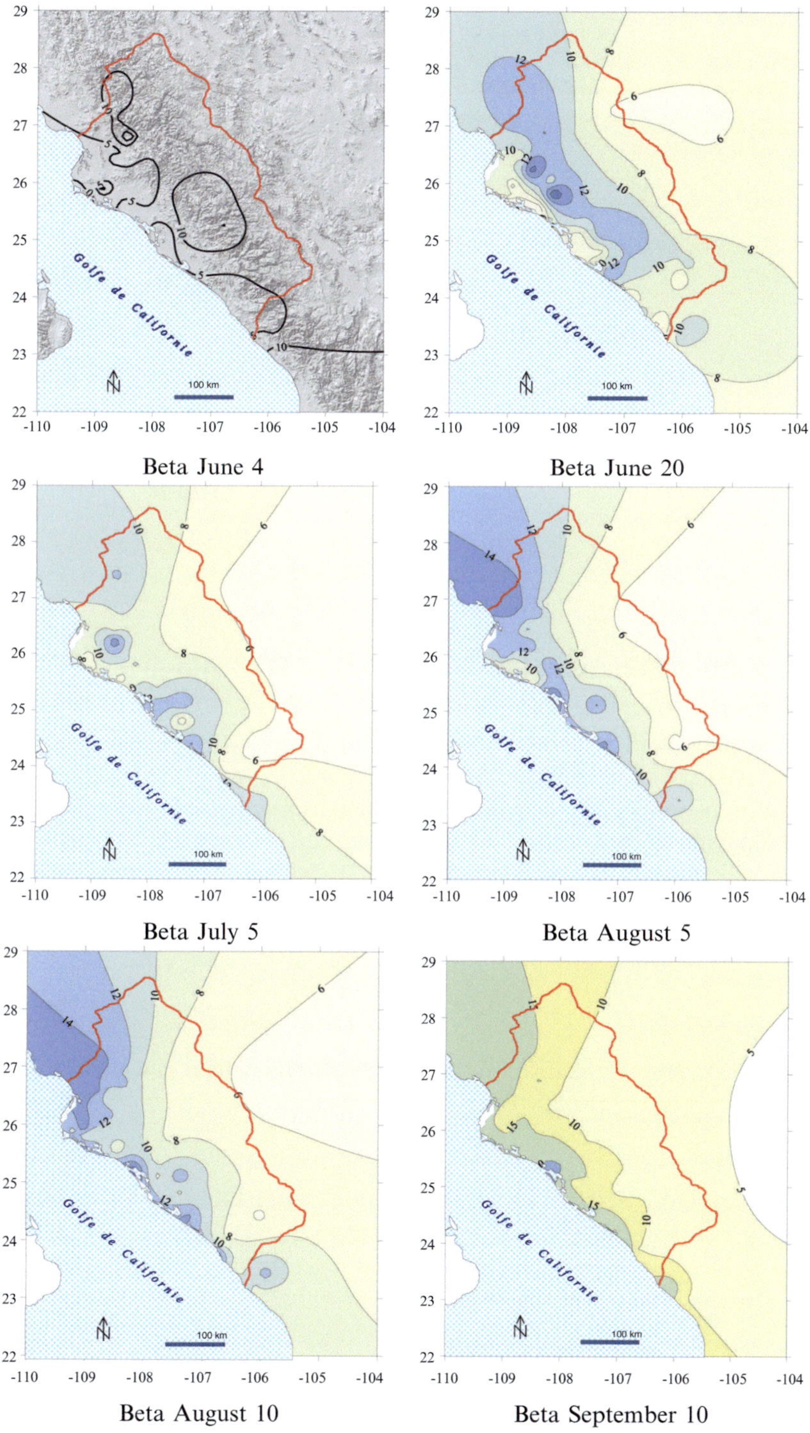

Fig. 14.8 Cartography of spatial evolution of mean depth of rain per event (β)

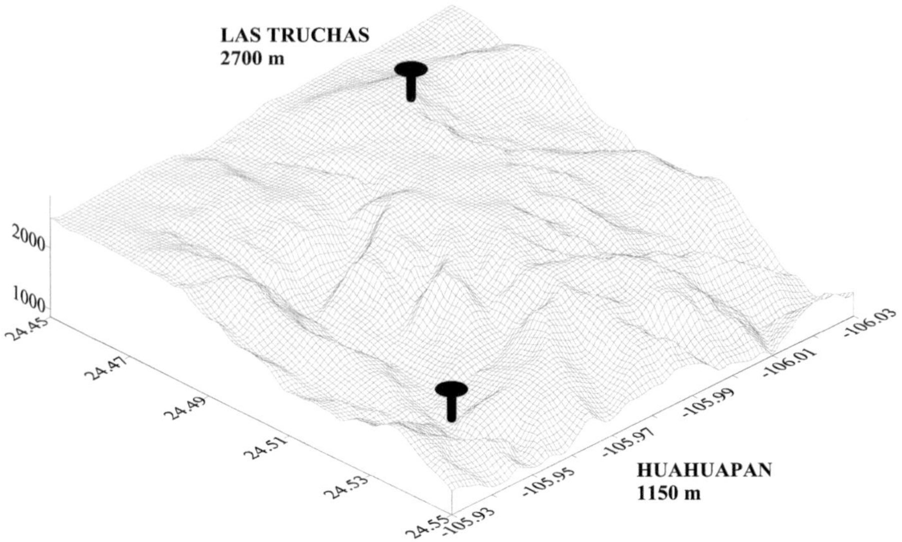

Fig. 14.9 Schematic comparison of location of stations: Las Truchas and Huahuapan (7-km distance)

14.7 Estimation and Representation of Hydrological Risk

Type I extreme value distributions (or Gumbel) are used frequently to model extreme events. This distribution is derived from a non-enclosed exponential-type distribution (Kite 1988). The type I distribution, based on the exponential distribution of a parameter, can be derived as follows:

$X_{\max}$ is defined as the maximum value of an independent series of variables ε_i with a cumulative probability distribution given by $P(x) = P(\varepsilon_i \leq y)$:

$$P(X_{\max} \leq y) = P(\varepsilon_1 \leq y,\ \varepsilon_2 \leq y,\ \ldots, \varepsilon_n \leq y) \tag{14.24}$$

$$P(X_{\max} \leq y) = [F(y)]^n \tag{14.25}$$

Taking the extreme of the distribution $P(y)$ as an exponential law, we obtain:

$$F(y) = 1 - \beta \mathrm{e}^{-y/\beta} \tag{14.26}$$

Based on Eq. (14.25), if we accept that $\ln(\alpha n)$ is a normalized constant, we get $P[X_{\max} \leq y + \ln(\beta n)] = [P(y + \ln(\beta n))]^n$ and from Eq. (14.26):

$$P[y + \ln(\beta n)] = 1 - \beta \mathrm{e}^{-[y+\ln(\beta n)]} \tag{14.27}$$

$$P[X_n \leq y + \ln(\beta n)] = \left\{1 - \beta \mathrm{e}^{-[y+\ln(\beta n)]}\right\}^n;$$
$$P[X_n \leq y + \ln(\beta n)] = [1 - \mathrm{e}^{-y}/n]^n$$

If $n \to \infty$:

$$\lim_{n\to\infty} P[X_n \leq y + \ln(\beta n)] = \lim_{n\to\infty}[1 - \mathrm{e}^{-y}/n]^n$$
$$\lim_{n\to\infty} P[X_n \leq y + \ln(\beta n)] = \mathrm{e}^{-\mathrm{e}^{-y}} \tag{14.28}$$

This is a reduced form of type I probability distribution. Taking the reduced variable $y = \beta(x - x_0)$, where β is a scale parameter and x_0 is a measure of the central tendency, the probability distribution type I can be written as:

$$P(x) = \mathrm{e}^{-\mathrm{e}^{-\beta(x-x_0)}} \tag{14.29}$$

According to the procedure proposed by Lebel and Laborde (1988), there are several applications of the scale parameter of the Gumbel distribution. Slimani and Lebel (1987) found that this distribution has a fit similar to that of an exponential distribution for the maximum monthly events. Following this same idea, Tapsoba (1997) proposes that the number of events recorded from a day j_1 to a day j_2 follow a Poisson distribution with parameter λ (sum of the days included within these two dates). If Hx_{j1-j2} is the maximum depth of rainfall recorded each year, during the periods between the day j_1 and j_2, $P\left(Hx_{j1-j2} \leq Hx\right)$, this depth can be obtained as:

$$P\left(Hx_{j1-j2} \leq Hx\right) = \sum_{0}^{\infty} \frac{\lambda_T^* \, \mathrm{e}^{-\lambda_T^*}}{i\,!} \left[1 - \mathrm{e}^{\left(-\frac{Hx}{\beta}\right)}\right] \tag{14.30}$$

Simplifying, we get:

$$P\left(Hx_{j1-j2} < 0\right) = 0 \tag{14.31}$$
$$P\left(Hx_{j1-j2} \leq Hx\right) = \mathrm{e}^{\left[-\lambda_T^* \, \mathrm{e}^{(-Hx/\beta)}\right]} \tag{14.32}$$

The previous expression is the Gumbel distribution for positive values. On the other hand, if we consider that the events are independent and that the average number of events within a considered time interval is constant, then the binomial and Poisson distributions can be used to evaluate the hydrological risk. If the occurrence of events follows a Poisson distribution and the magnitudes of the events follow a type I distribution, the risk in design events will be a function of parameter λ.

The scale parameter of the Gumbel distribution, fitted to the maximum annual events of the month of August, at the Tamazula weather station, is almost the same as that obtained when fitting the leak distribution to the daily rainfall sample data in the month of August for that same recorded period (Fig. 14.10). This figure shows the comparison of direct fit of the Gumbel distribution to the maximum monthly values ($\beta = 13.3$) and of the law inferred from the leak distribution to daily rain ($\beta = 12.3$). The study of the scale parameter of the Gumbel distribution must be used as a risk index, associated with extreme rain. For the annual maximum series, (37 years) of the month of August, from the fit to the Gumbel distribution, $\beta =$

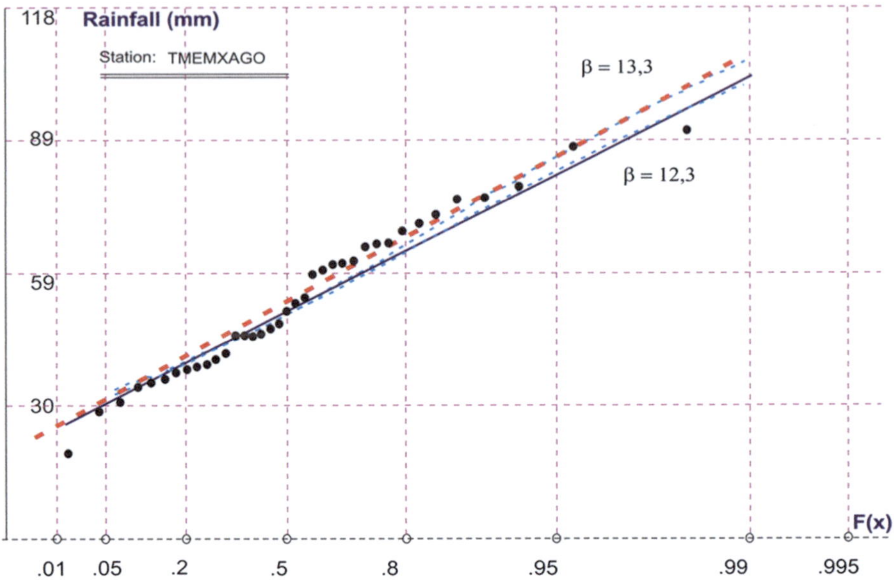

Fig. 14.10 Comparison of fit with scale parameters for the analysis of maximum rainfall for the month of August in Tamazula station

13.28 and $x_0 = 45.76$. The scale parameter is noticeably the same as for the analysis of daily rainfall in the month of August, using the leak distribution: $\beta = 12.214$ and $\lambda = 0.654$. The latter is another indirect confirmation of the good ability of the leak distribution to represent rainfall regime in North Mexico.

In effect, the leak distribution is supported by the property that the number of events recorded during a period of fixed duration T follows a Poisson distribution of parameter λ_T. Therefore, it can be affirmed that the scale parameter of the leak distribution is equal to the scale parameter of the Gumbel distribution; consequently, the spatial distribution of parameter β represents the risk of extreme precipitation occurrence in a given region. Similarly to Eqs. 14.33, 14.34, and 14.35, if X_T is the maximum annual rainfall over this duration T, we get:

$$P(X_T \leq y) = \sum_{i=0}^{\infty} \frac{\lambda^i e^{-\lambda}}{i\,!} \left[1 - e^{\left(\frac{-y}{\beta}\right)}\right]^i \tag{14.33}$$

simplifying:

$$P(X_T \leq y) = e^{\left[-\lambda e^{(-y/\beta)}\right]} \tag{14.34}$$

$$P(X_T < 0) = 0 \tag{14.35}$$

Figures 14.11 and 14.12 show the precipitation regime of region 10 for August. The spatial distribution of scale parameter β corresponds to an extreme rainfall risk map, whereas the map of parameter λ_{31} represents a drought risk map.

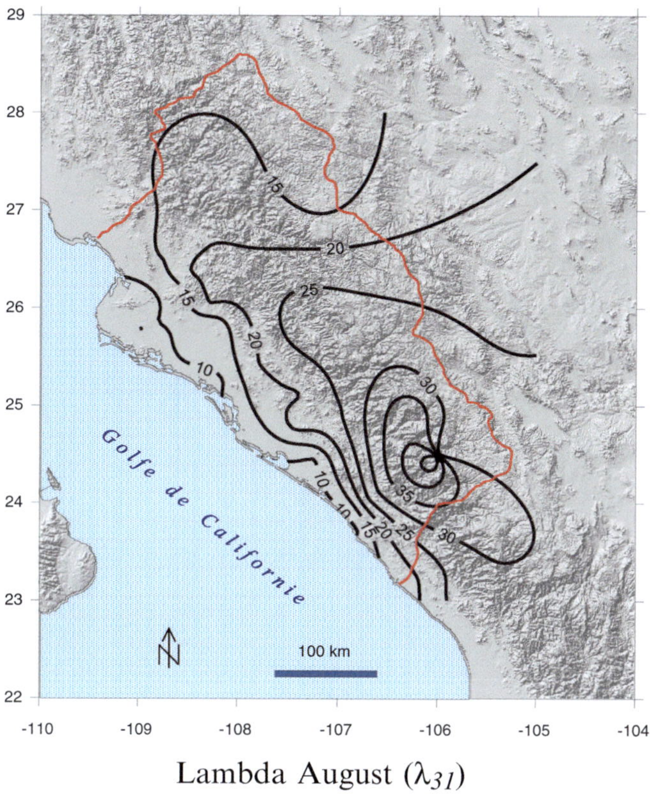

Fig. 14.11 Interannual mean number of events per day for the month of August, risk of drought for August in region 10

The graphic representation of the scale parameter of the Gumbel distribution allows the analysis of the data included in the historical precipitation series. Hydrologic region 10 is an example of a typical area exposed to extreme events occurring mainly in coastal areas. It is worth emphasizing that the same is obtained when analyzing the risk of maximum rainfall (Fig. 14.12).

This map of risk of extreme rain shows a more humid sector located near the Southern Coast region, which decreases toward the north. Likewise, the influence of the topography of the Sierra Madre is observed, which represents a transition between humid tropical climate in the south and semiarid climate in the north. If we follow the 20-mm isoline (Fig. 14.11), for example, it is evident that it delimitates the mountainous area almost perfectly.

Regarding the map of risk of drought, on the contrary, we observe a distribution which is analogous to that of the mean annual isohyets. The isolines are more variable near the coast, and maximum occurrence is observed over the mountain, mainly in the region's southern portion.

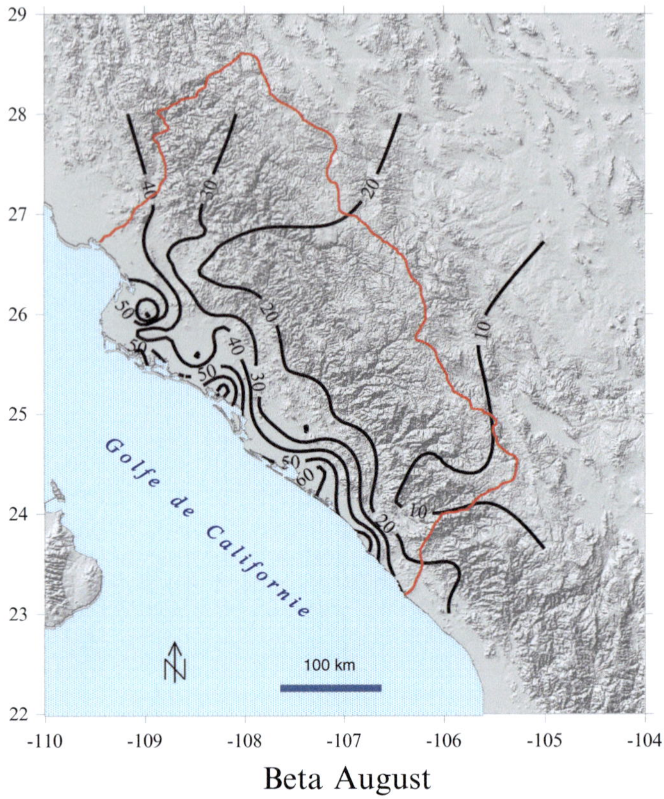

Fig. 14.12 Interannual mean depth in mm of rain per event, risk of extreme rain for August in region 10

14.7.1 Parameter λ of the Leak Distribution as a Drought Index

The standardized precipitation index (SPI) is one of the indexes used worldwide to evaluate and monitor drought (McKee et al. 1993; Edwards and McKee 1997). In general, this index is calculated for time intervals between 3 and 48 months, and it allows identifying and classifying possible areas that are susceptible to experience negative impacts due to drought. The methodology for the calculation of SPI is based on fitting a series of historical records of total monthly precipitation to an incomplete gamma distribution and the transformation of the resulting data to the standard normal distribution function. It is worth noting that the deduction of the leak distribution using an exponential distribution is a particular case of the so-called family of gamma distributions. In this context, the SPI represents the number of standard deviations that each precipitation record deviates from the historical average. Precipitation records higher than the historical average of the corresponding month will have positive SPI values (presence of humidity), whereas precipitation records below the historical average will produce negative SPI values (lack of humidity).

In general terms, it can be said that a drought event begins when a marked tendency of continuous negative SPI values are observed through time. Subsequently, the drought event ends when the SPI value reaches positive values. As can be intuited, the estimation of drought indexes requires the use of numerous variables. Among the principal variables, we can mention precipitation, evapotranspiration, flows in natural streams, levels of lakes, aquifer levels, etc. Although other simplified drought indexes have been proposed (Pita 2003), up to this date no work has explicitly included the number of rainy days (intuitively included in monthly precipitation) so as to estimate a drought index. The present application indicates another advantage of the leak distribution by establishing a relationship between the location parameter of this law and the SPI. Based on the fact that the duration of a drought event can be defined in relation to its detection, from its initial stages to its final stage, through time, it seems appropriate to introduce the variable corresponding to the number of rainy days. This way, we propose using the map of drought risk presented in Fig. 14.11 and comparing it with the traditional drought index SPI (Fig. 14.13) calculated for region 10 for the year 1968, estimated as a "normal" year in the North Mexico region (Nouvelot and Descroix 1996). The two figures show the same behavior; however, the spatial representation of the SPI over

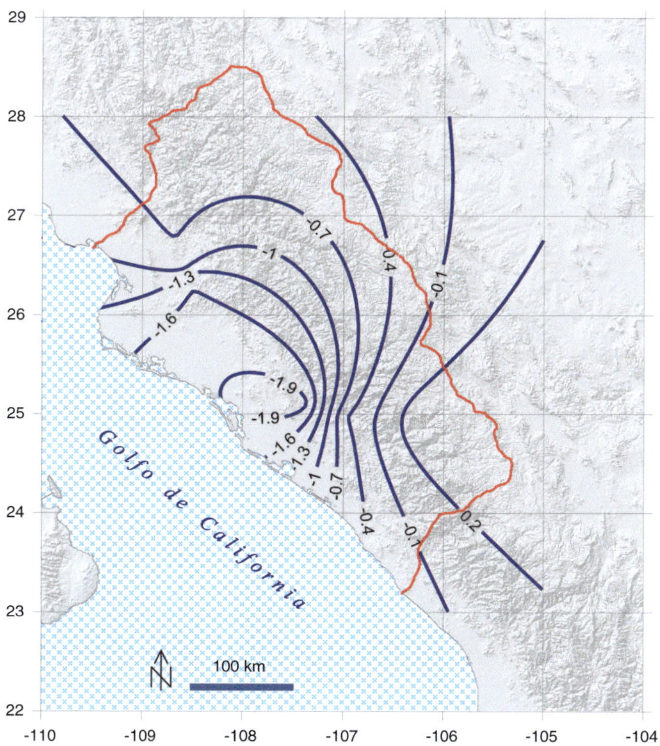

Fig. 14.13 SPI estimated according to traditional procedure. Values for August 1968 in region 10

region 12 seems to have no relationship whatsoever with the region's relief and topography, whereas the spatial distribution of λ reflects an orographic influence in the area (Fig. 14.11). We propose fitting the SPI values using a relationship that will allow considering the effect of the area's topography. Thus, the SPI-λ relationship is obtained, which will allow estimating a modified SPI which will be denoted as SPI*. This relationship, for the case of hydrologic region 10, has the form:

$$SPI^* = 1.734 - \frac{55.6303}{\lambda} \quad (14.36)$$

where:

λ is the number of events within a period of time T.
SPI* is the drought index, estimated according to the number of rainy days.

Figure 14.14 shows the spatial distribution of the modified SPI*; a clear definition of areas and a better relationship with the orography of the Sierra Madre can be observed. Finally, it can be affirmed that the modified SPI* reproduces in a reliable manner the spatial distribution of parameter λ, already defined previously as a drought risk index.

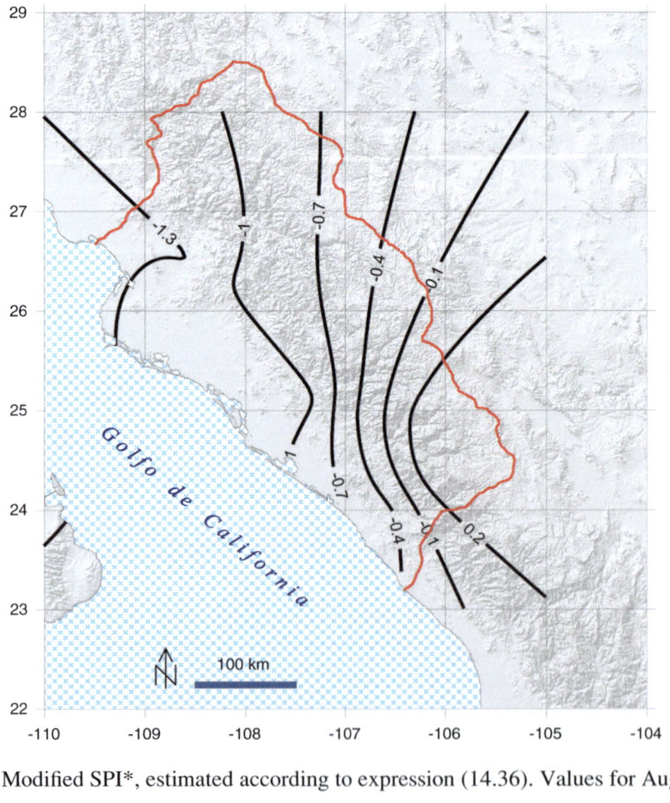

Fig. 14.14 Modified SPI*, estimated according to expression (14.36). Values for August 1968 in region 10

14.8 Conclusions

Rain in Northern Mexico is, without doubt, a complex phenomenon to study; however, the results of the application of the leak distribution have allowed a complete knowledge of the phenomenon. The leak distribution shows several advantages. First of all, it allowed determining the spatial and temporal distribution of rain, by obtaining the two distribution parameters (λ or number of rainfall events within a period of time T and β or average depth of a rainfall event). The simplicity of these parameters allows finding simple relationships with their geographic and physical environment.

It is shown that the proposed model is significantly consistent with the description of changes associated with extreme events, because it allows identifying the normal rainy season in a reliable manner, as well as the effects of the hurricane season, through the evolution of parameter λ. It is observed that at the end of September, λ decreases due to the ending of the rainy season; however, in some stations an increase in precipitation is observed, caused by extreme events, in a band that crosses hydrologic region 10 from south to north, following a direction almost parallel to the coast.

The graphic representation of the scale parameter (β) of the Gumbel distribution is the adequate tool to explore the information contained in an historical precipitation series. Hydrologic region 10 shows a typical behavior of a region affected by extreme events, because the main risks occur in the coastal areas and in the area of high rainfall. Finally, it is considered that the proposed methodology and the use of the leak distribution adapt to the hydrological risk estimation needs and allow obtaining a deeper knowledge of the regional climatology.

In relation to the comparison between the SPI and the number of rainy days, it can be observed that parameter λ alone represents acceptably a drought risk and that it allows obtaining, through a simple relationship, the SPI. Finally, it can be affirmed that the modified SPI*, related to the spatial distribution of parameter λ of the leak distribution, allows estimating in a faster and more reliable manner a drought index, which is generally hard to estimate and, especially, to interpret. It is worth mentioning that these aridity indexes are usually estimated through different physiographic or climatologic characteristics of the region, for example, potential of evaporation (Arora 2002).

References

Abi-Zeid I, Parent É, Bobée B (2004) The stochastic modeling of low flows by the alternating point processes approach: methodology and application. J Hydrol 285(1–4):41–61

Arora VK (2002) The use of the aridity index to assess climate change effect on annual runoff. J Hydrol 265(1–4):164–177

Babusiaux C (1969) Etude statistique de la loi de fuites. Thèse 3eme cycle, Faculté des sciences de Paris

Bacchi B, Becciu G, Kottegoda N (1994) T. Bivariate exponential model applied to intensities and durations of extreme rainfall. J Hydrol 155(1–2):225–236
Calder IR (1986) A stochastic model of rainfall interception. J Hydrol 89(1–2):65–71
Cameron D, Beven K, Tawn J (2000) An evaluation of three stochastic rainfall models. J Hydrol 228(1–2):130–149
Coles S, Pericchi LR, Sisson S (2003) A fully probabilistic approach to extreme rainfall modeling. J Hydrol 273(1–4):35–50
Edwards D, McKee T (1997) Characteristics of 20th century drought in the United States at Multiple Time scales. Climatology Report No. 97–2. Colorado State University, Department of Atmospheric Science, Paper No. 634, 155 p
Escalante C, Reyes L (2002) Técnicas estadísticas en hidrología. Facultad de Ingeniería, Universidad Nacional Autónoma de México, UNAM, México. 298 p
Gutiérrez-López MA (2003) Modélisation stochastique des régimes pluviométriques a l'échelle régionale pour la prévision des crues au Nord-Mexique. Doctoral thesis, Institute National Polytechnique de Grenoble, Grenoble, France. http://www.lthe.fr/PagePerso/boudevil/THESES/gutierrez_lopez_03.pdf.
Gutiérrez-López A, Lebel T, Descroix L (2002) Statistical analysis for modelling the hydrological risk in Northern Mexico. International Association for Hydraulic Research (IAHR) Hydraulic and Hydrological Aspects of Reliability and Safety of Hydraulic Structures, St. Petersburg
Gutiérrez-López A, Lebel T, Mejía R (2005) Estudio espacio-temporal del régimen pluviométrico en la zona meridional de la República Mexicana. Revista Ingeniería Hidráulica en México, IMTA XX(1):57–65
Hughes RL (2003) On detecting anomalous behaviour in runs. J Hydrol 278(1–4):253–266
Istok JD, Boersma L (1989) A stochastic cluster model for hourly precipitation data. J Hydrol 106(3–4):257–285
Kite GW (1988) Frequency and risk analyses in hydrology, Rev ed. Water Resources Publications, Littleton, 257 p
Le Barbé L, Lebel T (1997) Rainfall climatology of the HAPEX-Sahel region during the years 1950–1990. J Hydrol 188–189(1–4):43–73
Le Barbé L, Lebel T, Tapsoba D (2002) Rainfall variability in West Africa during the years 1950–1990. J Clim 15(2):187–202
Lebel T, Laborde J (1988) A geostatistical approach for areal rainfall statistics assessment. Stoch Hydrol Hydraul 2:245–261
Lebel T, Le Barbé L (1997) Rainfall monitoring during HAPEX-Sahel. 2. Point and areal estimation at the event and seasonal scales. J Hydrol 188–189(1–4):97–122
López-Segovia L, Villaseñor-Alva J, Vaquera-Huerta H (2002) Dos pruebas de bondad de ajuste para procesos de Poisson no homogéneos. Revista Agrociencia 36:703–712
McKee T, Doesken N, Kleist J (1993) Drought monitoring with multiple time scales. American Meteorological Society, 9th conference on applied climatology, pp 233–236
Nouvelot JF, Descroix L (1996) Aridité et sécheresse du Nord-Mexique. Revue Trace México 30:9–25
Onof C, Wheater HS (1994) Improvements to the modelling of British rainfall using a modified Ryom Parameter Bartlett-Lewis Rectangular Pulse Model. J Hydrol 157(1–4):177–195
Önöz BY, Bayazit M (2002) Troughs under threshold modeling of minimum flows in perennial streams. J Hydrol 258(1–4):187–197
Pandey GR, Nguyen V-T-V (1999) A comparative study of regression based methods in regional flood frequency analysis. J Hydrol 225(1–2):92–101
Pita M (2003) Un nouvel indice de sécheresse pour les domaines méditerranéens. Application au bassin du Guadalquivir (sud-ouest de l'Espagne). Publications de l'Association Intenrationale de Climatologie, vol 13, Nice, pp 225–234
Saporta G (1990) Probabilités, analyse des données et statistique, Editions. Technip, Paris, 193 p. ISBN 2-7108-0565-0

Seguis L (1989) La pluviometrie au Togo, Caracterisation Agronomique. Orstom Institut Francais de Recherche Scientifique pour le Developpement en Cooperation. Centre de Lomé. pp 63

Sharma TC (1996) Simulation of the Kenyan longest dry and wet spells and the largest rain-sums using a Markov model. J Hydrol 178(1–4):55–67

Singh P, Kumar N (1997) Effect of orography on precipitation in the western Himalayan region. J Hydrol 199(1–2):183–206

Sivapalan M, Blöschl G (1998) Transformation of point rainfall to areal rainfall: intensity-duration-frequency curves. J Hydrol 204(1–4):150–167

Slimani M, Lebel T (1987) Comparison of three methods of estimating rainfall frequency parameters according to the duration of accumulation. Hydrologic frequency modeling. In: Singh VP (eds) Proceedings of the international symposium on flood frequency and risk analyses. Louisiana State University, Springer, Baton Rouge, pp 277–291, May 1986

Tapsoba D (1997) Caracterisation evenementielle des regimes pluviometriques Ouest Africains et de leur recent changement. Th. D., Université de Paris XI (Orsay)

Wilks DS (1998) Multisite generalization of a daily stochastic precipitation generation model. J Hydrol 210(1–4):178–191

Wotling G, Bouvier C, Danloux JY, Fritsch JM (2000) Regionalization of extreme precipitation distribution using the principal components of the topographical environment. J Hydrol 233(1–4):86–101

Part IV
IWRM and Water Governance: Climate Change, Social, Economic, Public Health and Cultural Aspects

Chapter 15
Water Resources Management for Sustainable Environmental Public Health

Shimelis Gebriye Setegn

Abstract Water is essential for life and social and economic development. Its exploitation and use must be well planned and managed in a sustainable manner. In many developing countries, lack of adequate, clean, and safe water, pollution of aquatic environments, and the mismanagement of natural resources are still major causes of environmental health problem and mortality. With a human population that is continuing to grow, the management of water resources will become of vital importance. In order to accommodate more growth, sustainable freshwater resource management will need to be included in future development plans and implementations. One of the major environmental issues of concern to policy-makers is the increased vulnerability of surface and groundwater quality. Furthermore, the main challenge for the sustainability of water resources is the control of water pollution. To understand the sustainability of the water resources, one needs to understand the impact of future land use and climate changes on the water resources. Providing safe water and basic sanitation to meet the millennium development goals will require substantial economic resources, sustainable technological solutions, and courageous political will. A balanced approach to water resources development, on the one hand, and controls for the protection of water quality, on the other hand, is required for sustainability of water resources and environmental public health. In addition to providing improved water and sanitation services, we must ensure that these services provide safe drinking water; adequate quantities of water for health, hygiene, agriculture, and development; and sustainable sanitation approaches to protect health and the environment.

Keywords Water • Sanitation • Hygiene • Water and health • Water quality • Waterborne disease • Drinking water • Water pollution • Water scarcity

---

S.G. Setegn (✉)
Department of Environmental and Occupational Health and Global Water for Sustainability Program (GLOWS), Florida International University, 11200 SW 8th Street, Miami, FL 33199, USA
e-mail: ssetegn@fiu.edu

S.G. Setegn, M.C. Donoso (eds.), *Sustainability of Integrated Water Resources Management*, DOI 10.1007/978-3-319-12194-9_15

15.1 Introduction

Water is crucial and an essential life-sustaining element to the existence and success of human being. The human population is expected to grow from 6.7 billion to 9.2 billion people by 2050 (US Census Bureau 2009 as cited by IOM, 2009). With a human population that is continuing to grow, the management of water resources will become of vital importance. In order to accommodate more growth, sustainable freshwater resource management will need to be included in future development plans and implementations. The main challenge for the sustainability of water resources is the control of water pollution. To understand the sustainability of the water resources, one needs to understand the impact of future land use and climate changes on the natural resources.

In the USA, groundwater is one of the nation's most important natural resources that provide drinking water for more than one-half of the population (Solley et al. 1998) as cited by Twarakavi and Kaluarachchi (2006). It is also the source of much of the water used for irrigation. Even though groundwater resources appear to be ample, spatial availability of groundwater varies at large (Alley et al. 1999). One of the major environmental issues of concern to policy-makers is the increased vulnerability of groundwater quality. Twarakavi and Kaluarachchi (2006) propose a methodology to address sustainability of groundwater quality considering land use changes, aquifer vulnerability to multiple contaminants and public health risks. The study divided the methodology to quantify GWQ into the three steps: (1) identify the major contaminants in the study area (carcinogens and noncarcinogens), (2) estimate the vulnerability of groundwater to different contaminants, and (3) group the different vulnerability estimates of multiple contaminants based on toxicology endpoint.

Africa poses particular challenges to providing safe, accessible water for its rapidly growing population. The majority of the countries lack access to safe water (UNICEF 2006). In sub-Saharan Africa, water resources are scarce and water availability may be seasonal. According to JMP (2008), 28 % of the population of sub-Saharan Africa defecates in the open, and an additional 23 % uses "unimproved" sanitation facilities that "do not ensure hygienic separation of human excreta from human contact."

This review chapter addresses the different water-related disease, global perspective and challenges in water and sanitation, impact of water quantity and quality on human health, and integrated water resources management for sustainable public health. The paper emphasis how water quantity and quality are vital for sustainability of environmental public health. It gives an overview summary of the relationship between water and health that help policy- and decision-makers to take careful understanding of the different components and make appropriate planning and management of water resources and the environment.

15.2 Water-Related Infections: Direct Health Effects

While water quality affects transmission rates of many water-related diseases, water availability also plays a significant role in the spread of infectious diseases. Changes in water flow or quality, which can influence the population dynamics of vector species that transmit infectious diseases and intermediate hosts for microbial pathogens, also influence the prevalence and transmission dynamics of infectious diseases.

White et al. (1972) made relationship between water supplies and health in developing countries and developed the following classification of disease transmission routes that structures and clarifies information critical to effectively target interdisciplinary intervention efforts to the health effects of water and sanitation.

Waterborne Diseases Waterborne diseases may result when pathogenic organisms including but not limited to viruses (e.g., hepatitis A and hepatitis E), parasites (e.g., *Giardia*, *Cryptosporidium*), and bacteria (e.g., *Shigella*, *Campylobacter jejuni*, *Escherichia coli*, *Salmonella* spp. and *Vibrio cholerae*) present in feces or urine from human and animal waste contaminate water supplies, and this water is subsequently used for drinking or food preparation without adequate treatment.

Water-Washed Diseases (The pathogen is spread from person to person due to lack of water for hygiene, as it occurs in diarrheal diseases, scabies, and trachoma). Person-to-person transmission of certain skin and eye infections is favored by inadequate hygiene conditions, which frequently result from lack of access to water (Cairncross and Valdmanis 2006). These water-washed diseases include shigellosis, a bacterial infection affecting the intestinal tract; trachoma; and scabies, a skin infection caused by the microscopic mite *Sarcoptes scabiei*.

Water-Based Diseases (The pathogen is transmitted to humans through contact with and infection, multiplication in, and excretion from aquatic intermediate hosts, as occurs in the diseases schistosomiasis and dracunculiasis (guinea worm)). Aquatic intermediate hosts are involved in the transmission of water-based diseases such as schistosomiasis and dracunculiasis. Although associated with inadequate sanitation, schistosomiasis is limited to environments favorable to the snail in which the larval pathogen develops into a stage that is released from the snail into freshwater and can penetrate intact human skin. WHO (2008) indicated that schistosomiasis was introduced into both Mauritania and Senegal when the damming of the Senegal River created a less salty environment that allowed snails to flourish.

Water-Related Insect Vectors (The pathogen is carried and transmitted by insects that breed in or bite near water). Water-related vector-borne diseases include mosquito-borne diseases such as malaria and dengue fever. Flies that breed in water transmit the parasitic worm that causes onchocerciasis (river blindness); the tsetse fly, which bites near water, spreads African trypanosomiasis (sleeping sickness).

According to Cairncross and Valdmanis (2006), the prevention of waterborne disease transmission requires improvements in water quality, whereas water-washed transmission is interrupted by improvements in the water availability and hence the quantity of water used for hygiene. Water supply may affect water-based transmission or water-related insect vectors of disease, though that will depend on the life cycle of the parasite involved and the preferred breeding sites and behavior of the vector.

15.3 Water, Sanitation and Health: Global Challenges

Worldwide, over 1.1 billion people lack access to improved water supplies and 2.6 billion people lack adequate sanitation, which is more than 35 % of the world's population (Prüss-Üstün et al. 2008; WHO/UNICEF 2012; Moe and Rheingans 2006). Regions with the lowest coverage of "improved" sanitation in 2006 were sub-Saharan Africa (31 %), Southern Asia (33 %) and Eastern Asia (65 %). In 2006, 7 out of 10 people without access to improved sanitation were rural inhabitants.

Prüss-Üstün et al. (2008) indicated that according to the World Health Organization, unsafe water, inadequate sanitation, and insufficient hygiene account for an estimated 9.1 % of the global burden of disease and 6.3 % of all deaths. In developing countries, nearly half of the human populations have infections or diseases associated with inadequate water supply and sanitation (Bartram et al. 2005).

IOM (Institute of Medicine) (2009) indicated that the lack of access to and availability of clean water and sanitation has had devastating effects on many aspects of daily life. While poverty has been a major barrier to gaining access to clean drinking water and sanitation, access to and the availability of clean water is a prerequisite to the sustainable growth and development of communities in many parts of the developing world. Water availability, water quality, and sanitation are fundamental issues underlying infectious disease emergence. While water quality affects transmission rates of many water-related diseases, water availability also plays a significant role in the spread of infection. Changes in water flow or quality, which can influence the population dynamics of vector species that transmit infectious diseases and intermediate hosts for microbial pathogens, also influence the prevalence and transmission dynamics of infectious diseases (Institute of Medicine 2009).

Africa poses particular challenges to providing safe, accessible water for its rapidly growing population. The majority of the countries lack access to safe water (UNICEF 2006). In sub-Saharan Africa, water resources are scarce and water availability may be seasonal. According to JMP (2008), 28% of the population of sub-Saharan Africa defecates in the open, and an additional 23 % uses "unimproved" sanitation facilities that "do not ensure hygienic separation of human excreta from human contact."

Moe and Rheingans (2006) reviewed five major challenges to providing safe water and sanitation on a global basis: (1) contamination of water in distribution systems, (2) growing water scarcity and the potential for water reuse and conservation, (3) implementing innovative low-cost sanitation systems, (4) providing sustainable water supplies and sanitation for megacities, and (5) reducing global and regional disparities in access to water and sanitation and developing financially sustainable water and sanitation services.

The UN millennium development goals (MDG) aim to reduce by half the proportion of people without sustainable access to safe drinking water and basic sanitation by the year 2015. Lack of access to improved drinking water is still a serious problem in large portions of Asia where an estimated 675 million people are without improved drinking water sources (WHO and UNICEF 2004). In sub-Saharan Africa, only 36 % of the population has access to basic sanitation (WHO and UNICEF WHO 2004; Moe and Rheingans 2006). Meeting the MDGs for water and sanitation in the next decade will require substantial economic resources, sustainable technological solutions, and courageous political will.

In addition to providing "improved" water and "basic" sanitation services, we must ensure that these services provide safe drinking water; adequate quantities of water for health, hygiene, agriculture, and development; and sustainable sanitation approaches to protect health and the environment.

15.3.1 Foundations of Health: Water Supply Sanitation and Hygiene

With the importance of water supply and sanitation to health, it is necessary to consider the status of these services globally and regionally. Those without access to an adequate and safe water supply and appropriate sanitation are those most at risk from waterborne diseases. Access to safe drinking water and adequate sanitation is a recognized universal human need. According to Pruss-Ustun et al. (2008) as cited by Bartram and Cairncross (2010), globally, around 2.4 million deaths could be prevented annually if everyone practiced appropriate hygiene and had good, reliable sanitation and drinking water. These deaths are mostly of children in developing countries from diarrhea and subsequent malnutrition and from other diseases attributable to malnutrition.

Reliable and safe water for household purpose prevents diarrhea, guinea worm, waterborne arsenicosis, and waterborne outbreaks of diseases such as typhoid, cholera, and cryptosporidiosis. Much of the impact of water supply on health is mediated through increased use of water in hygiene. For example, hand washing with soap reduces the risk of endemic diarrhea and of respiratory and skin infections, while face washing prevents trachoma and other eye infections (Chant 2008; Cairncross and Kolsky 1997).

Environmentally caused mortality and malnutrition have substantial economic costs. Cairncross and Kolsky (1997) indicated that in Ghana and Pakistan, for example, the indirect effect on child mortality of environmental risk factors mediated by malnutrition adds more than 40 % to the cost of directly caused child mortality.

Rural water supply and sanitation are essentially simple engineering but much less simple sociology, and there needs to be a good and sustained program of hygiene education so that people with an improved water supply and improved sanitation know how to use them to maximize the benefits to their health (Fewtrell et al. 2005). Many studies have reported the results of interventions to reduce illness through improvements in drinking water, sanitation facilities, and hygiene practices in less developed countries.

Global water for sustainability (GLOWS) program at the Florida International University has been implementing water sanitation and hygiene project in different parts of the world particularly East and West Africa. The last 10 years of intervention resulted in an improved water supply, sanitation, and hygiene condition of different countries. It was observed that interventions such as improved hygiene practices, improved sanitation, improved access to water, and improved water quality significantly fight against water-related diseases. Many studies have been performed to evaluate these interventions and compare their effects in reducing the incidences of diarrhea and other diseases, especially for children.

15.4 The Impact of Water Quality on Health

Drinking water sources contaminated by sewage contain not only fecal bacteria and potentially pathogenic microorganisms but also organic material and ammonia, both of which exert substantial chlorine demands. Pathogenic microorganisms in drinking water, the leading causes of diarrhea, and chemical impurities, which are growing threats in many developing countries, have drawn a lot of attention in public health and other related fields.

Improved water quality has been shown to play a substantial role in reducing diarrhea and mortality in various countries (Cutler and Miller 2005; Arnold and Colford 2007; Kremer et al. 2009). Furthermore, there are other diseases that can be caused by chemical impurities in water such as toxic metals and inorganic or organic compounds. Some studies also indicated that there is strong link between water pollution and digestive cancers in many parts of the world.

There is still serious impact of inadequate and unsafe water supply on human health in many countries and not only in developing ones. Waterborne diseases, such as diarrhea, are responsible for more than 3 million deaths per year. Vector-borne diseases, such as malaria and schistosomiasis, are still rampant in the tropics causing severe human suffering and economic losses. Chemical hazards related to water are not only due to the presence of toxic substances, such as pesticides or heavy metals, but also due to excessive or deficient amounts of natural substances such as fluoride or iodine (Helmer 1999).

The interactions between water and human health are indeed complex. Human health may be affected by the ingestion of contaminated water, either direct or through food, and by the use of contaminated water for purposes of personal hygiene and recreation. Health hazards are also associated with various industrial and agricultural applications. The contamination of water by viruses, pathogenic bacteria, and other parasites can occur either at the water source itself or during conveyance of the water from source to consumer. In many developing countries, water in rivers, ponds, and canals is used for a variety of purposes, washing clothes, the disposal of human excreta, the ceremonial washing of persons and so on. As a result, it becomes highly polluted and, therefore, an important vehicle for the domestic transmission of infections and infestations.

Helmer (1999) indicated that water and its impact on health can be categorized according to the mode of disease transmission. There are basically six such classifications: waterborne diseases, water-contact diseases, vector-borne diseases, water-hygiene diseases, excreta-related diseases, and poisoning and disabilities associated with chemical contamination. In total, it is estimated that more than 3 million people per year die from waterborne or hygiene-related diseases due to unsafe drinking water, unclean domestic environments, and improper excreta disposal. The morbidity and mortality rate of water-related diseases can be reduced through the provision of a safe drinking water supply and adequate sanitation.

Waterborne diseases have long been a scourge of humanity. A drinking water supply contaminated either by excreta or wastewater containing organisms of enteric disease, or by an individual infected by such organisms, can easily spread the contaminated water throughout the whole or part of the water distribution system, resulting in an explosive outbreak of disease affecting hundreds or perhaps thousands of people. Such outbreaks, typically of typhoid fever, cholera, or shigellosis, have occurred many times. Infectious hepatitis and cryptosporidiosis can be added to this list of waterborne diseases. Dracunculiasis, or guinea worm disease, is the one disease which is only transmitted through drinking contaminated water, and it only affects humans. It is, therefore, the only disease which can be eliminated through water supply improvements alone.

Hygiene-related diseases: Diarrhea, as a consequence of poor hygiene, is one of the most common conditions in developing countries, its effect being greatest among children during their first 5 years of life and enhanced by the severe malnutrition which so frequently coexists. Pathogens causing diarrheal diseases are of fecal origin, conveyed by food or drink or by direct person-to-person transmission. Improvements in water quality, water availability, and utilization and in adequate excreta disposal facilities are considered essential elements in any remedial program. Many epidemiological studies have confirmed the importance of safe and adequate water supplies and improved hygiene behavior in curbing diarrhea and similar waterborne disease.

Chemical hazards: Perhaps, the most important element-deficiency problem associated with drinking water globally is that of endemic goiter and cretinism, both of which are linked to dietary iodine deficiency. For other elements, there is only a narrow range of concentrations in water within which beneficial effects are found. Excess of fluoride (above about 1.5 mg per liter) leads to dental fluorosis and skeletal fluorosis at higher levels. Arsenic is a toxic and carcinogenic element which is sometimes naturally present in groundwaters, although most drinking waters have concentrations below the WHO guideline value. There are areas with very high concentrations with well-documented cases of chronic arsenic poisoning such as in southern Taiwan, Chile, Mexico, China, and West Bengal, India. It has been estimated that some 30 million people may be at high risk from arsenic exposure in Bangladesh and India.

Most often, however, chemical contamination of water supplies arises from industrial, agricultural, or other man-made sources. Nitrate pollution of drinking water is a good example. A certain amount of nitrates in water may come from natural sources, but excess is likely to occur in water as a result of sewage discharges or farm effluents, or the use of nitrogenous fertilizers on the land. Helmer (1999) indicated that nitrates in drinking water may cause infantile methemoglobinemia. Agriculture is responsible for the serious pollution of many of the world's aquifers due to intensive use of nitrogen-rich fertilizers and of pesticides. In many places, the leaching of these substances into the ground leads to nitrate concentrations in abstracted drinking water which by far exceed the WHO guideline values.

15.5 Water Quantity and Human Health

Having enough water for drinking and hygiene purposes promotes better health and well-being. Without essential access to safe drinking water, humans, not to mention animals and plants, cannot survive. Adequate water is also a necessity for food production since it may benefit people's health through the prevention of malnutrition and thus enable them to more readily recover from illness. Good sanitation, which is critical to human health, also depends on adequate water supply to ensure the safe disposal of human waste and reduce disease and death. Water quantity may also have substantial impacts on human health by adversely changing water quality. Overabstraction of surface water may reduce the flow and thus the self-purification capability of rivers, lakes, and reservoirs. Similarly, in the case of groundwater, abstraction may bring about strong hydraulic gradients and thus the formation of preferential flow paths, which could reduce the efficacy of attenuation processes and increase the concentrations of contaminants. In coastal areas, seawater intrusion may correspondingly take place. Furthermore, changes in groundwater levels ensuing from abstraction may also change the subsurface environment, e.g., redox conditions, and therefore induce mobilization of contaminants such as heavy metals.

Surface water and groundwater are traditionally regarded as separate entities with respect to water resources management, whereas surface water may interact with groundwater in many situations. The water exchange between surface water and groundwater may recharge either of them, depending on the difference between the altitudes of the water surface of the former and the water table of the latter. As a result, withdrawal of water from either of them may deplete water in the other. Moreover, this interaction may also cause changes in water quality in either of them.

Water resources development projects often present modifications to the environment and conditions conducive to the breeding of vectors which are the agents of disease including malaria, schistosomiasis, lymphatic filariasis, and a number of arboviral diseases such as Japanese encephalitis in the irrigated rice-growing areas of Southeast Asia. All water resources development schemes, such as irrigation, hydroelectric power generation, navigation, and water supply, have to be designed and implemented with due regard to health if risks are to be minimized. For malaria and schistosomiasis, the most serious problems occur in sub-Saharan Africa. Schistosomiasis is by far the most notorious acute infection associated with irrigation development in Africa. It is primarily on dams in sub-Saharan Africa. It is a major constraint to economic development of the rural areas of the developing world.

15.6 Integrated Water Resources Management and Public Health

The freshwater that humans have direct access to is found in lakes, rivers, and reservoirs, which only accounts for 0.266 % of the total freshwater resources. Water crisis seemingly becomes inevitable in such a changing world in terms of the variation of global environment, the ever-growing population, and the serious environmental pollution. To cope with the water crisis in both quantity and quality, frameworks for source water management regarding both surface water and groundwater have been proposed all around the world, such as Integrated Water Resources Management (IWRM), Water Framework Directive (WFD) of the European Communities, and Watershed Planning and Management (WPM) and Source Water Assessment and Protection (SWAP) programs of the USA (Sun et al. 2011). With the development of water and wastewater treatment technology, exploitation of alternative water sources, including desalination, rainwater harvesting, and water reuse, has become a global practice to satisfy this thirsty world.

Source water quantity and quality are affected by a great variety of factors, natural, anthropogenetic, local, regional, or global. Therefore, management of source water calls for a holistic, integrated, and collaborative approach. A holistic perspective should be held in mind that water quantity and quality are related and should be considered together, while different water bodies, either surface and ground or inland and costal, may also be connected and should be regarded in the context of the whole hydrological cycle. Management of source water is also a scientific and technical issue in an extremely complex and mixed system across natural environment and human society, and hence, it needs an integration of multidisciplinary work, including meteorology, geography, geology, hydrology, hydraulics, ecology, chemistry, biology, demography, economics, sociology, and technology. Moreover, source water management, in the social aspect, requires a collaborative framework within which the concerned regions, sectors, and users should be involved, such as land use planning, agriculture, forestry, flood management, industry, tourism, and recreation.

IWRM, as the Global Water Partnership defined, is the process of promoting the coordinated development and management of water, land, and related resources, to maximize the resultant economic and social welfare in an equitable manner without compromising the sustainability of vital ecosystems. IWRM aims to support countries in their efforts to tackle specific water challenges, e.g., water scarcity, waterborne diseases, floods, droughts, and access to water and sanitation, and thus sustain their development to achieve the goals such as poverty alleviation, food security, economic growth, and ecological conservation. However, IWRM is not just about managing physical resources; it also requires and promotes the positive changes in water governance regarding the enabling environment, institutional roles, and management instruments. IWRM systems should, therefore, not only be responsive to changes among its development process, for example, between projected goals and decision-makers' willingness, but also be capable of adapting to new economic, social, and environmental conditions and to changing human values over a long-term implementation.

One of the major environmental issues of concern to policy-makers is the increased vulnerability of groundwater quality. Another issue of equal interest is the sustainability of natural resources for future generations. To understand the sustainability of the natural resources such as water in general, one needs to understand the impact of future land use changes on the natural resources.

15.7 Conclusions

Providing safe water and basic sanitation to meet the millennium development goals (MDGs) will require substantial economic resources, sustainable technological solutions, and courageous political will.

The effects of water shortages and water pollution have been felt in both industrialized and developing countries, and it will be necessary to transcend international and political boundaries to meet the world's water needs in a sustainable manner that will conserve and preserve this common resource. In the last few decades, national and international organizations from both the public and private sectors have come together to tackle global issues in water and sanitation.

Water is essential for life and social and economic development. Its exploitation and use must be well planned and managed. Lack of adequate safe water, the pollution of the aquatic environment and the mismanagement of resources are still major causes of ill-health and mortality, particularly in the world's developing countries. Efforts to combat the detrimental effects of water pollution and environmental mismanagement have to be addressed within the framework of water resources development. Schemes for the harnessing of water are designed to benefit humanity through increased crop and livestock production, improved industrial output, better communications, and increased energy availability for development. A balanced approach to water resources exploitation for development, on the one hand, and controls for the protection of health, on the other, are required if the benefits of both are to be realized without avoidable detrimental effects manifesting themselves.

Diarrheal disease is one of the leading causes of morbidity and mortality in less developed countries, especially among children under 5 years of age (Boschi-Pinto et al. 2008). Adequate sanitation can help prevent endemic diarrhea and intestinal helminthiases, giardiasis, schistosomiasis, trachoma, and numerous other globally important infections. There needs to be a good and sustained program of hygiene education so that people with an improved water supply and improved sanitation know how to use them to maximize the benefits to their health. All of these are important for health, development and social well-being. However, if the full potential of health benefits is to be derived from these schemes, health must be involved from their earliest planning stage, and adequate health and environmental impact studies must be performed.

Different studies indicated several classification of water-related diseases that structures and clarifies information critical to effectively target interdisciplinary intervention efforts to the health effects of water and sanitation. The prevention of waterborne disease transmission requires improvements in water quality, whereas water-washed transmission is interrupted by improvements in the water availability and hence the quantity of water used for hygiene. Water supply may affect water-based transmission or water-related insect vectors of disease that will depend on the life cycle of the parasite involved and the preferred breeding sites and behavior of the vector.

Meeting the millennium development goals for water and sanitation in the next decade will require substantial economic resources, sustainable technological solutions, and courageous political will. The challenge is to mobilize the political will to implement water resource development programs which cater in an equitable manner for the various demands on water, the finite resource. At the forefront of these considerations must be the sustained protection and promotion of human health.

In addition to providing "improved" water and "basic" sanitation services, we must ensure that these services provide safe drinking water; adequate quantities of water for health, hygiene, agriculture, and development; and sustainable sanitation approaches to protect health and the environment.

A great number of governments and international organizations have launched water-related programs and interventions all over the world as an effective way to improve people's health and welfare. But the challenges to overcome the impacts will be very high. An integrated approach should be designed to decrease the alarming impact of water quality, chemical impurities, and other water pollutions.

References

Alley WM, Reilly TE, Frank OL (1999) Sustainability of groundwater resources. US Geological Survey Circular 1186, 1p

Arnold B, Colford J (2007) Treating water with chlorine at point-of-use to improve water quality and reduce diarrhea in developing countries: a systematic review and meta-analysis. Am J Trop Med Hyg 76(2):354–364

Bartram J, Cairncross S (2010) Hygiene, sanitation, and water: forgotten foundations of health. PLoS Med 7(11):e1000367. doi:10.1371/journal.pmed.1000367

Bartram J, Lewis K, Lenton R, Wright A (2005) Focusing on improved water and sanitation for health. Lancet 365(9461):810–812

Boschi-Pinto C, Velebit L, Shibuya K (2008) Estimating child mortality due to diarrhoea in developing countries. Bull World Health Organ 86:710–717

Cairncross S, Kolsky PJ (1997) Letter: water, waste and well-being. Am J Epidemiol 146:359–360

Cairncross S, Valdmanis V (2006) Water supply, sanitation, and hygiene promotion. In: Jamison DT, Breman JG, Measham AR, Alleyne G, Claeson M, Evans DB, Jha P, Mills A, Musgrove P (eds) Disease control priorities in developing countries, 2nd edn. Oxford University Press, New York

Chant R (2008) The role of water, hygiene and sanitation in neonatal mortality. MSc dissertation. London School of Hygiene & Tropical Medicine, London

Cutler D, Miller G (2005) The role of public health improvements in health advances: the 20th century United States. Demography 42(1):1–22

Fewtrell L, Kaufmann RB, Kay D, Enanoria W, Haller L, Colford JM Jr (2005) Water, sanitation, and hygiene interventions to reduce diarrhoea in less developed countries: a systematic review and meta-analysis. Lancet Infect Dis 5:42–52

Helmer R (1999) Water quality and health. Environmentalist 19:11–16

Institute of Medicine (2009) Global issues in water, sanitation, and health: workshop summary. The National Academies Press, Washington, DC

IOM (Institute of Medicine) (2009) Microbial evolution and co-adaptation: a tribute to the life and scientific legacies of Joshua Lederberg. The National Academies Press, Washington, DC

JMP (Joint Monitoring Programme for Water Supply and Sanitation) (2008) Progress on drinking water and sanitation: special focus on sanitation. UNICEF/WHO, New York/Geneva

Kremer M, Miguel E, Null C, Van Dusen E, Zwane A (2009) Measuring diarrhea: quantifying Hawthorne effects in frequently collected data. Working Paper, University of California, Berkeley

Moe CL, Rheingans RD (2006) Global challenges in water, sanitation and health. J Water Health 4:41–57. IWA Publishing 2006

Prüss-Üstün A, Bos R, Gore F, Bartram J (2008) Safer water, better health: costs, benefits and sustainability of interventions to protect and promote health. World Health Organization, Geneva

Solley WB, Pierce RR, Perlman HA (1998) Estimated use of water in the United States in 1995. US Geological Survey Circular 1200, 71pp

Sun F, Chen M, Chen J (2011) Integrated management of source water quantity and quality for human health in a changing world. Encycl Environ Health 254–265

Twarakavi NKC, Kaluarachchi JJ (2006) Sustainability of ground water quality considering land use changes and public health risks. J Environ Manag 81(2006):405–419

UNICEF (United Nations Children's Fund) (2006) Progress for children: a report card on water and sanitation. UNICEF, New York

White GF, Bradley DJ, White AU (1972) Drawers of water. University of Chicago Press, Chicago

WHO (World Health Organization) (2008) Initiative for vaccine research: schistosomiasis, http://www.who.int/vaccine_research/diseases/soa_parasitic/en/index5.html. Accessed 10 Aug 2008

WHO, UNICEF (2004) Water supply and sanitation global assessment year 2000 report. Joint monitoring programme for water supply and sanitation

WHO, UNICEF (2012) Progress on drinking water and sanitation: 2012 update. External web site icon United States: WHO/UNICEF joint monitoring programme for water supply and sanitation

Chapter 16
Vulnerability and the Probability of Households Having Access to Water in Locations with Extreme Weather in Mexico City

Armando Sánchez-Vargas

Abstract We carry out a statistical analysis to estimate the probability of households having access to water and identify the most vulnerable people in counties with extreme weather in Mexico City. We use a methodology that combines the use of spatial, climate, and household survey data. Our results suggest that locations in 10 out of 16 counties in Mexico City are currently affected by extreme conditions and in addition show a lower probability of having water access at home. From the 8.8 million people living in Mexico City, about 3,142,660 are living in areas with decreasing mean rainfall over time, "scarce rainfall zones," and approximately 1,500,100 in areas with an increasing mean temperature over time, "high temperature zones." Only five counties are currently affected by scarce rainfall and high temperature at the same time; which implies that there are around 508,840 highly vulnerable people in Mexico City affected by both types of extreme conditions. We also find that by coincidence, the odds of having water at home are much lower for people living in six counties with extreme conditions and their odds are reduced between 20 and 30 %. Such counties also have high poverty levels, so water scarcity conditions are reinforcing the vulnerability of the population. Knowing the precise spatial location and the number of affected people will definitively contribute to improve the implementation of public policies and the effectiveness in the use of resources devoted to problems like water access in the face of extreme events.

Keywords Water access • Extreme weather • Probability models

A. Sánchez-Vargas (✉)
Institut for Economics Research, UNAM, Mexico City, Mexico
e-mail: armando_sanchez123@hotmail.com

S.G. Setegn, M.C. Donoso (eds.), *Sustainability of Integrated Water Resources Management*, DOI 10.1007/978-3-319-12194-9_16

16.1 Introduction

Extensive research work regarding extreme weather and its association to the water sector have been carried out worldwide. However, very few studies have investigated the household's vulnerability, in terms of water access, in urban locations with extreme weather (Ludi 2009). So, we aim at carrying out a statistical analysis to estimate the probability of households having reduced access to water and at the same time living in counties with more extreme climate in Mexico City to identify the most vulnerable people. In order to do so, we use an integrated methodology that combines the use of spatial, climate, and household survey data. Our study is carried out by answering the following questions: Is there any difference in the odds of households having access to water between counties under more extreme and regular climate conditions in Mexico City? How can we identify the most vulnerable people? In this context, another interesting issue would be to determine whether more extreme climate explains the differences in the odds of having access to water for the different counties. The answer to this question will require an additional level of analysis in relation to the link between the climate in the various counties and that at the sources of the water supply for the city. Since we are interested in the household's vulnerability, we do not address such topic here.

In order to conduct our analysis, we construct two groups of comparable households that are similar in all their observable characteristics, except that one group has already been exposed to more extreme weather conditions (treatment group) and the other group has not (control group). Subsequently, we compute the discrepancy in the probabilities of having access to water as the mean differences of household water access binary indicators between the defined comparable groups. Such pairs of comparable groups are constructed through econometric techniques, instead of using randomization. We use a quasi-experimental technique called propensity score matching (PSM), which pairs similar groups of individuals based on the scores resulting from a nonlinear probability model of participation in the treatment or control group (probit model). Once we have such groups (i.e., people living in affected and not affected counties), we can compute the difference in their respective average probabilities of having access to water, which, in turn, are obtained from household survey data.

More specifically, our proposed methodology consists of the following steps: First, we spatially identify the counties, households, and hydrometric and climate stations in Mexico City. We also obtain a long span of daily time series records of precipitation and temperature, registered by climate stations, in the localities under study. Second, based on such time series, we estimate 15 extreme climate indicators to identify the specific counties that have strong evidence of more extreme weather conditions (Zhang et al. 2005, 2011). Third, we use such indicators to map zones facing extreme events, by constructing contour lines, to spatially locate the people who are living under more extreme conditions and the ones who are not. Finally, once we have identified the affected and non-affected people spatially, we construct comparable groups of households, at the county level, and calculate the mean differences in their probabilities of having water access in the household. In order to compute such probabilities, we use binary indicators on water access inside the households, provided by the Mexican income-spending survey (ENIGH). A preliminary version of this methodology has already been used to study the well-being of poor families in the counties of Mexico City (Sánchez et al. 2011).

The results of our approach suggest that 10 out of 16 counties in Mexico City might be facing much lower probabilities of having access to water, in zones affected by extreme conditions, compared to the ones in zones with regular climate conditions. We make this comparison for two types of extreme weather conditions: "scarce rainfall" and "high temperature." In what follows we define "scarce rainfall" and "high temperature" as statistically significant trends in the mean levels of precipitation and temperature over a long time span in the locations of interest. According to our estimates, from the 8.8 million people living in Mexico City, there are approximately 3,142,660 people living in areas with scarce rainfall and around 1,500,100 in areas with high temperature. There is also evidence that only six counties are currently affected by both phenomena: Azcapotzalco, Coyoacán, Miguel Hidalgo, Tlalpan, Xochimilco, and Álvaro Obregón. This implies that there might be around 508,840 people in Mexico City affected by both extreme conditions.

In terms of water accessibility, the odds of having water at home are lowered between 20 and 30 % per month in five counties which experience extreme climate events: Azcapotzalco, Álvaro Obregón, Iztapalapa, Tlalpan, and Xochimilco. These have a total population of around 936,420 individuals. In addition, these counties have high poverty levels and are considered among the most vulnerable in Mexico. Our analysis suggests that water scarcity might be playing a role in reinforcing the vulnerability of such locations.

Finally, our results should influence the scope of the public policies intended to adapt to climate change effects at Mexico City. In fact, it is clear that such counties and specific people would need some kind of support to maintain their access to water. Knowing the specific spatial location and the number of affected people will definitively contribute to improve the design and implementation of coordinated public policies and the effectiveness in the use of resources devoted to problems like water access.

16.2 Methodology

16.2.1 Study Area

The focus of this study is Mexico City. It covers 1,495 km^2 and has 8.8 million people according to the 2010 population census (Fig. 16.1). More specifically, we investigate the water access of the households in the 16 different counties of

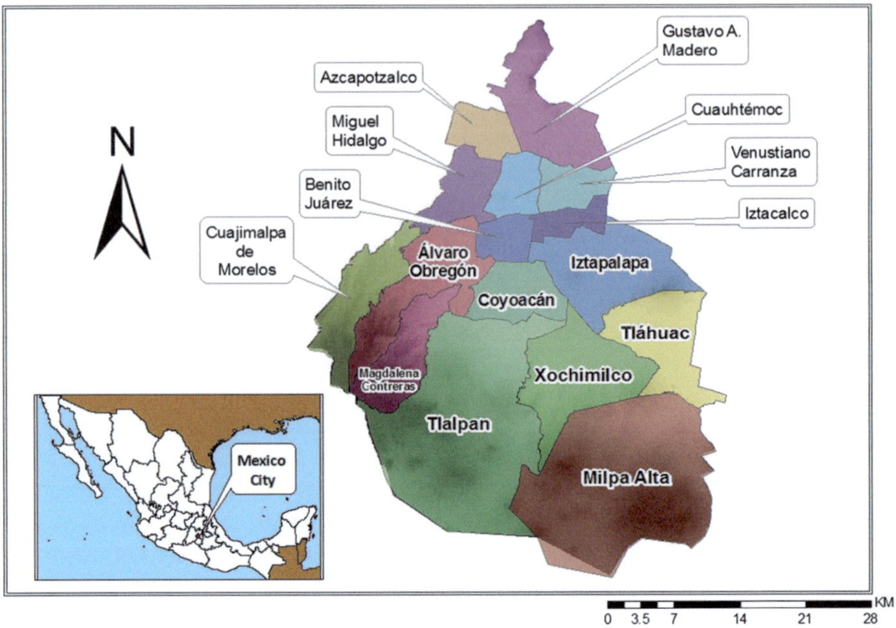

Fig. 16.1 Mexico City map including its 16 counties

Mexico City. The upper and middle sections of the city are basically human settlements, and the lower sections are considered agricultural and protected areas. The annual average temperature in Mexico City is around 16 °C, and the annual total precipitation varies from 600 to 1,200 mm.

16.2.2 A Statistical Analysis to Determine the Probability of Having Access to Water at the County Level in the Face of Extreme Weather

In this section, we discuss our methodology to estimate the probability of having low water access at the county level by integrating spatial and household data. We also analyze spatial and climate data. We try to answer the following questions: How do the odds of households having access to water differ for counties subject to extreme weather conditions compared to locations under regular climate conditions in Mexico City? How can we identify the most vulnerable people? In order to define such simulation, we statistically construct and spatially locate two groups of comparable households, by using quasi-experimental techniques. Both groups would be similar in all their observable characteristics, except that the so-called treatment group has been affected by extreme weather and the "control" group has been not. For this stage of our analysis, we ignore differences in income levels. More specifically, to construct the groups, we propose to use the technique known as matching based on propensity scores (PSM, propensity score matching) (Rosenbaum and Rubin 1983) at the local level by combining the use of spatial, climate, and household survey data. Once we define these groups that indicate the specific people living in affected and not affected counties, we can compute the difference in their respective average probabilities of having access to water, based on indicators obtained from household survey data. The geographic information system (GIS) is useful for the definition of the spatial limits, for the areas affected by extreme weather, as well as for the identification of the population living in such regions. Our methodology can be divided into six steps: (1) data preparation, (2) estimation of extreme events indicators, (3) spatial identification of zones affected by extreme events and spatial identification of the affected population by extreme weather conditions, (4) statistical construction of comparable groups inside the zones, (5) estimation of the different probabilities and vulnerability assessment, and (6) validation of the results. The flowchart of our methodology is shown in Fig. 16.2.

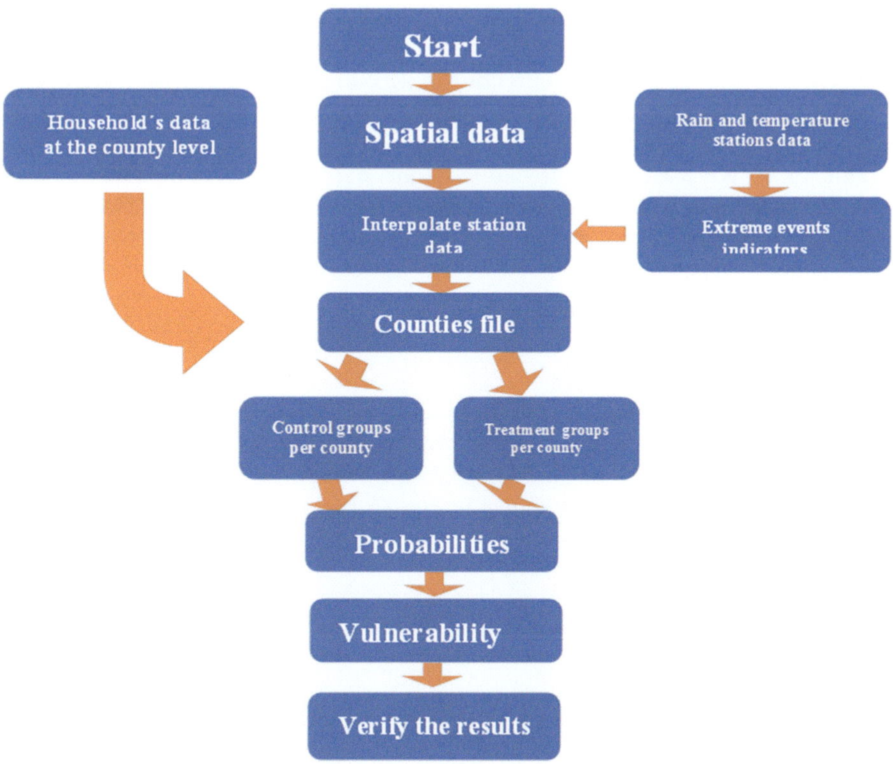

Fig. 16.2 Methodology flowchart

Table 16.1 Variables used in analysis

Variable	Data source
Mexico City map	The National Institute of Statistics and Geography (INEGI)
Hydrometric stations	The National Commission for Knowledge and Use of Biodiversity (CONABIO)
Climatic data	National Meteorological Service (SMN)
Hydrometric data	The Water System of Mexico City (SACM)
Household data	The National Institute of Statistics and Geography (INEGI)

16.2.3 Data Preparation

In order to construct our statistical analysis, we require three types of datasets: First, a spatial dataset to locate the sixteen counties in Mexico City. Second, large-scale datasets for rainfall and temperature; such daily time series records are generated by hydrometric and climatic stations. Finally, household water access data at the county level. A county-based dataset, including variables in the three preceding categories, was completed and processed to carry out our simulation. The data preparation implied to map the counties, their respective households, and the hydrometric and climatic stations. We downloaded the boundaries of the counties and the stations from the National Institute of Statistics and Geography (INEGI) of Mexico and the National Commission for Knowledge and Use of Biodiversity (CONABIO). Household survey data were also obtained from INEGI. The climatic time series were provided by the National Meteorological Service of Mexico (Table 16.1). We used GIS software to define and locate the comparable groups needed for our exercise. The specific characteristics of the data are described in the following sections.

16.2.3.1 Spatial Data

For the purpose of this analysis, the spatial units are the 16 counties of Mexico City. We obtained the counties' information from the web page of the National Institute of Statistics and Geography (INEGI) of Mexico. We use an area which includes the 16 counties of Mexico City, as shown in Fig. 16.1.

16.2.3.2 Climate Data

Climate data used in our simulation consists of daily rainfall and temperature data. Such variables were obtained from 70 working hydrometric stations scattered around Mexico City (Fig. 16.3). The time span of the daily data records for these two variables covers a varying range, starting in 1952. However, we included in our analysis only the 49 stations that had at least 25 years of daily data. The daily rainfall

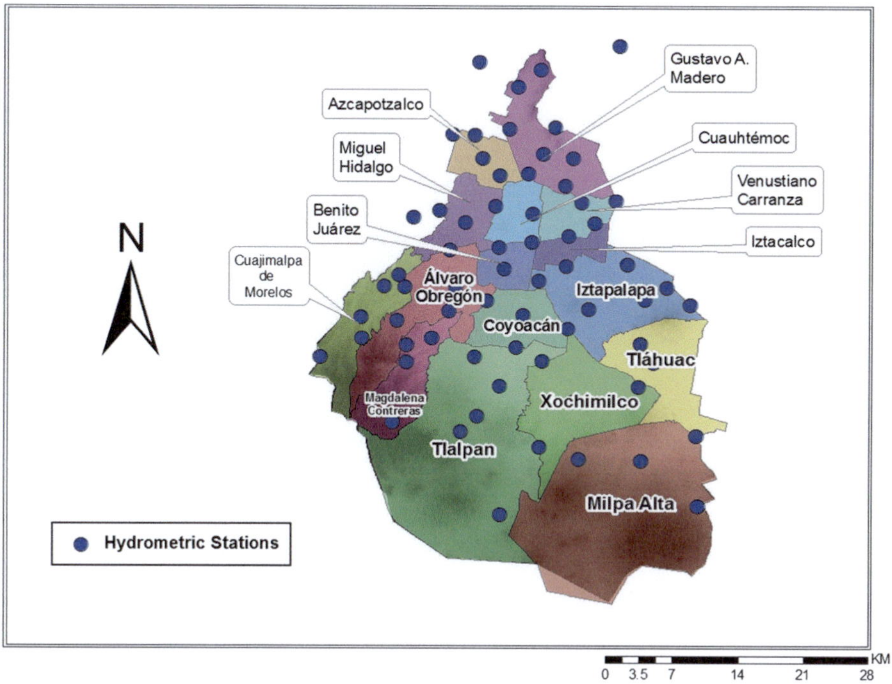

Fig. 16.3 Hydrometric stations in Mexico City

records from these stations were almost complete; there were very few missing points. Missing data values were replaced statistically by daily averages of the same calendar day obtained for the previous years where there were existing data values. The spatial location of these 49 stations seems to be fairly representative of the rainfall and temperature across the city, since they are scattered homogeneously in the populated areas around Mexico City. The continuous and complete time series of temperature were obtained from 61 stations distributed across Mexico City (Fig. 16.4). The time span for both time series covers a daily range of at least 25 years, from 1986 to 2011. So, the observed rainfall and temperature records have sufficient observations to get statistically robust evidence of extreme climate in the areas.

16.2.3.3 Household Survey Data

The final variable in our study is the probability of having daily access to water at home. We construct such variable based on the household's income-spending survey, which includes an indicator that shows whether the household received water at home every day. Thus, we construct a binary indicator which takes the value of one if the household i at county j has access to water every day and zero otherwise according to the survey results. Such indicator will be used to estimate the probability of having access to water by using a probabilistic model (probit model).

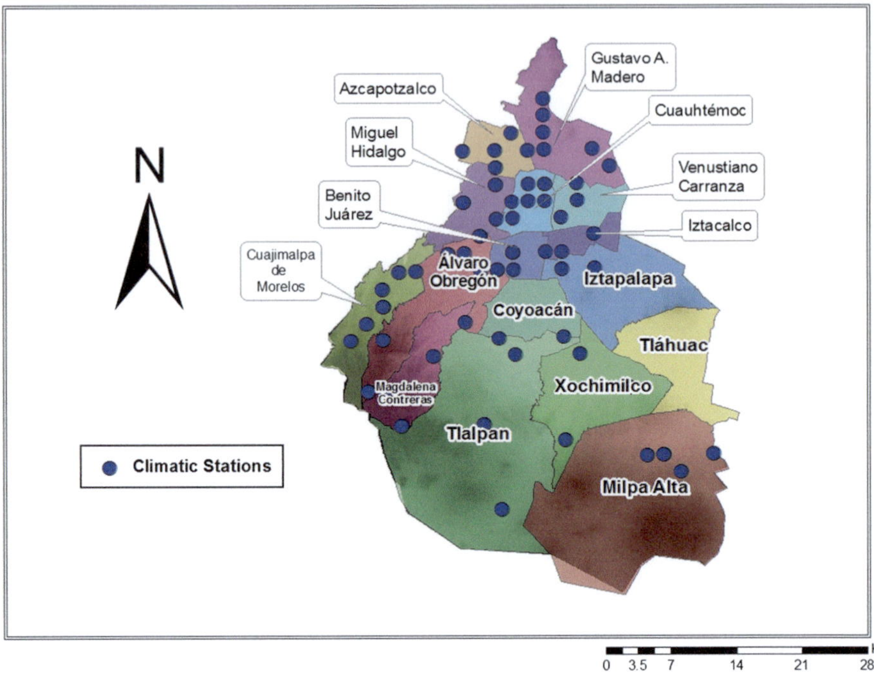

Fig. 16.4 Climatic stations in Mexico City

16.2.4 Estimation of Extreme Weather Indicators

Our methodology implies to have some evidence that extreme climate events are already affecting some of the counties of Mexico City and to identify such regions spatially to estimate the different probabilities per county later on. We choose to identify two types of anomalous climate parameters in Mexico City: "scarce rainfall" and "high temperature." As we have said above, these events correspond to statistically significant trends in mean levels of rainfall and temperature over a long time series span. In order to do so, we construct a series of indicators tracking signs of the aforementioned extreme conditions. In this case we use the ones proposed by the World Meteorological Organization. The indicators used are a selected set of extreme weather-event indicators suggested by such organization and are not an exhaustive group of all the existing climate change indicators. The hydrometric and climate data provided by the hydrometric and climatic stations allow us to estimate a series of 15 indices to spatially identify the regions which are facing frequent extreme weather conditions, such as scarce rainfall and high temperature (Figs. 16.5 and 16.6).

The hydrometric and climatic stations were located using information provided by the National Biodiversity, Ministry of Environment, and the daily temperature and precipitation time series data was provided by the National Meteorological

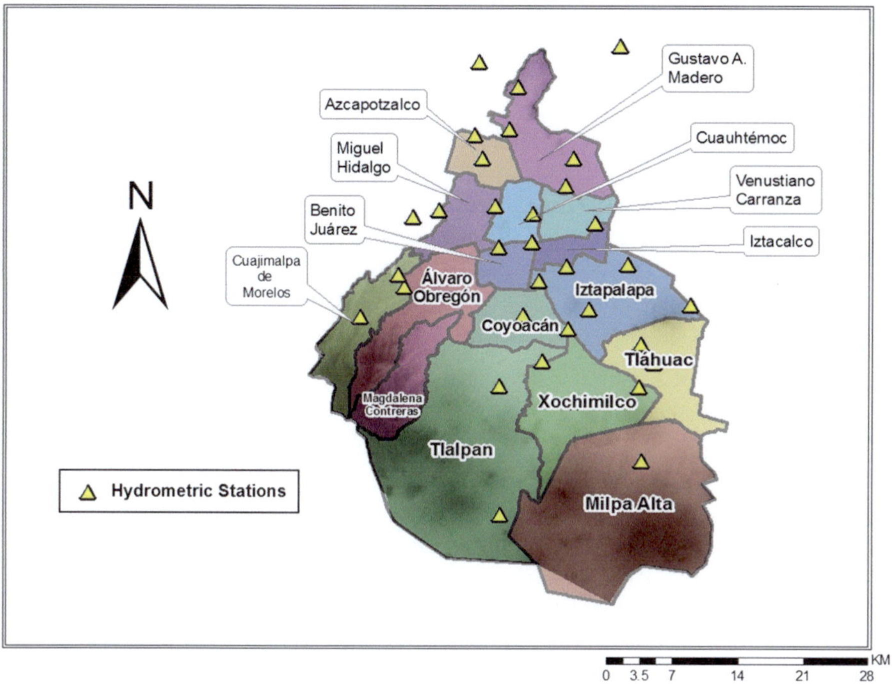

Fig. 16.5 Hydrometric stations showing evidence of scarce precipitation

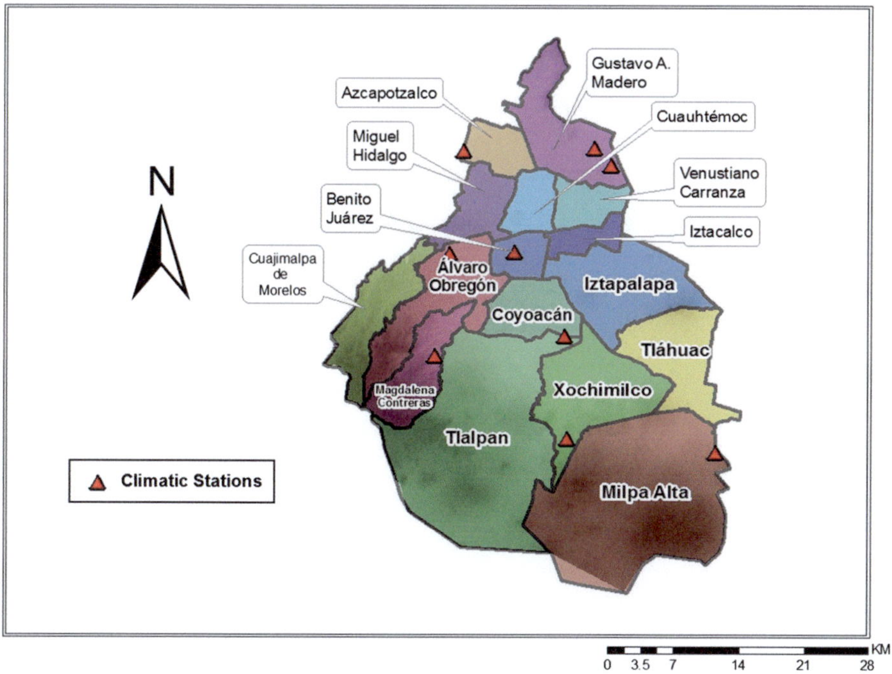

Fig. 16.6 Climate stations showing evidence of high temperatures

System of Mexico, starting in 1952. The time series of rainfall and temperature from the located stations were processed to estimate 15 temperature indices that might provide evidence of extreme events using the climatic program "RClimDex 1.0" (Zhang and Yang 2004). This software computes a series of extreme-condition indicators and runs regressions of such indicators against a linear trend; if most of the trends are statistically significant, then we can say that we have evidence of extreme weather (i.e., scarce rainfall or high temperature) in the areas.

16.2.5 Spatial Identification of Zones Affected by Extreme Weather Conditions

This step consists of using the estimated indicators and the software GIS to geographically identify zones with extreme conditions. We do so by constructing contour lines to spatially locate the places where there might be people currently living under extreme climate conditions and others who are not. If most of the indices in any station show significant climatic trends, then we create a variable to identify it, and based on it we interpolate the climatic data getting level curves associated to both levels of scarce precipitation and high temperature (Figs. 16.7 and 16.8).

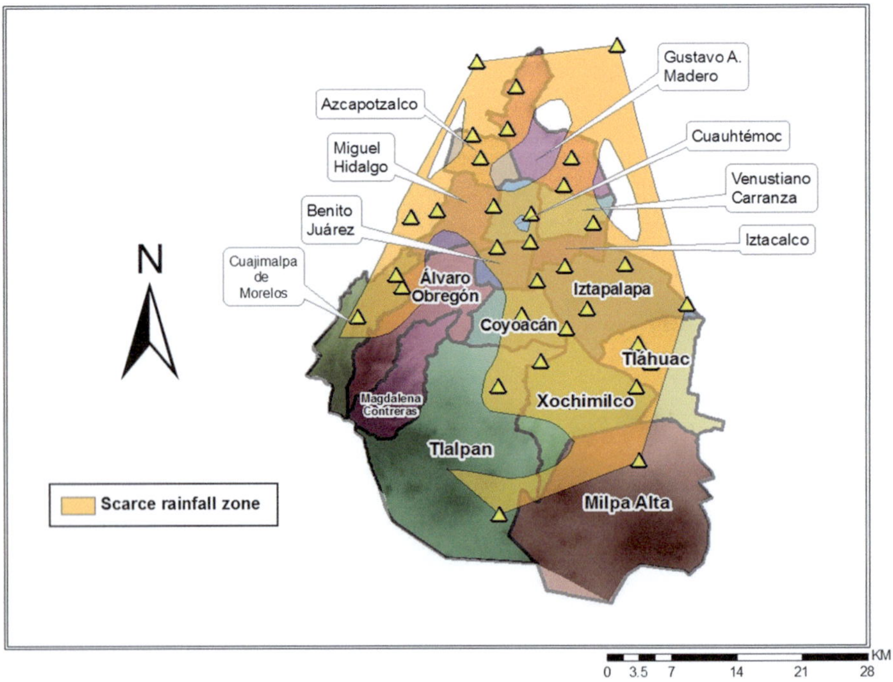

Fig. 16.7 Zones of scarce precipitation events

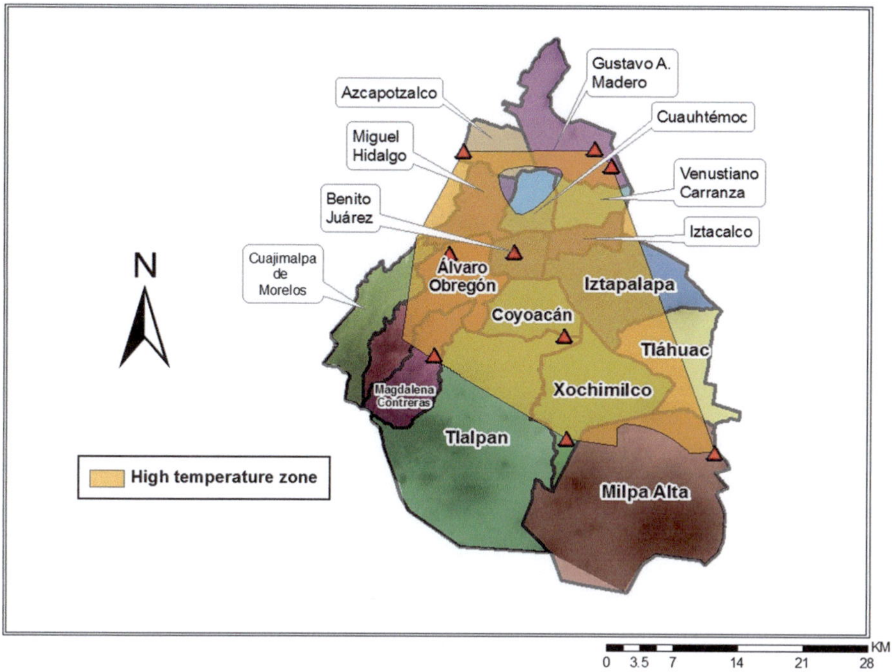

Fig. 16.8 Zones of high temperature

16.2.6 Spatial Identification of the Affected Population by Extreme Events

In the next step, we spatially locate and statistically identify the affected population by estimating the probability of households having reduced access to water at home in locations with extreme weather conditions in Mexico City and then identifying the most vulnerable people living in such areas. To do so, we overlap the zones with evidence of extreme weather events and the census areas, and we can then locate the people who are living in climatic vulnerable zones (Figs. 16.9, 16.10, and 16.11).

16.2.7 Construction of Comparable Groups and Estimation of Probabilities

The fifth step implies to estimate the probability of having daily water access in the households for the different counties of Mexico City (Sánchez et al. 2011). Such analysis can be applied at very different spatial scales; once we have the whole dataset, we carry out the propensity score matching technique for the different counties. In order to identify the probabilities of interest, we construct comparable groups: one of which is subject to extreme weather conditions and other is not.

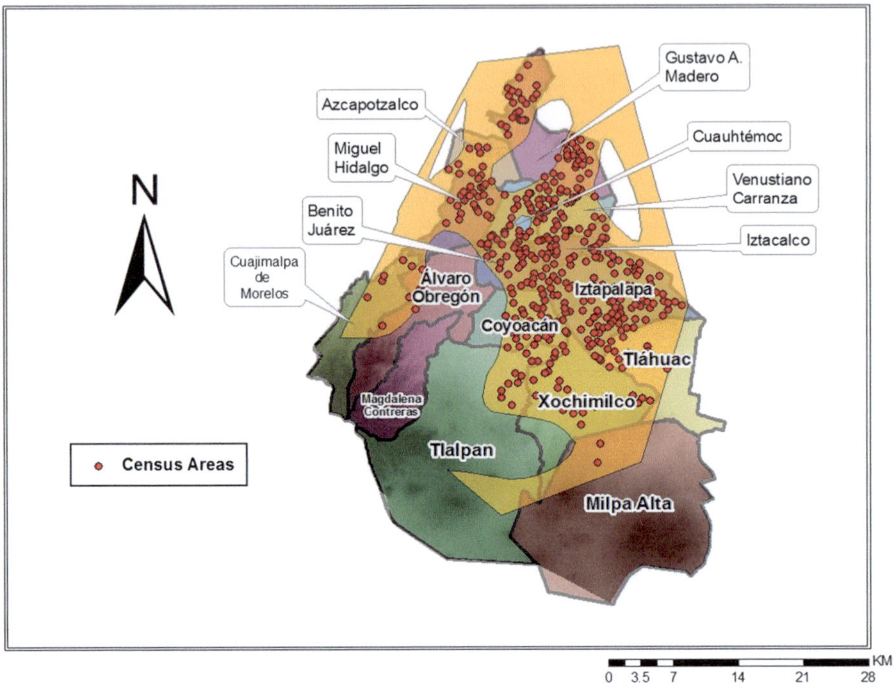

Fig. 16.9 Distribution of the population in zones of scarce rainfall

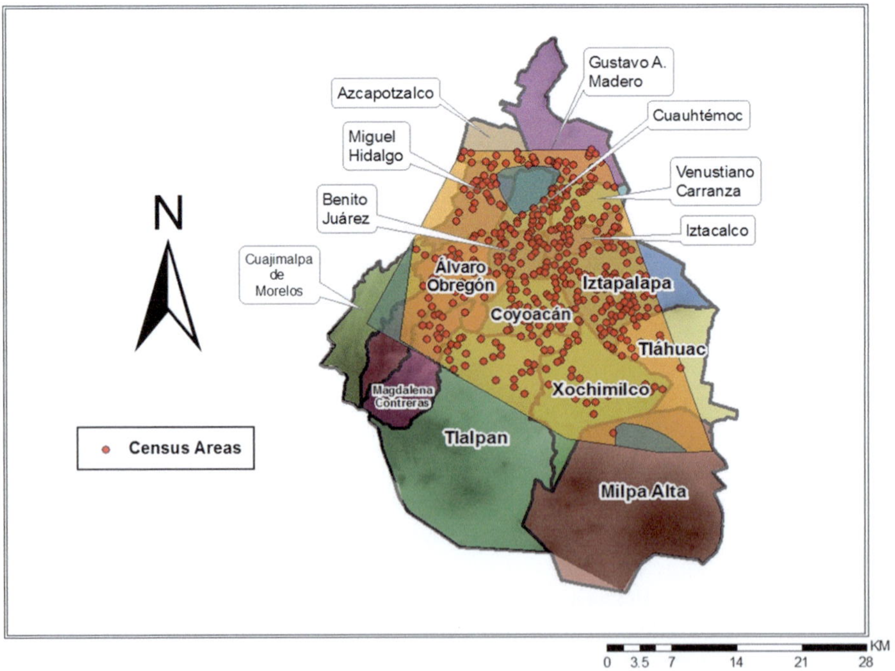

Fig. 16.10 Distribution of the population living in zones of high temperature

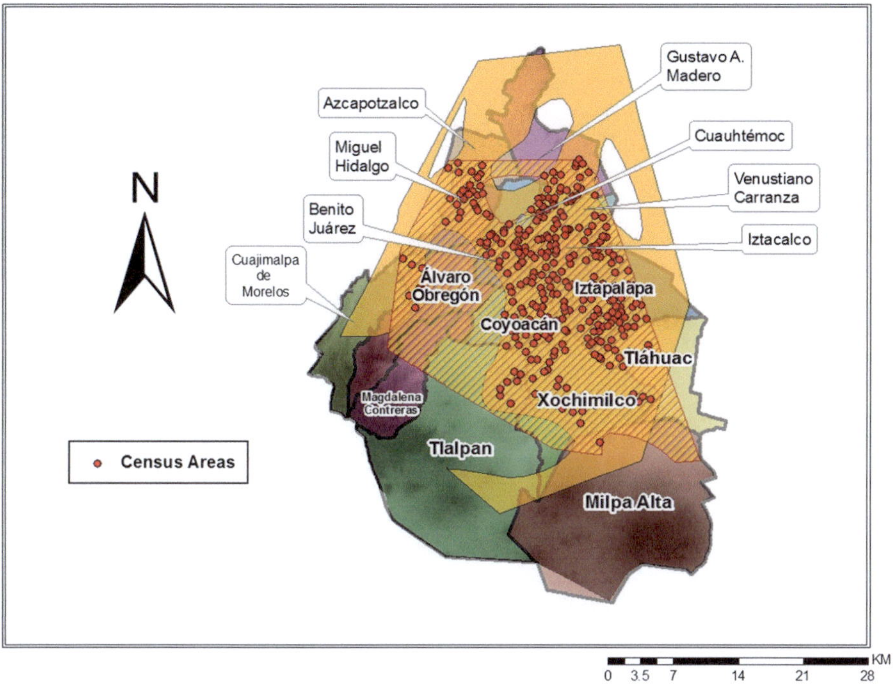

Fig. 16.11 Distribution of the population living in zones of scarce rainfall and high temperature

When we have these comparable groups, we can compute the discrepancies in the probabilities as the difference in mean outcomes between the affected and non-affected groups. It is worth mentioning that we construct such groups through econometric techniques, instead of using randomization, to ensure that there are no systematic differences in the observed characteristics between those groups. More specifically, we propose to use a matching process based on propensity scores. Such scores, obtained by using a nonlinear probability econometric model (probit model), allow us to pair potential affected individuals and non-affected individuals. We can then estimate a probability comparison by choosing a group of household living in areas which are currently facing extreme conditions and a similar group of households living in areas without such a problem (i.e., probability of having water access at home).

16.2.8 Model Validation

In order to verify the empirical validity of our results, we carry out diagnostic checks on the key assumptions that are made in the estimation and corroborate that the model specification is appropriate and that the results do not suffer from bias. A common diagnostic check used is to plot and compare the distributions of the propensity scores for the treated and untreated groups to visually check the overlapping condition and to see if the matching is able to make the distributions of both groups more similar. The distributions of the propensity scores, before and after the matching, for our aggregated models for Mexico City are plotted in Figs. 16.12 and 16.13. The rest of the graphs for all the counties are reported in the Appendix.

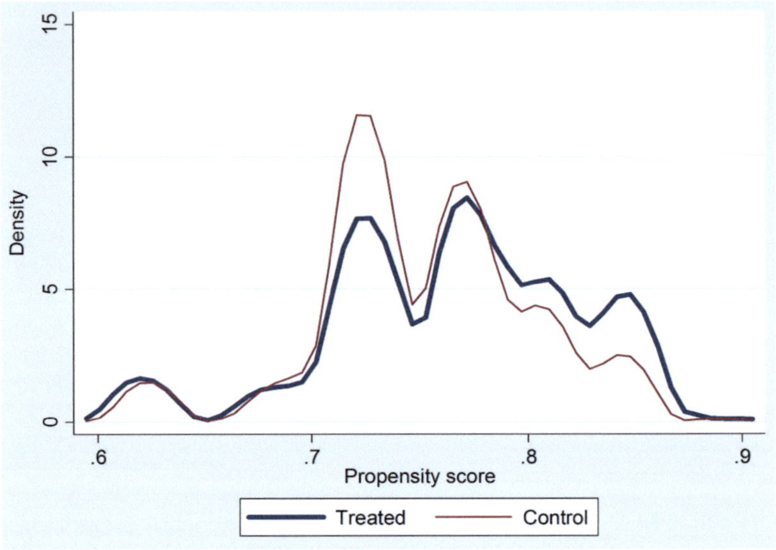

Fig. 16.12 Propensity score distribution for scarce rainfall

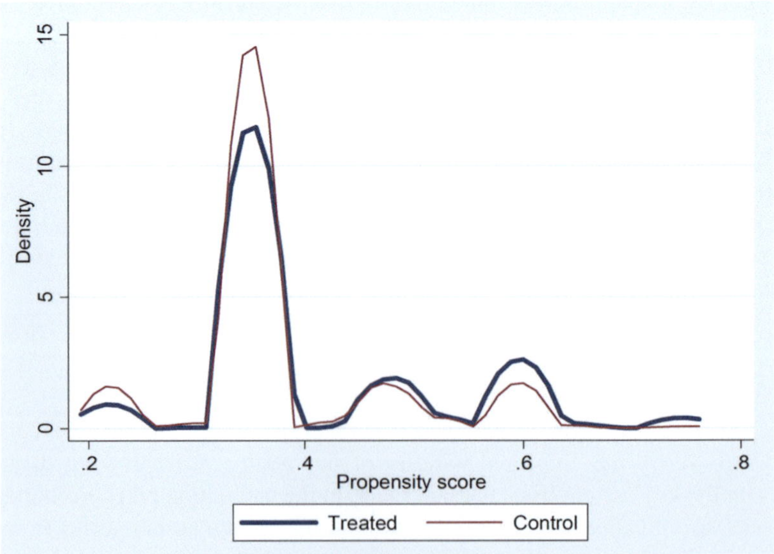

Fig. 16.13 Propensity score distribution for high temperature

Visual inspection suggests that the densities of the propensity scores are similar after matching for the groups of people living in the counties in Mexico DF (*distrito federal*). The whole set of tests for every county reported in the Appendix confirms that the same applies for the rest of the counties. Some other statistical *t*-tests are reported as a way to validate the equality of the two distributions. We report such statistics in the Appendix.

16.3 Results and Discussion

In the previous section, we described our methodology to determine the odds of households having water access for the counties of Mexico City that suffer extreme weather conditions in relation to the ones that are not facing such climate conditions (scarce rainfall and/or high temperatures). We not only seek to identify the affected regions but also the affected people. It is worth mentioning that this methodology might be applied for other types of extreme conditions as well. Our main results suggest that the following ten counties are currently facing scarce rainfall, high temperatures, or both (Table 16.2).

Table 16.3 shows the simulated differences in the probability of households having access to water for ten counties in Mexico City, which are currently suffering of scarce rainfall and/or high temperatures, in relation to households in counties with regular weather conditions. In general, we find that there are lower probabilities of having water access at home in 10 out of 16 counties. It is worth noticing that the results are highly heterogeneous for the different counties of Mexico City.

Table 16.3 suggests that 10 out of 16 counties in Mexico City might be affected by extreme weather conditions and have a lower probability of having water access. Specifically, we refer to the counties of Azcapotzalco, Coyoacán, Gustavo A. Madero, Iztapalapa, Álvaro Obregón, Miguel Hidalgo, Tlalpan, Xochimilco, Cuauhtémoc, and Venustiano Carranza. According to our estimates, from the 8.8 million people living in Mexico City, around 3,142,660 are living in areas with scarce rainfall and about 1,500,100 in areas with high temperature. There is also

Table 16.2 Counties suffering scarce rainfall or high temperature

Counties	Scarce rainfall	High temperatures
Azcapotzalco	✓	✓
Coyoacán	✓	✓
Gustavo A. Madero	✓	
Iztapalapa		✓
Álvaro Obregón	✓	✓
Tlalpan	✓	✓
Xochimilco	✓	
Cuauhtémoc	✓	
Miguel Hidalgo	✓	✓
Venustiano Carranza	✓	

Table 16.3 Differences in the probability of households having access to water for the counties with extreme weather conditions in Mexico City compared to the households with regular weather conditions

	Scarce rainfall				High temperature			
Counties	Normal*	Logit	Kernel	Radius	Normal	Logit	Kernel	Radius
Azcapotzalco	−0.216 (0.077)***	−0.216 (0.077)***	−0.184 (0.073)***	−0.189 (0.071)***	−0.081 (0.027)***	−0.081 (0.024)***	−0.073 (0.022)***	−0.069 (0.017)***
Coyoacán	−0.078 (0.015)***	−0.078 (0.015)***	−0.078 (0.015)***	−0.078 (0.015)***	−0.094 (0.014)***	−0.094 (0.014)***	−0.094 (0.014)***	−0.094 (0.014)***
Gustavo A. Madero	−0.042 (0.030)	−0.044 (0.030)	−0.062 (0.022)***	−0.072 (0.012)***				
Iztapalapa					−0.306 (0.045)***	−0.306 (0.045)***	−0.279 (0.040)***	−0.268 (0.031)***
Álvaro Obregón	−0.217 (0.171)	−0.217 (0.171)	−0.289 (0.098)***	−0.201 (0.069)***	0.000 (0.167)	0.000 (0.167)	−0.132 (0.119)	−0.133 (0.118)
Tlalpan	−0.095 (0.072)	−0.101 (0.072)	−0.093 (0.058)*	−0.094 (0.038)***	−0.268 (0.311)	−0.135 (0.286)	−0.025 (0.156)	0.442 (0.027)***
Xochimilco	−0.274 (0.326)	−0.0215 (0.322)	−0.123 (0.253)	0.223 (0.040)***				
Cuauhtémoc	−0.064 (0.019)***	−0.064 (0.019)***	−0.065 (0.020)***	−0.064 (0.019)***				
Miguel Hidalgo	−0.069 (0.020)***	−0.069 (0.020)***	−0.070 (0.020)***	−0.069 (0.020)***	−0.020 (0.010)***	−0.020 (0.017)***	−0.0003 (0.017)	0.058 (0.010)***
Venustiano Carranza	−0.055 (0.013)***	−0.055 (0.013)***	−0.055 (0.013)***	−0.055 (0.013)***				

*Differences in probabilities obtained with three different methods: normal, logit, kernel, and radius
Counties with no significant differences in probabilities are not reported
Standard errors reported between parentheses
**Statistically significant at the 5 % level of confidence
***Statistically significant at the 1 % level of confidence

evidence that only five counties are currently affected by both phenomena: Azcapotzalco, Coyoacán, Álvaro Obregón, Tlalpan, and Miguel Hidalgo. This implies that there are around 508,840 vulnerable people in Mexico City affected by both types of extreme conditions.

On the other hand, the odds of having water at home are much lower for people living in six of the counties with extreme conditions, where the odds are reduced between 20 and 30 %. Specifically, the odds are reduced in five counties with scarce rainfall – Azcapotzalco (−0.216), Álvaro Obregón (−0.217), Iztapalapa (−0.306), and Xochimilco (−0.274) – which have a total population of around 936,420 individuals. These results are worrying since the probabilities of receiving water at home for these counties vary from 74 to 50 %, respectively, so reductions of about 20 or 30 % might imply a serious impact on the well-being of such families. Besides, these ten counties have high poverty levels and are considered among the most vulnerable ones in Mexico, so our results suggest that water scarcity might be playing a role in reinforcing vulnerability processes in such counties. Besides, among the counties with high temperature, Iztapalapa and Tlalpan are having a significant negative difference in the probability of having access to water, −0.306 and −0.268, respectively. So, it is clear that such counties would need some kind of support to maintain water access sustainability. These results are important since knowing the spatial location and the number of affected people will definitively contribute to improve the design and implementation of coordinated public policies and the effectiveness in the use of resources devoted to problems like water access combined with extreme climate around the world.

16.4 Conclusions

We analyze the probability of households having water access in the counties with more extreme climate of Mexico City. Our aim is to identify vulnerability to water access at the household and county level. We do so by combining climate, spatial, and household survey information in the context of a probability model that compares the odds of having access to water of affected and non-affected people. This methodology allows us to estimate not only the probabilities but also to identify the most vulnerable people in zones with more extreme climate and develop focused public policy recommendations in the face of extreme weather. Our results suggest that 10 out of 16 counties in Mexico City are facing much lower probabilities of having access to water in zones affected by extreme weather, compared to the locations with regular climate conditions. Specifically, we refer to people located in Azcapotzalco, Coyoacán, Gustavo A. Madero, Iztapalapa, Álvaro Obregón, Miguel Hidalgo, Tlalpan, Xochimilco, Cuauhtémoc, and

Venustiano Carranza. From the 8.8 million people living in Mexico City, about 3,142,660 are living in areas with scarce rainfall and approximately 1,500,100 in areas with high temperature. Empirical evidence shows that only five counties are currently affected by both phenomena: Azcapotzalco, Coyoacán, Álvaro Obregón, Tlalpan, and Miguel Hidalgo. This implies that there are around 508,840 highly vulnerable people in Mexico City affected by both types of extreme conditions. We also find that coincidentally the odds of having water at home are lower for people living in these zones with more extreme climate conditions; the odds are reduced between 20 and 30 %, compared to zones with regular climate conditions; such areas have a vulnerable population of around 936,420 individuals. These counties have high poverty levels and are considered among the most vulnerable ones in Mexico, so our results suggest that water scarcity conditions might be reinforcing the vulnerability of the population in such counties. Knowing the spatial location and the number of affected people will definitively contribute to improve the design and implementation of public policies and the effectiveness in the use of resources devoted to problems like water access in the face of extreme weather events.

Appendix

Scarce Rainfall

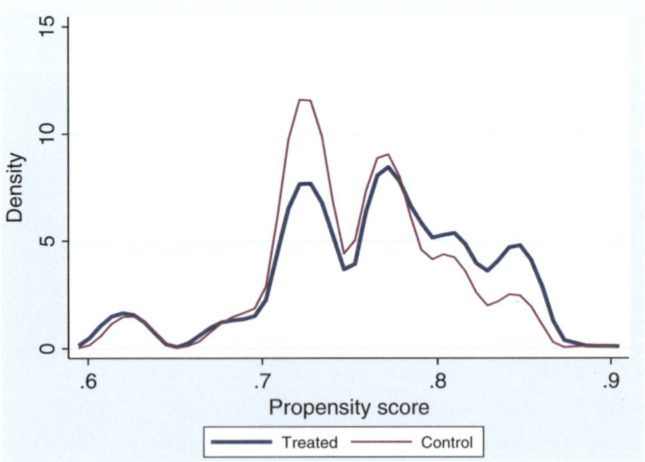

Fig. 16.14 Propensity score distribution scarce rainfall

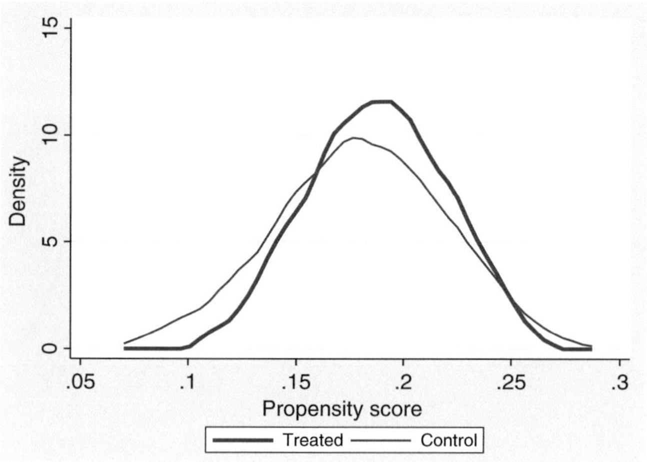

Fig. 16.15 Propensity score distribution scarce rainfall at Azcapotzalco county

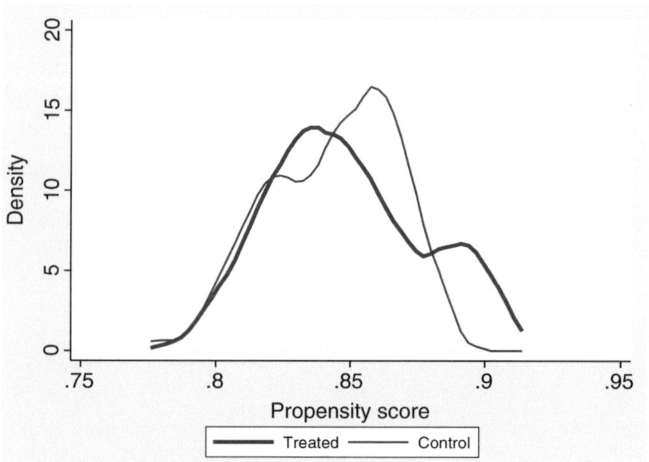

Fig. 16.16 Propensity score distribution scarce rainfall at Coyoacán

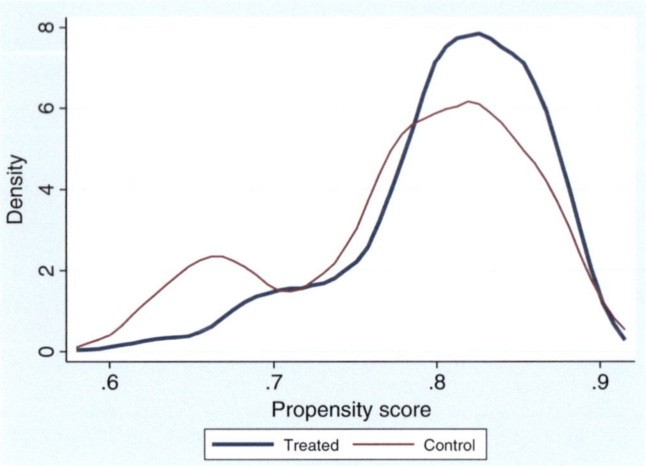

Fig. 16.17 Propensity score distribution scarce rainfall at Gustavo A. Madero county

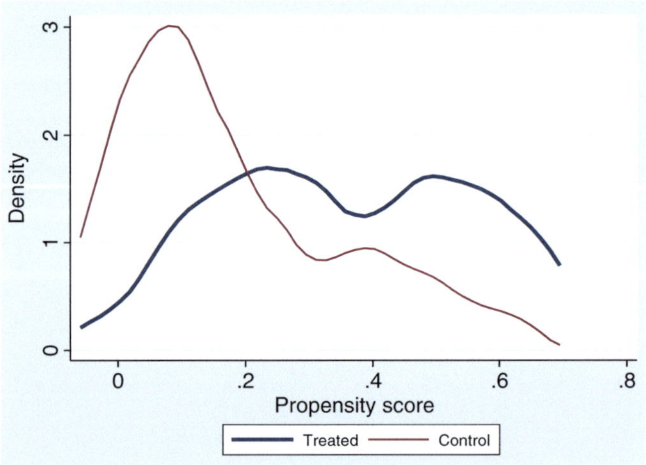

Fig. 16.18 Propensity score distribution scarce rainfall at Álvaro Obregón county

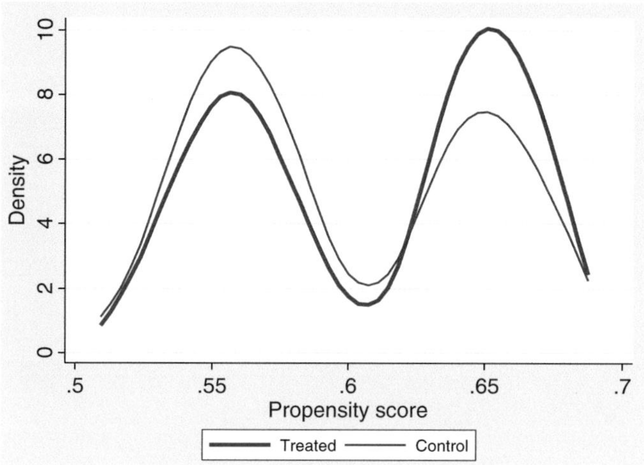

Fig. 16.19 Propensity score distribution scarce rainfall at Tlalpan county

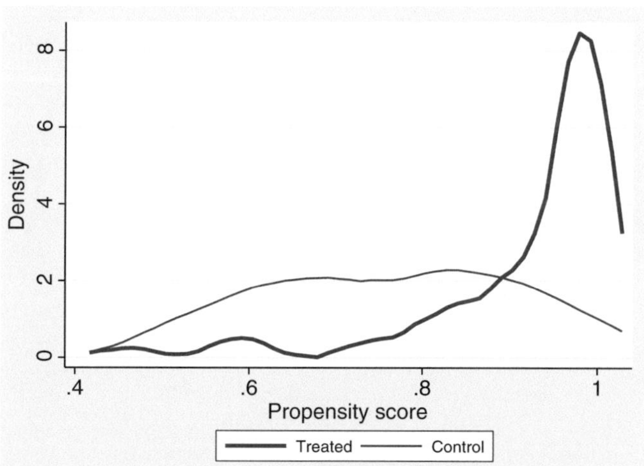

Fig. 16.20 Propensity score distribution scarce rainfall at Xochimilco county

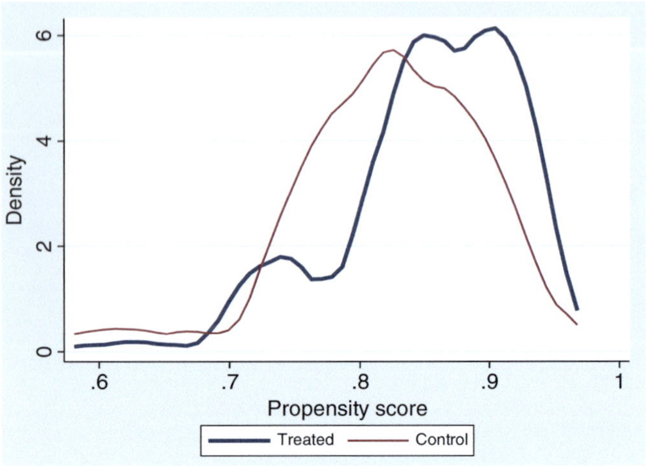

Fig. 16.21 Propensity score distribution scarce rainfall at Benito Juárez county

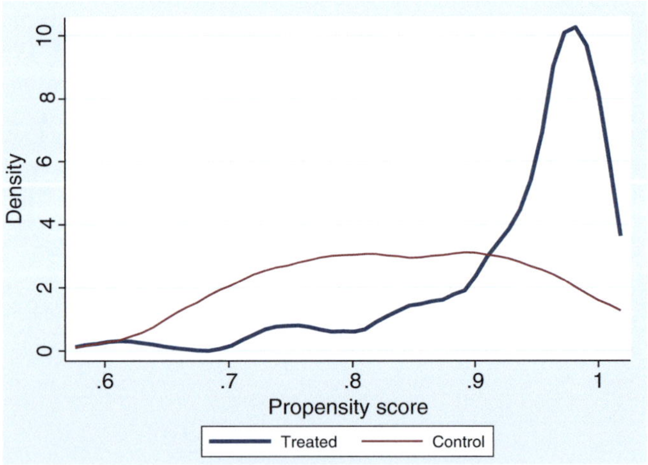

Fig. 16.22 Propensity score distribution scarce rainfall at Cuauhtémoc

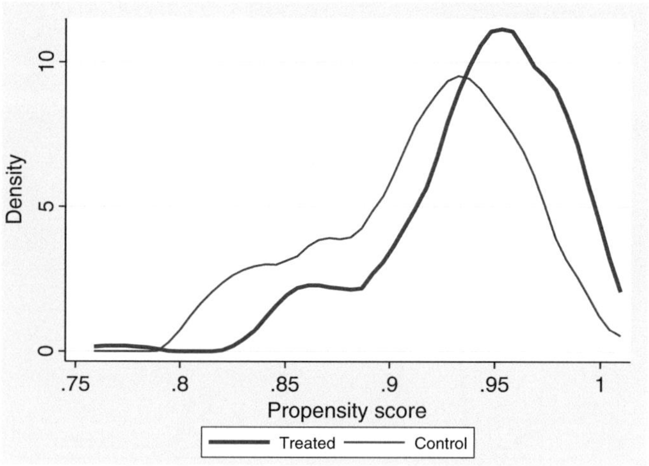

Fig. 16.23 Propensity score distribution scarce rainfall at Miguel Hidalgo county

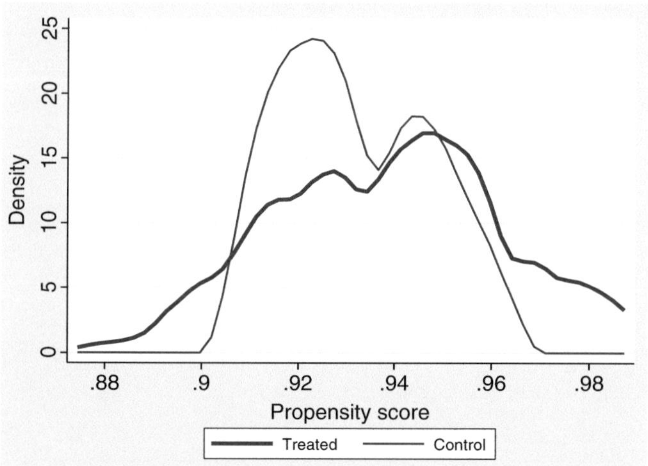

Fig. 16.24 Propensity score distribution scarce rainfall at Venustiano Carranza

High Temperature

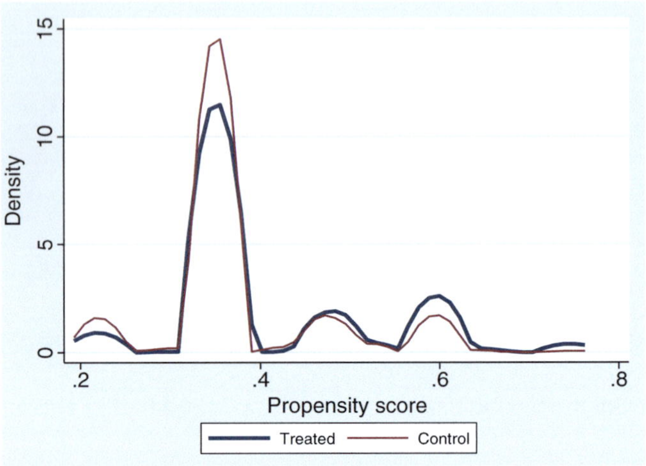

Fig. 16.25 Propensity score distribution high temperature

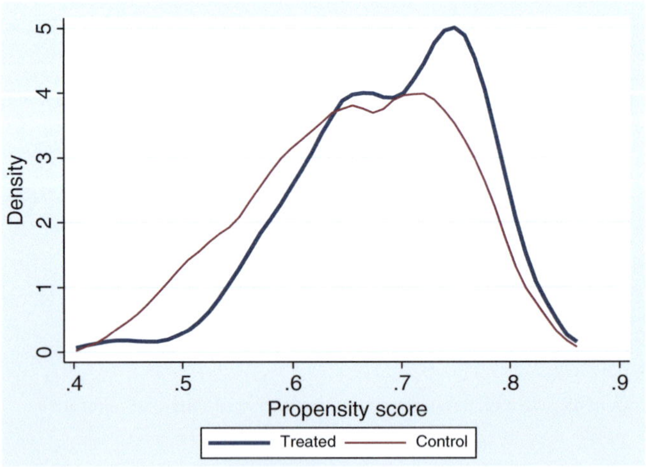

Fig. 16.26 Propensity score distribution high temperature at Azcapotzalco county

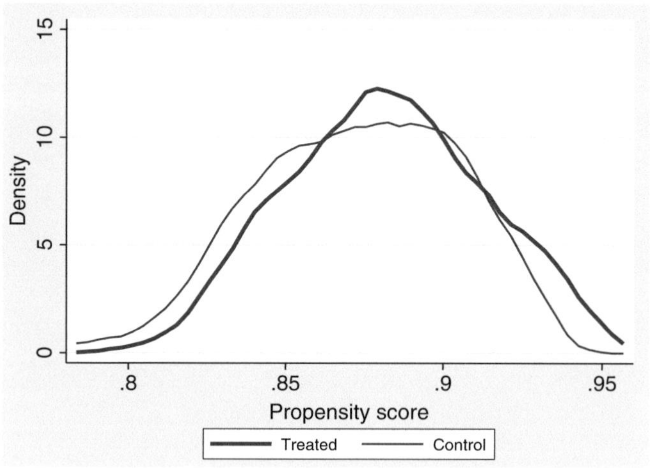

Fig. 16.27 Propensity score distribution high temperature at Coyoacán county

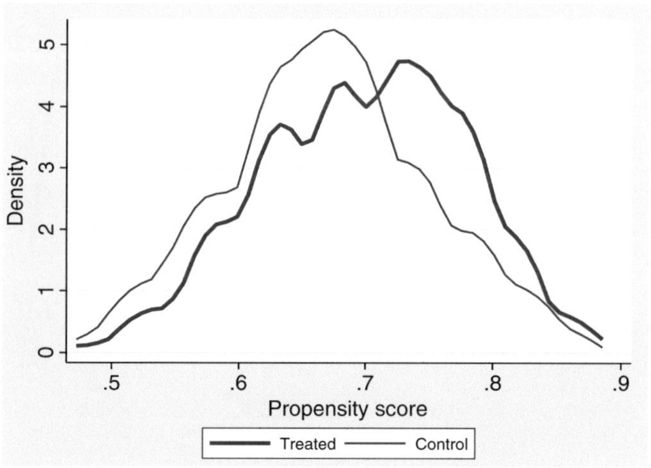

Fig. 16.28 Propensity score distribution high temperature at Iztapalapa county

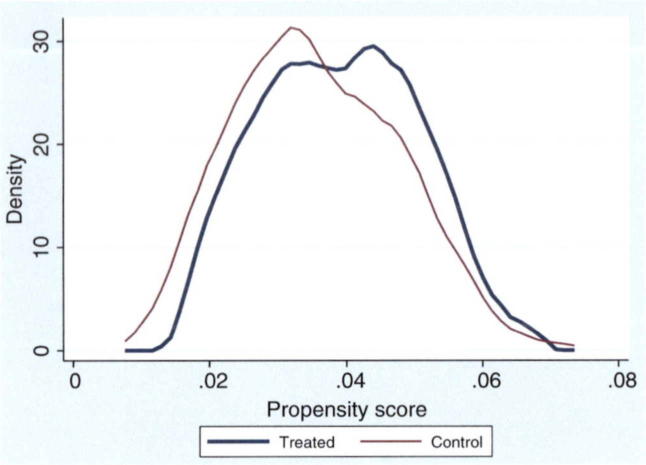

Fig. 16.29 Propensity score distribution high temperature at Álvaro Obregón county

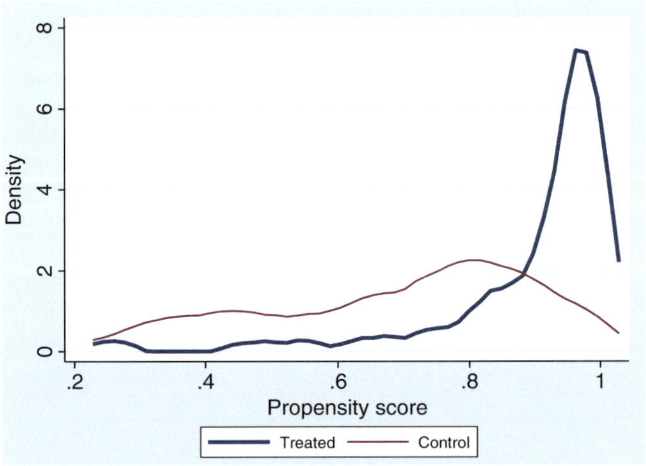

Fig. 16.30 Propensity score distribution high temperature at Tlalpan county

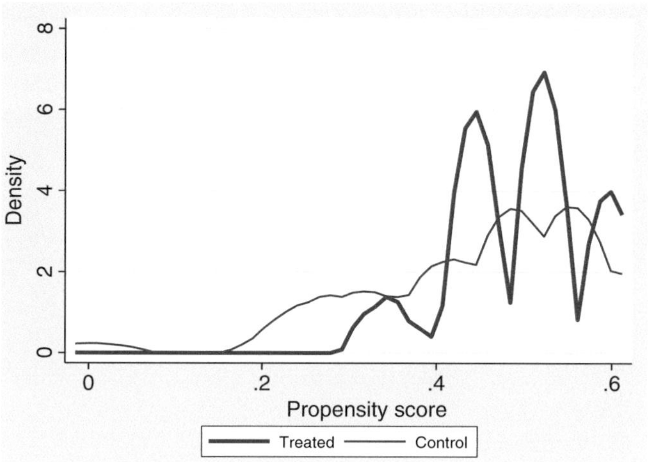

Fig. 16.31 Propensity score distribution high temperature at Miguel Hidalgo county

Scarce-Rainfall T-*test*

Table 16.4 Propensity score statistical *t*-test scarce rainfall

Variable	Unmatched / Matched	Mean: Treated	Mean: Control	% Bias	% Reduct |bias|	*t*-test: *t*	*t*-test: $p > \|t\|$
Sex	Unmatched	.48529	.5	−2.9		−0.12	0.904
	Matched	.48529	.5	−2.9	0.0	−0.02	0.983
Age	Unmatched	32.665	33.667	−4.1		−0.19	0.848
	Matched	32.665	33.667	−4.1	0.0	−0.03	0.973
Work type	Unmatched	.54464	.57143	−5.3		−0.19	0.846
	Matched	.54464	.57143	−5.3	0.0	−0.03	0.973
Household size	Unmatched	4.5074	3.7778	47.1		1.53	0.127
	Matched	4.5074	3.7778	47.1	0.0	0.26	0.797

Table 16.5 Propensity score statistical *t*-test scarce rainfall at Azcapotzalco County

Variable	Unmatched / Matched	Mean: Treated	Mean: Control	% Bias	% Reduct |bias|	*t*-test: t	*t*-test: $p>\|t\|$
Sex	Unmatched	.43243	.46108	−5.7		−0.32	0.753
	Matched	.43243	.35135	16.2	−183.1	0.71	0.482
Age	Unmatched	33.405	36.461	−14.6		−0.78	0.438
	Matched	33.405	36.243	−13.6	7.1	−0.61	0.545
Work type	Unmatched	.46667	.50714	−8.0		−0.40	0.690
	Matched	.46667	.57576	−21.6	−169.5	−0.86	0.395
Household size	Unmatched	4.1351	4.3832	−15.5		−0.74	0.460
	Matched	4.1351	3.973	10.1	34.6	0.49	0.625

Table 16.6 Propensity score statistical *t*-test scarce rainfall at Coyoacán county

Variable	Unmatched / Matched	Mean: Treated	Mean: Control	% Bias	% Reduct |bias|	*t*-test: *t*	*t*-test: $p>\|t\|$
Sex	Unmatched	.48265	.45614	5.3		0.37	0.713
	Matched	.48265	.53312	−10.1	−90.4	−1.27	0.204
Age	Unmatched	37.117	39.088	−9.1		−0.64	0.525
	Matched	37.117	38.331	−5.6	38.4	−0.71	0.477
Work type	Unmatched	.53184	.39216	28.1		1.83	0.068
	Matched	.53184	.29617	47.5	−68.7	5.79	0.000
Household size	Unmatched	4.1987	3.8421	21.3		1.34	0.181
	Matched	4.1987	3.9905	12.4	41.6	1.52	0.128

Table 16.7 Propensity score statistical *t*-test scarce rainfall at Gustavo A. Madero county

Variable	Unmatched / Matched	Mean: Treated	Mean: Control	% Bias	% Reduct |bias|	*t*-test: *t*	*t*-test: $p>\|t\|$
Sex	Unmatched	.45814	.48447	−5.3		−0.60	0.549
	Matched	.45814	.4551	0.6	88.4	0.11	0.912
Age	Unmatched	35.116	33.925	5.6		0.64	0.522
	Matched	35.116	32.187	13.8	−146.0	2.47	0.014
Work type	Unmatched	.50536	.54135	−7.2		−0.75	0.456
	Matched	.50536	.61914	−22.7	−216.1	−3.73	0.000
Household size	Unmatched	4.1598	4.8758	−35.9		−4.30	0.000
	Matched	4.1598	4.2283	−3.4	90.4	−0.69	0.493

Table 16.8 Propensity score statistical *t*-test scarce rainfall at Álvaro Obregón county

	Unmatched	Mean			% Reduct	t-test	
Variable	Matched	Treated	Control	% Bias	|bias|	*t*	*p* > |*t*|
Sex	Unmatched	.41304	.46324	−10.1		−0.59	0.557
	Matched	.41176	.29025	24.3	−142.1	1.04	0.301
Age	Unmatched	28.065	33.662	−27.9		−1.64	0.103
	Matched	27.647	30.716	−15.3	45.2	−0.65	0.516
Work type	Unmatched	.68571	.54054	29.9		1.52	0.132
	Matched	.61538	.42718	38.7	−29.6	1.37	0.176
Household size	Unmatched	5.8261	3.8382	132.0		7.79	0.000
	Matched	5.1765	5.0647	7.4	94.4	0.41	0.685

Table 16.9 Propensity score statistical *t*-test scarce rainfall at Tlalpan county

	Unmatched	Mean			% Reduct	t-test	
Variable	Matched	Treated	Control	% Bias	|bias|	*t*	*p* > |*t*|
Sex	Unmatched	.46497	.56311	−19.7		−1.55	0.123
	Matched	.46497	.46497	0.0	100.0	0.00	1.000
Age	Unmatched	36.554	36.359	0.9		0.07	0.944
	Matched	36.554	35.389	5.4	−498.0	0.50	0.620
Work type	Unmatched	.51515	.64706	−26.8		−1.92	0.056
	Matched	.51515	.55556	−8.2	69.4	−0.66	0.510
Household size	Unmatched	3.6624	3.7184	−4.2		−0.33	0.740
	Matched	3.6624	3.6943	−2.4	43.2	−0.22	0.827

Table 16.10 Propensity score statistical *t*-test scarce rainfall at Xochimilco county

	Unmatched	Mean			% Reduct	t-test	
Variable	Matched	Treated	Control	% Bias	|bias|	*t*	*p* > |*t*|
Sex	Unmatched	.43791	.3125	25.7		0.96	0.337
	Matched	.43791	.11111	66.9	−160.6	6.86	0.000
Age	Unmatched	32.092	39.75	−33.9		−1.38	0.170
	Matched	32.092	15.366	74.1	−118.4	7.58	0.000
Work type	Unmatched	.52033	.61538	−18.9		−0.65	0.517
	Matched	.52033	.47727	8.5	54.7	0.49	0.626
Household size	Unmatched	4.0196	4.375	−20.9		−0.87	0.385
	Matched	4.0196	4.9346	−53.9	−157.5	−6.42	0.000

Table 16.11 Propensity score statistical *t*-test scarce rainfall at Benito Juárez county

Variable	Unmatched / Matched	Mean Treated	Mean Control	% Bias	% Reduct |bias|	*t*-test *t*	*p* > |*t*|
Sex	Unmatched	.41791	.58333	−33.1		−1.50	0.135
	Matched	.41791	.42537	−1.5	95.5	−0.12	0.902
Age	Unmatched	36.134	43.5	−36.0		−1.54	0.127
	Matched	36.134	41.425	−25.8	28.2	−2.07	0.039
Work type	Unmatched	.58407	.78261	−43.2		−1.79	0.075
	Matched	.58407	.79545	−46.0	−6.5	−3.68	0.000
Household size	Unmatched	3.5075	3.6667	−8.2		−0.45	0.655
	Matched	3.5075	2.9179	30.3	−270.3	2.95	0.003

Table 16.12 Propensity score statistical *t*-test scarce rainfall at Cuauhtémoc county

Variable	Unmatched / Matched	Mean Treated	Mean Control	% Bias	% Reduct |bias|	*t*-test *t*	*p* > |*t*|
Sex	Unmatched	.45455	.25	42.9		1.37	0.171
	Matched	.45455	.24675	43.6	−1.6	3.90	0.000
Age	Unmatched	36.026	57.667	−119.4		−3.62	0.000
	Matched	36.026	39.091	−16.9	85.8	−1.45	0.149
Work type	Unmatched	.65152	.33333	65.6		2.20	0.029
	Matched	.65152	.31169	70.0	−6.8	6.08	0.000
Household size	Unmatched	3.3766	3.1667	18.3		0.49	0.622
	Matched	3.3766	3.6883	−27.2	−48.5	−2.50	0.013

Table 16.13 Propensity score statistical *t*-test scarce rainfall at Miguel Hidalgo county

Variable	Unmatched / Matched	Mean Treated	Mean Control	% Bias	% Reduct |bias|	*t*-test *t*	*p* > |*t*|
Sex	Unmatched	.49367	.4	18.4		0.57	0.568
	Matched	.49367	.5443	−9.9	45.9	−0.90	0.369
Age	Unmatched	36.215	28.8	32.5		1.02	0.311
	Matched	36.215	35.443	3.4	89.6	0.34	0.736
Work type	Unmatched	.5625	.57143	−1.7		−0.05	0.963
	Matched	.5625	.4	31.5	−1720.0	2.66	0.008
Household size	Unmatched	4.2975	3.4	45.3		1.13	0.262
	Matched	4.2975	3.6582	32.3	28.8	3.00	0.003

Table 16.14 Propensity score statistical *t*-test scarce rainfall Venustiano Carranza county

Variable	Unmatched Matched	Mean Treated	Control	% Bias	% Reduct |bias|	*t*-test *t*	$p > \|t\|$
Sex	Unmatched	.48529	.5	−2.9		−0.12	0.904
	Matched	.48529	.55882	−14.5	−400.0	−1.72	0.086
Age	Unmatched	32.665	33.667	−4.1		−0.19	0.848
	Matched	32.665	41.195	−34.6	−751.9	−3.40	0.001
Work type	Unmatched	.54464	.57143	−5.3		−0.19	0.846
	Matched	.54464	.43229	22.2	−319.4	2.29	0.022
Household size	Unmatched	4.5074	3.7778	47.1		1.53	0.127
	Matched	4.5074	4.1176	25.1	46.6	2.92	0.004

High-Temperature **T-*test***

Table 16.15 Propensity score statistical *t*-test high temperature

Variable	Unmatched Matched	Mean Treated	Control	% Bias	% Reduct |bias|	*t*-test *t*	$p > \|t\|$
Sex	Unmatched	.4712	.48131	−2.0		−0.20	0.839
	Matched	.4712	.48131	−2.0	0.0	−0.05	0.957
Age	Unmatched	35.445	35.215	1.0		0.10	0.917
	Matched	35.445	35.215	1.0	−0.0	0.03	0.978
Work type	Unmatched	.55414	.54286	2.3		0.21	0.837
	Matched	.55414	.54286	2.3	0.0	0.05	0.956
Household size	Unmatched	4.5393	4.5935	−2.3		−0.23	0.818
	Matched	4.5393	4.5935	−2.3	0.0	−0.07	0.946

Table 16.16 Propensity score statistical *t*-test high temperature at Azcapotzalco county

Variable	Unmatched Matched	Mean Treated	Control	% Bias	% Reduct |bias|	*t*-test *t*	$p > \|t\|$
Sex	Unmatched	.45946	.46429	−1.0		−0.09	0.925
	Matched	.45946	.44595	2.7	−180.0	0.33	0.742
Age	Unmatched	29.75	37.129	−35.3		−3.54	0.000
	Matched	29.75	29.321	2.1	94.2	0.26	0.791
Work type	Unmatched	.53879	.47899	11.9		1.06	0.290
	Matched	.53879	.54673	−1.6	86.7	−0.17	0.867
Household size	Unmatched	5.0608	4.5357	22.8		2.14	0.033
	Matched	5.0608	4.8885	7.5	67.2	0.90	0.366

Table 16.17 Propensity score statistical *t*-test high temperature at Coyoacán county

Variable	Unmatched / Matched	Mean: Treated	Mean: Control	% Bias	% Reduct |bias|	*t*-test: *t*	*t*-test: $p > \|t\|$
Sex	Unmatched	.50461	.45763	9.4		0.68	0.499
	Matched	.50461	.52765	−4.6	51.0	−0.68	0.498
Age	Unmatched	36.124	39.797	−17.4		−1.26	0.207
	Matched	36.124	36.18	−0.3	98.5	−0.04	0.968
Work type	Unmatched	.5122	.48148	6.1		0.42	0.674
	Matched	.5122	.40199	21.9	−258.8	3.09	0.002
Household size	Unmatched	3.9747	3.5424	27.5		1.88	0.060
	Matched	3.9747	3.8548	7.6	72.3	1.13	0.257

Table 16.18 Propensity score statistical *t*-test high temperature at Iztapalapa county

Variable	Unmatched / Matched	Mean: Treated	Mean: Control	% Bias	% Reduct |bias|	*t*-test: *t*	*t*-test: $p > \|t\|$
Sex	Unmatched	.47297	.51	−7.4		−0.61	0.540
	Matched	.47297	.45045	4.5	39.2	0.48	0.635
Age	Unmatched	34.414	36.19	−8.1		−0.68	0.494
	Matched	34.414	32.865	7.1	12.7	0.80	0.427
Work type	Unmatched	.54497	.58537	−8.1		−0.61	0.540
	Matched	.54497	.6	−11.1	−36.2	−1.07	0.283
Household size	Unmatched	4.4144	3.86	36.0		2.97	0.003
	Matched	4.4144	4.3784	2.3	93.5	0.25	0.805

Table 16.19 Propensity score statistical *t*-test high temperature at Álvaro Obregón county

Variable	Unmatched / Matched	Mean: Treated	Mean: Control	% Bias	% Reduct |bias|	*t*-test: *t*	*t*-test: $p > \|t\|$
Sex	Unmatched	.53333	.48642	9.2		0.36	0.722
	Matched	.53333	.53333	0.0	100.0	−0.00	1.000
Age	Unmatched	26.667	32.832	−33.2		−1.18	0.240
	Matched	26.667	26.267	2.2	93.5	0.07	0.948
Work type	Unmatched	.36364	.54142	−35.4		−1.16	0.246
	Matched	.36364	.66667	−60.4	−70.4	−1.46	0.160
Household size	Unmatched	4.7333	4.4667	20.7		0.59	0.559
	Matched	4.7333	4.6	10.4	50.0	0.41	0.687

Table 16.20 Propensity score statistical *t*-test high temperature at Tlalpan county

Variable	Unmatched Matched	Mean Treated	 Control	 % Bias	% Reduct |bias|	*t*-test *t*	 $p > \|t\|$
Sex	Unmatched	.46547	.58	−22.9		−1.51	0.131
	Matched	.46547	.87087	−81.2	−254.0	−12.29	0.000
Age	Unmatched	33.634	25.58	42.0		2.59	0.010
	Matched	33.634	37.384	−19.6	53.4	−2.98	0.003
Work type	Unmatched	.49281	.55556	−12.5		−0.71	0.480
	Matched	.49281	.86624	−74.3	−495.1	−10.71	0.000
Household size	Unmatched	4.2793	5.16	−49.3		−3.05	0.002
	Matched	4.2793	2.7147	87.5	−77.6	12.29	0.000

Table 16.21 Propensity score statistical *t*-test high temperature at Miguel Hidalgo county

Variable	Unmatched Matched	Mean Treated	 Control	 % Bias	% Reduct |bias|	*t*-test *t*	 $p > \|t\|$
Sex	Unmatched	.4712	.48131	−2.0		−0.20	0.839
	Matched	.4712	.03665	86.8	−4200.7	11.23	0.000
Age	Unmatched	35.445	35.215	1.0		0.10	0.917
	Matched	35.445	37.209	−8.0	−666.9	−0.89	0.376
Work type	Unmatched	.55414	.54286	2.3		0.21	0.837
	Matched	.55414	1	−89.3	−3851.6	−7.58	0.000
Household size	Unmatched	4.5393	4.5935	−2.3		−0.23	0.818
	Matched	4.5393	2.9686	66.9	−2798.4	7.22	0.000

Acknowledgments This paper would not have been possible without the financial support of SEDESOL-CONACYT's project No. 166140, PAPIIT project No. 301715 and the PAPIME project No. 304914. We are Grateful for the financial support of the BIARI program at Brown University and the GLOWS program at Florida International University. We are also grateful to the PINCC project: "Implementar un manejo integrado de biodiversidad y servicios ambientales en comunidades indígenas pobres de Oaxaca: Un enfoque de paisaje"

References

Ludi E (2009) Climatic change, water and food security, ODI Background Notes, 8p

Rosenbaum P, Rubin D (1983) The central role of the propensity score in observational studies for causal effects. Biometrika 70(1):41–55

Sánchez A, Gay C, Estrada F (2011) El cambio climático y la pobreza en el Distrito Federal, Investigación Económica 70(278):45–74. http://www.redalyc.org/pdf/601/60121367004.pdf

Zhang X, Yang F (2004) RClimDex (1.0) user manual, Climate Research Branch, Environment Canada, 23p

Zhang X, Hegerl G, Zwiers FW et al (2005) Avoiding inhomogeneity in percentile-based indices of temperature extremes. J Clim 18(11):1641–1651

Zhang X, Alexander L, Hegerl G, Jones P, Tank A, Peterson T, Trewin B, Zwiers F (2011) Indices for monitoring changes in extremes based on daily temperature and precipitation data. WIREs Climatic Change 2(6):851–870. doi:10.1002/wcc.147

Chapter 17
Climate Change and Households' Willingness to Pay for Protecting High Quality Water and Its Provision in a Small Basin at Ecuador

Diana del Cisne Encalada-Jumbo and Armando Sánchez-Vargas

Abstract In this article we aim at eliciting households' willingness to pay (WTP) for protecting water quantity and quality, by restoring a small basin, in the face of climate change in Ecuador. To do so, we carry out a discrete choice experiment based on a representative survey of 248 users of the water system that depends on such basin; this type of methodology is often used in the economics literature to predict consumer choice and prices. Our results suggest that about 62.1 % of the respondents are willing to pay to secure water quantity and quality at home by carrying out a restoration plan of the basin to cope with climate extreme events. Households' willingness to pay for restoring the basin and securing water quality and quantity is about 1.24 and 0.5 dollars, respectively. These payments would be charged in the water bill in the form of monthly local taxes. Our findings also suggest that households' climate change perceptions have a significant effect on the willingness to pay. These results might constitute important inputs for policy makers to take decisions on the value of water and on the amount of investments needed to adapt to climate change.

Keywords Willingness to pay • Choice experiment • Climate change • Water quality • Water quantity • Water supply attributes

17.1 Introduction

Climate change has posed new challenges for all the countries around the world (FAO 2008). Weather variability affects hydrological cycles and, in consequence, the access to water is becoming a serious concern of developed and underdeveloped countries (UNESCO-WWAP 2009). In fact, the empirical evidence has shown that

D.C. Encalada-Jumbo (✉)
Department of Economics, UTPL, Loja, Ecuador
e-mail: dencalada1@utpl.edu.ec

A. Sánchez-Vargas
Institut for Economics Research, UNAM, Mexico City, Mexico
e-mail: sva@economia.unam.mx; asanchez@vt.edu

S.G. Setegn, M.C. Donoso (eds.), *Sustainability of Integrated Water Resources Management*, DOI 10.1007/978-3-319-12194-9_17

extreme events, linked to climate change, might affect the quantity and quality of water available in rivers, lakes, and underground reservoirs which, in turn, might generate water scarcity at the household level, affecting people's well-being (Sánchez et al. 2011; Bates et al. 2008).

In order to ameliorate such problems, a huge amount of public investments for the water sector are needed around world (Briscoe et al. 1990). Given the scale of such expenses, monetary contributions from water consumers would be needed to pay for financing such infrastructure for adaptation purposes (Veronesi et al. 2013). In this paper we aim at estimating a household's willingness to pay (WTP) to secure water quantity and quality, by investing in the restoration of a deteriorated small basin at Catacocha Town in Ecuador (NCI 2010), to counteract the effect of extreme climate events. This type of analysis is important since water supply systems, in small basins, will have to cope with climate change and design adaptation strategies to maintain and finance high quality water provision and reduce the social, ecological, and health risks of climate change (Gregersen and Arnbjerg 2011).

To do so, we carry out a discrete choice experiment (following Aizaki and Nishimura 2008) to elicit households' willingness to pay to secure high quality water and its provision by implementing restoration activities in a small basin that supplies freshwater to Catacocha Town at Ecuador. More specifically, our experiment consists in asking the town's households about their willingness to pay considering different payment plans, with different attributes, to restore the basin which might help to improve water quantity and quality in an environment of extreme events.

Such exercise is based on a representative sample survey from a population consisting of all the water users in Catacocha Town in Ecuador. Our representative sample consists of 248 interviews from a population of 2,160 water users. Our survey collects households' information on the level of climate change perception (Gunatilake and Tachiiri 2012), the degree of concern about water provision and its quality, and the costs for restoring the watershed. Respondents were asked to choose the preferred option among a set of three alternative policies to restore the basin (plan A, plan B, or neither of them), which might help to avoid the reduction of water quantity and quality in the face of climate change. Each policy option differs from the others in the following attributes: (1) prediction probability of extreme events, (2) average number of occurrences of water supply reduction in the basin, (3) number of times water will be unavailable at home, (4) water quality, (5) ecological risk for the basin, and (6) value of water (introduction of a local tax for the basin's protection) (Berrens et al. 2004).

Our statistical results suggest that the users of the water system at Catacocha strongly value the quantity and quality of water, and, in consequence, they are willing to pay for the restoration of the small basin, in the face of climate change. Specifically, we find that about 62.1 % of the respondents are willing to pay an annual local tax of about 1.24 dollars per month to protect the basin and reduce the possibility of having less access to high quality water; such an amount is not negligible given that the average water bill per month is around 6 dollars. We also find that the respondents are willing to pay an annual local tax of about 0.47

dollars per month to protect the basin and avoid having a reduced provision of water at home. Our findings also suggest that a higher probability of extreme climate events and the variation of the number of extreme events also have a statistically significant and positive effect on a household's willingness to pay for restoring the basin. These results might constitute a good source of information to support public policy makers on how to deal with climate change in the water sector for small basins.

17.2 Methodology

17.2.1 Study Area

The small basin of interest provides water to the town of Catacocha and covers 391.1 ha. The town has around 6,617 people, according to the 2010 population census from the National Institute of Statistics and Census of Ecuador. The average education is secondary level. The upper and middle sections of Catacocha are basically human settlements, and the lower sections are considered agricultural areas. The annual average temperature in the area is around 18 °C, and the average annual total precipitation is about 831.79 mm.

17.2.2 A Choice Experiment to Estimate the Willingness to Pay for High Quality Water and Its Provision in a Small Basin at Ecuador

Our contingent valuation methodology to elicit people's willingness to pay consists in carrying out a discrete choice experiment (Zwerina et al. 1996), which is often used in the economic literature to predict consumer choices and prices. It would allow us to quantify the households' welfare change associated with variations in the quantity, quality, price, or other attributes of water. In this type of experiments, respondents of a survey are often shown alternative options of a "public policy" they might be able to "purchase," (acquire its implementation) by paying a tax, in the near future. Then, the survey's respondents are asked to pick their most preferred option. In our case, the offered policy is a restoration plan, which implies that households will decide to accept or reject a water tax that might help to secure appropriate water attributes in the face of climate change. The options offered to the households differ from one another in the levels taken by two or more of the attributes of the goods (policies), such as the quality and quantity of water associated to different restoration levels of the basin. The attributes are not "purchased" separately, but come in a bundle, which also includes the cost of the particular option. We expect that the results of such experiment reveal the number of respondents who are willing to pay to conserve a specific water quantity and quality by carrying out some of the restoration plans or neither or them. We will also determine the average willingness to pay for protecting the basin, which can be the basis for establishing a local tax per household in the water bill to finance the project. Finally, our methodology will show what factors are the most important for the WTP of the water users in the town of Catacocha, including the climate change perception. All these results might constitute important inputs for policy makers to take anticipated decisions on the value of water in this small basin. Our methodology can be divided into four steps: (1) survey instrument and data collection, (2) discrete choice experiment, (3) econometric model and estimation of households' willingness to pay, and (4) validation of the results.

17.2.3 Survey Instrument and Data Collection

To collect the data needed for our experiment, we implemented a representative stated preference household's survey at Catacocha Town. We selected respondents randomly from the whole set of water users of such town (water users census), consisting of about 2,160 households. Such users are statistically representative of the population under study; the age range of the interviewed head of households varies between 18 and 94 years of age. Our questionnaire was pretested with 21 respondents and modified to make questions clearer. The final survey was administered by visiting 324 randomly selected houses. However, only 248 water users were interviewed since 76 heads of households were not located at home. This implies a response rate of about 76.54 %, which is much better than the range usually found for stated preference surveys. The average age of the respondents is 56 years, 36 % of them are women and 63.7 are men. A summary of the sociodemographic characteristics of our sample is presented in Table 17.1.

Our survey was organized in three sections. The first section focused on the demographic and social characteristics of the respondents. The second section characterized the water supply system in Catacocha Town. This section is also focused on the respondents' climate change perceptions; we got that information by asking them whether they noticed any long-term changes in temperature and/or rainfall in the area and whether they were concerned about the impacts of climate extreme events. It is worth mentioning that our sample shows that the climate change perception in the area is very high, since 88 % declared to have noticed

Table 17.1 Characteristic of the respondents

Variable	Sample average	Town's average
Age (years)	55,99	54
Female sex rate (%)	36,30	52,51
Household size	4,8	3,8
Household's annual income (USD)	7,216.82	7,170.24

temperature changes and 98 % of the respondents perceived long-term changes in precipitation in the last 10 years. Finally, we also asked for the average concern of households about climate extreme events. Respondents suggested that they are very concerned about the effects of climate change on the water sector, around 84.27 % of the sample. This descriptive result suggests that households might be willing to take some important decisions to avoid climate change impacts.

17.2.4 Discrete Choice Experiment

According to Louviere et al. (2000), an appropriate choice experiment has to fulfill two requirements: (1) the survey respondent should be asked to make a discrete choice between two or more alternatives in a choice set (i.e., select one of two alternative "goods" or "policies"), and (2) the alternatives presented for choice should be constructed by using an experimental design that changes one or more attributes within and/or between respondents. Thus, we design a choice experiment in which we allow households to choose between different options of the event of interest: "policies to protect the watershed to preserve water quantity and quality in the face of climate change." In order to identify the combination of attributes and options in the choice sets, we implemented a fractional factorial experimental design (Louviere et al. 2000; Adamowicz et al. 2011). The attributes of the policies to be "purchased" by households in this study are presented in Table 17.2 below. We will define each policy plan by the combination of six attributes: (1) prediction probability of extreme events, (2) average number of occurrences of water supply reduction in the basin, (3) number of times water is unavailable at home, (4) water quality, (5) ecological risk, and (6) price of the policy (a local tax for the basin's protection).

Table 17.2 Potential policy choices attributes for each adaptation policy

Attributes	Levels
Price of the policy	2 USD/family per month
	8 USD/family per month
Prediction probability of extreme events	10 %
	20 %
Average number of occurrences of extreme events	5 in 48 years
	10 in 48 years
Decrease in the water supply at home	0 h per month
	300 h per month
Water quality	Little contaminated
	Very contaminated
Ecological risk for the basin	Low
	High

Table 17.3 Example of choice question for the extreme event: scarce quantity of water in the basin

	A	B	C
Extreme events prediction probability	10 %	10 %	I do not choose any option
Average number of occurrences of extreme events	5 low flows in 48 years	10 low flows in 48 years	
Decrease in water supply at home	300 h per month	0 h per month	
Water quality	Low and contaminated	Low and contaminated	
Ecological risk	Low	Low	
Price of the policy	$ 2	$ 8	

The levels of such attributes are reliable since they were computed with the help of experts from the municipal services and from a local NGO called Nature and Culture International NCI. The "prediction probability" describes the confidence in the prediction of occurrences of extreme events. The level of the "average number of occurrences of extreme events" implies that lower levels correspond to an improvement of the situation and higher levels to more severe climate change conditions. Water availability was chosen as an attribute representing water safety in the area. Quality was chosen as an attribute to provide an indication of having access to healthy water at home. We chose a quality range with only two possibilities: good and bad quality. The ecological risk was chosen as the latent attribute of the basin in the face of climate change. Finally, the "price" of the policy was chosen as an attribute to provide a realistic comparison of water access and to allow for a monetary valuation of the whole set of water attributes. We selected a price range with only two levels, 2 dollars or 8 dollars, because these options represent realistic limits of monthly water taxes needed to pay for protecting the basin. With these options and their corresponding attributes, the respondents were then asked to choose the most preferred among three possible options, resulting from different combinations of attributes (Table 17.3).

17.2.5 Econometric Model and Estimation of Households' Willingness to Pay

Choice modeling implies to estimate a regression model that contains both a deterministic and random utility component. In this context, utility is a measure of consumer's welfare resulting from the consumption of a good or service. The deterministic component captures the influence of attributes known by the researcher, and the random component contains the influence of nonobservable attributes. The utility of choice j for consumer i is described as follows:

$$U_{ij} = V_{ij} + e_{ij} \tag{17.1}$$

where V_{ij} is the indirect observable utility of individual i and the attributes of alternative j. The indirect observable utility of individual i and alternative j can also be defined as follows:

$$V_{ij} = \beta' X_{ij} \tag{17.2}$$

where X_{ij} is a vector of the attributes of alternative j. We now use a conditional logit model to estimate the probability of individual i choosing alternative j as follows:

$$P_{ij} = \exp\left(\beta' x_{ij}\right) / \sum_j \exp\left(\beta' x_{ij}\right) \tag{17.3}$$

Where β is a parameter vector of interest. The X vector includes all the variables (attributes) considered in this study: (1) prediction probability of extreme events, (2) average number of occurrences of water supply reduction in the basin, (3) number of times water is unavailable at home, (4) water quality, (5) ecological risk, and (6) price of the policy (a local tax for the basin's protection). We choose such attributes, since we hypothesize that the policy's "purchasing" decision will depend on them.

This final utility model to be estimated can be specified as follows:

$$\begin{aligned} V_{ij} = {} & \beta_1' \, \mathrm{COS}_J + \beta_2' \, \mathrm{PPR}_j + \beta_3' \, \mathrm{OEV}_j + \beta_4' \, \mathrm{DSE}_j + \beta_5' \, \mathrm{CAA}_j \\ & + \beta_6' \, \mathrm{REC}_j + \mathrm{CONST} \end{aligned} \tag{17.4}$$

where COS denotes the price of the policy presented in the choice set, PPR denotes the prediction probability of extreme events, OEV is the average number of occurrences of extreme events, DSE is the decrease in the water supply at home, CAA is water quality, and REC is the ecological risk. Price was chosen as an attribute to provide a realistic estimation of the willingness to pay for a restoration plan, which includes different levels of quality and quantity of water. We selected two possible prices (2 dollars and 8 dollars) because they represent realistic limits of restoration prices that might deliver different levels of restoration. It is assumed that consumers are rational individuals and the water tax acceptance decision is based on constrained utility maximization, as reflected in the indirect utility function.

17.2.6 Validation of the Results

The validity of the willingness to pay responses can be verified by showing that the willingness to pay correlates in predictable ways with socioeconomic variables and ensuring that the usual regression statistics are appropriate. So, we estimate different models to check the robustness of the results and we include the sociodemographic characteristics of the respondents, and in particular we change the socioeconomic variables.

17.3 Empirical Results

Our survey results reveal that about 62.10 % of the respondents are willing to pay an annual local tax to secure water quality and quantity. Table 17.4 presents the estimated coefficients of a conditional logit model that shows the determinants of the willingness to pay.

The chi-squared statistic shows that all model variables are jointly statistically significant. The value of the adjusted R squared is around 0.13. Even more, the coefficients of our model display the expected signs. In other words, as predicted by the theory, the policy price coefficient (COS) is negative and statistically significant (−0.104), indicating that an increasing price decreases the probability of a consumer purchasing the policy. The PPR and OEV coefficients are positive and statistically significant at the 1 % level (0.206 and 0.0605) and indicate that increasing extreme events raise the probability that a consumer will be willing to pay for a policy to restore the basin. The coefficient of DSE is positive and is statistically significant at the 1 % level, indicating that a decreasing amount of water at home increases the probability of a consumer purchasing a policy, by paying a tax, to protect the basin (0.582). The CAA coefficient is positive and is statistically significant at the 1 % level, which also suggests that access to higher quality water

Table 17.4 Conditional logit estimates

Variable	Definition	Coefficient[a]	Standard error
COS	Price	−0.104*	0.0138
PPR	Prediction probability of extreme events	0.206*	0.0949
OEV	Average number of occurrences of the event	0.065*	0.0894
DSE	Decrease in the water supply at home	0.582*	0.0818
CAA	Water quality	1.544*	0.0870
REC	Ecological risk	0.100*	0.1058
ASC	Constant	−0.859*	0.1420
	n (Choice Set)	5,208	
	L (,O)	−1,900,028	
	L (β)	−1,670,446	
	Chi squared (d.f. = 7)	471.0	
	ρ^2	0.121	
	Adjusted ρ^2	0.128	

Note: "*" denotes statistically significant at the 1 % level
[a]It means 99 % confidence

increases the probability of a consumer purchasing the policy. Respondents are also willing to pay more money to avoid high ecological risks for the basin.

Changes in consumers' economic welfare arising from the availability of high quality water and quantity are computed following the method suggested by Blamey et al. (2000). Such method computes economic welfare as the compensating variation associated with a change in the level of a particular attribute. The change in economic welfare C is calculated as:

$$C = -1/\mu\left[\ln\left(e^{v^{i0}} - \ln e^{v^{i1}}\right)\right] = -1/\mu\left[V^{i0} - V^{i1}\right] \quad (17.5)$$

where μ = the marginal utility of money (the price coefficient is used to represent the marginal utility of money), V^{i0} = the indirect utility function in a baseline situation (i.e., PPR = 0, OEV = 0, DSE = 0, CAA = 0, REC = 0), and V^{i1} = the indirect utility function when higher quality water is available (i.e., PPR = 0, OEV = 0, DSE = 0, CAA = 1, REC = 0). Applying formula (17.5), we find that the respondents are willing to pay an annual local tax of about 1.24 dollars per month to protect the basin and reduce the possibility of having less access to high quality water; such amount is not negligible given that the average water bill per month is around 6 dollars. We also find that the respondents are willing to pay an annual local tax of about 0.47 dollars per month to protect the basin and avoid having a reduced provision of water at home. Finally, we should emphasize that a higher probability of extreme climate events and the number of extreme events also have a statistically significant and positive effect on a household's willingness to pay for restoring the basin. All these results might constitute a good source of information to support public policy designers on how to deal with climate change in the water sector for small basins.

17.4 Conclusions

We estimate the willingness to pay of the town's consumers to maintain water quantity and quality and, at the same time, to adapt to climate change by restoring a deteriorated small basin that supplies the town with freshwater. In order to do so, we offer different restoration options to water users in order to maintain the communal supply sources, and we ask them their WTP. More specifically, we offer them to restore and protect the superficial sources of water in the basin. The actions might include the restoration of the current water supply sources. In this study we find a relatively high willingness to pay, not only to secure the household's own water provision but also to maintain a good quality of it, by restoring a basin that supplies water to Catacocha Town, in an environment of climate change. We carried out a survey among 248 respondents in the area. Our results show that households have perceived climate change trends in the area and are also worried about their impacts on the water sector which has generated concerns about adaptation polices to restore the basin and secure water quantity and quality. We find that about 62.1 % of the respondents are willing to pay a higher annual local tax of about 1.24 dollars per month in order to protect the basin and secure water quality and almost half a dollar to secure water provision at home. To the best of our knowledge, there are no studies that have elicited the willingness to pay for a restoration policy of a small basin to protect the supply of water and its quality at the household level. Our results reveal that Catacocha's households value in particular the quality of water, and priority should be given to an upgrade of the basin that provides the water. Interestingly enough, we also find that the uncertainty in the scenarios does affect the willingness to pay of people for water quality and quantity. We find that the number of extreme events has positive impact in the determination of the WTP, which implies that climate change perception is pushing forward households' WTP. These results are important to support the policy makers' decisions on how to deal with the emerging risks of climate change in the water sector and where to set priorities.

Acknowledgments This paper would not have been possible without the financial support of SEDESOL-CONACYT's project No. 166140 and the PAPIT project No. 301313. We are grateful for the financial support of the BIARI program at Brown University and the GLOWS program at Florida International University. We are also grateful for the financial and academic support of the Universidad Técnica Particular de Loja (UTPL), the General Direction of Research and Technology Transfer (CITTE), and the Economics Department of UTPL.

References

Adamowicz W, Dupont D, Krupnick A, Zhang J (2011) Valuation of cancer and microbial disease risk reductions in municipal drinking water: an analysis of risk context using multiple valuation methods. J Environ Econ Manag 61:213–226

Aizaki H, Nishimura K (2008) Design and analysis of choice experiments using R: a brief introduction. JSAI 17(2):86–94. http://www.jstage.jst.go.jp/

Bates BC, Kundzewicz ZW, Wu S, Palutikof PJ et al (2008) Climate change and water. Technical paper of the Intergovernmental Panel on Climate Change. IPCC Secretariat, Geneva. 210pp

Berrens RP, Bohara AK, Smith HC, Silva CL, Weimer DL et al (2004). Information and effort in contingent valuation surveys: application to global climate change using national internet samples. Elsevier B.V. 47(2):331–363. http://www.elsevier.com/locate/jeem

Blamey RK, Bennett JW, Louviere JJ, Morrison MD, Rolfe JC (2000) A test of policy labels in environmental choice modelling studies. Ecol Econ 32(2):269–286

Briscoe J, Furtado P, Griffin C et al (1990) Toward equitable and sustainable rural water supplies: a contingent valuation study in Brazil. Int Bank Reconstr Dev/World Bank 4(2):115–134. http://www.jstor.org/stable/3989925

FAO (2008) Climate change and food security: a framework document. Summary. FAO, Rome

Gregersen IB, Arnbjerg NK (2011) Decision strategies for handling the uncertainty of future extreme rainfall under influence of climate change. Water Sci Technol 66(2):284–375. doi:10.2166/wst.2012.173

Gunatilake H, Tachiiri M (2012) Willingness to pay and inclusive tariff designs for improved water supply services in Khulna, Bangladesh. ADB South Asia working paper series, no. 9. p 25. Asian Development Bank. http://beta.adb.org/sites/default/files/wtp-tariff-designs-water-supply-ban.pdf

Louviere J, Hensher D, Swait JD, Adamowicz V (2000) Stated choice methods: analysis and applications. Cambridge University Press, Cambridge, p 420

NCI (Nature and Culture International) (2010) Vegetation map of the province of Loja, scale 1:25000

Sánchez A, Gay C, Estrada F (2011) El cambio climático y la pobreza en el Distrito Federal. Investigación Económica 70(278): 45–74. http://www.redalyc.org/pdf/601/60121367004.pdf

UNESCO-WWAP (2009) Climate change and water. United Nations World Water Assessment Programme. www.unesco.org/water/wwap

Veronesi M, Chawla F, Maurer M, Lienert J et al (2013) Climate change and the willingness to pay to reduce ecological and health risks from wastewater flooding in urban centers and the environment. Ecol Econ 98:1–10

Zwerina K, Huber J, Kuhfeld WF (1996) A general method for constructing efficient choice designs. SAS technical support documents, TS-650D, pp 48–67. http://www.sas.com/service/techsup/tnote/tnote_index6.html

Chapter 18
Shared Waters of the South Caucasus: Lessons for Treaty Formation and Development

Ryan B. Stoa

Abstract The Kura-Aras River Basin is the largest and most critical water resource in the South Caucasus. As the primary source of freshwater for Georgia, Armenia, and Azerbaijan and a significant source for Turkey and Iran, the basin is at the center of a complex geopolitical region. The Kura-Aras has not been managed under a cooperative management treaty since the fall of the Soviet Union and remains one of the most significant watercourses ungoverned by a transboundary agreement. Tense relations between neighbors, as well as ambitious development plans and economic priorities, have pushed international cooperation over the basin to the fringes of the region's agenda. In this chapter, the Kura-Aras River Basin is examined in order to identify factors inhibiting, and opportunities to promote, cooperation. While a multilateral, basin-wide treaty appears unrealistic given extreme levels of diplomatic discord, opportunities exist to move toward a cooperative management framework through bilateral agreements.

Keywords Kura-Aras River Basin • South Caucasus • Treaty formation • Multilateral • Bilateral • Water resources management • International development

18.1 Introduction

On May 6, 2014, the presidents of Turkey, Georgia, and Azerbaijan met in Tbilisi to discuss pressing economic issues in the region, with the objective of strengthening cooperation over trade, transportation, and energy. The meeting highlighted the countries' intent "to expand the involvement of all new countries in regional projects, as their involvement will ensure the stability in the Caspian-Black Sea and Mediterranean regions" (Babayeva 2014). The stated inclusiveness in regional projects, however, seems not to extend to conspicuously absent Armenia, which has a long-standing diplomatic feud with Azerbaijan. Nor, it seems, do the projects

R.B. Stoa (✉)
Florida International University, College of Law and Global Water for Sustainability Program – GLOWS, Miami, FL 33199, USA
e-mail: rstoa@fiu.edu

S.G. Setegn, M.C. Donoso (eds.), *Sustainability of Integrated Water Resources Management*, DOI 10.1007/978-3-319-12194-9_18

include cooperation over the Kura-Aras River Basin. That too is conspicuous, as the basin is one of the most significant watercourses on earth ungoverned by a framework cooperation agreement.

The Kura-Aras River Basin comprises the Kura River, which flows from the mountains of Turkey to the Caspian Sea (via Georgia, Azerbaijan, and Iran), and the Aras River, which begins in Turkey and comprises various borders between Turkey, Armenia, Azerbaijan, and Iran, before joining with the Kura in Azerbaijan. It is the largest source of freshwater in the region and the primary water source for Georgia, Armenia, and Azerbaijan. In total, the basin encompasses approximately 188,200 km^2 (Campana et al. 2012), while the total population of basin states exceeds 167 million (World Bank 2014). Kura-Aras water is a key input in several economic sectors, including energy, agriculture, and municipal water supply.

Despite the vital role the Kura-Aras River Basin plays in the South Caucasus, it is not cooperatively managed. On the contrary, its management scheme can be characterized as flowing downstream – it is subject to the regulations and water policies of one country until its waters enter a downstream jurisdiction, at which point a new regulatory framework is applied. Partly as a result, the basin and its dependents are suffering. Agricultural pollutants are degrading water quality, allocation decisions are piecemeal and poorly monitored, and countries' insecurities are heightened by upstream threats to expected flow levels. Meanwhile, little is being done in the region to systematically quantify groundwater supplies or analyze their interaction with surface waters.

Transboundary water agreements are not a panacea to the many challenges of water resources management. They cannot ensure that all water needs will be met or that conflict will not arise over scarce resources in the future. But the strength of water treaties is their ability to create a mechanism for cooperation and facilitate information sharing, impact mitigation, and future planning by providing assurances to basin states of their respective rights and duties. Bilateral or multilateral water treaties are especially beneficial due to the nascent state of international water law treaties and custom.

Those benefits would appear to be needed in the South Caucasus, where the basin's water resources are not as clean or sustainably managed as they could be, largely due to a lack of cooperation between countries. It would also calm the fears of downstream countries that are vulnerable to upstream allocation decisions. And yet, a Kura-Aras River Basin treaty or cooperative management framework does not exist and does not appear likely in the near future. Given the importance of the basin in the region's economic growth, it is important to understand why cooperation is limited: first, because the obstacles to cooperation can be targeted if they are identified and, second, because the obstacles to cooperation in the South Caucasus may manifest themselves elsewhere in the future and the lessons of one region can inform another.

The South Caucasus has a long history, replete with armed conflict, political transitions, and long-standing regional tensions. As a result, transboundary water resources management cannot be studied in isolation and instead must be understood as one piece of a very complex puzzle. There are likely to be many subtle factors or interacting dynamics that help explain why a transboundary agreement is not in place. This chapter will focus on two.

First, there is a diplomatic deficit between countries on a macroscale that inhibits cooperation in many areas, one of which is shared waters. Diplomatic relations – or a lack thereof – between Armenia and Azerbaijan, for example, have thus far precluded negotiations over cooperative management of the Aras River or a multilateral Kura-Aras River Basin agreement. Second, countries of the South Caucasus are undergoing a post-Soviet period of economic growth, where natural resources are valued for their short-term economic gains and not appreciated for their long-term economic potential. Accordingly, the perceived value of a treaty – whether multilateral or bilateral – that may inhibit short-term consumption of water resources is low, while the value of water-intensive sectors like energy and agriculture is high. While many other factors contribute to complex geopolitical dynamics in the region, these two – diplomatic relations and economic growth strategies – are central obstacles to a basin-wide or bilateral agreement.

18.2 From Intranational to International: The Kura-Aras Management Revolution

The Kura-Aras River Basin of the South Caucasus lies at the intersection of Europe and Asia. For centuries its people have been religiously, politically, and ethnically diverse. Herodotus wrote in the fifth century BC that the region had the "loftiest of mountain ranges. . .inhabited by many different tribes" (King 2010). Even a modern history of the South Caucasus can fill and has filled volumes. From the Persian Campaign of Peter the Great in 1722 to the Russo-Georgian war in 2008, the region has experienced the rise and fall of the Soviet Union, World Wars I and II, and myriad independence movements. The Kura-Aras, consequently, has rarely enjoyed a stable management regime.

For much of the twentieth century, however, the basin was governed by bilateral treaties between the USSR and Turkey and the USSR and Iran. The bilateral agreements were focused on border waters (waters that form the border between two countries) and so were limited in scope. Nonetheless, the countries agreed to share waters equally, while establishing joint commissions tasked with managing common resources (Newton 2007). While the treaties were likely a positive development for the region's water resources, they were not as remarkable as a basin treaty would be in the twenty-first century. First, Turkey and Iran were (and are) not as reliant on waters of the Kura-Aras River Basin as Georgia, Armenia, and Azerbaijan are today. Second, contemporary understandings of environmental degradation and sustainable resource management challenges, as well as water quality and quantity monitoring technologies, were not available. Therefore, it is likely the countries were not attuned to the extent to which management challenges – such as agricultural runoff, groundwater depletion, and ecological needs –

plagued the basin. Finally, the scope of the treaties was limited to border waters, preempting the need for a complex multilateral agreement.

Nonetheless, creating a mechanism for cooperation is a step toward joint management. Had an environmental crisis arisen in the basin, as it did in Chernobyl, the commissions could have cooperatively addressed the implications for water resources. In that sense, the Soviet era water management regime for the Kura-Aras was a significant, if modest, achievement. With the fall of the Soviet Union and fragmentation of the region into Georgia, Armenia, and Azerbaijan – as well as several autonomous or semiautonomous states – the bilateral treaties dissolved. In their place, national water strategies emerged to harness the vast potential of the basin's water resources. Armenia, Georgia, and Azerbaijan each have a complex web of national laws, policies, and institutions that regulate water resources (Vener 2006).

To some extent the national approaches have enabled the countries to mollify agricultural, municipal, energy, and industrial needs. Further advancements, however, such as flood control, sediment management, and hydropower development, would benefit from a cooperative strategy (Vardanyan and Volk 2014). The basin is suffering from mounting challenges necessitating collective action. Waters are becoming severely polluted from urban, agricultural, and industrial runoff; nearly 80 % of the countries' wastewater is deposited into the basin's surface waters (Campana et al. 2012). In addition, water is not allocated efficiently: Azerbaijan consumes 58 % of its renewable water resources, while Armenia and Georgia consume 28 and 5 %, respectively (Campana et al. 2012). Further work could be done to better understand the extent, quality, and hydrological dynamics of groundwater deposits as well.

While cooperation in the basin has been low, the record of basin states in the development of an international water law regime has been mixed. None of the five basin states are parties to the United Nations Watercourses Convention, and only Azerbaijan is a party to the United Nations Economic Commission for Europe Water Convention (UN Treaty Database 2014). On the other hand, Armenia, Azerbaijan, Georgia, and Turkey are members of the Council of Europe. Further integration with the European Union would require the countries to meet minimum standards of the EU Water Framework Directive, including integrated water management policies and spatial planning measures. The Directive requires basin-level management of water resources (Frederiksen et al. 2008), which is not being applied to the Kura-Aras.

Today the Kura-Aras River Basin remains one of the most vulnerable transboundary watercourses lacking a cooperative management agreement (De Stefano et al. 2012). Following the collapse of the Soviet Union, no joint mechanism has emerged to fill the void left by bilateral treaties with Turkey and Iran. National water policies have developed to harness the economic potential of the basin's water resources, while international efforts to promote integrated water management and cooperation over water resources will push the countries away from individualistic approaches. But given the potential benefits of joint management, stakeholders must recognize and overcome the hurdles to its adoption.

18.3 Conflict and Diplomatic Discord: Dim Hopes for Basin-Wide Cooperation

Despite their newfound status as independent nations, Armenia, Azerbaijan, and Georgia were not ushered into a period of peace and prosperity in the 1990s. On the contrary, the fall of the Soviet Union left the countries in disarray. From the ashes of a centralized command economy, political institutions had to be created, infrastructure updated, and economic growth strategies developed. While the Soviet model pushed hydropower and large-scale irrigated agriculture to their limits, the water needed for those sectors was not being sustainably managed (van Harten 2002), and the region's entire natural resource management strategy had to be rethought.

At the same time, independence did little to preclude armed conflict. Most importantly for the Kura-Aras, Armenia and Azerbaijan's long-standing dispute over the Nagorno-Karabakh region erupted and remains a central obstacle to cooperation in the basin. The Nagorno-Karabakh, comprised predominantly of ethnic Armenians, was designated by the Soviet Union as an autonomous region of Azerbaijan. Between 1992 and 1994, however, Armenia and Azerbaijan waged war over the region, culminating in a cease-fire brokered by Russia and maintained by the Organization for Security and Cooperation in Europe (OSCE) (Vardanyan and Volk 2014). Today the region remains in dispute, acting as an autonomous state but recognized internationally as part of Azerbaijan (OSCE Minsk Group 2008).

The situation is problematic for the Kura-Aras in part because it lies within the basin but also because the conflict has strained relations between countries in the region. Armenia and Azerbaijan, primarily, are technically still at war, with closed borders and severed diplomatic relations (Cecire 2012). Without the ability to engage Armenia and Azerbaijan in joint talks, there is little to no hope that a basin-wide management framework can be seriously discussed, much less agreed upon. The conflict has larger implications for the region than two feuding rivals, however. Turkey has also cut diplomatic ties with, and closed its border to, Armenia (in solidarity with Azerbaijan), while Armenia's ties with Iran and Russia make Azerbaijan's relations with those powers tense (Cecire 2012).

Georgia has managed to stay above the fray concerning the Nagorno-Karabakh, but it too has a predominantly ethnic Armenian region, Javakheti, whose autonomy is supported by Armenia (Campana et al. 2012). Meanwhile, Russia provides support to Abkhazia and South Ossetia, autonomous regions of Georgia. And Russia's recent actions in the Ukraine threaten to cover the entire South Caucasus in a blanket of insecurity (Assenova 2014).

The regional rivalries and alliances have not prevented countries from cooperating in all areas. As mentioned above, Turkey, Georgia, and Azerbaijan cooperate on many levels, not least of which are pipelines transporting oil and gas reserves from the Caspian Sea to Turkey via Azerbaijan and Georgia. There is evidence that dialogue over shared waters is taking place between technical experts as well (Vardanyan and Volk 2014). For purposes of executing a multilateral water agreement over the Kura-Aras River Basin, however, prospects are nonexistent if diplomatic relations aren't in place to begin with. There is some potential for shared water resources to provide an impetus for conflict resolution in some cases, but the South Caucasus demonstrates that cooperation potential is minimized when relationships between basin states are so strained that negotiations are a nonstarter. Under the circumstances it appears unlikely that the importance of transboundary water resources management will overcome extreme levels of diplomatic discord.

18.4 Bilateral Agreements: Oasis or Mirage?

Recognizing the futility of a basin-wide agreement, regional and international stakeholders have turned their gaze toward the promise of bilateral agreements. Most notably, talks between Georgia and Azerbaijan have been taking place for several years in an attempt to create a bilateral treaty governing the Kura River. There is reason for optimism: first, because relations between Georgia and Azerbaijan are relatively harmonious (the countries have joint energy and transportation projects and depend on one another for bilateral trade flows) (Ministry of Foreign Affairs of Georgia 2010) and, second, because the Kura River is of great importance to both countries. Georgia relies on the Kura for municipal water supply, as well as ambitious agricultural irrigation and hydropower development plans. The Kura, similarly, is Azerbaijan's primary source of fresh drinking water (Campana et al. 2012). A cooperative management agreement would provide clarity to both countries regarding their rights and duties over shared water resources, while providing certainty regarding the quantity and quality of water available.

The opportunity has been recognized by foreign donors and the international community. The United States Agency for International Development (USAID), United Nations Development Program (UNDP) and Global Environmental Facility (GEF), North Atlantic Treaty Organization (NATO), EU, World Bank, UNECE, and OSCE have all made efforts to promote cooperation (Vener 2006). For the past several years, the UNECE and OSCE, in particular, have supported dialogue and negotiations between Georgia and Azerbaijan in order to develop a Kura River agreement. As recently as January 16, 2014, bilateral consultations have taken place in order to refine the text of an agreement. In May 2014, the author of this article provided advisory opinions to Georgian government agencies regarding rights and duties imposed by the agreement. While it is still under consideration, there are outstanding concerns that must be overcome for the agreement to be executed.

The draft "Agreement between the Republic of Azerbaijan and Georgia on Cooperation in the Field of Protection and Sustainable Use of the Water Resources of the Kura River Basin" is a relatively straightforward document. Most of the controversial or operational elements of the agreement – such as dispute resolution, notification procedures, transboundary impact assessment, and coordination – are vested in a joint commission. The rest of the agreement's provisions mirror customary international water law principles such as reasonable and equitable utilization, impact mitigation, intergenerational equity, prior notification and consultation of planned measures, and the precautionary principle. In essence, the agreement is light on firm obligations and heavy on principles and further action.

Because so many powers are conferred on the commission, the agreement functions mostly as a mechanism for further cooperation. Even the commission's composition and rules of procedure are yet to be determined. For these reasons, the agreement may appear unthreatening and a modest step toward cooperation. However, three obstacles lie in the path of treaty formation.

First, as the upstream country, Georgia does not perceive the agreement as necessary or a strong promotion of its interests. This is nothing new; tensions between upstream and downstream countries are a fundamental dynamic international water laws must address (Stoa 2014). In this case, Georgia enjoys the privileges of clean and bountiful surface water flows. It is concerned that the agreement may curtail its ability to develop hydropower facilities and expand irrigated lands or require costly wastewater cleanup costs. In addition, notification and consultation procedures are seen as unilaterally burdensome. Less obvious are the benefits the agreement would provide.

Those concerns may be overblown. First, the agreement itself does not impose substantive requirements on either Georgia or Azerbaijan. Water allocations and specific pollution control measures are to be discussed and determined by the joint commission, requiring Georgia's approval. Second, procedural requirements – such as the 6-month grace period for consideration of planned measures by another party – are firmly entrenched as customary international water law norms. In other words, the obligation already exists. Finally, there are material benefits to a cooperative agreement with Azerbaijan. While Georgia is the upstream country in terms of surface waters, both countries can tap into the Kura-Aras River Basin's groundwater stores. The agreement's hydrological jurisdiction covers both surface and groundwaters of the basin, providing Georgia with a mechanism to cooperatively monitor and manage shared groundwaters. In addition, while Georgia can exercise leverage over Azerbaijan as the upstream country, Georgia is downstream of Turkey, where the Kura River originates. Taking a hard-line position with Azerbaijan may place Georgia in an awkward position should Turkey seek to follow suit.

The second challenge is that while perceived benefits may be unreasonably skeptical, sustainable and integrated water resources management is not afforded priority over other economic sectors in the region. With the support of the World Bank, the Ministry of Agriculture of Georgia aims to irrigate up to 300,000 ha, placing further demands on water resources (Shotadze 2011). In addition, Georgian elections in 2012 handed power to a coalition with ambitious plans to install hydroelectric facilities throughout the Kura basin. These efforts are supported by foreign donors like USAID who see the benefits of clean energy security (Deloitte 2012). Azerbaijan's economic growth, meanwhile, is heavily focused on exploiting oil reserves. The oil sector in Azerbaijan accounts for over 60 % of overall GDP (Hasanov 2013). While Azerbaijan has a strong interest in controlling upstream pollution in the Kura River, it is unlikely to push the issue when harmonizing the oil and gas relationship with Georgia is so critical to Azerbaijan's continued economic growth.

While water resources management may not receive as much attention as agriculture or energy, it is a key input in both sectors. Water resources of the Kura River play a large role in Georgia's agricultural and energy development plans. A bilateral agreement can provide certainty regarding the quality and availability of those resources for the government, industry, and investors. Similarly, Azerbaijan's reliance on oil makes the economy vulnerable to price fluctuations (Hasanov 2013). Diversification is likely to require secure water supplies in order to expand the agricultural sector, while Azerbaijan's population of nearly ten million needs a dependable source of clean freshwater (CIA 2014). The need for a bilateral agreement may not be making headlines, but it is present nonetheless.

18.5 Conclusion

It is unlikely that a multilateral, basin-wide agreement governing the Kura-Aras River Basin will be executed anytime soon. While the benefits await, transboundary water resources management has not, in this case, overcome extreme levels of diplomatic discord between countries in the region. The disconnect between Armenia and Azerbaijan, in particular, makes agreement formation all but impossible. Still, cooperation can occur without an agreement, and knowledge sharing, information exchange, and technical collaborations are known to take place between technical experts (Vardanyan and Volk 2014). Ultimately, however, an agreement will be needed to establish rights and duties and reinforce customary international principles and norms. As challenges mount, effective management will require direct and collaborative action. Until long-standing political and ethnic tensions subside, however, prospects for an agreement appear slim.

Still, there is merit to bilateral agreement formation. As the primary basin countries of the Kura River, Georgia and Azerbaijan's efforts to create a bilateral treaty are commendable and encouraging. Although concerns exist for both countries, they can be overcome with a holistic understanding of international water law rights and duties and the strategic importance of secure water resources. This presents an opportunity for international experts and foreign donors to facilitate contemporary and nuanced understandings of integrated water resources management and the mechanisms that support it.

As yet, the Kura-Aras River Basin lacks a multilateral or even bilateral management agreement. While the need may exist, diplomatic discord and competing economic priorities have pushed treaty formation to the fringes of the region's agenda. Bilateral agreements that manage segments of the basin have the most promise and may provide a piecemeal, step-by-step approach to basin-wide management. Until then, the Kura-Aras River Basin's people and ecosystems will suffer from the limitations of fragmented water management.

References

Assenova M (2014) Ukrainian crisis sparks worries in the South Caucasus. Retrieved June 23, 2014, from http://www.jamestown.org/regions/thecaucasus/single/?tx_ttnews%5Btt_news%5D=42268&tx_ttnews%5BbackPid%5D=642&cHash=3028e39fcd269a18f143acb708024d0d#.U4-hRvldXIU

Babayeva J (2014) Azerbaijani, Georgian, Turkish presidents call for expansion of trilateral co-op (UPDATE). AzerNews. Retrieved June 12, 2014, from http://www.azernews.az/azerbaijan/66769.html

Campana ME, Vener BB, Lee BS (2012) Hydrostrategy, hydropolitics, and security in the Kura-Araks basin of the South Caucasus. J Contemp Water Res Educ 149:22–32. doi:10.1111/j.1936-704X.2012.03124.x

Cecire M (2012) Azerbaijan-Armenia tensions: regional risks, policy challenges. Retrieved June 23, 2014, from http://www.worldpoliticsreview.com/articles/12046/azerbaijan-armenia-tensions-regional-risks-policy-challenges

CIA (2014) The World Factbook. Retrieved June 23, 2014, from https://www.cia.gov/library/publications/the-world-factbook/geos/aj.html

De Stefano L, Duncan J, Dinar S, Stahl K, Strzepek K, Wolf A (2012) Climate change and the institutional resilience of international river basins. J Peace Res 49(1):193–209

Deloitte (2012) U.S. Federal Government case study Georgia Hydropower Investment Promotion Project 2010–2012. Retrieved June 23, 2014, from http://www.deloitte.com/view/en_US/us/Insights/Browse-by-Content-Type/Case-Studies/US-Federal-Government-Case-Studies/856db061e88da310VgnVCM2000003356f70aRCRD.htm

Frederiksen P, Mäenpää M, Hokka V (2008) The water framework directive: spatial and institutional integration. Manag Environ Qual Int J 19(1):100–117

Harten MV (2002) Europe's troubled waters – a role for the OSCE: the case of the Kura-Araks. Redirecting.... Retrieved June 23, 2014, from http://heinonline.org/HOL/Page?handle=hein.journals/helsnk13&collection=journals&page=338#347

Hasanov F (2013) Dutch disease and the Azerbaijan economy. Communist Post-Communist Stud 46(4):463–480

King C (2010) The ghost of freedom: a history of the Caucasus (1. issue as an Oxford Univ. Press paperback ed.). Oxford University Press, Oxford

Ministry of Foreign Affairs of Georgia (2010) Retrieved June 23, 2014 from http://www.mfa.gov.ge/index.php?lang_id=ENG&sec_id=265

Minsk Group (2008) Statement of the co-chairs of the OSCE Minsk Group. Retrieved June 23, 2014, from http://www.osce.org/mg/49564

Newton J (2007) Case study of transboundary dispute resolution: the Kura-Araks basin. Case studies | Water Conflict Management and Transformation at OSU. Retrieved June 12, 2014, from http://www.transboundarywaters.orst.edu/research/case_studies/Kura_Arkas_New.htm

Shotadze M (2011) Technical report 1. Rapid National Assessment February 2011. Retrieved June 23, 2014, from http://www.globalwaters.net/wp-content/uploads/2012/12/Technical-Report-1-Repid-National-Assesemnet-of-Legal-Policy-and-Institutional-Settings.pdf

Stoa R (2014) International water law principles and frameworks: perspectives from the Nile River basin. In: Melesse AM, Abtew W, Setegn SG (eds) Nile River basin: ecohydrological challenges, climate change and hydropolitics. Springer International Publishing, Cham, pp 581–595

UN Treaty Database (2014) https://treaties.un.org/Pages/ViewDetails.aspx?src=UNTSONLINE&tabid=2&mtdsg_no=XXVII-12&chapter=27&lang=en#Participants and https://treaties.un.org/Pages/ViewDetails.aspx?src=TREATY&mtdsg_no=XXVII-5&chapter=27&lang=en

Vardanyan M, Volk R (2014) Transnational cooperation over shared water resources in the South Caucasus: reflections on USAID interventions. In: Nakayama M, Weinthal E, Troell J (eds) Water and post-conflict peacebuilding (Book 3). Earthscan, Oxford

Vener BB (2006) The Kura-Araks basin: obstacles and common objectives for an integrated water resources management model among Armenia, Azerbaijan, and Georgia. Retrieved June 12, 2014, from http://repository.unm.edu/bitstream/handle/1928/2591/master-PPOriginal.pdf?sequence=1

World Bank (2014) Data. Retrieved June 12, 2014, from http://data.worldbank.org/indicator/SP.POP.TOTL

Chapter 19
Basin Comanagement Plans – A Participative Approach to Water Governance: A Case Study in Honduras, Central America

Claudia Cecilia Lardizabal

Abstract This case study addresses the governance and institutionality of integrative and participative basin management in the micro-basin of the Valle de la Soledad located on the outskirts of the Honduran capital Tegucigalpa. The initiative comprises four basic stages: (1) identification of existing laws and state institutions, (2) knowledge of local actors' capacities to protect and administer water resources, (3) determination of water conflicts and steps for its possible solution, and (4) determination of factors and actors in favor and against water protection. The process allowed to develop municipal capacities, participative and efficient basin, and land use management programs that benefited local communities, particularly small and medium farmers, as well as to develop financial sustainability mechanisms.

Keywords Comanagement • Conflict resolution • Basin management • Water governance

19.1 Introduction

Global water problems are likely to increase in severity, rendering existing approaches inadequate to deal with such issues (Dellapena et al. 2013). The Caribbean is not exempt from such problems. According to Cashman (2012), due to increase in and on water demand, reordering of institutions in the Caribbean is under way by altering their institutional roles in order to achieve adequate water governance. In the majority of democratic countries, water-related decisions are left to government officials; however, in many parts of the world, water users have additional possibilities to provide input on such decisions before implementation. The involvement of water users and stakeholders in decision making can produce fairer and increasingly sustainable results (Susskind 2013).

C.C. Lardizabal (✉)
Department Francisco Morazan, National Autonomous University of Honduras, Tegucigalpa, Honduras
e-mail: claudia_lardizabal@unah.edu.hn

S.G. Setegn, M.C. Donoso (eds.), *Sustainability of Integrated Water Resources Management*, DOI 10.1007/978-3-319-12194-9_19

For more than 30 years, the Tropical Agricultural Research and Higher Education Center (CATIE) has been working in the comanagement of watersheds. According to Kammerbauer et al. (2010), the comanagement approach is based on the need to ensure drinking water availability through participative actions and decisions. The following chapter reviews a successful case study of comanagement for the Rio Soledad basin in Valle de Angeles, Honduras, where the involvement of water users, stakeholders, and decision makers in comanaging the watershed has proved to be an effective collaborative management scheme of water resources. First, a definition and reason for the use of the comanagement approach is provided, followed by the key elements involved and an overview of the legal framework in place and conflict resolutions that have taken place. Furthermore, the chapter attempts to identify the factors that help achieve the social, administrative, and financial sustainability of the comanagement approach.

19.1.1 Definition

According to Prins (2009), comanagement refers to the conjugation of wills, capacities, and responsibilities of a series of actors that play different roles in a particular basin, who together through interaction should develop the desired results. A comanagement model should result from a series of participative processes that will provide the foundation for a sustainable and operative action that should carry out the management of the watershed. Comanagement plans should present a foundation based on a participative approach among actors that can be leaders of key sectors such as municipalities, social organizations, or any other with influence over the area (Bucardo 2007). This approach is a shared management process based on experimental observation and attentive consideration of action results, feedback, and action readjustments through analysis (Faustino and Jiménez 2005).

Faustino and Jimenez (2005) also propose that the action research and learning alliances are a fundamental support for the implementation of adaptive comanagement in river basins. This process seeks to strengthen the governability of the local political structure in its role as articulator and authority of actors and interest groups. The model starts from the impact in water quality and quantity as the ultimate goal of the basin management as well as an indicator of the efficiency of the processes and a solid foundation due to the convergence of diverse interests and conflict resolutions regarding water (Kammerbauer et al. 2010).

Adaptive comanagement is referred to as the adaptation of intervention activities in accord to the particular characteristics of a basin as well as the modification of strategies, methodologies, and actions in order to advance effectively toward the programs' vision (Prins 2009).

19.1.2 Why Comanagement

The base hypothesis for the comanagement model is that basin conflicts are so complex and demanding that no one actor by itself can effectively address the conflicts. Therefore, it is necessary to bring together wills, capacities, responsibilities, and resources of a relevant group of actors through a basin organism. In other words, this new and viable organism should be the critical factor in order to obtain a successful basin management (Prins 2009). Many municipalities have formulated strategic, management, territorial, and protected area plans, among others, that are not being implemented. Comanagement addresses the need for implementing formulated plans rather than discarding and producing new ones.

19.1.3 Key Elements

The central purpose of the comanagement model is to regulate water use and extractions in order to guarantee continuous water supply and quality. The model seeks to promote measurable positive changes that improve the population's quality of life. The hard part is finding change tendencies since hydrological cycles are influenced by a number of external factors that require the dedication of time and constant analysis for a certain amount of time to establish direct correlations between cause and effect (Kammerbauer et al. 2010).

In order to find a middle point in situations involving multiple actors and sectors who possess similar capacities and wills, *conversation platforms* are mandatory. These platforms bring together actors with different competencies and roles in order to unite efforts toward developing joint planning and monitoring activities in the watershed. Furthermore, these platforms will depend highly on the steering capabilities of individuals, the adequate articulation among them, and the disposition of addressing conflicts of interest. Finally, the communication during these conversation processes has to be of high quality in order for them to succeed (Prins 2009).

Comanagement is directly related to governability (ASDI CATIE a, 2008). Actions agreed during the process are flexible and adaptable to the changing conditions of the area and to the various actors involved ranging from the local to the regional scales. One of the necessary conditions is convergence mechanisms (Kammerbauer et al. 2010) among key players in order to ventilate latent conflicts in search for common solutions. On a local level, different structures can undertake these key functions, contrary to a higher level, such as municipality, where some sort of formalization process must take place. Another condition is the equal and representative participation of all actors with disregard for sex, race, economic status, or other in order to attain efficiency and credibility.

A comanagement plan is based on a collective territorial vision (Kammerbauer et al. 2010); this is not an official plan but rather a flexible tool that affects agreements between different organizations. The common agenda is based on organizational needs in order to achieve higher efficiency, equity, and legitimacy in the agreed actions, arrangements, and practices.

In order to implement the agenda, minimum financial requirements must be met. These can be autonomous and decentralized funding or contributions from each institutional actor. A very effective way to guarantee process sustainability is the creation of a common initial fund with clearly defined rules and mechanisms aimed to strengthen the fund. In addition, the acquisition of government funds through forward-thinking mechanisms, such as payment for environmental services, is a must in order to achieve local governability (Kammerbauer et al. 2010).

Due to the high uncertainty that characterizes systems where a social factor is involved, it is necessary to implement feedback mechanisms through collective processes that include technical as well as popular and ancestral knowledge (Bucardo 2007). The multiple perspectives of the different actors will allow to develop innovative and unique approaches. For example, in the Rio Soledad basin, one of the main experiences was the resolution of water conflicts (Prins 2009).

19.2 Study Area

The Soledad River micro-basin is mostly located in the Valle de Angeles County with the rest of the area distributed between the Central District, Santa Lucia, and San Antonio de Oriente at approximately 22 km from Tegucigalpa to the Francisco Morazán state (Fig. 19.1). The micro-basin covers an area of 5,542 Ha and is located in the upper Choluteca River basin (lat 14° 7′ 15″ to 14°11′22″ North, lon 87°0′13″ to 87°5′40″ East). The mean annual rainfall of the area ranges from 1,500 to 2,500 mm; the average annual temperature is 18 °C, and average relative humidity is 84 %. In Honduras, according to Argeñal (2010), two very distinct seasons can be distinguished; from May to October, the rainy season takes place, and from November to April, the dry season occurs.

The soil of the basin is of medium fertility, humid, and mainly of forest vocation. Most of the soils in the watershed have developed from sedimentary rocks. Such soils include Chimbo,[1] Chandala, Espariguat, and El Naranjito. These soils tend to be deep to very deep depending on the altitude, with the exception of the Chimbo soils that are shallow and severely erodible. In general, soils of sedimentary origins have good drainage and medium to fine texture with a defined structure. Its parental material corresponds to conglomerates and red sand and in lesser proportions of calcilutites.

In the lower parts of basin where there are gentle slopes, agriculture can take place (Barahona 2006). Most of the basin, approximately 4,357.50 Ha, is covered

[1] Chimbo, Chandala, Espariguat, and El Naranjito are Honduran soil series (Simmons Honduran Soil Study).

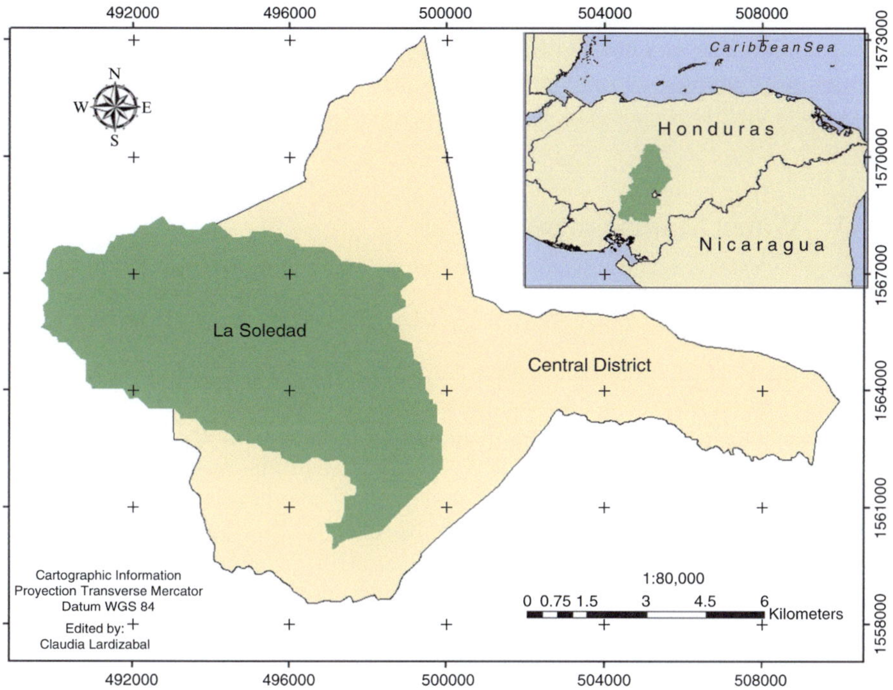

Fig. 19.1 Rio Soledad basin, Valle de Angeles, Honduras

with pine forests, and in a lesser percentage of broad leaf and mixed forest, also there is a smaller part (175.7 Ha) dedicated to traditional agriculture which comprises mainly of subsistence production crops such as grains.

Hydrologically, the Rio Soledad micro-basin belongs to the Yeguare River subbasin, which in turn is part of the Choluteca River upper subbasin. The Rio Soledad basin is very mountainous, with approximately 70 % of its area consisting of steep slopes with no rocks and only 24 % corresponding to relatively flat and soft slopes.[2] The natural resources of the basin are sensitive to accelerated land degradation due to the intensity of land use changes for agricultural production purposes, as well as inadequate production practices. Land cover is constantly being diminished, there is a reduction in water producing areas, and this is directly related to water resources degradation along with intensive agrochemical use in crop production and practices related to mineral extraction which was one of the main economic activities of the area (PREVDA 2008).

The Valle de Angeles municipality has an approximate population of 13,400 inhabitants of which 44.6 % can be considered urban and 55.4 % rural. According to PNUD (2013), illiteracy in Valle de Angeles is 19 %, with the biggest illiteracy problems in the rural area. The watershed has an extension of 55.47 km^2 and a

[2] Soft slopes refer to topography with less than 5 % inclination.

population density of 116 persons per square kilometer. Depending on the number of inhabitants per area and the low illiteracy percentage, there are different ways that the development and environmental situation of the basin can be addressed (PREVDA 2008).

19.3 Water Management Policies

19.3.1 Legal Framework

The Honduran legislation has a wide legal framework for the protection of natural resources, dating from 1902 when concessions and natural resources were legislated. Municipalities are given the responsibility to monitor the areas in which water sources are located. There are 14 laws regarding water and watershed management (Table 19.1). However, only a few of these have a direct relationship with the management of the resources: the Law for National Water Use was approved in 1927 and is still in use, the General Environmental Law (1993), the Framework Law for Potable Water and Drainage (2003), the General Water Law (2009), and the Forest Law approved in 1972 and reviewed in 2007 in which watershed management is addressed (Ley 98-2007).

The Law for National Water Use was the legal instrument that addressed water resources. The General Water Law is based on it and elaborates on water resource management, including the creation of the National Water Resources Council which is the institutional entity in charge of regulating the policies regarding water resource use and management. However, there are several other institutions and entities that provide technical and operational support (Fig. 19.2).

19.4 Local Actors

The Valle de Angeles has a 3-level organizational chart that includes the municipal corporation, the mayor, and five departments: land registry, treasury, auditory, municipal justice, and the UMA.[3] However, there is no specific entity to handle drinking water. Drinking water is under the Department of Municipal Justice (HYTSA 2005). Water system maintenance is usually handled by plumbers, while water boards are in charge of the administration of the water system (Aguilar et al. 2008). Water boards have their legal background in the Framework Law for Potable Water and Drainage and are set up in three parts: the user assembly, which is the top authority since it expresses the will of its members; the board of directives that is comprised of seven members; and a support committee. However, in none of

[3] UMA Municipality Environmental Unit (acronym in Spanish).

Table 19.1 Honduran legal framework for water resources

Law	Decree	Date	Observations
Law for National Water Use	137	April 9th 1927	
Forestry Law	85	February 10th 1972	Revised in 2007 where watershed management was included
Law for agricultural reform and other dispositions	170	December 30th 1974	
Municipalities Law	134-90	November 7th 1990	
Health Code	65-91	August 8th 1991	
Law for the Modernization and Development of the Agricultural Sector	31-92	March 19th 1992	
General Environmental Law	104-93	June 6th 1993	
Law for forestry incentives, reforestation, and forest protection	163-93	September 22nd 1993	
Law for the Environmental and Natural Resources, Attorney General's Office	134-99	September 17th 1999	
Law for Sustainable Rural Development	12-2000	March 30th 2000	
Framework Law for Potable Water and Drainage	118-2003	September 29th 2003	
Territorial Ordainment Law	180-2003	November 28th 2003	
General Water Law	181-2009	December 14th 2009	

the communities is the support committee organized, and according to law if the committees are not organized, the organizations are incomplete.

According to Aguilar et al. (2008), despite the importance of the water boards' work and the fact that they manage approximately 50 % of the total amount of water that is captured in the dams that comprise the potable water system, none of them have an annual plan. Overall decisions are taken on a daily basis and according to system necessity or the day to day. In general, most of the water boards have mentioned of the absence of general assemblies as their main issue.

Water boards have also mentioned that some of their main problems are related to the lack of system maintenance and also water source protection and chlorination system maintenance deficiencies. In terms of social aspects, the central issue is the indifference of the assembly, the lack of collaboration, and in some cases the board of members does not fulfill its responsibilities.

However, despite a generalized indifference from the general population, some organizations in the area provide support to the Valle de Angeles municipality (Table 19.2).

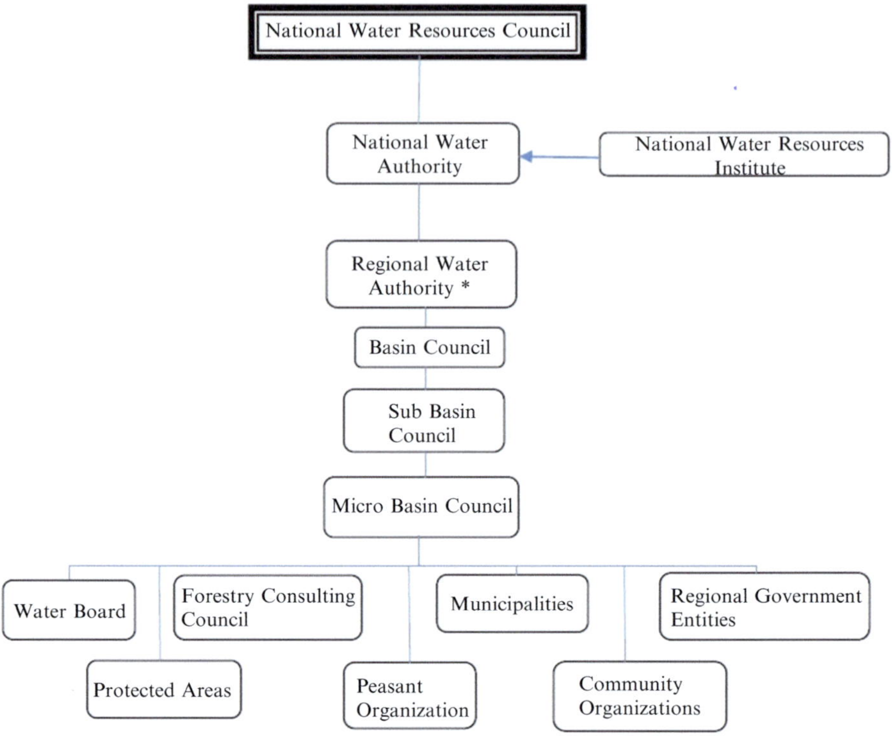

Fig. 19.2 National Water Resources Council organization chart

Table 19.2 Supporting institutions of the Valle de Angeles municipality

Institution	Type of support	Legal framework
State Finance Ministry	Financial and technical support and infrastructure construction	Signed financing agreement
Friends of the Valley Foundation	Infrastructure construction	Signed agreement
FOCUENCAS II-CATIE	Support in water source protection and rural bank implementation	Signed agreement through the Basin Council
AFE-COHDEFOR (now Forestry Conservation Institute, ICF)	Technical support in forest inspection	Signed agreement, joint work with the UMA
Friends of La Tigra National Park (AMITIGRA)	Infrastructure construction	Signed agreement
FOCUENCAS	Technical support in natural resources management	Signed Agreement
Ministry of Natural Resources and Environment	Conflict resolution in water resources issues	N/A
Governance and Justice Ministry	Technical assistance	Defined by law

Source: Data from Aguilar et al. (2008), water management in Valle de Angeles, Honduras

19.5 Conflict Resolution

Diverging interest and perspective conciliation can be achieved through diverse methods of communication, transactions, and compensations which can be accomplished through binding contracts or payment of water ecosystem services (Prins 2009). In this section, different cases per type of conflict will be discussed. Also, the steps toward their solution will be listed.

Of the 22 water supply systems in Valle de Angeles, 13 have some sort of conflict. Conflicts in the Rio Soledad basin can be classified in six types which allow prioritization of actions and the illustration of cases important in determining key elements toward good water administration and management.

19.5.1 Soil Use

It is common to have a problem related to agricultural activities in areas where water capture infrastructure for potable water is present. A common element in conflicts arising from these situations is that such problems are not perceived as one, but are rather considered latent problems. In the Rio Soledad basin, such problem can be found in the communities of San Francisco, Las Martitas, and Jocomico (Aguilar et al. 2008).

In the San Francisco case, problems presented themselves when agricultural activities reached the upper part of the creek and surrounding areas of the reservoir that supplies water to the town center of Valle de Angeles. The area inhabitants would deposit trash and biological human and animal waste in the field which in turn began contaminating the water collected by the dam (Aguilar et al. 2008).

In order to correct the problem, first the water recharge areas for that particular catchment were delimited. Next, the specific problems were identified, in this case source of the human and household waste. In turn, the situation of such contamination was reported to the municipality's Commission of Health and Environment which resulted in onsite visits to verify the contamination. This leads to the elaboration of a budget for the purchase of materials needed to protect the area. Such budget included mainly the fencing of the catchment. The fencing turned out to be a very easy solution. After its construction, no contamination in subsequent visits was found. In hindsight, this problem was easily solved due to the fact that the land, where the conflict was taking place, was property of the municipality, and therefore no permits were regulated to build the fence. However, environmental education programs on the importance of water production areas to capacitate the inhabitants of nearby towns are necessary (Aguilar et al. 2008).

19.5.2 *Ownership of Water Sources*

This type of conflict is better exemplified by the case of the Chiquistepeque community. The intake supplying water to this community is located near a road. Since it is close to a transit area, it was exposed to contamination. This was a multipurpose intake in which the structure served for providing both potable water to the community and as a drinking place for animals. According to Reyes (2006), water analyses were done, and the results demonstrated contamination due to organic matter. A decision to fence the reservoir was made. However, a particular person (offender) of the community went ahead and fenced a part of the reservoir but built a direct access. He sustained his action alleging that his right to water usage of the reservoir was being violated and a conflict for the property of the water source arose.

In order to find a solution to the conflict, several steps had to be taken. First, a complaint was filed in the municipality citing the abuses of the offender. Subsequently, the municipality intervened and summoned both parties to explanatory meetings. However, the first two meetings had to be suspended due to the absence of the offender.

As an alternative to the offender's absence, the municipality in turn visited the conflict area, managing to meet with both sides and hear their positions. As a result, an approach began and a date for a second meeting was defined; during this meeting, legal documents of land ownership were presented and possible solutions were discussed over the course of two more meetings. Finally, during the last encounter between the affected community, the offender and the municipality, an agreement was reached to move the water intake further upstream to land property of the community, and in doing so both parties would have access to the water (Aguilar et al. 2008).

19.5.3 *Access to the Water Source*

For more than 30 years, the community called "Bordo de las Martitas" has sheltered more than 35 families, which are permanently supplied with water from a source located in a private property. Traditionally, the water has been collected by hoses, but recently the water board decided to install permanent piping. Simultaneously, the property was sold and the new owner denied the permit since he planned to divide the area into lots. This conflict came about because of the plan to build an urbanization project which challenged the prior acquired rights of a community.

In this case, the parties involved are the community and the land owner who plans to develop a housing and tourism project which depends on this water source. However, he offers to cooperate with justifying alternative mechanisms in order to solve the water problem in the area. On the other hand, the community demands respect to its common law right to water and desire to improve its water supply by building a pipeline all the way to the community water tank (Aguilar et al. 2008). Despite the fact that the conflict began in 2007, there is yet a solution to be found since current law states guidelines for contamination, service abuse, or bad quality cases but not about access to water resources. There are a couple of possible solutions; one option is to declare the area an area of priority for water production through a municipality regulation, but this would mean that according to law the land would be expropriated. A second option would have to guarantee the water supply to the aforementioned project (Aguilar et al. 2008).

19.5.4 Water Quality

This problem usually presents itself when agricultural activity happens near water catchment zones and thus contaminates. A common element in these cases is that neither producer nor water users attribute any importance to the matter. For water quality, the case study will be the Jocomico community. The water source is on private property, and very close by upstream of the subbasin, there is a vegetable plantation on which agrochemicals are used. Despite the danger of the situation, water users have made no formal complaint for the possible illnesses as a result of agrochemical-contaminated water. This water use conflict is not perceived by the owner of the plantation nor by the community; as a matter of fact, there is a good relationship between the involved parties that could prove to be of benefits in future conflict resolutions (Aguilar et al. 2008).

19.5.5 Water Availability

This basin is characterized by the small amount of water in catchment zones which in turn doesn't allow a proper supply to the communities. In some cases, water availability is scarce as in El Chaguitillo and Los Lirios communities where water supply is supplemented by hose connections to other smaller water sources. In other cases, water demand is higher than water supply due to population growth like in La Esperanza where the water system was built to supply 50 users, but the current users are above 250 (Aguilar et al. 2008).

Stream flow varies according to seasonality, which is an important information that the board of directives of the water board needs to use in order to plan distribution and calculate the total number of users. In the case of Chaguitillo, bad decision making has influenced the problem, as previous local councils subscribed agreements with the neighboring municipality of Santa Lucia to supply water with no foresight to future supply problems of the local communities. These agreements were not analyzed, socialized, or agreed with the local communities (Aguilar et al. 2008). The solution process has involved two types of parties, the community and institutions working on water issues (AMITIGRA,[4] PRRACAGUA[5]). This conflict has taught the community to assume other strategies to increase water supply. In addition, institution support and guidance have also contributed to promote a collaborative management of the conflict.

In La Esperanza, because of the disproportional population growth, water provisioning is insufficient. Currently, water rationing is in place; houses receive water 12 h a day. Alternatives to increase water supply are being sought such as water extraction from adjoining areas such as Buena Vista.

19.5.6 Deforestation

Deforestation conflicts usually arise in water sources. In two cases, the areas are close to residential areas where firewood is needed to cook and in other areas in communal land with no protection or preservation activities in place. This problem is difficult to control since it usually involves low-income families that collect firewood for food preparation. These situations require integral actions in which ecologic stoves are a good solution since they require less wood and provide shorter cooking times. In addition, the local council has to incorporate in its municipal politics the protection of water production areas and the promotion of alternative cooking mechanisms that do not include firewood as well as environmental education regarding water protection (Aguilar et al. 2008).

19.6 Good Water Management: Supporting Actors and Factors

There is a diverse internal dynamic as well as external factors that affect water management in the basin. There are favorable factors that promote good water management such as a positive attitude, an extensive legal framework, and the presence of an active basin council (ASDI CATIE b, s.f.). However, adverse factors can also be found such as the inexistence of a local policy regarding water, water conflicts, and land titles provided by National Agricultural Institute.

Some interested parties in the basin such as restaurant owners and the Catholic Church have yet to announce their participation; such actors are considered potential collaborators in favor of water management, especially the Catholic Church which reaches all areas of the municipality and also addresses natural resources conflicts in their regular meetings (Aguilar et al. 2008).

---

[4] Friends of La Tigra National Park, nongovernmental organization.

[5] Honduran Rural Aqueduct, Wells and Basic Drainage Project.

Contrary to many watersheds in Honduran territory where land use is over- or underutilized, 98.49 % of the Valle de La Soledad basin has an adequate land use category and only 1.5 % in the improperly used category. The aforementioned data, point to a positive potential regarding the execution of actions directed to establish compensations to land users in the basin that will allow adequate water management (PREVDA 2008).

In order to have a clearer picture regarding positive and negative forces affecting water management, a table comparing both has been elaborated (Table 19.3).

19.7 Results: Local Implementation Strategies

All scenarios are changing and processes have constant readjustments. Problems and conflicts are gradually addressed in order to look for solutions that benefit all the players involved. This allows negative factors to be transformed into positive factors through the implementation of strategies. Additionally, in participative processes such as comanagement activities, the actors involved widen their knowledge and learn the importance of protecting the basin (GWP 2009).

19.7.1 Basin Council

With the help of *FOCUENCAS*[6] and *FOCUENCAS II*,[7] the participation of local actors, as well as other programs, a Basin Council for the Valle de la Soledad Watershed was organized. Internal problems of Basin Council are now being addressed, such as changes in land use through binding contracts. These contracts specify the agreement between municipalities and farmers to halt the advance of farming activities in water production areas in order to protect water sources (Aguilar et al. 2008). Basin councils tend to have better results when its operation is decentralized and base committees are formed in the communities which in turn during the process allowed the creation of bonds among the inhabitants of the different ecological levels in the basin (Prins 2009).

---

[6] Program for the strengthening of local capacities in watershed management and natural disaster prevention.

[7] Program for innovation, learning, and communication in the integrated adaptative management of watersheds.

Table 19.3 External and internal factors that influence positively or negatively on water management

In favor		Against	
External factors	Internal factors	External factors	Internal factors
General Water Law regulates conservation, protection, and adequate water management	Awareness and local politics will be for good water management	Disperse institutional framework regulation, administration, information, planning and policy-making roles	Lack of a local water policy
National Territorial Ordainment Law establishes basins as areas under special regimen and basin councils as integration mechanisms	Presence of a mixed local basin council that gathers actors from all different sectors	Lack of coordination and knowledge of activities among state entities	No local regulations or bylaws for water management
Natural Resources and Environment Ministry supports in establishing vinculating contracts	Comanagement plan that addresses water management and establishes the protection of water sources and recharge areas	Municipality unaware of land title granting of water-producing areas by INA	No municipal information and communication system in order to convey news or messages
General Direction of Water Resources in charge of the measurement, conservation, and evaluation of water resources as well as the authorization of its use	Availability of qualified local human capital	Urbanization of water producing areas due to the uncontrolled expansion of the nation's capital Tegucigalpa	No information on water management and care available
Private organizations support such as AMITIGRA, CATIE, and VIDA Foundation which support and promote good water management activities	Base water organization is promoting and strengthening capacitation and participation in water management to different inhabitants	Vegetable production increase due to higher demand from Tegucigalpa, causing a demand for more farming land diminishing water production priority areas	Little municipal capacity for water resources management, organization structure lacks a water management department as well as logistical support
	Growth and organization of water boards; leading conservation, administration, and protection activities in rural areas		No economic resources available to water boards
	Adequate local practices for water protection and usage		Not all local organizations participate actively in water management

(continued)

Table 19.3 (continued)

In favor		Against	
External factors	Internal factors	External factors	Internal factors
	Implementation of fencing practices to protect water sources		Deforestation of water source areas
	Positive land use and land use practices change such as agroforestry		Land use conflicts in water source areas
	Binding contracts between municipalities and farmers in order to protect water sources		Dispersion of efforts and resources

Sources: (Aguilar et al. 2008), (Law 181-2009)

19.7.2 Comanagement Plan

In order to make decisions, develop strategies, and establish structured directives with technical validity, several principles and criteria are to be taken into account (Fig. 19.3). According to Faustino and Jimenez (2005), comanagement plans are based on the premise "it is not about making new plans"; rather the purpose is to implement existing ones that are of interest to the community regarding natural resources management. The main reason of comanagement plans as an alternative is due to the increasing trend found in multiple municipalities in which strategic, management, territorial, protected areas, and other types of plans have been formulated and very few have been implemented. This in turn brings the question, why haven't they been implemented, when the plans have been elaborated through participative methodologies and respond to strong needs of the community? One of the considerations regarding these aspects could be the need for management processes and actions in a participative and collaborative way in order to achieve the resources and means necessary to implement these actions.

19.7.3 Critical Factors

Critical factors that influence the viability and sustainability of the watershed management processes[8] can be identified, such as adequate management of the complexity of the issue, the adaptation of the processes to the peculiarities of the basin, the construction of a common agenda, and the development of capacities to

[8] Processes refer to all the actions regarding watershed management such as conservation of water sources and land use among others.

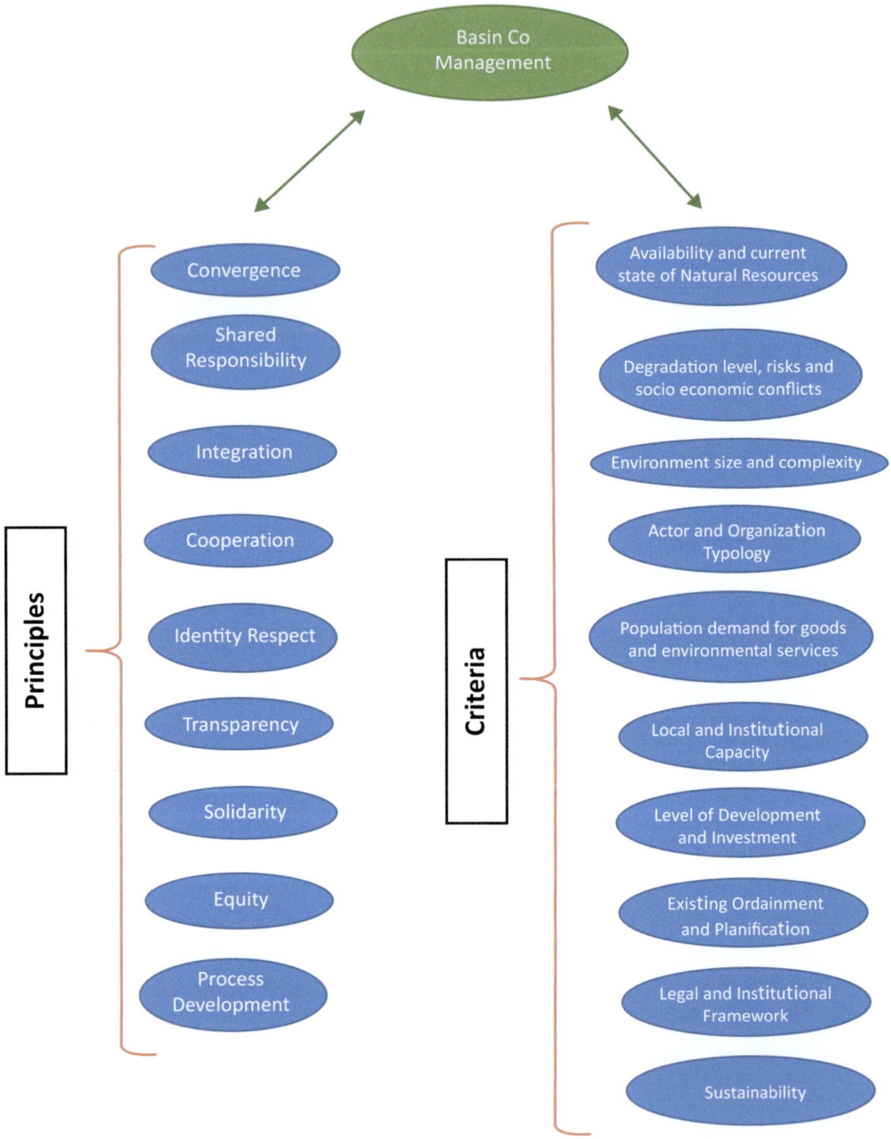

Fig. 19.3 Comanagement principles and criteria (Source: Faustino and Jimenez 2005)

conciliate common and personal interest, as well as support from the civil and public sectors. Also, an adequate balance between public and private sectors as well as role definition of each actor that will in turn lead to the construction and appropriation of work instruments such as rules, procedures, tasks, and responsibilities can be identified (Prins 2009).

19.7.3.1 Sustainability

In order to fulfill activities oriented toward sustainable management, it is necessary for national and even local organizations to be able to manage and procure technical, financial, and administrative resources. Also, the strengthening of local capacities is needed in order to adequately manage their available human and financial resources in the most efficient way without losing focus of the integral approach required for natural resources management (Espinal 2007).

Current financial mechanisms for watersheds include international cooperation funding, private sector contributions, PSE,[9] binding contracts for commercial activities, natural resources exploitation, and contamination canons.[10] However, these mechanisms have no legal framework or specific regulations; only municipal bylaws and private agreements between parties are legally binding.

The Valle de Angeles municipality does not have the necessary resources to develop continuous conservation and management activities in the basin; therefore, it is necessary to generate local resources for these activities. However, the basin generates diverse goods and services to the local and external population (Tabora 2004). Additionally, 43.5 % of its territory forms are part of La Tigra National Park which is one of the main water sources for the Honduran capital Tegucigalpa, as well as to other communities and industries surrounding the watershed (Espinal 2007).

One of the purposes of a comanagement is the creation of an environmental fund that is managed by the local actors. In the case of the Soledad River micro-basin, the fund was created, but also a novel approach was implemented through the use of loans granted from the environmental fund using binding contracts conditioned by an environmental conduct code between the person receiving the loan and the Basin Council. This strengthens the capacities of the Basin Council and the municipality to generate appropriate mechanisms for the use and control of the fund in order to stimulate watershed management impact on water quality and quantity (Espinal 2007).

19.8 Conclusions

Good water management and empowerment of protection activities in basins are the main goal of many organizations. The Soledad River basin case addressed in this chapter allowed to identify key elements for water management such as the national and regional laws of Honduras orienting water management and municipal

[9] PSE, Ecosystem Service Payment referring to any payment received for the conservation, sustainable use, or other of the natural resources of a particular area.

[10] Contamination canons refer to the "fee" paid by companies or institutions whose activities generate some sort of contamination; fees paid are typified by the General Natural Resources Law of Honduras.

bylaws and the identification of certain communities that already have good water management and use which can also serve as examples for future municipal bylaws and strategies.

The three components of municipality plumbers and UMA[11] can condensed into a single water management department in order to coordinate and maximize efforts. Collaborative efforts between the municipality and the Basin Council resulted in effective strategies and actions in different areas of the basin that allowed to resolve otherwise difficult conflicts in an appropriate way that was beneficial to all parties; these collaborations should be strengthened.

Basin councils and water boards are vital to good water management in order to promote participative actions and processes. In general, experiences generated through pilot programs such as the one described in this chapter are excellent as examples when addressing future conflicts in other areas.

Acknowledgments I acknowledge the FOCUENCAS II project for making available all their publications and Dr. Jorge Faustino and Dr. Francisco Jimenez from the Tropical Agricultural and Higher Education Center (CATIE) for their revision and inputs to this chapter.

References

Aguilar O, Prins C, Faustino J y Madrigal R (2008) La Gestion del Agua en Valle de Angeles, Honduras: Elementos claves para la protección y buen aprovechamiento del agua. ASDI CATIE. Technical report No. 360, Turrialba

Argeñal F (2010) Variabilidad Climatica y Cambio Climatico en Honduras. Tegucigalpa

ASDI, CATIE (a) (2008) La Cogestión de Cuencas Hidrográficas en América Central. Turrialba

ASDI CATIE (b) (s.a.) Innovación, aprendizaje y comunicación para la gestión adaptativa de cuencas. In FOCUENCAS

Barahona J (2006) Caracterizacion Paisajistica y Definicion de Lineamientos de Manejo de Montana Grande: zona de recarga de las cuencas de La Soledad, Las Canas y El Cobre, Valle de Angeles y Santa Lucia. Tesis de Licenciatura. Universidad Nacional Agricola, Catacamas

Bucardo E (2007) Escalamiento de la Co gestión de cuencas del rio Jucuapa a las micro cuencas de la parte alta del rio viejo Jinotega, Nicaragua. Tesis de Maestría. Centro Agronómico Tropical de Investigación y Enseñanza, Turrialba

Cashman A (2012) Water policy development and governance in the Caribbean: an overview of a regional progress. Water Policy 14:14–30

Dellapena J, Gupta J, Wenjing L, Schmidt F (2013) Thinking about the future of global water governance. Ecol Soc 18(3):383–391

Espinal E (2007) Mecanismos de financiamiento para el manejo y cogestión de la micro cuenca del Rio de la Soledad, Valle de Ángeles, Honduras. Tesis de Maestría. Centro Agronómico Tropical de Investigación y Enseñanza, Turrialba

Faustino J, Jiménez F (2005) Guía para la Elaboración de Planes de Co Gestion de Cuencas. CATIE, Turrialba

[11] Municipal Environment Unit.

Global Wáter Program (GWP) (2009) Experiencias de Agua Potable y Saneamiento con enfoque de gestión integrado de recursos hídricos en Honduras. KARES, Tegucigalpa

HYTSA (Estudios y Proyectos SA) (2005) Politicas Municipales para los Proyectos de Agua Potable y Alcantarillado sanitario en 16 municipalidades de Honduras. Segundo informe. Tegucigalpa, 47 p

Kammerbauer H et al (2010) Modelo de Cogestión Adaptativa de Cuencas Hidrográficas. Recursos Naturales y Ambiente No. 59–60:117–122

Ley No. 98-2007 (2008, de 26 de Febrero) Ley Forestal, Áreas Protegidas y Vida Silvestre. Diario Oficial La Gaceta No. 31,544

Ley No. 104-93 (1993, del 30 de Junio) Ley General del Ambiente. Diario Oficial La Gaceta No. 27,083

Ley No. 118-2003 (2003, de 8 de Octubre) Ley Marco del Sector de Agua Potable y Saneamiento. Diario Oficial La Gaceta. No. 30, 207

Ley No. 181-2009 (2009, de 14 de Diciembre) Ley General de Aguas. Diario Oficial La Gaceta No. 32,088

PNUD (2013) Informe Desarrollo Humano. PNUD, New York

PREVDA-HON (2008) Diagnostico participativo de la microcuenca del rio de La Soledad. CATIE, Tegucigalpa

Prins C (2009) Experiencias y Reflexiones sobre la Gobernanza de los Recursos Naturales en las Subcuencas de Aprendizaje de Honduras y Nicaragua por el Programa FOCUENCAS II del CATIE. In Webinar Desarrollo Rural Territorial y Gobernanza de los Recursos Naturales: Reflexiones de los Andes Mayo 2009

Reyes K (2006) Analisis del estado de las fuentes de agua para consumo humano y funcionamiento de los acueductos rurales en la cuenca del Rio La Soledad, Honduras. (Tesis Maestria). Centro Agronomico Tropical de Investigacion y Enseñanaza. Turrialba

Susskind L (2013) Water and democracy: new roles for civil society in water governance. Int J Water Resour Dev 29(4):666–677

Tabora F (2004) Desarrollo de un modelo de fondo ambiental para el manejo y conservación de la recursos naturales de una microcuenca de Honduras. Com Tec Recursos Naturales y Ambiente

Chapter 20
Integrating Local Users and Multitiered Institutions into the IWRM Process

Ryan H. Lee, Lauren Herwehe, and Christopher A. Scott

Abstract Participation among stakeholders and tiered institutions in a collaborative policymaking process is essential to IWRM's stated goals of securing water for people in a manner that reconciles economic efficiency, social equity, and environmental sustainability. Using the energy-water nexus and case examples from Tajikistan and Mexico, we define a mechanism by showing that poorly articulated multitiered institutional arrangements coupled with failure to generate truly participatory interaction of stakeholders lead to water insecurity. In the case examples, we found that the livelihoods of vulnerable populations are threatened when users experience water insecurity that is created or exacerbated when tiered institutions neglect users' signals by failure to respond with actions that promote sound resource management or mitigate livelihood threats. Water and livelihood security would be improved by adaptive actions targeted at user-defined causes of water insecurity and coordination between local resource users and institutions at multiple levels. Our results are a diagnostic tool that can be used to identify one cause that, among a possible multitude, contributes to water insecurity. Institutions and decision-making among stakeholders will be an explicit component of the human capacity to respond with programs, policies, and actions able to deal with the dual pressure on water resources posed by climate change and heightened demand while reconciling economic efficiency, social equity, and environmental sustainability. Institutions that operate at the intersection of local users and state and non-state actors have the greatest chance of inducing IWRM solutions if the tiered nature of linkages is expressly accounted for and used to adaptive advantage.

R.H. Lee (✉)
Arid Lands Resource Sciences, University of Arizona, Tucson, AZ, USA
e-mail: rhlee@email.arizona.edu

L. Herwehe
School of Geography and Development, University of Arizona, Tucson, AZ, USA
e-mail: laurenherwehe@email.arizona.edu

C.A. Scott
Udall Center for Studies in Public Policy and School of Geography and Development,
University of Arizona, Tucson, AZ, USA
e-mail: cascott@email.arizona.edu

S.G. Setegn, M.C. Donoso (eds.), *Sustainability of Integrated Water Resources Management*, DOI 10.1007/978-3-319-12194-9_20

Keywords IWRM • Multitiered institutions • Policy • Water-energy nexus • Water governance • Participatory approach

20.1 Introduction

In the twenty-first century, freshwater resources are becoming increasingly constrained as demand for water increases and drought diminishes supply (Overpeck and Udall 2010; Hoerling and Kumar 2003; IPCC 2001a; Easterling et al. 2000). Climate change is predicted to negatively alter global (IPCC 2001b) and basin (Setegn et al. 2011) hydrologic cycles, stream flows, and water availability. Rapid urbanization and modern economic development compete against each other for water resources, heighten demand for freshwater, and contribute in combination to the creation of a water-scarce world (Scott et al. 2013; Hightower and Pierce 2008). Integrated water resources management (IWRM) is "a process, which promotes the coordinated development and management of water, land and related resources in order to maximize the resultant economic and social welfare in an equitable manner without compromising the sustainability of vital ecosystems" and emphasizes integrated, basin-wide governance and public participation (Global Water Partnership 2000). IWRM is viewed as the water management and governance paradigm best suited for "securing water for people" while reconciling economic efficiency, social equity, and environmental sustainability (Global Water Partnership 2000).

The major challenge faced by IWRM is its effective implementation in the field (Rahaman and Varis 2005). There is no clear consensus on how to weigh priorities or ensure their realization (Molle et al. 2008). IWRM lacks a "universal blueprint" and is instead articulated as a set of principles to be put into practice. In some respects this is a strength of the paradigm but also risks that deficient or untargeted applications of the principles leave IWRM to become an empty buzzword used to pursue business as usual (Molle 2008a) or a panacea that fails to remedy problems by virtue of a misdiagnosis that obscures and ignores particular solutions or aspects of the IWRM process (Ostrom 2007). When IWRM is oriented toward technical outcomes (optimality/water productivity, adequacy of supply, irrigation efficiency, and meso- and macroscale impacts) while ignoring the "enabling environment," it risks falling short of the paradigm's intentions, covering for business as usual or failing to diagnose local-scale consequences and impacts (Mehta 2008).

This chapter will focus on how government and communities should shape the enabling environment, the vertical (e.g., local- to mesoscale, user to government) or horizontal (e.g., user to user, organization to organization) arrangements, and the interplay among multitiered institutions (laws, policies, and organizations that operate across jurisdictional levels). The enabling environment is the institutional capacity to actually achieve IWRM's stated goals, particularly "securing water for people" which can be understood as reducing vulnerability due to resource access or scarcity. Participation is a core IWRM principle that affects the enabling environment, is "more than consultation," and is put into practice when "stakeholders at all levels of the social structure have an impact on decisions at different levels of water management (Global Water Partnership 2000, p.16). The emphasis on participation as a core IWRM principle is important, but without defining a specific mechanism to generate truly participatory interaction among stakeholders, the principle is at risk of becoming an empty buzzword and the opportunity to induce IWRM solutions lost. This chapter will justify why participation is important to any well-built IWRM architecture by (a) using the water-energy nexus as a vehicle to define a specific participation mechanism and understand the importance of user inclusion and (b) providing evidence that insecurity results when mesoscale governance institutions (state, regional) are unlinked or unresponsive to feedbacks from vulnerable local water users within the institution's jurisdiction.

20.1.1 The Water-Energy Nexus Is an IWRM Paradigm with Lessons About Water Governance, Institutions, and the Participatory Approach

Analysis of river-basin management reveals that governance is inherently a political process and about access to, and the allocation of, a contested and scarce resource (Molle 2008b, 2009). The political process can be responsible for socially constructed scarcity, business as usual, and exclusion. Little improvement for water governance is possible without rebalancing decision-making power and the empowerment of the community at large. Therefore, participation in IWRM, like community-based conservation, can be understood as governance that starts from the ground up and involves networks and linkages across various levels of organization (Berkes 2007). Linkages across networks and various levels of organization are a good foundation but do not advance our understanding much beyond the description articulated by the Global Water Partnership nor of how participation functions as an actual mechanism to fulfill IWRM goals. The challenge is to build upon this foundation and become more descriptive and focused in our understanding of integration between users and institutions so that we are able to diagnose water governance failures according to symptoms. To do this we need to know what participation and IWRM mean for local resource users.

Studies into the water-energy nexus provide a convenient vehicle to travel beyond the foundation laid by statements about integration across resources, sectors, and institutions in order to reach a useful, multidimensional framework that will support efforts at structured analysis of water governance case studies. The energy-water nexus is a variant of IWRM: it views energy and water as inextricably linked because energy supply involves the use of water, while water supply requires energy. As the concept has grown from an operational tool used for optimizing resource use according to input-output to a joint-resource management and policy paradigm, it has been found that policy, decision-making, and institutional

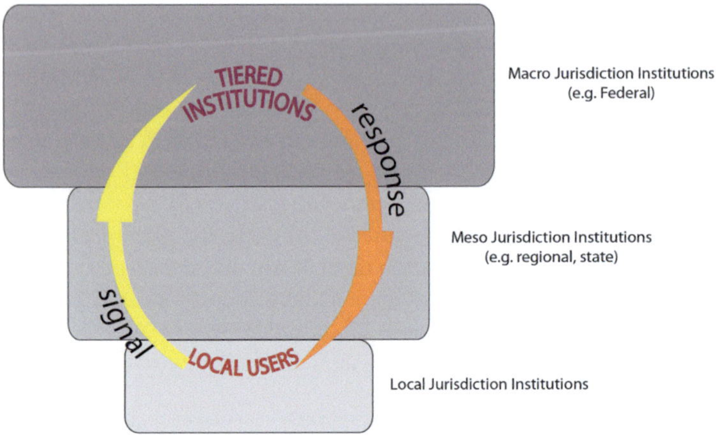

Fig. 20.1 Feedback loop between users and multitiered institutions

relationships play an important role in the effectiveness of joint energy-water management and governance. Conceptually the energy-water nexus is concerned with integration across resources (energy and water), but in order to shape policy and decisions that impact economic and environmental security, the energy-water nexus must involve coordination of multitiered institutional arrangements (Scott et al. 2011). With respect to coupled energy-water management, institutions across horizontal and vertical scales must be sufficiently connected to another in order to coordinate synchronous governance that brings water management and policy closer in line with energy policy. Central to synchronous governance are multitiered institutions and the extent to which they are able to manage and respond to local environments, impacts, and users (Fig. 20.1), i.e., "reconcile spatial scales." In a multitiered institutional system, resource users are within the jurisdiction of local, regional, state, and federal authorities. While it is possible that an individual can responsibly manage a resource themselves, it is typical that governance defaults to an institution found within the tiered arrangement of laws, policies, and organizations. Individual users, however, experience scarcity, lack of access, and practices that influence demand. User experiences are scaled horizontally when the experience becomes a pattern across multiple users and therefore a feedback that *ought* to be scaled vertically and registered at higher institutional tiers in order to invoke an adaptive response – a change in governance, management, or policy. The adaptive response by a local, regional, or federal authority is transmitted downward through the multitiered institutional arrangements and back to resource users in order to correct or mitigate against their experience of scarcity or misuse, thereby completing the feedback loop.

Participation in IWRM is concerned with (a) coordination of tiered institutions and also (b) how stakeholder involvement enables the emergence, generation, and implementation of adaptive actions that preserve the security of core resources and in turn promote the well-being within human communities. In the energy-water nexus, this second institutional dimension is termed collaborative policymaking and involves public decision-makers, private initiative, and a range of stakeholders. It is with this second dimension, collaborative policymaking, that we begin to move into the internal mechanics of how participation, users, and management interact in order to elicit adaptive actions and resource governance across scales, sectors, and institutions. Collaborative policymaking is needed in order to counter special interest group's influence over energy and water policy or effectively create and implement policy (Scott and Pasqualetti 2010). The range of participating stakeholders and their degree of participation have important equity implications. For example, in many basins in the United States, water is delivered under legally binding agreements or withdrawn under issued permits, and making water available to the energy sector would likely decrease water availability for another sector, such as agriculture (Carter 2010). Trade-offs with equity implications can even occur within the same sector: a federal program in Sonora, Mexico, which aimed to improve electrical and water-use efficiency by sharing the cost to upgrade electrical and irrigation technologies has been largely ineffective because the program's criteria favor large commercial growers to the exclusion of small-scale farmers (Scott and Pasqualetti 2010).

These examples from the water-energy nexus reveal that, like property rights among resource users (Schlager and Ostrom 1992), the policymaking process can also produce situations of unequal exclusion, alienation, access, and withdrawal (Fig. 20.2). In the IWRM process, participation by all resource users in creating policy outcomes, regardless of position and associated bundle of rights, is crucial to water governance. Exclusion, alienation, access, and withdrawal affect the function and effectiveness of the multitiered institutions that have jurisdiction over a particular basin. Water is fundamental to life and livelihood and all stakeholders within a basin are water resource users. All require, de facto, rights of access and withdrawal. However, for environmental, climatic, proximity, or institutional reasons, access and withdrawal among users are unequal to the point that some users experience scarcity and therefore livelihood insecurity. Livelihood insecurity due to differences in access underscores the necessity to involve all water users in the creation and management of water policy. According to societal and environmental determinants of scarcity and access, users can be categorized according to the strength of their connection to multitiered institutions (i.e., likelihood of institutional response to their signal, strength of involvement in collaborative policymaking) and the water resource (i.e., water security, experience of scarcity, vulnerability due to changes in climate and water availability). Users in Tier One, with best access to water resources and institutions, are the most secure. Users in Tier Two experience a weak-strong connection to water resources, and institutions (strong water, weak institution/weak water, strong institution) are less secure than users found in Tier One. Users in Tier Three are the most vulnerable and insecure because they have weak access to water resources and ability to invoke an institutional response or participate in setting policy outcomes. "Securing water for people" in a manner that reconciles economic efficiency, social equity, and

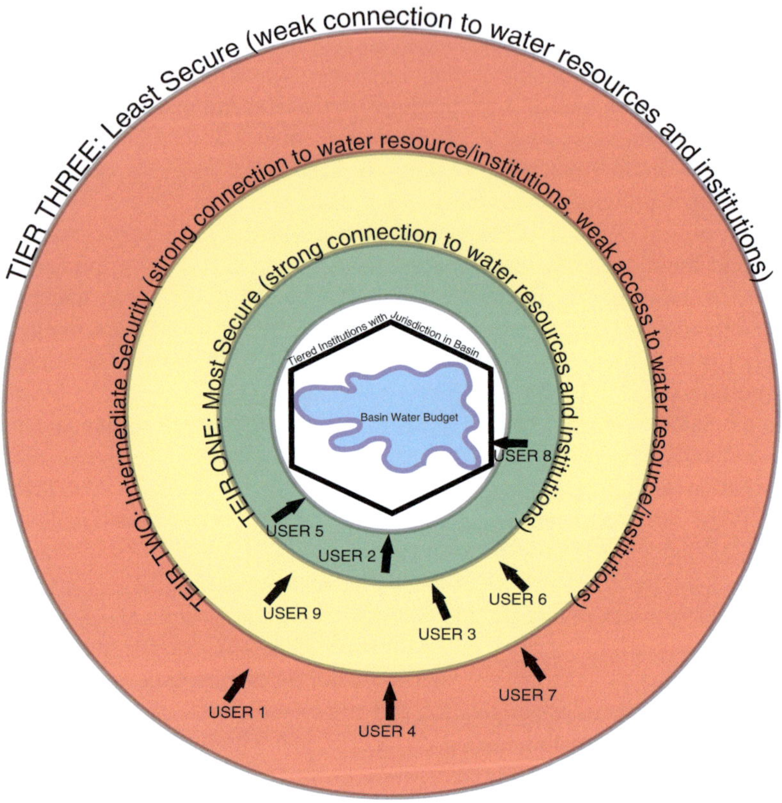

Fig. 20.2 Basin water users grouped according to strength of connection to water resource and tiered institutions

environmental sustainability requires users from all three tiers to participate in the collaborative, feedback-response policymaking process.

20.1.2 Diagnosing a Cause of Water Scarcity: Exclusion of Local Resource Users from Collaborative Policymaking

Insights from the water-energy nexus have led us through the rather vast, formless "enabling environment" into a foundation and structure built by linkages, political processes, multitiered institutional arrangements, collaborative policymaking, and internal processes about feedback-response loops, user exclusion, and access. However, we have not yet arrived at an actual mechanism that can be used to justify participation's importance or diagnose causes of water insecurity. The energy-water nexus leads us to the mechanism we seek by reconsidering the feedback-response loop between users and multitiered institutions and combining it with our insights about users who have weak access to resources and collaborative policymaking. Specifically, we consider what happens when linkages with users, multitiered institutions, and collaborative policymaking processes fail to materialize or function.

A feedback loop between users and multitiered institutions is the basic process whereby governance between polity and citizenry is supposed to function. This same feedback-response process is also responsible for coordinating multitiered institutions in a manner that aims to reconcile local needs with meso- or macroscale institutions by bringing local-scale signals (e.g., scarcity, access, mismanagement) and policy objectives in line with one another. Institutional arrangements can produce positive as well as negative results for environmental governance (Young et al. 2008), and in order to derive an actual mechanism whereby IWRM can prove effective at addressing water problems, we are interested in the negative results – what causes failure in multitiered institutional systems? In the energy-water nexus, user needs and practices are not always congruent or in agreement with regional, national, or global policy strategies and objectives, nor is there always a clear path regarding meso- or macroscale responses. In the arid southwest United States, individual consumer practices raise carbon emissions and water consumption and act against national climate mitigation efforts because rising temperatures and variable water availability increase per capita water and energy consumption for cooling, irrigation, and other uses (Scott et al. 2011). In the United States in general, regional and local scales of water resources and how they are managed often complicate federal water-related actions, especially complicating matters when energy's water demand must also be considered (Carter 2010). In these examples local practices compromise or complicate meso- or macroscale management efforts. However, what happens when the opposite is true: meso- or macroscale institutions are incapable of responding to local signals as well as at fault for water insecurity among users. For example, in the United States, local adoption of new technologies that can reduce energy production's freshwater use is hindered by irregular coordination between federal institutions, the United States Geological Survey and Energy Information Administration, especially in regard to compilation and dissemination of water-energy data to policymakers and industry analysts (Mittal 2009).

Both failure to respond to feedbacks and institutionally caused scarcity have not been linked into an actual IWRM mechanism that can be used to direct specific courses of action among researchers, policymakers, and practitioners. In order for IWRM to succeed and break from business as usual, the components for action must exist in place and be cohesive in a manner that produces adaptive action or a blueprint that provides specific, mechanistic instructions that permit us to evaluate what is missing and potential cause of dysfunction. Institutional failure to respond can be caused by user exclusion from collaborative policymaking and ultimately precludes the emergence, generation, and implementation of adaptive actions. IWRM is business as usual when participation is biased, unequal, and exclusive, especially when implementation excludes or ignores vulnerable users who have weak access to water and influence over policy setting. A diagnostic mechanism to evaluate water insecurity takes shape when unequal, exclusive participation and weak feedback-response loops that produce institution-caused scarcity or non-useful responses are evaluated in connection with the most vulnerable users within a basin. Therefore, one mechanism that causes water resource management to become integrated among stakeholders in a manner that reduces water insecurity is when a basin's most vulnerable users are sufficiently linked with meso- and macroscale institutions so that (a) tiered institutions respond to user-defined experiences of scarcity, livelihood insecurity, and other threats to well-being and (b) management and policy outcomes from tiered institutions result from collaboration and consensus among all basin users. Hence, participation happens in practice, not just theory, when signals from a basin's most vulnerable users are linked and received by tiered institutions in order to generate collaborative, consensual responses.

20.2 Two Case Studies: Methodology

Participation is important because the livelihoods of vulnerable populations are threatened when users experience water insecurity which is created or exacerbated when tiered institutions neglect users' signals or respond to feedbacks with actions that fail to promote sound resource management for sustained livelihoods. The participatory approach defined by specific mechanisms is required in order to diagnose for and ultimately correct these particular system failures. Without integration between users and tiered institutions (by sector and scale), attempts to develop IWRM are truncated at specific scales or administrative levels. Two case studies, one from a water-rich region in Khatlon province, Tajikistan, and another from an arid region in Sonora, Mexico, will be presented to support these claims. These case studies will demonstrate that water management issues and IWRM implementation challenges are larger than just the water-energy problem or integration across resources and instead extend into policy setting, institutions, and decision-making.

Both presented case studies tell a story about the challenges posed to IWRM to promote social-ecological security when resource users and tiered institutions fail to integrate in a manner that produces adaptive actions that mitigate livelihood insecurity due to water scarcity. Although research in Khatlon province, Tajikistan, and Sonora state, Mexico, was conducted independently, the methods used in each case study are similar in their focus on water users in the context of variable or scarce supply, organizational forms, and vertical and horizontal linkages (as outlined above) and tiered institutions and local resource users that are not sufficiently integrated to enable adaptive actions that reduce water insecurity. The two case studies address, as introduced by the water-energy nexus, the conceptual focus of this chapter: that without concerted emphasis on integration between local resource users and tiered institutions in order to enable adaptive actions, attempts to

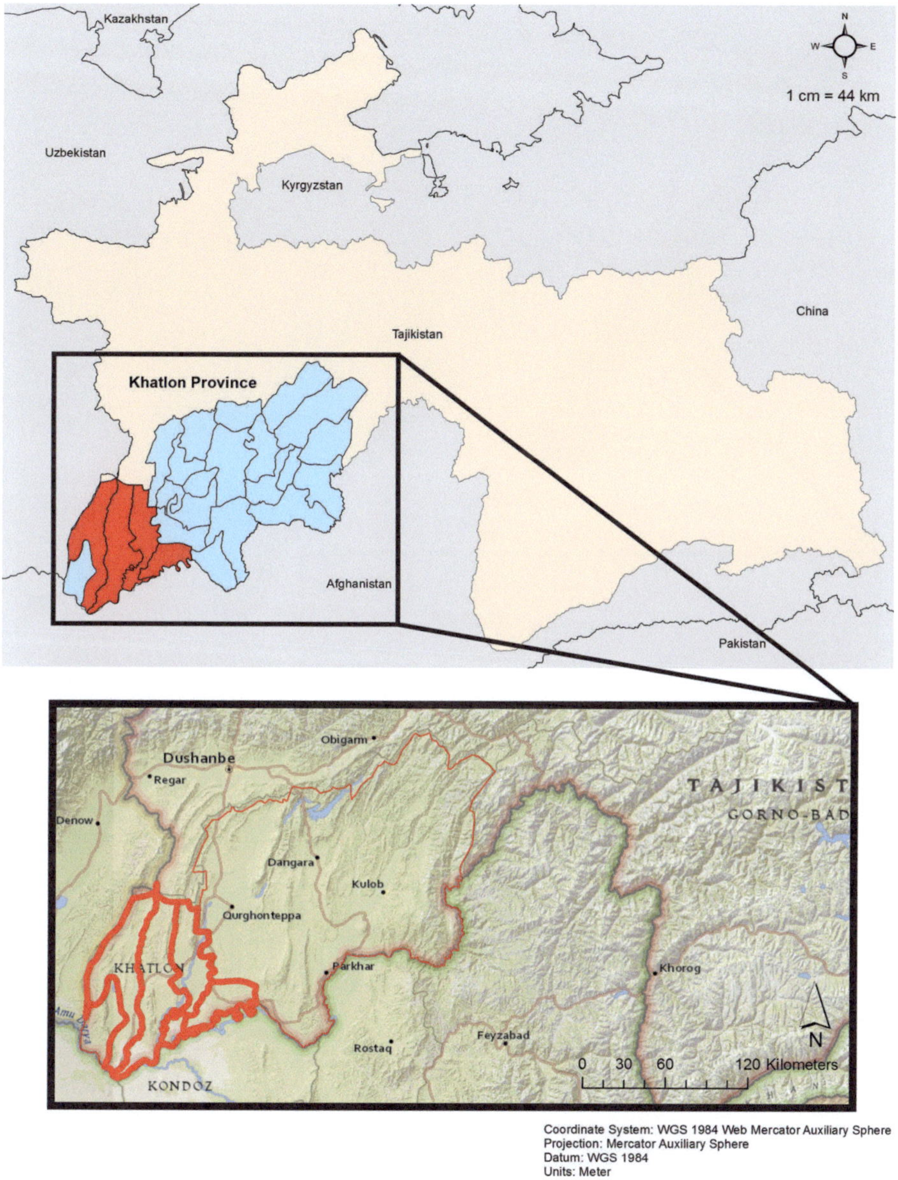

Fig. 20.3 Study area districts within the Khatlon province, Tajikistan (Feed the Future 2012)

develop IWRM are truncated at specific spatial or administrative scales. The case from four districts within the Khatlon province, Tajikistan (Fig. 20.3), contributes to the policy and institutional dimensions of the water-energy nexus while also grounding the theoretical IWRM participation problem highlighted above with an actual process that leads to institution-caused scarcity. The second case Rayón,

Fig. 20.4 Location of Rayón, Sonora, Mexico

Sonora, Mexico (Fig. 20.4), moves us away from the water-energy nexus vehicle and instead illustrates complications between local resource users and tiered institutions under global-change-type drought. Both cases deal with agrarian communities and illustrate that policy setting, institutions, and decision-making must also be considered when seeking water management solutions and overcoming IWRM implementation challenges.

In both cases chain-referral (i.e., "snowball") and convenience sampling were used to recruit survey respondents in order to administer semi-structured interviews. Follow-up interviews with key respondents were used to add depth to survey results in both cases. In Tajikistan 79 total respondents from four different districts (Shahrituz, Qabodion, Jilikul, and Qumsangin) were surveyed using four different sets of interview questions, each targeted at four stakeholder types: Water Users Association leaders, Water User Association members, farmers not members of Water User Associations, and provincial mayors. In Rayón, Sonora, Mexico, a total of 15 respondents were surveyed who held roles ranging from elected or appointed officials to community leaders. Numerous follow-up interviews were conducted with key respondents, commonly long-standing appointed officials or community leaders. In order to standardize across responses from the two cases, a common vocabulary will be used to describe survey results. The word "few" will be used when specific results are found in 1–20 % of all responses for that particular case study. The word "some" will be used to characterize specific results from 21–50 % of all responses, the word "most" to characterize 51–70 % of all responses, "majority" to characterize 71–99 % of all responses, and "all" when a particular result is found in 100 % of all responses.

20.3 The Water-Rich Case: Khatlon, Tajikistan

Four districts (*khukumats*) located within the Khatlon province (*oblast*) in southern Tajikistan comprise the study area for this case. Like the majority of Tajikistan's land area, Khatlon is classified as semiarid with an average annual precipitation of 140 mm (Yang et al. 2006; Abdullaev and Akbarzaheh 2010). However, situated within the Vakhsh and Kofarnion River basins, it possesses significant amounts of available surface water. Tajikistan itself is characterized by an abundance of water, with 55 % of the Aral Sea basin's total flow originating within its borders (Melikyan and Ghukassyan 2011). With 12,706 m^3 per capita, the country's renewable water resources far exceed the UN definition of water scarcity (1,000 m^3 per capita) (Varis and Rahman 2008; United Nations Economic Commission for Europe 2007). Only 7 % of Tajikistan's land area is arable because the mountains that serve as the headwaters for the country's vast water resources cover the other 93 % of the country's land area. (Melikyan and Ghukassyan 2011). Despite limited arable land, 75 % of the population in Tajikistan is employed in agriculture (USAID 2012). Most of the country's agricultural activity occurs in the lowlands of Khatlon, which possesses 49 % of Tajikistan's total sown area and accounts for 45 % of the nation's agricultural gross domestic product (Zvi and Sedik 2008). Of Tajikistan's total 7.6 million population in 2008, 2.6 million live in Khatlon (World Bank 2013). Contributing two-thirds of the nation's total harvest, Khatlon province is

Tajikistan's main cotton grower (World Bank 2013). In addition to cotton, farmers in the study area grow wheat and, to a lesser extent, vegetables and orchards.

Central Asia's abundant water resources were an important cog in the former Soviet Union's economic engine. The region's water resources and climate provided an ideal environment for growing cotton on collective farms (*kolkhozes*) and state farms (*sodhozes*), large enterprises that consisted of many specialized workers on any single farm. The collapse of the Soviet Union began a transition of collective and state farms into *dekhan* farms, midsized privately owned farms that average 20 ha and account for 60 % of all agricultural land in Tajikistan. Individuals who had previously played only a specialized role in the agricultural system became responsible for all aspects of farm management, in particular, allocating water resources and maintaining irrigation infrastructure. Both are tasks that were previously managed by centralized state entities. In the late 1990s, the World Bank, the US Agency for International Development (USAID), and the Asia Development Bank began funding and facilitating the creation of Water User Associations (WUAs) as a method of assisting farmers with the shift to decentralized water management and encouraging user participation[1] (Lam 2010). WUAs primarily function to serve the water management needs of the country's *dekhan* farms (Lam 2010). Donor organizations focused their resources on organizing farmers at the grassroots level. As such, while much attention was given to the internal structure and organization of WUAs themselves, the broader federal-, state-, and district-level institutional environments in which WUAs function under remained unchanged. In this manner, WUAs were created without coherence or coordination among farmers, WUAs, donors, and state and federal institutions so that tiered institutional water management policies could be brought into line with local practices and WUA strategies. This is an example of failure to coordinate between institutional scales and, as demonstrated below, leads to interagency mismanagement of resources core to local resource users' livelihood.

Most of the irrigation infrastructure found within the four Khatlon districts are in poor condition and have not been repaired since before the Soviet Union's collapse. Pumps, installed as early as the 1960s, either break regularly or are permanently broken. The large cement-lined canals built in the Soviet era are found broken and cracked, and the dirt canals are filled with silt and vegetation and require excavation in order to restore their function. A few farmers and WUA chairmen explained that excavating dirt canals requires heavy equipment that is near impossible to find. For the majority of respondents, the lack of financial resources to fix irrigation infrastructure and disagreement between farmers and government regarding who is responsible for repairing irrigation infrastructure are at the foundation of impediments that prevent seasonal crop irrigation and therefore jeopardize livelihoods. For most respondents the poor condition of irrigation infrastructure had a stronger

[1] All Water User Associations that were surveyed in the Khatlon province were created by Winrock International/USAID's Water Users Association Support Program and Family Farming Program between 2005 and 2011.

influence on water reliability than environmental or management factors. A majority of farmers responded that they experience water shortages during the irrigation season and have lost crops due to water shortage, and in general water delivery is highly unpredictable. Irrigation infrastructure from the Soviet era was built and geographically organized for singular, centrally managed, large-scale farms, not the smaller, individually managed farms that are now characteristic of Tajikistan. The condition and spatial arrangement of irrigation infrastructure have led to a host of contemporary problems related to water use, access, management, competition for water resources, salinization, waterlogging, and erosion. These problems have led to inequitable and unpredictable supply of water resources. In order to combat poor water access due to decaying, unsatisfactorily arranged infrastructure, a few WUA chairmen reported that farmers have broken and removed gates in order to receive more water than their neighbor and that new irrigation instruments were stolen. A majority of farmers responded that competition for water resources is the cause of conflicts with other farmers. Access and scarcity have limited the capacity of farmers to devise or undertake adaptive actions. All responded that if water access is more predictable, they would diversify the types of crops they grow. The above concerns voiced by famers regarding the poor physical condition of infrastructure are the types of feedbacks from local users that tiered institutions should use to inform a response and tailor policies aimed at mitigating livelihood insecurity.

Water shortage and unpredictable delivery are caused by the physical condition of infrastructure as well as institutional factors. Integration between water users, WUAs, and state organizations is disorganized and leads to livelihood insecurity by causing weak water access and delivery among Khatlon farmers. Khatlon farmers in the four districts surveyed pay two separate water-use fees: an annual irrigation service fee paid to the *Vodhoz* (district water department created during the Soviet era) and an annual, biannual, or monthly WUA membership fee. All irrigation infrastructures are owned by the *Vodhoz*. The majority of farmers, WUA chairmen, and municipal officials responded that their district's water department has the most power over water access in a community. The majority of farmers noted that the *Vodhoz* actually offers nothing in the way of infrastructure repair in return for paid irrigation service fees. Key respondents report that livelihood insecurity is heightened because most farmers are unable to pay their irrigation service fee and are therefore fiscally in debt to their district's *Vodhoz*. The *Vodhoz's* failure to reprieve farmers' weak access to water while concurrently causing farmers to become indebted is an institutional failure that also fouls farmers' relationship with another tiered institution, the WUA. Key respondents reported that farmers are hesitant to pay WUA membership fees because they already pay a fee to their *Vodhoz* and do not receive any benefit in return. WUA chairmen, WUA members, and municipal officials explained repeatedly that WUAs are ineffective because they do not have adequate financial resources to fix and maintain irrigation infrastructure. Most farmers replied that either they do not pay their WUA fee or pay it but do not receive any benefit from paying. Often WUAs are unable to regulate water resources sufficiently so that only farmers that pay their membership fees receive water. A few WUA chairmen reported that they did not cut off water access to farmers who did not pay their WUA membership fee because without water, those farmers would lose their crop and be placed further into debt. Farmers are often already in debt to private investors and suppliers and therefore without liquid capital to actually pay for water. Attempts at coordination between WUAs and their *Vodhozes* in order to address weak and unpredictable access to water resources among farmers have failed. A few WUA chairmen were unsuccessful with their requests that their Vodhozes transfer responsibility for infrastructure repair and fee collection to each chairman's WUA. In instances where responsibility for irrigation service fee collection was transferred to a WUA, the WUA was taken to court over debts owed to their *Vodhoz*. The issue of livelihood insecurity due to water fees not improving water access in Khatlon province is an example of poorly articulated tiered institutions failing to receive or respond to feedbacks generated by local water users. The lack of adaptive responses targeted at these user-defined causes of water insecurity is the product of malfunctioning within the feedback-response loop. This loop undergirds the collaborative policymaking process and requires participation from both secure and insecure users in order to function properly. Poorly articulated multitiered institutional arrangements coupled with failure to truly engage a range of Khatlon stakeholders in the policymaking process serve to erode farmer livelihood and the capacity of provincial water institutions. This is illustrated in greater depth below.

Farmers in the four Khatlon districts are drowning in a cycle of debt. Often unable to repay investors and suppliers, farmers are forced to meet their debt obligation by re-upping at a higher debt burden with the same investor or supplier, plunging them deeper into a debt cycle. Some farmers do not pay their WUA membership fees because they view it as another financial contract that only serves to plunge them further into debt. Livelihood insecurity caused by weak water access and debt between farmers and tiered institutions is exacerbated when state organizations are indebted to another state organization. In an example of broken coordination between water and energy management institutions, a few municipal mayors and WUA chairmen reported their *Vodhoz* uses collected irrigation service fees to pay off the department's debt to the district energy department. Without debt payments the energy department will stop electricity service to the water department, and farmers will be unable to irrigate with electric pumps. A few farmers reported having irrigation water only a few months per year because their *Vodhoz* was indebted, and electricity rationing in the region constrained access to water resources. In other instances, a few farmers reported paying both an irrigation service fee to their *Vodhoz* and a pump-operation fee to their district energy department. In another example of dissonance between local resource users and tiered institutions, one WUA chairman reported that their regional WUA normally collects electricity fees for the region, but in 2012 the *Vodhoz* collected them instead.

The creation of livelihood insecurity and weak access to irrigation water in Khatlon's four provinces is twofold: (a) eroding anachronistic irrigation infrastructure mechanically restricts water access and (b) discord between energy and water resource agencies institutionally restricts water access because interagency debt causes disruption in electrical service and, ultimately, water delivery. Attempts to improve livelihood security by renovating irrigation infrastructure are obstructed because the funds that a *Vodhoz* should use for this purpose are instead used to service the debt to the district energy department. Unfortunately the feedback loop between local users and tiered institutions in Khatlon province's four districts only serves to compound problems by passing the consequences of debt burden onto local farmers instead of enabling the emergence, generation, and implementation of adaptive actions able to repair infrastructure, improve water access, and reduce insecurity. Tiered institutions in the Khatlon province's four districts are creating water scarcity where it need not exist. The plight of local users has not generated an adaptive response nor has an effort been made to alleviate the dual pressures on livelihood caused by infrastructural and institutional failure. Admitting institutional failure and then including WUA members and chairmen in the dialogue regarding reforms to reconcile user and interagency discord might enable both users and agencies to escape from the circle of debt that threatens to drown them.

20.4 The Arid Region Case: Rayón, Sonora, Mexico

Rayón, Sonora (29.7170° N, 110.5830° W), is found in the Mexican portion of the Sonoran Desert and is located in a narrow valley alongside the once perennially flowing Rio San Miguel, a tributary to the larger Río Sonora. The Sonoran Desert's precipitation pattern is regulated by the North American Monsoon, resulting in a bimodal precipitation distribution that produces winter and summer rainfall. The intensity of precipitation and drought brought by the North American Monsoon is determined by dynamic interaction between Pacific basin atmospheric and oceanic conditions (Loik et al. 2004; Comrie and Glenn 1998; Leathers et al. 1991). The Pacific Decadal Oscillation (PDO) and El Niño Southern Oscillation (ENSO) combine to drive the cyclical switching between wet and dry periods, strong or weak monsoons, and other alternations to temperature and precipitation patterns. Rainfall volume and timing are variable, especially at fine spatial scales (Comrie and Broyles 2002), but generally winter precipitation is widespread and associated with frontal systems, while summer rainfall from thunderstorms is each year's precipitation maximum and more localized and intense (Sheppard et al. 2002; Adams and Comrie 1997). The reason precipitation timing and volume are variable in the Sonoran Desert is because its topography is also highly complex with mountain ranges, landscapes, and elevation changes. Rayón is located west of the Sierra Madre Occidental, and the majority of ranchers report that the clouds that bring the summer monsoons arrive through the valley from the southeasterly direction, confirming NA Monsoon models (Fawcett et al. 2002; Adams and Comrie 1997). As is common of arid zones, evaporative demand exceeds precipitation because on an annual basis, the potential for the atmosphere to evaporate water is always greater than the amount of precipitation delivered (Lauenroth and Bradford 2009).

Rayón has approximately 1,500 residents (Secretaria de Gobernación 2013), and the primary economic activity is ranching interspersed with a few who farm non-pasture crops. Most are smallholders farming on 5 ha or less, while others are able to support stocking rates of 600 cattle or more. Since its founding in 1628, agriculture has been the town's primary economic activity, relying on seasonal rains and the Rio San Miguel's surface waters to water and flood irrigate their crops. In the mid-twentieth century, diesel and then electric pumps were permitted access to groundwater by drilling wells. In 2012 groundwater and surface water are used to grow crops or pasture. Farmers and ranchers in Rayón use dirt canals, cement-lined canals, tube/pipe, or some combination of each to deliver irrigation water to their parcels. The majority of ranchers responded that the climate has changed in the past years. Results varied, but ranchers report the time range for their observed climatic change is anywhere from 40 to 2 years prior to the 2012 monsoon. The majority of reported changes are similar to those expected to result from climate change: less annual precipitation, increased temperatures, and tree die-off. Global-change-type drought and, as alleged by one key respondent, overpumping groundwater by users upstream of Rayón's municipality have combined to cause landscape and hydraulic changes that in turn negatively impact livelihood. The San Miguel River located within the Rayón municipality first ran dry when the uppermost reach lost perennial surface water flow in 1970. The section at the town's entrance, whose reputation among ranchers and townspeople is that this section has the shallowest depth-to-bedrock level within the entire municipality, causing subsurface flows to be pushed toward the surface and transformed into surface flows, went dry for the first time in anyone's memory in 2010. In 2012 the entire reach within the Rayón municipality was without flow and a key respondent and another respondent who previously served as president of the surface water users' association,[2] reported that many of the surface water-fed fields went unplanted that year. A key respondent and former head of the town's potable water rated each individual well in the municipality according to its water level after the 2012 summer monsoon. According to this rating system, 69 % of the municipality's wells do not have enough water for the next growing season because they are dry (21 %) or have water but not enough to last an entire season (48 %). After the 2013 summer monsoon, the same respondent estimates the number of wells without enough water for the next growing season to be higher than the previous year, closer to 75–80 %. It is normal that well water levels fluctuate throughout the year – dropping during times of use and then recuperating after summer or winter rains. One respondent noted that declines in naturally available forage led to clearing mesquite forests in order to plant pasture crops to supplement cattle herds. Another key respondent noted that in 2012, his rain-fed buffel grass field yielded plants half their normal size. Two key

[2] The surface water users' association is formally known as the Rayón *ejido*. An *ejido* is a system of communal agricultural property rights unique to Mexico and given to a group of people (typically landless farmers) by the Mexican government. The system was created by the Constitution of 1917 and eliminated as a constitutional right in 1991.

respondents, "field judges" responsible for permitting cattle movements in and out of the municipality,[3] and a state agriculture official (SAGARPA),[4] citing drought, noted that cattle stocking rates have gradually decreased over the past 6 years.

The causes of water scarcity due to drought in Rayón are likely both human and naturally caused. Responses from tiered institutions, however, are relatively muted and do little to allay the fears of a majority of those interviewed that the sustained drought will lead to livelihood loss. The National Water Commission (CONAGUA) is responsible for permitting new wells and deepening existing ones in addition to education programs about consumption, conservation, and compliance. Some responded that CONAGUA regulation has generally had a positive effect on water management, but one key respondent, a "field judge," noted that the problem is not water use in Rayón but instead the combined water use of the entire Rio San Miguel watershed causing the water table to drop. The "field judge" elaborated and explained the mechanism whereby combined water use within the watershed becomes a problem, especially for downstream users – upstream more and more wells are being drilled and at deeper levels. The agricultural ministry, SAGARPA, is the organization that ranchers contact when they are experiencing water scarcity. SAGARPA has a drought support program in place, but discussions with both "field judges" and the former head of potable water revealed that ranchers view this program as largely ineffective. Ranchers must wait until they are actively experiencing drought before they can apply, often finding they will not receive any support or the support they do receive is for actions that, at the earliest, will not have any effect until the next year. The application process and late or undelivered funds can further delay program benefits (Acuña 2013).

All ranchers in Rayón belong to a ranching association and in times of drought turn to the ranching association for immediate assistance because the response is quicker and actions are more effective at relieving pressures on livelihood. SAGARPA officials at the regional office hold the view that ranchers in Rayón are experiencing an agricultural drought – the stocking rates are too high for available resources and need to be lowered. The majority of Rayón ranchers counter that they are experiencing hydrological drought: the average stocking rate in the municipality is relatively small, 32.79, the median, 17, even smaller (SAGARPA 2010), and to alleviate drought, most ranchers view improved water supply or techniques to more efficiently use available supply as essential.

Ranchers have experienced sustained symptoms of global-change-type drought (less precipitation, higher temperatures, tree die-off, perennial river drying) and drought caused by overextended resources (depleted wells, not enough surface water). In the context of feedbacks between resource users and tiered institutions, ranchers' responses indicate that tiered institutions, e.g., CONAGUA and SAGARPA, have failed to address the real problem of drought and water scarcity

---

[3] These federally appointed officials are officially termed *Juez de Campo* or "Field Judge."

[4] Secretariat of Agriculture, Livestock, Rural Development, Fisheries, and Food is a unit of Mexico's Federal Executive branch.

in Rayón with institutional actions. The dissonance between SAGARPA officials and ranchers regarding the cause of drought (agricultural vs. hydrological) illustrates the disconnect between water users and tiered institutions and how this situation fails to elicit a programmatic, long-term plan to mitigate drought impacts on livelihood among ranchers with high, medium, and small stocking rates. The observation by the "field judge" that the cause of drought is due to combined groundwater use throughout the Rio San Miguel basin indicates a need to manage water resources in a coherent, basin-wide manner, an institutional action that has yet to emerge. Seemingly because feedbacks between local users and tiered institutions go unregistered and vulnerable, downstream users are not consulted or asked to participate in designing programs meant to mitigate vulnerability caused by global-change-type drought or overuse. A coherent, basin-wide response is further hindered by dissonance and failed institutional coordination between the agricultural (SAGARPA) and water (CONAGUA) ministries. In times of water scarcity due to drought, ranchers contact the agricultural ministry, who responds directly to the ranchers, but fails to relay ranchers' drought experience to the water ministry, who is ultimately responsible for articulating a basin-wide water plan to manage local- and basin-level growth and resource use. The failure to produce a coherent, basin-wide management plan is caused by dissonance between water users in the Rio San Miguel watershed and tiered institutions responsible for its administration. This is a clear example of poorly articulated multitiered institutional arrangements coupled with failure to generate truly participatory interaction of stakeholders, as alluded in the introduction above. We return to this point in the conclusions. The failure to produce such a plan has consequence for residents within the basin and the state's capital city, Hermosillo, Sonora. Rapid urban growth in Hermosillo is occurring in the context of fixed or declining water resources availability and urban-rural conflicts over water (Pineda-Pablos et al. 2012). The Rio San Miguel is a main tributary to the Rio Sonora River that, in addition to groundwater, serves as Hermosillo's primary water source. As one Rayón respondent noted, a dry Rio San Miguel also means less water for Hermosillo. Lack of coordination between water users within the Rio San Miguel basin and tiered institutions hinder attempts at IWRM within the basin as well as planning for growth and resource use at interbasin scales.

20.5 Conclusion: IWRM, Local Resource Users, Multitiered Institutions, and Policy

This chapter demonstrates the importance of coordination between local resource users and institutions at multiple levels in order to facilitate a collaborative policymaking process aimed at identifying and implementing short- and long-term adaptive actions to manage water, environment, and livelihood security. Water security is predicted to be one of the most pressing resource challenges of our time, and all management strategies – institutional and technical – must be on the table in order to mitigate the threats water scarcity poses to humans and the environment. Adaptive management is a specific IWRM mechanism that can be used to evaluate potential causes of water insecurity and aid practitioners when they seek to overcome water management challenges. Coordination between local users and tiered institutions to facilitate collaborative policymaking is not a panacea; rather it is a diagnostic tool that can be used to identify one cause that, among a possible multitude, contributes to water insecurity. Ecological and social issues are intrinsically interwoven to form a complex pattern of challenges to livelihood and environmental security as well as potential policy and management options suited to support livelihood and ecosystems.

Our examples show that under a range of conditions (from arid to water abundant), poorly articulated integration between users and tiered institutions can limit IWRM development, fail to mitigate threats caused by water scarcity, and compound livelihood degradation by causing scarcity itself. Under the discordant integration presented by each case, short-term coping actions were curtailed or nonexistent and desperately needed long-term, cross-scale adaptive strategies and policies which are impossible given each case's present institutional condition. Though the presented cases are regional, the outcomes from failure to mitigate livelihood degradation due to water scarcity are real, even in water-rich areas, and particularly troubling given that climate change and increased competition and demand for water are predicted to become more prevalent among the Earth's communities. Localized instances of global-change-type symptoms and water scarcity may one day become standard meso- and macroscale conditions.

Institutions and decision-making among all stakeholders will be an explicit component of our capacity to respond with programs, policies, and actions able to deal with the dual pressure on water resources posed by climate change and heightened demand while also reconciling economic efficiency, social equity, and environmental sustainability. Cohesion and cooperation between local users and tiered institutions is one institutional mechanism of many that can, and *ought* to, be used. In the short and long term, collaborative policymaking between local users, especially the most vulnerable, and tiered institutions will best ensure that IWRM is able to deliver on its promises and remain a sustainable alternative to business-as-usual-caused water governance failures.

The specific mechanisms by which water security impacts economics, equity, and environment are a general class of problem that concerns people today and in the next decades will continue to need to be addressed. Addressing water scarcity and its impacts is a process and will require continual evaluation and fundamental restructuring of how stakeholders and institutions create collaborative policymaking and remain cohesive throughout the process (Varady et al. 2013). Collaborative policymaking is faced with social and geographical challenges that threaten its implementation as a means to create alternatives to business as usual. There are no simple, clear-cut answers or protocols to prevent business as usual nor some formula based on a combination of actions that will cause water insecurity to disappear. However, we should continue to pursue participatory and collaborative processes (e.g. Varady et al. 2013). In the cases presented (arid, water rich, coupled resource), exclusion and dissonance between local resource users and tiered institutions caused real failures that have real impacts on people's lives. We suggest that this should renew and strengthen our commitment to participation and the collaborative process and that evaluating and facilitating coordination between local users, especially the most vulnerable and tiered institutions – in particular those responsible for administering at basin, district, and municipal levels – is a good place to start.

In sum, we have raised the need for a systems thinking, adaptive approach to water insecurity challenges. Institutions that operate at the intersection of local users and state and non-state actors have the greatest chance of inducing IWRM solutions if the tiered nature of linkages (local level, meso-level, and macro-level) is expressly accounted for and used for adaptive advantage.

Acknowledgments The authors gratefully acknowledge the support of the National Science Foundation (NSF); Coupled Natural-Human Systems grant DEB-1010495; the Inter-American Institute for Global Change Research grant SGP-CRA #005 (supported by NSF grant GEO-1138881); the Fulbright Program and Mercy Corps Tajikistan, especially John Ross; the Tinker Foundation; the National Science Foundation's Pan-American Advanced Studies Institute on Adaptive Energy-Water Management in the Americas (NSF grant OISE-1242209); and the University of Arizona's Udall Center for Studies in Public Policy, for making this research possible. Special thanks to Dr. Shimelis Setegn, Dr. Alan Navarro, Katherine Curl, and Lily House-Peters.

References

Abdullaev K, Akbarzaheh S (2010) Historical dictionary of Tajikistan. Scarecrow Press, Lanham

Acuña D (2013) Reportan Ganaderos Beneficios Por Lluvias. El Imparcial, 10 July 2013, Online edition, sec. Local

Adams DK, Comrie AC (1997) The North American monsoon. Bull Am Meteorol Soc 78(10): 2197–2213. doi:10.1175/1520-0477(1997)078<2197:TNAM>2.0.CO;2

Berkes F (2007) Community-based conservation in a globalized world. Proc Natl Acad Sci 104(39):15188–15193, http://www.pnas.org/content/104/39/15188.short

Carter NT (2010) Energy's water demand: trends, vulnerabilities, and management. Congressional Research Service, Washington, DC

Comrie AC, Broyles B (2002) Variability and spatial modeling of fine-scale precipitation data for the Sonoran Desert of south-west Arizona. J Arid Environ 50(4):573–592. doi:10.1006/jare.2001.0866

Comrie AC, Glenn EC (1998) Principal components-based regionalization of precipitation regimes across the Southwest United States and Northern Mexico, with an application to monsoon precipitation variability. Clim Res 10(3):201–215, http://www.int-res.com/abstracts/cr/v10/n3/p201-215/

Easterling DR, Meehl GA, Parmesan C, Changnon SA, Karl TR, Mearns LO (2000) Climate extremes: observations, modeling, and impacts. Science 289:2068–2074

Fawcett PJ, Stalker JR, Gutzler DS (2002) Multistage moisture transport into the interior of Northern Mexico during the North American summer monsoon. Geophys Res Lett 29(23): 2094. doi:10.1029/2002GL015693

Feed the Future (2012) Tajikistan Fact Sheet. www.feedthefuture.gov

Global Water Partnership Technical Advisory Committee, Agarwal A, de los Angeles MS, Bhatia R, Chéret I, Davila-Poblete S, Falkenmark M et al (2000) Integrated water resources management, TAC Background Papers. Global Water Partnership, Stockholm
Hightower M, Pierce SA (2008) The energy challenge. Nature 452(7185):285–286, http://www.nature.com/nature/journal/v452/n7185/full/452285a.html
Hoerling M, Kumar A (2003) The perfect ocean for drought. Science 299(5607):691–694. doi:10.1126/science.1079053
Intergovernmental Panel on Climate Change (2001a) Climate change 2001: synthesis report, contribution of working groups I, II, and III to the third assessment report of the intergovernmental panel on climate change. Cambridge University Press, New York
Intergovernmental Panel on Climate Change (2001b) Climate change 2007: synthesis report, contribution of working groups I, II, and III to the fourth assessment report of the intergovernmental panel on climate change. IPCC, Geneva
Lam S (2010) Assessment of water user association support program
Lauenroth WK, Bradford JB (2009) Ecohydrology of dry regions of the United States: precipitation pulses and intraseasonal drought. Ecohydrology 2(2):173–181. doi:10.1002/eco.53
Leathers DJ, Yarnal B, Palecki MA (1991) The Pacific/North American teleconnection pattern and United States climate. Part I: regional temperature and precipitation associations. J Clim 4(5):517–528. doi:10.1175/1520-0442(1991)004<0517:TPATPA>2.0.CO;2
Loik ME, Breshears DD, Lauenroth WK, Belnap J (2004) A multi-scale perspective of water pulses in dryland ecosystems: climatology and ecohydrology of the western USA. Oecologia 141(2):269–281. doi:10.1007/s00442-004-1570-y
Mehta A (2008) Integrated water resource management assessment report: Wakal River Basin, India – The enabling environment (Water Related Policies, Laws and Institutional Mechanism). Project Study Series. Global Water for Sustainability Program, Florida International University, North Miami
Melikyan L, Ghukassyan H (2011) Poverty and social impact assessment: energy sector in Tajikistan
Mittal AK (2009) Energy-water nexus: improvements to federal water use data would increase understanding of trends in power plant water use. Report to the Chairman, Committee on Science and Technology, House of Representatives GAO-10-23. United States Government Accountability Office, Washington, DC. http://www.gao.gov/products/GAO-10-23
Molle F (2008a) Nirvana concepts, narratives and policy models: insights from the water sector. Water Altern 1(1):131–156, http://www.water-alternatives.org/index.php?option=com_docman&task=doc_download&gid=20%3E
Molle F (2008b) Why enough is never enough: the societal determinants of river basin closure. Int J Water Resour Dev 24(2):217–226. doi:10.1080/07900620701723646
Molle F (2009) Water, politics and river basin governance: repoliticizing approaches to river basin management. Water Int 34(1):62–70. doi:10.1080/02508060802677846
Molle F, Mollinga PP, Meinzen-Dick R (2008) Water, politics and development: introducing water alternatives. Water Altern 1(1):1–6, http://www.water-alternatives.org/index.php?option=com_docman&task=doc_download&gid=14
Ostrom E (2007) A diagnostic approach for going beyond panaceas. Proc Natl Acad Sci 104(39):15181–15187, http://www.pnas.org/content/104/39/15181.short
Overpeck J, Udall B (2010) Dry times ahead. Science 328:1642–1643
Pineda-Pablos N, Scott CA, Wilder M, Salazar-Adams A, Díaz-Caravantes R, Brito L, Watts C, Moreno JL, Oroz L, Neri C (2012) Chapter 5: Hermosillo. In: Wilder M, Scott CA, Pineda-Pablos N, Varady RG (eds) Moving forward from vulnerability to adaptation: climate change, drought, and water demand in the urbanizing Southwestern United States and Northern Mexico/Avanzando Desde La Vulnerabilidad Hacia La Adaptación: El Cambio Climático, La Sequía, y La Demanda Del Agua En Áreas Urbanas Del Suroeste de Los EEUU y El Norte de México. Udall Center for Studies in Public Policy, The University of Arizona, Tucson

Rahaman MM, Varis O (2005) Integrated water resources management: evolution, prospects and future challenges. Sustain Sci Pract Policy 1(1):15–21, http://sspp.proquest.com/archives/vol1iss1/0407-03.rahaman.html

Schlager E, Ostrom E (1992) Property-rights regimes and natural resources: a conceptual analysis. Land Econ 68(3):249–262, http://www.jstor.org/stable/10.2307/3146375

Scott CA, Pasqualetti MJ (2010) Energy and water resources scarcity: critical infrastructure for growth and economic development in Arizona and Sonora. Nat Resour J 50:645, http://heinonlinebackup.com/hol-cgi-bin/get_pdf.cgi?handle=hein.journals/narj50&section=34

Scott CA, Pierce SA, Pasqualetti MJ, Jones AL, Montz BE, Hoover JH (2011) Policy and institutional dimensions of the water–energy nexus. Energy Policy 39(10):6622–6630. doi:10.1016/j.enpol.2011.08.013

Scott CA, Meza FJ, Varady RG, Tiessen H, McEvoy J, Garfin GM, Wilder M, Farfán LM, Pablos NP, Montaña E (2013) Water security and adaptive management in the arid Americas. Ann Assoc Am Geogr 103(2):280–289. doi:10.1080/00045608.2013.754660

Secretaria de Gobernación (2013) Distribución de La Población Por Tamaño de Localidad, 2010. Instituto Nacional Para El Federalismo y El Desarrollo Municipal, Sistem Nacional de Información Municipal, 8 Oct 2013. http://www.snim.rami.gob.mx/#

Secretariat of Agriculture, Livestock, Rural Development, Fisheries and Food (SAGARPA) (2010) Livestock census – Rayón, Sonora, Mexico

Setegn SG, Rayner D, Melesse AM, Dargahi B, Srinivasan R, Wörman A (2011) Climate change impact on agricultural water resources variability in the Northern Highlands of Ethiopia. In: Melesse AM (ed) Nile River basin. Springer, Dordrecht, pp 241–265, http://dx.doi.org/10.1007/978-94-007-0689-7_12

Sheppard PR, Comrie AC, Packin GD, Angersbach K, Hughes MK (2002) The climate of the US Southwest. Clim Res 21(3):219–238, https://webspace.utexas.edu/wsc226/GeoPapers/Climate/Sheppard2002_US-SW-Climate.pdf

United Nations Economic Commission for Europe (2007) Our waters: joining hands across borders: first assessment of transboundary rivers, lakes and groundwaters. United Nations Publications, Geneva

USAID (2012) Feed the future Tajikistan fact sheet. United States Agency for International Development. http://feedthefuture.gov/sites/default/files/resource/files/ftf_factsheet_tajikistan.pdf

Varady RG, Scott CA, Wilder M, Morehouse B, Pablos NP, Garfin GM (2013) Transboundary adaptive management to reduce climate-change vulnerability in the western U.S.–Mexico border region. Environ Sci Pol 26:102–112. doi:10.1016/j.envsci.2012.07.006

Varis O, Rahman MM (2008) Central Asian waters: social, economic, environmental and governance puzzle. Water and Development Publications – Helsinki University of Technology, Helsinki

World Bank Poverty Reduction and Economic Management Unit Europe and Central Asia (2013) Tajikistan: reinvigorating growth in the Khatlon Oblast

Yang S, Ding F, Ding Z (2006) Pleistocene chemical weathering history of Asian arid and semi-arid regions recorded in loess deposits of China and Tajikistan. Geochimica 70:1695–1709

Young OR, King LA, Schroeder H (2008) Institutions and environmental change: principal findings, applications, and research frontiers. MIT Press, Cambridge, MA

Zvi L, Sedik D (2008) The economics of land reform in Tajikistan

Chapter 21
The Environmental Regulatory Shift and Its Impact on Water Resources Management in Latin America

Juan Bautista Justo and Liber Martín

Abstract The new environmental paradigm that has gained momentum in recent decades has had a concrete impact on water resources management in Latin America. From a critical perspective, this chapter aims to identify and review ten different aspects of the new scenario. In doing so, it points to two main conceptual transitions: firstly, the transformation of the conceptual basis of water management and, secondly, the regulatory power shifts over water sources. The authors conclude that a strategy on water governance that does not take into account those contrasts exposes itself to failure.

Keywords Environmental regulatory • IWRM • Latin America • Environmental paradigm

21.1 Introduction

There has been a significant change in the approach to water management over the last few decades. As a result of the detection of the adverse natural and social consequences of the economic paradigm centered on the productive role of water, an environmental vision has gained ground in most countries.[1]

Under the economic paradigm, water was considered almost exclusively as an input in the production process of goods and services, a mere object of which to extract the maximum economic benefit. This one-dimensional approach was impacted by

---

[1] By "paradigm," we refer to decision models that have a prior status to rules and influence decisions. Thus, the term is used to identify the pattern of presuppositions that guide human actions in a certain time and place.

J.B. Justo (✉)
Universidad Nacional del Comahue, Neuquén, Argentina
e-mail: juanjusto@speedy.com.ar

L. Martín
CONICET/Universidad Nacional de Cuyo, Mendoza, Argentina
e-mail: libermartin@gmail.com

S.G. Setegn, M.C. Donoso (eds.), *Sustainability of Integrated Water Resources Management*, DOI 10.1007/978-3-319-12194-9_21

sustainability requirements, giving way to the implementation of a catalog of principles aimed at avoiding the destruction of ecosystems by human intervention.[2]

These changes have not come about by chance. They are the result of the "water crisis," namely, a growing scarcity of adequate water, increasing demand, climate change, overpopulation, pollution, unequal distribution, multiple uses, and new technologies, among other factors that have had a direct impact on legislation and water management. All these aspects have led to growing conflicts between social groups that compete for the ownership and the use of water resources, shaping a scenario significantly more complex than that in which the formulation of the traditional water management model was developed.

The environmental paradigm is based on an idea of complex interaction of elements within a system that takes into account the individual and collective, present and future, effects of human actions. It is a systemic conception that gives prominence to the cumulative effect of each action.

The whole theoretical structure of modern Western culture was built on the basis of the individual. That rational, independent, and self-interested subject, only linked to others through exchange relationships (of civil powers or commodities), is the main character of the economic paradigm. That rational, independent, and self-interested subject kingdom was in property rights, the only possible incentive for progress.

After the revolutions of the eighteenth century, the new legal order sought to free the individual from all "external" bondage in order to allow him to unfold in different fields (especially the economic). This purpose led to a system centered on the idea of rights held by independent and rational individuals pursuing their own goals in competition. Nature happened to be one of the objects bound to appropriation by those unstoppable subjects, and the usufruct of the commons has been the trend since.[3] On the other hand, the environmental approach demands a collective look,[4] starting to recognize the environment as a common.

The environmental paradigm is a direct result of the negative consequences of that individualistic view, especially by underestimating the collective effects of the economic actions. To that end, the idea of externality has played a key role in building the environmental paradigm, because it has forced us to make visible the many shared consequences of each singular behavior. Usually ignored, this aspect was incorporated into both the economic analysis and the law and helped to abandon the State neutrality in respect to this allocation of costs.

In this context, the chapter analyzes two aspects of the impact of the new paradigm in Latin America's way of understanding water resources that are not normally addressed in the various articles which appear every year[5] on the subject,

---

[2] That is, prevention, precaution, intergenerational equity, sustainability, or cooperation.

[3] Horkheimer, Max, *Eclipse of Reason*, New York, Oxford University Press, 1947, pp. 97 y ss.

[4] Arnold, Craig A., "The Reconstitution of Property: Property as a Web of Interests", *Harvard Environmental Law Review*, Vol. 26, Cambridge, MA, 2002, p. 283.

[5] Vgr. Martinez-Santos, P. Aldaya, Maite M., Llamas, M. Ramón (Eds), *Integrated Water Resources Management in the 21st Century: Revisiting the paradigm*, CRC Press, 2014. Specifically referred to Latin America; see the Special Issue IWRM in Latin America, *International Journal of Water Resources Development* Vol. 24, Issue 1, 2008.

such as, the transformation of the conceptual basis of water management and the regulatory power shifts over water sources.

These issues are particularly relevant because they are directly linked not only to the legal structure of the water regime but also to the behavior of public agencies, courts, and civil society in relation to water resources. Each of the ten analyzed factors influences the way in which those actors perform their role in water management, and the chapter intends to show the main changes in their decisions as a result of the current prevalence of the environmental view. Basically, each one of the topics studied here shows how the paradigm's change reflects in the daily actions of water governance makers.

In regard to the transformation in the fundamentals of water management models in the region, we'll study some examples of this phenomenon that have a common ground in the response to the deficits of the economic paradigm in order to address sustainability challenges.

First, we'll analyze how the new conceptualization of water resources as collective goods is intended to cope with the shortcomings of the water's ownership model. In the same vein, the principle of intergenerational equity aims to reverse the negative effects of the tendency of modern societies to favor their current interests at the expense of future generations. Furthermore, the recognition of access to water and sanitation as a human right tends to close the gap that the commoditization of water has produced in accessibility and universality matters. In terms of public authority and police power, the result is that the foundations of the regulatory capacity of the State over water has moved from ownership to the duty to protect internationally recognized human rights, thereby varying both the fundamentals and the limits of the powers of public authorities.

The second aspect, shifts in the regulatory power holders, expresses the institutional implementation of the substantial changes seen above. In fact, globalization of the legal order has involved an unprecedented transfer of regulatory capacities from nation States to the international arena, something that has internally strengthened the role of courts and fostered centralizing tendencies in the territorial distribution of power. In the new scenario, the State no longer has the monopoly of water resources management, because civil society is gaining a considerable role in this area. However, neither the State nor civil society is the sole recipient of all duties: corporate responsibility opens a new range of subjects to whom rights holders may demand some performance requirements to ensure the sustainability of human intervention in the context of water management.

In this context, Latin America follows global trends in the evolution of water management, but also presents particularities, like the rise of specific water conflicts, the expansion of environmental judicial activism, the existence of the inter-American human rights system, the colonial reminiscences, the conceptualization of collective goods and indigenous population, and the ongoing debate of the water rights, among other elements.

The success of the new public policies to manage the growing conflict around water resources depends on addressing these changes. If those policies continue resting on a scenario that has evolved dramatically, their ability to generate a real and sustainable framework of water governance will be limited. Therefore, observing, at least briefly, the new picture is of paramount importance.

21.2 The Transformation of the Conceptual Basis of Water Management

21.2.1 The Foundations of Water Resources Management: From Ownership to Collective Goods

In Latin America, the legal regime of water resources was structured around water ownership, with its core in the distinction between public and private waters.[6] The protection of common goods was embodied in the idea of appropriation by the State, as an exception to the rule of private property.[7]

[6] The influential French Civil Code of 1804 endorsed this distinction. Public waters were those which were considered to be "navigable" or "floatable" and belonged to the public or national domain. Their use required a government permit or authorization. Private waters, which were those located below, along or upon privately owned land, could be freely utilized subject to certain limitations of a statutory nature such as servitudes and rights of way. The right to use such private waters, both surface and subsurface, derived from land ownership which recognized the owner's right to use at pleasure the water existing upon his land without any limitation.

[7] As it has been explained, in continental Europe, the Roman legal model of *res communes*, defined the open to free use as a natural right belonging to every citizen, was progressively abandoned in favor of a state ownership model as a derogatory regime compared to private ownership. This choice necessarily led to the establishment of administrative systems of water rights, whereby private property rights on water are debarred and public authorities enjoy the exclusive power of water allocation through temporary and revocable permits. At first, the right to use water resources vested in public ownership did not entail a previous grant of concession. Water resources vested in public ownership were open to the public and of free use for the citizens (for watering, cattle drinking, washing, fishing, and other domestic uses) as far as such uses did not hamper the public uses (navigation, timber floating, irrigation, or other industrial or productive uses) assigned to the public water resources. When new ways of water use became possible, thanks to technological innovation, water power became scarce and contended (Cavallo Perin, R. and Casalini, D., "Water Property Models as Sovereignty Prerogatives: European Legal Perspectives in Comparison", *Water* 2010, 2, 429–438). Thus, the private law framework of regulation proved to be unsuitable for an efficient allocation of water rights.

Following the French model,[8] the system was organized over the primacy of property rights and the private use of public goods through administrative entitlements (Míguez Núñez 2008). In the formative period of most of the countries of the region, water laws were aimed at a central function: to provide security to the holders of rights to use water and coordinate their different purposes,[9] always favoring the economic usage of natural resources, even at the expense of environmental and social functions (Martin 2010b).

In that commoditized vision of natural resources underlay the idea that property rights played a civilizing role that allowed humanity to move from a late-developmental stage – marked by the collective ownership of the native American people – to an upper one focused on the goods of trade.

Adam Smith proclaimed, for instance, that "The four states of society are hunting, pasturage, farming, and commerce... The age of commerce generally succeeds that of agriculture. As men could now confine themselves to one species of labor, they would naturally exchange the surplus of their own commodity for that of another of which they stood in need."[10]

Private property was, thus, the key for progress of civilization. "By breaking the grip on power of elites wedded to old regimes and re-ordering incentives to induce dynamic efficiency, it was possible to move an entire society forward in history –that is, along the civilizational timeline marked out by Europe's historical

---

[8] European conceptions of water and water law have strongly influenced the development of formal water regimes around the world through the two principal European legal traditions: the civil law and the common law. That influence is a result of the colonizer eurocentrism that began with the conquest of America. The French Revolution inaugurated what decolonial thinking has called the "Second Modernity," in which Europe is attributed a global civilizing mission that is a continuation – secularization by – of the Christianizing mission begun in the sixteenth century (First Modernity). The political philosophy of the Enlightenment was one of the main vehicles for Western Europe global hegemony and the civil codes one of its main instruments. *See* Dussel, Enrique, "Europe, Modernity and Eurocentrism," *Nepantla*, Vol. 1(3), Duke University Press, Durham (NC), 2000; Escobar, Arturo, "Mundos y conocimientos de otro modo: el programa de investigación de modernidad/colonialidad Latinoamericano," *Tabula Rasa*, Vol. 1, Bogotá, 2003; Lander, Edgardo (ed.), *La colonialidad del saber: eurocentrismo y ciencias sociales*, Buenos Aires, CLACSO, 2000; Mignolo, Walter, *Local Histories, Global Designs: Coloniality, Subaltern Knowledges and Border Thinking*, Princeton, Princeton University Press, 2000; Quijano, Aníbal, "Coloniality of Power, Ethnocentrism, and Latin America," *Nepantla*, Vol. 1(3), Duke University Press, Durham (NC), 2000; De Sousa Santos, Boaventura, *Toward a New Legal Common Sense. Law, globalization, and emancipation*, Londres, Butterworths, 2002; *La globalización del derecho: los nuevos caminos de la regulación y la emancipación*, Bogotá, ILSA Ediciones Universidad Nacional de Colombia, 1998; Grosfoguel, Ramón, "Descolonizando los universalismos," Castro-Gómez, Santiago y Grosfoguel, Ramón (eds.), *El giro decolonial*, Bogotá, Pontificia Universidad Javeriana, Universidad Central y Siglo del Hombre Editores, 2007.

[9] Martín, L. "Cuando el río suena... el establecimiento de una nueva matriz disciplinar para el derecho argentino de aguas", *Revista Electrónica del Instituto de Investigaciones "Ambrosio L. Gioja"* – Año IV, Número 5, 2010a.

[10] Smith, A. *Lectures on jurisprudence* 459–60 (R.L. Meek et al. ed.) (Oxford University Press ed. 1978.

experience."[11] On the contrary, collective rights over resources were a synonym of an inferior stage of civilization (native Americans) where no incentive exists for the improvement in productivity of real property.

This approach envisaged the holder of property rights as the "virtuous man" upon which the new institutions should be built. In the new ideology, law was conceived as a regulator of a society formed by citizens whose economic aspiration was the establishment of private property (Locke 1698). The regulation of water did not escape this logic and that's why even State's power over water had to resort to ownership.

Facing the traditional division between public and private goods, in recent years – and as a result of the advance of the environmental paradigm – a new category of goods has been gaining ground. Collective goods do not belong to the State or private parties exclusively and are insusceptible of division or appropriation. In their case, the regulation capacities are derived from their communal status and the need to protect fundamental rights associated with them. Conceptually, ownership is not the key factor in the management of this kind of goods.

It is no longer possible to speak of exclusive property of the State in water resources. Between the public and private field, there is a social sphere where we place the collective goods that belong to the whole community.

Thus, from the traditional binary structure (division between public and private goods), collective goods introduce a third sphere – the social one – which gives rise to a unitary category (common assets managed in trust by the State).

Paradoxically, this "new" approach implies – for the traditional and colonial view – a kind of move backwards, a return to collective management schemes banished by the conquest of America.

21.2.2 The Subject of Water Resources: From the Individual of the Present to the Community of the Future

The traditional economic vision led to a system where the public ownership of the resource and the possibility of appropriation of water for exclusive uses were the key factors. Until the late twentieth century, that interaction between individual appropriation and public needs was always thought within a single generation. The consequences of human intervention in the medium and long term were not taken into account.

The belief in the infinite nature of water resources led, in effect, not to address the inter-temporal variables of human interventions on them. This idea was combined with the difficulties of the contractualist philosophical-political theory (a trend that prevailed in the making of rule of law) to include past, present, and

[11] Purdy, J, "Property and Empire: The Law of Imperialism in Johnson v. M'Intosh", *George Washington Law Review*, 2007, 330–31.

future generations within the same ethical community with corresponding duties and obligations.

In the context of the contractualist conceptions, it was complex to extend the scope of ethics to a community of past, current, and future generations. This is because the conclusion of contracts has been understood as made between current agents and in terms of reciprocity, disconnecting the present generation of its historical insertion. The most direct consequence of such ideas is the artificial view of man as an isolated subject and a synchronous idea of commutative justice, which considers that offsets occur within a cohort and at the same time.[12]

The process of environmental destruction that began with the industrial revolution changed dramatically the static approach referred above. We speak now of an ethical link between generations that adds dynamism to the analysis and forces us to look upon the needs of people whose parents have not been born. We have been forced to acknowledge a tendency of each generation to give priority to its wishes even at the expense of the rights of those who are yet to come. That is why protection is necessary by way of the rights of future generations.

As a result, we have moved from a system in which individuals were the essential subjects interacting with the public interest in a static way (current generation) to a regime where the community, current and future, has the main role. The economic vision omitted any consideration of the connection between generations, which were left within their own temporal context. The exploitation of water resources was not limited by principles of intergenerational justice, and therefore, restrictions related to sustainability were nonexistent.

By contrast, today the notion of sustainable development prevails, understood as development that meets the needs of present generations without compromising the ability of future generations to meet their own needs. In the new model, the actors in water management have changed. Its main character is no longer the temporarily isolated individual but a generationally connected community, bound by ethical and regulatory linkages. In this regard, the authorities and private actors face tangible limits in resource management in order to ensure appropriate use and enjoyment of the environment by the present and future generations.

21.2.3 Legal Title: From Individual Rights to Diffuse Interests

Under the context in which Latin American water law was sanctioned, only individual interests, qualified as property rights, enjoyed full legal protection. In the absence of concession, the law did not recognize the individuals a right of this kind in relation to public waters, so there was no title to demand the preservation of

---

[12] Ost F. and van Hoecke M., "Del contrato a la transmisión. Sobre la responsabilidad hacia las generaciones futuras", *Doxa*, 22, 1999.

the resource. The endorsement of the so-called collective goods collides with that conception, because it emphasizes the recovery of the communitarian side of rights and in doing so produces fractures in a legal system designed to protect individual property rights (Ferrajoli 2004; Macpherson 1978).

The shift towards the environmental approach has extended the range of rights under State supervision, adding *diffuse rights over collective goods*. These collective goods belong to the whole community, are indivisible, and do not support any exclusion. There is no individual ownership over the good, and no single entitlements are at stake. They simply do not belong to the individual sphere but rather to the social one. Clearly, water resources enter this category making it necessary to reformulate the institutions that govern them, in so far as they were based on the opposite assumption.

Under the economic paradigm, only the State – as the owner of water – and the holder of administrative rights to use it (concessions) had a voice on water management. The concept of diffuse rights over collective goods broke that restriction. The key factor for influencing is no longer in ownership. Rather, any person who shows interest, as a community member, in protecting those collective goods is authorized to defend the water resources. This means a sweeping extension of the list of people with weight to fall on the management of water resources.

21.2.4 Conceptualization of the Resource: From Commodity to Human Right

Although the right of access to water has long been recognized at national level in many forms, its consideration as a human right involves, also in this case, a paradigm's change. That is so because giving water this classification means subjecting it, and its related activities, to a normative system, the one of human rights treaties, that limits the scope of action of the State.

Human rights covenants have a number of unique features that differentiate them from traditional treaties.[13] Basically, this type of convention establishes a new legal order – and not just reciprocal engagements –[14] consisting of a series of objective obligations that are binding between States even without proof of the involvement of a national entity or individual and must be secured by a joint action of States

[13] See generally Teitel, Ruti, *Humanity's Law*, Oxford University Press, 2011.

[14] This is so to the extent that the making of a treaty of this nature does not respond to a negotiation under guidelines of reciprocity, a compromise of competing interests, but to the existence of common challenges, the defense of human rights, for whose achievement a joint effort is needed. After the ratification of a human rights treaty, there are no areas of state activity exempt from the network of commitments and guarantees. The system exists "to recognize rights and freedoms to the people and not to empower states to do so" (IACHR, AO-7/86, *Exigibilidad del derecho de rectificación o respuesta – artículos 14.1, 1.1 y 2 Convención Americana sobre Derechos Humanos*, 1986, para. 24. Malan, Koos, "The nature of human rights treaties: Minimum protection agreements to the benefit of third parties," *De Jure*, 2008, 82).

(collective enforcement).[15] At the same time, human rights obligations are grounded on constitutive and substantive norms representing the adherence to a normative system, not on an exchange of rights and duties.[16]

Human rights treaties, therefore, provide a comprehensive legal system for all areas of government activity.[17] In this line, the United Nations Sub-Commission on the Promotion and Protection of Human Rights has affirmed the "centrality and primacy of human rights obligations in all areas of governance and development."[18]

At present, there is considerable consensus that access to clean water is an essential human right protected by international law. This is confirmed in the resolution 64/292 of the General Assembly of the United Nations and the General Comment 15 of the United Nations Committee on Economic, Social, and Cultural Rights (CESCR).[19] Directly related to the management of water resources, the later established the following guidelines derived from the human right to water and sanitation (HRWS):

(a) *Legal nature*: Water is a limited natural resource and a public good fundamental for life and health.[20]
(b) *Allocation criteria*: Priority in the allocation of water must be given to the right to water for personal and domestic uses. Priority should also be given to the water resources required to prevent starvation and disease, as well as water required to meet the core obligations of each of the Covenant rights.[21] However, CESCR noted the importance of ensuring sustainable access to water resources for agriculture in order to realize the right to adequate food. State parties should ensure that the allocation of water resources facilitates access to water for all members of society. Inappropriate resource allocation can lead to discrimination that may not be overt.[22]
(c) *Environmental consequences of the HRWS*: The right to water contains both freedoms and entitlements. The freedoms include the right to maintain access to existing water supplies necessary for the right to water, and the right to be

[15] ECHR, *Ireland v. United Kingdom*, 1978, para. 239; *Loizidou v. Turkey*, 1995, para. 70; *Mamatkulov & Askarov v. Turkey*, 2005, para. 100; IACHR, AO-1/82, *"Otros Tratados" Objeto de la Función Consultiva de la Corte (artículo 64 Convención Americana sobre Derechos Humanos)*, 1982, para. 24; *Ivcher Bronstein v. Peru*. 1999, para. 42.

[16] Provost, Rene, "Reciprocity in Human Rights and Humanitarian Law," *British Yearbook of International Law*, 1994, Vol. 65, p. 386.

[17] Human rights *erga omnes* obligations "are grounded not in an exchange of rights and duties but in an adherence to a normative system" (Provost, René, "Reciprocity in Human Rights and Humanitarian Law," supra, 386).

[18] UN, "Human Rights as the Primary Objective of Trade, Investment and Financial Policy," Doc./E/CN.4/Sub.2/RES/1998/12.

[19] CESCR, General Comment N° 15 (2002), *"The right to water (arts. 11 and 12 of the International Covenant on Economic, Social and Cultural Rights),"* E/C.12/2002/11.

[20] CESCR, GC 15, para. 1.

[21] CESCR, GC 15, para. 6.

[22] CESCR, GC 15, para. 14.

free from interference, such as the right to be free from contamination of water supplies.[23] State parties should ensure that natural water resources are protected from contamination by harmful substances and pathogenic microbes. Likewise, State parties should monitor and combat situations where aquatic ecosystems serve as a habitat for vectors of diseases wherever they pose a risk to human living environments.[24]

(d) *Resource's management guidelines*: The elements of the right to water must be *adequate* for human dignity, life, and health, in accordance with Arts. 11.1 and 12 of the ICESCR. The adequacy of water should not be interpreted narrowly by mere reference to volumetric quantities and technologies. Water should be treated as a social and cultural good and not primarily as an economic good. The manner of the realization of the right to water must also be sustainable, ensuring that the right can be realized for present and future generations.[25]

(e) *Disadvantaged groups*: Whereas the right to water applies to everyone, State parties should give special attention to those individuals and groups who have traditionally faced difficulties in exercising this right, including women, children, minority groups, and indigenous peoples (WALIR 2002). In particular, State parties should take steps to ensure that, inter alia, (1) women are not excluded from decision-making processes concerning water resources and entitlements, (2) rural and deprived urban areas have access to properly maintained water facilities, and (3) indigenous peoples' access to water resources on their ancestral lands is protected from encroachment and unlawful pollution. States should provide resources for indigenous peoples to design, deliver, and control their access to water.[26]

(f) *International cooperation*: To comply with their international obligations in relation to the right to water, States have to respect the right to water in other countries. International cooperation requires States to refrain from actions that interfere, directly or indirectly, with the right to water in other countries. Any activities undertaken within the State party's jurisdiction should not deprive another country of the ability to enforce the right to water for persons in its jurisdiction.[27]

(g) *Regulatory implications of the State's duties related to HRWS*: Violations of the obligation to respect HRWS follow from the State party's interference with the right to water, including, inter alia, pollution and diminution of water resources affecting human health. Violations of the obligation to protect follow from the failure of a State to take all necessary measures to safeguard persons within their jurisdiction from infringements of the right to water by third parties. This includes, inter alia, failure to enact or enforce laws to

[23] CESCR, GC 15, para. 10.

[24] CESCR, GC 15, para. 8.

[25] CESCR, GC 15, para. 11.

[26] CESCR, GC 15, para. 16.

[27] CESCR, GC 15, para. 31.

prevent the contamination and inequitable extraction of water. Violations of the obligation to fulfill occur through the failure of State parties to take all necessary steps to ensure the realization of the right to water. Examples include, inter alia, insufficient expenditure or misallocation of public resources which results in hampering the right to water by individuals or groups, particularly the vulnerable or marginalized, and failure of a State to take into account its international legal obligations regarding the right to water when entering into agreements with other States or with international organizations.[28]

(h) *Duty to reform water laws inconsistent with HRWS*: Existing legislation, strategies, and policies should be reviewed to ensure that they are compatible with obligations arising from the right to water and should be repealed, amended, or changed if inconsistent with ICESCR requirements.[29]

21.2.5 Regulation's Bedrock: From State's Ownership to the Duty to Protect Human Rights Related to Water Resources

The regulatory powers over water resources, once settled on public ownership, are today under the scope of the State's duty to protect human rights, such as HRWS or the right to a healthy environment.

The tendency to strengthen the regulation of resources – based initially on the shortage and the need to ensure equitable access – is accentuated today from a new perspective. It is no longer only about safeguarding the public domain but something more: to comply with the duty to protect the human right to a healthy environment, especially in light of the obligations assumed by States under universal and regional human rights treaties.

The European and American conventions on human rights lack a specific and express provision on environmental issues.[30] However, this vacuum has been remedied by jurisprudential integration, finding the right to a healthy environment in the right to life, in the American case, and in the right to private and family life in the European one.[31]

The Inter-American Court of Human Rights (IACHR) considers a healthy environment as a component of the international corpus juris derived from Art. 4 of the American Convention on Human Rights (ACHR), in relation to the general

[28] CESCR, GC 15, para. 44.

[29] CESCR, GC 15, para. 46.

[30] Except for art. 11 of the San Salvador Protocol.

[31] ECHR, *López Ostra v. Spain*, 1994, para. 51; *Guerra and others v. Italy*, 1998, para. 60; *Kyrtatos v. Greece*, 2003, para. 52; *Hatton and others v. UK*, 2003, para. 96; *Taşkin y otros v. Turkey*, 2004, para. 113; *Giacomelli v. Italy*, 2006 para. 76.

obligation to guarantee it contained in Art. 1.1 and the duty of progressive development contained in Art. 26. Also, Art. 11 of the Protocol of San Salvador endorses the right to a healthy environment.[32]

In that line, the IACHR has developed the guarantees required to face public or private actions that endanger the environment. While these criteria have been exhibited in relation to indigenous communities, we believe that they are generally applicable to environmental issues whenever we find anyone's human rights at stake. They are effective participation regarding any development, investment, exploration or extraction plan, reasonable benefit, and prior environmental and social impact assessment.[33] The European Court of Human Rights (ECHR) has also understood that participation requirements and environmental impact assessments are required generically by all affected groups, even if they are not indigenous communities.

The Inter-American Commission on Human Rights (IAComHR) has also stated that the exercise of the right to life and physical integrity is necessarily related to and in some ways dependent upon the environment.[34] For that reason, when pollution and environmental degradation pose a persistent threat to life and health for human beings, those rights are at stake. As a result, Member States must take positive steps to safeguard life and health, and the absence of regulation, inappropriate regulation, and lack of supervision in the implementation of the rules constitute a breach of its duties. That is now the real pillar of regulation.

21.3 Regulatory Power Shifts over Water Sources

The deep changes in the foundations of water management have resulted in new institutional dynamics. Many relationships are affected: not only between Nations but also the very checks and balances of a republic, the federal system, and even the link between public authorities and civil society.

21.3.1 Political Power: From Sovereignty to Globalization

Since World War II, Western countries have been engaged in a legalization process of international relations, or the internationalization of law, depending on the focus. This approach implies that areas which were traditionally destined for the national

[32] IACHR, *Caso Comunidad indígena Yakye Axa v. Paraguay*, 2005, para. 66.

[33] IACHR, *Case of the Saramaka People v. Suriname*, 2007, para. 129.

[34] IAComHR, *Informe sobre la situación de los derechos humanos en Ecuador. La situación de los derechos humanos de los habitantes del interior del Ecuador afectados por las actividades de desarrollo*. OEA/Ser.L/V/II.96, Doc. 10 rev. 1, 24 April 1999.

domain, and released within the scope of sovereign relationships, are increasingly regulated at an international level, with the subsequent limitation to State action in both spheres. We can see the same tendency within the field of human rights, commerce, or investments, among others, and its three typical characteristics are:

(a) The explicit establishment of minimum levels of action on the part of public authorities
(b) The prevention of States from invoking their national legal systems, their cultural traditions, or any other aspect of their national identity to justify their distancing from those minimum levels[35]
(c) The creation of international bodies – courts, groups of experts, or peers – with the ability to control the compatibility of governmental practices with those minimum levels of action and to declare a State's international liability in case of infringement

It is the presence – to varying degrees – of each and every one of these aspects which identifies the legalization process of international relations and allows us to distinguish three different stages within it[36]:

(a) *Precision* of rules, when we go from the ambiguity of political commitments to the clear definition of what is demanded, permitted, or prohibited by the establishment of minimum levels
(b) *Obligation*, when it is explicitly assumed – through the rule of inadmissibility of domestic law invocation and the principle of *pacta sunt servanda* (agreements must be kept) – that the undersigned are *legally* bound to a commitment from which they cannot step out and that their actions will be measured against that commitment
(c) *Delegation* of decision-making powers, when a third party is given the authority to interpret and apply the rules that derive from the commitment

The degree to which these three dimensions are present determines whether an international instrument is closer to "hard" or "soft" law. Therefore, the concept of legalization is gradable: from the ideal type, where the three dimensions are at a maximum, passing through "hard" legislation, where the three dimensions (or at least obligation and delegation) are high, to different forms of "soft" legalization that combine these attributes, and finally to the total absence of legalization as another ideal type. In environmental matters, the growing presence of instruments of "soft" and "hard" law gives account for the movement from sovereignty to globalization.

---

[35] Arts. 27 of the Vienna Convention.

[36] Abbott et al., 'The Concept of Legalization', 54 *INT'L ORG*. 401, 401 (2000); Koremenos et al., 'The Rational Design of International Institutions', 55 *INT'L ORG*. 761, 761–62 (2001); Boyle, 'Some Reflections on the Relationship of Treaties and Soft Law', 48 *INT'L L.Q*. 901, 901–13 (1999).

The common denominator of this movement is, thus, the disruption of the sources of law in each country and, especially, of the ability to monitor compliance with the obligations of States. That scenario requires taking account of new interpretive challenges, among which the competition between two contradictory trends highlights the *homogeneity*, derived from the persistent setting of courses of action common to all States that are given by the international order, and the *diversity*, embodied in the legal and cultural expressions of each country.

This trend has turned around the creation and enforcement law procedures – understood for centuries as the exclusive realm of sovereign entities – facing us with an evolutionary process characterized by the variation of the poles of production and application of the rules that bind the State with its citizens.

This new drift is based on the identification of a set of goals that transcend the will of the States – the historical starting point of the link between those – to settle on the collective protection of a range of interests that are considered universal. In this context, the contradiction between any domestic practice – including norms, judgments, or administrative acts – and international guidelines results in their nullity, regardless of their national hierarchy, their roots in the traditions of the country, or the acceptance by the community.

The power of international systems – which increasingly embrace environment and HRWS as one of their main issues – resides, in short, in advancing, under the banner of global interests, over any national standard, tradition, or practice deemed in violation of the rights fixed in treaties. This trend involves undoubtedly tensions with the cultural, social, and legal systems of each country and thereby significantly reduces the political influence of the State as a key notion of modernity (Harris and Roa-García 2013).

21.3.2 Separation of Powers: From the Preeminence of the Legislature to the Predominance of the Courts

Water laws in Latin America were part of the consolidation process of the nineteenth century republican State, in which civil codification played a central role. The leaders of the nascent States saw in the codes –especially the French one – a policy instrument ideal for perfecting the spirit of independence and ensuring political control. It represented a legal instrument, unitary and coherent, necessary to overcome the legal particularism of the previous period. The civil code, substituting the pluralism of old colonial sources, was aimed at strengthening national unity based on reason. The dominance of the law as an expression of the general will was a natural consequence of this construction, and in it the role of the courts was limited to the strict enforcement of the law's text.

Today, that legalistic model is giving way to the courts' role as guardians of enforceable human rights. Judges become a key piece to the enforcement of the rights related to the environment and water, and they must ensure minimum levels of protection in those fields, acquiring a growing influence in determining water policy.

Some of the samples of the new role of the courts are the following:

(a) *Class actions*

Latin American jurisprudence is facing, from a practical perspective, the challenges that the recognition of collective rights has created in the legal system. In the case of the environment, the individual approach results inefficient, since the protected good belongs to the whole community, is indivisible, and does not exclude anyone. That is why the capacity to demand certain performance – from the State or companies – includes now not only the affected individual but also the ombudsman and the NGOs that concentrate on the collective interest.

Today, a specific mechanism that attempts to overcome the substantial and procedural deficiencies of the individualistic approach is increasingly recognized: the collective actions that lead to rulings with *erga omnes* effect. This includes the ability of the courts to issue general mandates, something reserved to the legislature in the traditional scheme.

(b) *Conventionality control*

This type of scrutiny of government's practices is advancing rapidly in Latin America as a result of the consolidation of regional systems of human rights protection. It mainly consists of the verification of the compatibility of each State's conduct, whatever their origin or form, with the directives of the treaties that enshrine these rights. State parties must adapt their acts to human rights' standards, and the conventionality test intends to make sure this happens effectively.

Conventionality control embodies a fundamental objective of the international regimes of protection. Those models lay the rights' defense on two correlative levels: first, the spontaneous work of local authorities and, second, the subsidiary role of supranational bodies. The whole system relies on a central principle: the international jurisdiction of human rights is complementary to the internal one, and this feature determines that its principal actors are initially the national authorities. Subsidiarity implies that the first review of compliance must come from the local level, and therefore, each country must, at each step, verify that their decisions are compatible with the text and purpose of the covenants.

This kind of control is, thus, crucial to the integrity of the whole system, at the extent that – in accordance with the IACHR – it must be exercised ex officio[37] and with *erga omnes* effect.[38] As we can see, as far as national courts must compare the actions or omissions of the State with the human rights covenants with general effects, the judicial role has also expanded from this device.

---

[37] IACHR, *Almonacid Arellano y otros v. Chile,* 2006, para. 124; *Trabajadores Cesados del Congreso (Aguado Alfaro y otros) vs. Perú,* 2006, para. 128; *Cabrera García y Montiel Flores vs. México*, 2010, para. 225 y ss.

[38] IACHR, *Barrios Altos v. Perú,* 2001, para. 18; *La Cantuta v. Perú,* 2006, para. 188.

(c) *Pilot judgment procedure*

The ECHR has also shaped instruments to address rights violations caused by improper State practice affecting a large number of people, as it happens in most water-related issues. It has noticed that in a great number of cases, the violation does not arise from isolated incidents or due to the circumstances of a particular situation, but it is the result of a regulatory and administrative conduct of the authorities towards a group or class of persons. Such situations have been described as "systemic problems."[39]

In these scenarios, the ineffectiveness of individual resolutions has been underlined, both for those affected and for the court itself, so that the birth of the notion of a systemic problem resulted in a specific procedural response, known as "pilot judgment." When the court uses this mechanism, it indicates to the State the general measures to be taken to reverse the problem. If these measures are performed – and work – similar cases are "blacked out" of the lists of judgments pending in Strasbourg. This means that when a systemic problem is detected, ECHR issues a judgment with *erga omnes* effect, absorbing all related individual cases.

(d) *Public interest litigation*

This litigation model[40] seeks to change – through a court's case – the structural conditions of a situation which threatens or is contrary to any constitutional value.[41] The scale of the objectives that characterize these remedies often makes judicial intervention to entail the reorganization of government agencies. This transforms the figure as a relevant alternative for addressing problems related to water resources from the judiciary,[42] mainly because it involves (1) the negotiation between the parties to the dispute; (2) the continuous, provisional, and fluid features

---

[39] ECHR, *Broniowski v. Poland* ("Bug River" case), 2004, para. 189 y resolutive points 3 andy 4; *Lukenda v. Slovenia*, 2005, para. 93 and resolutive points 4 and 5; *Sejdovic v. Italy*, 2006, para. 119 and 120; *Hutten-Czapska v. Poland*, 2006, para. 231 y Friendly Agreement of April 2008; *Burdov v. Russia (N° 2)*, 2009, paras. 125 and 140 y resolutive points 7 and 8*; Kauczor v. Poland*, 2009, para. 62; *Rumpf v. Germany*, 2010, para. 59 y ss.; *Greens y M.T. v. UK*, 2010, paras. 117 and ss.; *Finger v. Bulgary*, 2011.

[40] Correa Montoya, L., "Litigio de alto impacto: Estrategias alternativas para enseñar y ejercer el Derecho", *Revista de Derecho*, N° 30, Universidad del Norte, Barranquilla, Colombia, 2008, p. 250.

[41] Linares, S., "El diálogo democrático entre las cortes y las instituciones representativas", *Revista Mexicana de Sociología*, 70, Vol. 8, N° 3, Universidad Autónoma de México, 2008, p. 510; Bergallo, P., "Justicia y experimentalismo: la función remedial del poder judicial en el litigio de derecho público en Argentina" en AA.VV, *SELA*, Del Puerto, Buenos Aires, 2005; Puga, M., "La realización de los derechos en casos estructurales: Las causas Verbitsky y Mendoza", *Universidad de Palermo*, Buenos Aires, 2008.

[42] Belski, M., in AA.VV *El litigio estratégico como herramienta para la exigibilidad del derecho a la educación*, Asociación por los Derechos Civiles, 2008, www.redligare.org/IMG/pdf/litigio_estrategico_educacion.pdf

of the judicial intervention; and (3) the transparency of the process of adoption of these measures.[43]

In Argentina, an emblematic sample of this new role of the courts is the *Mendoza* case,[44] of the Matanza-Riachuelo River Basin, conducted by the Supreme Court of that country since 2006. In July 2004, a group of neighbors in the Matanza-Riachuelo Basin sued before the Supreme Court the Federal Government, the Province of Buenos Aires, the Autonomous City of Buenos Aires, and 44 companies, seeking compensation for damages suffered as a result of contamination of the basin, the cessation of pollution, and environmental restructuring.

The Matanza-Riachuelo Basin is one of the most polluted in the country and one of the most degraded urban areas, environmentally and socially. It is located in the industrial belt of the Metropolitan Area of Buenos Aires, covering an area of 2,240 km^2, and its population reaches five million, a high percentage of whom live in slums, lacking water and sanitation services and with unsatisfied basic needs. The area involves 14 municipalities, three jurisdictions, and more than 22 agencies that apply 55 different regulatory frameworks that sometimes collide. Some 368,000 m^3/day of sewage is dumped in the river, of which only 5 % receive treatment. In addition, many pit latrines and septic tanks contaminate groundwater.

The Supreme Court took action in the case in 2006 and called on government authorities and businesses to provide information to diagnose the situation. Five months later, the National Congress created the Authority of the Matanza-Riachuelo Basin (ACUMAR), an interjurisdictional entity responsible for the sanitation of the basin. In July 2008, the Court delivered its ruling and established an intervention program that forced the ACUMAR to follow a schedule of measures that included the production and dissemination of public information, industrial pollution, and sanitation landfills' control in order to avoid contamination; the expansion of water sewer and storm drains works, carrying out an emergency health plan; and the adoption of a measurement system to evaluate the degree of fulfillment of the established objectives.

The Court delegated the implementation process to a federal judge of the affected area, which was responsible for overseeing compliance, and also created a collegiate body coordinated by the Ombudsman's Office along with NGOs in order to strengthen and enable citizen participation in the enforcement of the ruling. It also instructed the Auditor General's Office (*Auditoría General de la Nación*) to carry out a specific control over funds allocation and budget execution related to the

---

[43] Thea, F., "Hacia nuevas formas de justicia administrativa: Apuntes sobre el "Litigio Estructural" en la Ciudad de Buenos Aires", *LL* Sup. Adm 2010 (February), 2010-A, 309.

[44] CSJN, *Mendoza*, June 20 2006, *Fallos*, 329:2316; August 30 2006, *Fallos*, 329:3528; August 22 2007, *Fallos*, 330:3663. This case is just an example of this clear trend in the region that have been followed in Colombia, where a very significant judicial decision was held by the Estate Council in 2014 with respect to the Bogota river basin. Consejo de Estado, Sala de lo Contencionso Administrativo, Sección Primera, Acción Popular, Consejero Ponente: Dr. Marco Antonio Velilla Moreno, March 28th, 2014.

Integrated Sanitation Plan (PISA), to ensure transparency in the management of public funds.

The 2008 ruling forced ACUMAR to realize actions aimed at the expansion of water supply and sanitation. To do this, the entity had to be accountable for the progress of the expansion plan of the works of collection, treatment, and distribution of water. To this end, the judge imposed the authorities to submit a unique master plan for the provision of water and sewerage in all jurisdictions under a single operator. In all cases, compliance deadlines, funding sources, timelines, and budgets involved were detailed. In 2010, ACUMAR approved the Integrated Sanitation Plan (PISA), which is under implementation and involves a cumulative investment of about six billion dollars through 2011. In 2012, 258 industrial establishments were closed, and more than 1,000 contaminant businesses have taken steps to restructure their production practices to comply with environmental requirements.

In its ruling, the Court stated that the plan's goal (PISA) should be to improve the quality of life for residents and required the adoption of specific health programs to meet the needs of the population. Far from the traditional approach, in this structural litigation the law enacted by the legislature was secondary: there is no reference to the Civil Code, but mention is made of Integrated Water Resources Management (IWRM).

21.3.3 Territorial Distribution of Powers: From Isolated National Management to Transboundary Water Cooperation and from the Centrifugal Tendency (Federal) to the Centripetal (Unitary) One

As water resources are more overexploited and environmental water problems increase, conflicts arise but also transboundary cooperation develops. This is because environment and water do not respect neither borders nor private/public property boundaries.

Transboundary water conflicts and cooperation have a long history in Latin America, and transboundary water cooperation increases all over the world when it comes addressing significant problems. This can be seen in bilateral or multilateral agreements for specific basins, watercourses, or rivers, as well as in multilateral agreements that serve as a general framework for specific covenants.[45]

[45] Martin, L., Pinto, M. & Salinas Alcega, S., "Cooperation on Transboundary Waters in Latin American and the Caribbean," *Background Study Workshop on Transboundary Water Cooperation "Latin American and Pan-European Regions: Sharing Experiences and Learning from Each Other"* UNECE, 2013; Biswas, Asit K. (Ed.), *Managing Transboundary Waters of Latin America*, NY, Routledge, 2013.

Latin America and the Caribbean accounts for more than 35 international specific treaties regulating on shared waters. In respect to multilateral general agreement, the UNECE Water Convention, the NY Convention 1997, and the UE Water Framework Directive 2000/60/CE present sufficient evidence of how environmental principles increasingly apply to water.[46] The well-known principles of International Water Law – equitable and reasonable utilization, obligation not to cause significant harm, principles of cooperation, information exchange, notification, consultation, and peaceful settlement of disputes – are complemented by the specific one of protection and preservation of watercourses.

In line with these arguments, such transboundary water cooperation cannot refer to surface water exclusively and reaches also subsurface waters, whose interaction is crucial for a proper IWRM. This big step in transboundary water cooperation is well exemplified in Latin America. Through the UNESCO/OAS ISARM Americas Programme, transboundary aquifers were identified in the region, and as a result of this process, Argentina, Brazil, Paraguay, and Uruguay signed the Guarani Aquifer Treaty (2010). In parallel, the International Law Commission (ILC) has been working for the last years in the draft articles on the law of transboundary aquifers.

The many facts that challenge IWRM nowadays (intensive use, climate change, droughts, floods, conflicts, etc.) result in the increasing role of transboundary water cooperation and international water law applying for the scale of entire basin, river, or water course at the expense of national and local regulations (Engle et al. 2011; Molle 2009).

The second transformation to highlight with respect to the territorial distribution of power is a movement from the centrifugal tendency (federal) to the centripetal (unitary) one.

The environmental protection requires the fixing of minimum levels of performance in water management and thereby brings about a concentration of decision-making powers that impact heavily on those models of territorial distribution of authority that rest on the idea of decentralization.

Minimum levels of performance aim to grant a uniform environmental protection for the entire country, and they are increasingly internationally configured,

[46] For instance, the UNECE Water Convention that now allows the incorporation of non-UNECE Member States is one of the texts where the modern attempt of IWRM is articulated. To that end, it has established a set of obligations to the Member States: (a) to prevent, control, and reduce pollution of waters causing or likely to cause transboundary impact; (b) to ensure that transboundary waters are used with the aim of ecologically sound and rational water management, conservation of water resources, and environmental protection; (c) to ensure that transboundary waters are used in a reasonable and equitable way, taking into particular account their transboundary character, in the case of activities which cause or are likely to cause transboundary impact; and (d) to ensure conservation and, where necessary, restoration of ecosystems. In addition, it establishes specific principles like the precautionary principle, polluter pays principle, intergenerational equity and sustainability, and making management instruments available to the States.

either through specific conventions or by the inclusion of the concept of "healthy environment," along with HRWS, within the system of human rights protection. In any of the cases, the State cannot make use of internal rules – including those concerning the distribution of powers – to justify a breach of the commitments. This is summarized in that a State cannot plead its federal structure to avoid complying with an international obligation.[47]

As GC 15 states, where implementation of the right to water has been delegated to regional or local authorities, the State still retains the responsibility to comply with its covenant obligation.[48]

International engagement acts as a centripetal force towards the inside of the federation because the national State, as part of the human right treaties, assumes duties involving the unification of criteria and thereby centralizes a number of powers that internally may be the competency of the smaller political entities. The trend is, in other words, towards the accumulation of competences of the Nation as a condition for the fulfillment of its international obligations.[49] Conversely, federalism tends to scatter the power to be transferred to smaller entities, acting as a centrifugal force into the State. At the local level, such models require a significant fragmentation of decision-making that seems to go against their unification under the internationally bound head.

Progress towards global environmental protection finds in the denial of decentralization systems of political power another example of the paradigm shift. The minimum protection sets the direction towards uniformity, while decentralization does it towards diversity.

21.3.4 Water Protection's Holder: From State Exclusivity to Community Empowerment

The division between public and private waters is one of the many expressions of the dichotomy between public and private *purposes* that have been used historically to draw the line between State powers and individual rights.

This dichotomy has determined the configuration of relations between the State and civil society, to the point that the ability to act and the content of that activity have historically responded to a differentiation of the ends sought: The State action, through powers and subordination bonds, was considered dominant because it pursued *public interest*, while individual action, through guarantees and

---

[47] IACHR, *Garrido y Baigorria v. Argentina,* 1998, para. 46; AO-16/99, *El Derecho a la Información sobre la Asistencia Consular en el Marco de las Garantías del Debido Proceso Legal*, 1999, para. 139 y Resolutive Point 8; *Caso de las Penitenciarías de Mendoza*, 2006.

[48] CESCR, GC 15, para. 51.

[49] IAComHR, Informe N° 3/87, Caso 9647, *Estados Unidos, Pena de muerte a menores de edad*, paras. 62 y ss. ECHR, *Assanidze v. Georgia*, 2004, para. 146.

coordination bonds, was restricted in its ability to impact on community decisions, as its only end was the pursuit of a *private interest*. In other words, the possibility of influence in the collective social life was the absolute domain of the State as exclusive manager of public interest, while individuals or groups recognized only the possibility of participation within a circle of goals alien to the collective sphere.

Third-generation rights have faded that dividing line, as before them, the distinction between public and private interest has diluted greatly. Such circumstances should lead to changes in the participation schemes in the management of water resources. It is no longer possible to qualify the State as sole manager of the goals at stake. Rather, the new era needs a perspective of inclusion of new participants that also happen to embody the public interest that was previously untouchable heritage of State's authority.

The fracture of the dichotomy is perceived; then, from the two poles of the relationship, not only the public interest ceases to be State's property and acquires new managers – previously limited in their capacity for collective action by the individual content of their rights – but also private interest is crossed by a strong public intervention in relationships that were once considered limited to the spectrum of ties between individuals. It is no longer possible to consider that the environment involves only the public or private interest.

The change is profound and forces us to cope with dynamic responses. If we continue to understand environmental resources from the tenets designed according to the dichotomy between public and private interest, we will build ineffective protection mechanisms.

Moreover, it is essential to involve people in the management of water resources. Even when the need to reverse environmental deterioration is urgent, the efficacy of current protection policies is compromised to the extent that both the economic operators and the public authority still perceive the environment as an *outer limit*. Instead of promoting the cooperative management of collective goods, the legal system continues to operate from the omnipresence of rights hold by individuals guided by self-interest seeking to maximize profits and reduce losses. That is why the focus of environmental protection measures is still centered in *coercion*, i.e., the imposition of adverse consequences for the alteration of the ecosystem. The two main tools are *liability* for environmental damages and *sanctions* because it is understood that only through them harmful behaviors in this field can be stopped.

The problem with this view, which is transferred to the utility of the instruments it promotes, is their merely palliative role. While the law does not absorb a paradigm shift that allows people to *internalize the environment* and while we believe that its preservation is about inhibiting selfish subjects by imposing sanctions or remedial convictions, we'll be halfway in their protection, as these measures will simply be evaluated by the economic agents as just another barrier to their "real" objectives of profit maximization.

Responsibility and punishment are certainly necessary, but not sufficient. They still show the environment as an uncomfortable presence in front of the persecution of the tangible interests of the people. Therefore, we must resort to other institutional mechanisms to recover the collective approach of natural resources, and a first step towards this is the restitution of communal ties (Boelens et al. 2010; Graham 2011). From this view, other institutional arrangements arise, different from punishment and responsibility and more focused on prevention, participation, and, fundamentally, the rationality of cooperative action. This vision calls for encouraging contact between people and to give them a strong role in the decision-making processes related to the environment.

21.3.5 Accountability: From the Exclusive Role of the State to Corporate Responsibility

For a long time, strategies and initiatives for the protection of human rights focused almost exclusively on the role and responsibility of the State. However, as the global role of companies grew, their exposure in this area increased significantly. Since the mid-twentieth century, corporate responsibility for human rights violations has reached a rapid development which has caused its inclusion – especially from the 1990s – in the agenda of the United Nations and other global forums.

In June 2011, the Human Rights Council of the United Nations adopted a set of recommendations of the Special Representative of the Secretary General on human rights and transnational corporations and other business – Prof. John Ruggie – known as the "Guiding Principles on Business and Human Rights. Protect, Respect and Remedy."[50] These proposals are the outcome of a process that began in 2005 and up to date constitute the experience of debate on human rights and companies more evolved worldwide. In the management of water resources, the value of the Principles is to shed light on the role of business as a responsible subject for their preservation.

The initiative carried out by Ruggie was to mark a "principled pragmatism" aimed at achieving the greatest possible consensus on the part of NGOs, companies, and governments from the interplay of three fundamental principles:

(a) The State has a *duty to protect* against human rights abuses by third parties, including business enterprises, through appropriate policies, regulation, and adjudication. Such protection lies at the very core of the international human rights regime.
(b) Companies have the *responsibility to respect* human rights, which means that business enterprises should act with due diligence to avoid infringing on the rights of others and to address adverse impacts with which they are involved. That respect embodies the most basic social expectation regarding companies.
(c) It is necessary to improve victims' access to *effective redress*, judicial or extrajudicial, for even better coordinated efforts cannot totally prevent abuse.

[50] UN, HRC, A/HRC/RES/17/4, 2011.

The development of corporate responsibility in human rights matters opens a new field of debate, nonexistent under the classic economic view. Companies must perform in accordance with sustainability requirements.

21.4 Conclusions

The environmental paradigm has changed the reasons, ways, and subjects of water resources management. If in the past they were framed in a model centered on private property and individual rights, with legislative prevalence and trends towards decentralization, we face the shift to a model based on collective goods, human rights, intergenerational equity, and the leading role of courts in the solution of the increasing conflicts that the harnessing of water resources currently involves.

In this chapter, two main aspects of this change have been analyzed: first, the transformation of the basis of water management, showing how a series of principles that shaped the building of water law in Latin America have evolved from the economic and individualistic approach that prevailed in the last two centuries to new ways of environmental and social empowerment, and, secondly, the shifts in regulatory power over water show how the institutions responsible for implementing substantive principles also have new roles, leading to new dynamics marked by globalization and judicial dominance.

Some of the referred institutions and trends are still pending for consolidation, and its definition and content is not still clear at all, but a national or international strategy on water resources that does not take into account the many changes from the consolidation of the environmental paradigm exposes itself to failure. This is so because the context and the actors of water management have changed considerably.

Indeed, the effectiveness of any regulation first depends on the accuracy of the diagnosis on which it is formulated. If regulations are designed on false premises, their inefficiency is inevitable. This is especially true in Latin America, where the main problems of water governance are associated with the gap between the content of the legislation and its practical observance by administrative agencies, courts, and civil society.

The regional situation shows, in fact, that the institutional frameworks fail when they do not correlate with the dynamics of power in which they must operate. If one insisted on legislating based on a fictional scenario, the only result would be aggravating the current problems, which are closely linked to the lack of institutional capacity to implement the legal regime of each country. If the new government strategies do not consider reality accordingly, they will only relapse into that chronic deficit, leading to rising social conflicts around water resources.

The concept of governance applied to water refers to the social ability to mobilize energies consistently for sustainable development of water resources. This includes the ability to design socially accepted public policies for sustainable development and to make their implementation effective to the different stakeholders. For this to happen, the following three points are of utmost importance: (a) the degree of social agreement (implicit or explicit) about the nature of the water-society relationship, (b) the existence of consensus on the basis of public policies which express this relationship, and (c) the availability of effective management systems that enable the implementation and monitoring of policies, within a framework of sustainability.[51]

The environmental paradigm has turned the water-society relationship and thus the premise of governance referred to above. If the new initiatives for water management do not realize that change, the consensus and public policies necessary to achieve sustainable governance will not be possible. That's the whole point of the paradigm shift: to prevent the dissociation between the models of decision-making that are prior to the rules and the rules themselves.

References

Abbott KW (2000) The concept of legalization. 54 Int'l Org 401, 401

AO-16/99 (1999) El Derecho a la Información sobre la Asistencia Consular en el Marco de las Garantías del Debido Proceso Legal

Arnold CA (2002) The reconstitution of property: property as a web of interests. Harv Environ Law Rev 26:281–364, Cambridge, MA

Belski M (2008) in AA.VV El litigio estratégico como herramienta para la exigibilidad del derecho a la educación. Asociación por los Derechos Civiles. www.redligare.org/IMG/pdf/litigio_estrategico_educacion.pdf

Bergallo P (2005) Justicia y experimentalismo: la función remedial del poder judicial en el litigio de derecho público en Argentina, en AA.VV, SELA, Del Puerto, Buenos Aires

Boelens R, Getches D, Guevara A (eds) (2010) Out of the Mainstream. Water rights, politics and identity. Earthscan/Routledge, London/Washington, DC

Boyle AE (1999) Some reflections on the relationship of treaties and soft law. Int Comp Law Q 48:901–913

Cavallo Perin R, Casalini D (2010) Water property models as sovereignty prerogatives: European legal perspectives in comparison. Water 2(3):429–438

Correa Montoya L (2008) Litigio de alto impacto: Estrategias alternativas para enseñar y ejercer el Derecho, Revista de Derecho, N° 30. Universidad del Norte, Barranquilla

de Sousa Santos B (1998) La globalización del derecho: los nuevos caminos de la regulación y la emancipación". ILSA Ediciones Universidad Nacional de Colombia, Bogotá

de Sousa Santos B (2002) Toward a new legal common sense. Law, globalization, and emancipation. Butterworths, Londres

Dussel E (2000) Europe, modernity and eurocentrism. Nepantla 1(3):465–478, Duke University Press, Durham (NC)

Engle NL, Johns OR, Lemos M, Nelson DR (2011) Integrated and adaptive management of water resources: tensions, legacies, and the next best thing. Ecol Soc 16(1):19

Escobar A (2003) Mundos y conocimientos de otro modo: el programa de investigación de modernidad/colonialidad Latinoamericano, vol 1. Tabula Rasa, Bogotá

Ferrajoli L (2004) Derechos y Garantías. La ley del más débil. Trotta, Madrid

[51] Peña, Humberto and Solanes, Miguel 2003, La Gobernabilidad Efectiva del Agua en las Américas, un Tema Crítico, GWP SAMTAC.

Graham N (2011) Lawscape: property, environment and law. Routledge, Abingdon
Grosfoguel R (2007) Descolonizando los universalismos. In: Castro-Gómez S, Grosfoguel R (eds) El giro decolonial. Pontificia Universidad Javeriana, Universidad Central y Siglo del Hombre Editores, Bogotá
Harris LM, Roa-García MC (2013) Recent waves of water governance: constitutional reform and resistance to neoliberalization in Latin America (1990–2012). Geoforum 50:20–30
Horkheimer M (1947) Eclipse of reason. Oxford University Press, New York
IAComHR. Informe sobre la situación de los derechos humanos en Ecuador. La situación de los derechos humanos de los habitantes del interior del Ecuador afectados por las actividades de desarrollo. OEA/Ser.L/V/II.96, Doc. 10 rev. 1, 24 April 1999
Koremenos B (2001) The rational design of international institutions. Int Organ 55:761, 761–762
Lander E (ed) (2000) La colonialidad del saber: eurocentrismo y ciencias sociales. CLACSO, Buenos Aires
Linares S (2008) El diálogo democrático entre las cortes y las instituciones representativas. Revista Mexicana de Sociología, 70, vol 8, N° 3, Universidad Autónoma de México 487–539
Locke J (1698) Two treatises of Government: in the former, The false principles and Foundation of Sir Robert Filmer...; The latter is an Essay concerning The True Original, Extend, and End of Civil-Government, Edición digital basada en la de London, Printed for Awnsham and John Crullchill, at the Black Swan in de Pater – Noster- Row
Macpherson CB (ed) (1978) Property. Mainstream and critical positions. University of Toronto Press, Toronto
Martin L (2010a) Cuando el río suena... el establecimiento de una nueva matriz disciplinar para el derecho argentino de aguas. Revista Electrónica del Instituto de Investigaciones "Ambrosio L. Gioja" – Año IV, Número 5
Martin L (2010b) Derecho de Aguas. Estudio sobre el uso y dominio de las aguas públicas. Abeledo Perrot, Buenos Aires
Martinez-Santos P, Aldaya MM, Llamas MR (eds) (2014) Integrated water resources management in the 21st century: revisiting the paradigm. CRC Press, Leiden
Mignolo W (2000) Local histories, global designs: coloniality, subaltern knowledges and border thinking. Princeton University Press, Princeton
Míguez Núñez R (2008) Las oscilaciones de la propiedad colectiva en las constituciones andinas. Global Jurist 8(1)
Molle F (2009) River-basin planning and management: the social life of a concept. Geoforum 40:484–494
Ost F, van Hoecke M (1999) Del contrato a la transmisión. Sobre la responsabilidad hacia las generaciones futuras. Doxa 22:607–630
Provost R (1994) Reciprocity in human rights and humanitarian law. Brit Yearb Int Law 65:1–102
Puga M (2008) La realización de los derechos en casos estructurales: las causas verbitsky y mendoza. Universidad de Palermo, Buenos Aires
Purdy J (2007) Property and empire: the law of imperialism in Johnson v. M'Intosh. George Washington Law Rev 75:329–371
Quijano A (2000) Coloniality of power, ethnocentrism, and Latin America. Nepantla 1(3):533–580, Duke University Press, Durham (NC)
Smith A (1978) Lectures on jurisprudence. In: Meek RL et al (eds). Oxford University Press ed, 459–460
Teitel R (2011) Humanity's law. Oxford University Press, New York
Thea F (2010) Hacia nuevas formas de justicia administrativa: Apuntes sobre el "Litigio Estructural" en la Ciudad de Buenos Aires, LL Sup. Adm 2010 (February), 2010-A
UN. Human rights as the primary objective of trade, investment and financial policy. Doc./E/CN.4/Sub.2/RES/1998/12
VVAA (2008) Special issue IWRM in Latin America. Int J Water Res Dev 24(1):1–200
Water Law and Indigenous Rights (WALIR) (2002) Towards recognition of indigenous water rights and management. International WALIR Seminar, Wageningen, 4–8 March 2002. http://www.eclac.cl/drni/proyectos/walir/doc/walir1.pdf. 8 Oct 2012

Chapter 22
Environmental Provisions in the Constitutions of Uruguay and Argentina Affecting Water Resource Management

Maria Catalina Bosch and Maria Concepcion Donoso

Abstract This chapter identifies the similarities and differences between the existing constitutional and legislative environmental regimes in Uruguay and Argentina, with attention being paid to aspects that may be reflected in the management of water resources.

Emphasis in differences offers interesting avenues of thought to the scholar of comparative law that they constituted a single political entity from the discovery of the La Plata River, in 1516 up to 1828, when Uruguay became an independent state. The two countries share borders marked by a river and a vast estuary (the Uruguay and the La Plata Rivers) of enormous economic and environmental significance for both nations, which have agreed in entrusting their management to binational commissions. Uruguay and Argentina are founding members of the Common Market of the South.

Significant differences are noticeable in environmental constitutional approaches and provisions in each country. The impacts of the constitutional structure of the respective states upon the width of the environmental rights recognized in terms of the hierarchy of international treaties compared with that of the national constitution in each country.

Finally, two outstanding environmental water-related issues are invoked, namely, (1) the recent Argentina-Uruguay dispute on contamination of the Uruguay River by European pulp mills operating in the Uruguayan territory and the International Court of Justice decision thereupon and (2) the nature and legal effects of the Uruguayan October 2004 constitutional amendment on the country's water resources, heralded by *The UNESCO Courier* as unique ("A world's first") ("In an overwhelming majority vote, water was enshrined in Uruguay's constitution as public property – a world first." The UNESCO Courier, March, 2006 • ISSN 1993-8616) after a popular pronouncement with over 64 % of citizens' support.

M.C. Bosch (✉) • M.C. Donoso
Global Water for Sustainability Program, Florida International University, 3000 NE 151st Street, AC1- 267, Miami, FL 33181, USA
e-mail: catabosch81@hotmail.com; mcdonoso@fiu.edu

S.G. Setegn, M.C. Donoso (eds.), *Sustainability of Integrated Water Resources Management*, DOI 10.1007/978-3-319-12194-9_22

Keywords IWRM • Sustainability • Policy • International treaties • Uruguay • Argentina

22.1 Introduction

The point of departure for this paper consists of reviewing the concept of *environment* as defined and regulated by the Constitutions of Uruguay and Argentina, with the nature, scope, and hierarchy of environmental rights and obligations being paid particular attention. Next, the procedural, notional, and positive-law consequences of such constitutional provisions are examined, particularly in terms of active and passive legal standing for environmental protection claims, procedural guarantees for environmental rights, and the uniqueness of the "world-first" provisions of the Uruguayan constitutional reform on water resources. The effects in each country's domestic legal realm of the international environmental treaties and conventions adhered to by Argentina and Uruguay are next compared in aspects such as the hierarchy of international treaties as a source of law in either country and the domestic law impacts of the Common Market of the South (MERCOSUR) standards. Finally, the Argentina-Uruguay protracted environmental dispute around pulp mill operations on the Uruguay River and the International Court of Justice decision thereupon are also highlighted as a reminder of the need for reality to actually be reflected in abstract water and environmental law provisions – a requirement often remarked by the scholarly community as deserving keen attention by the Latin American and Caribbean (LAC) States.

22.2 Relevant Constitutional Provisions

Compared in this section are those constitutional stipulations that are most relevant to the issue of environmental regulation in the countries being considered, with water resources being the kernel in any discussion of this sort for two nations so plentifully endowed with such strategic assets.

Argentina introduced the environment issue in its Constitution with the 1994 amendment. The text is included under the heading "New Rights and Guarantees" and reads:

> Section 41. All inhabitants are entitled to the right to a healthy and balanced environment fit for human development in order that productive activities shall meet present needs without endangering those of future generations; and shall have the duty to preserve it. As a first priority, environmental damage shall bring about the obligation to repair it according to law.
>
> The authorities shall provide for the protection of this right, the rational use of natural resources, the preservation of the natural and cultural heritage and of the biological diversity, and shall also provide for environmental information and education.

> The Nation shall regulate the minimum protection standards, and the provinces those necessary to reinforce them, without altering their local jurisdictions.
>
> The entry into the national territory of present or potential dangerous wastes, and of radioactive ones, is forbidden.

The Uruguayan Constitution, in turn, included the matter in its Constitution in two stages. The first step, in 1996, consisted of adding the following provisions to Article 47 of its "Rights, Duties and Guarantees" chapter:

> Environmental protection is a matter of common interest. All persons should abstain from committing any act that may cause serious degradation, destruction or contamination of the environment. The law shall regulate this disposition and may provide sanctions for transgressors.

Subsequently, in 2004, by virtue of a constitutional referendum and as approved by over 64 % of the voters, Uruguay added the following text to Article 47 of its Constitution[1]:

> Water is a natural resource essential to life.
>
> Access to drinkable water and access to sewerage constitute fundamental human rights.
>
> 1) The national policy of Waters and Sewerage will be based on:
>
> a) the arranging of the territory, conservation and protection of the environment and restoration of nature.
> b) sustainable management, based on solidarity with future generations, of water resources, and preservation of the hydrologic cycle, which constitute matters of general interest. The users and the civil society, will take part in all instances of planning, management and control of water resources; with hydrographic basins established as basic units.
> c) the establishment of priorities for the use of water by regions, basins, or parts of them, being the first priority the supplying of drinkable water to populations.
> d) the principle that provision of drinkable water and sewerage services will have to be done placing reasons of social order before those of economic nature. Any authorization, concession or permission that in any manner infringes these principles will have to be stopped without any responsibilities for the State.
>
> 2) Superficial, as well as the underground waters, with the exception of rainwaters integrated into the hydrologic cycle, constitute an unitary resource, subordinated to the general interest, which will be part of the State's public domain as hydraulic public domain.
>
> 3) Public sewerage services, and public services of water for human consumption will be provided exclusively and directly by State legal persons.
>
> 4) The law, by the three fifths of the total of components of each Chamber, may authorize the supply of water to another country, when the latter lacks the necessary supplies and by reasons of solidarity.
>
> Also, Article 188 complements the above statements indicating,
>
> The provisions of this article will not be applicable to essential services of drinkable water and sewerage.
>
> *Transitory and Special provisions*
>
> The compensation that could arise because of the entering into force of these reforms will not generate any indemnization for lucrum cessans: only non-amortized investments will be refunded.

[1] Non-official translation by the NGO "Redes Uruguay", with necessary changes introduced by the authors.

22.3 Argentina-Uruguay Constitutional Text Comparisons[2]

22.3.1 Differences Stemming from the Structure of the State in Each Country

The fact that Argentina possesses a federal system of government (even though states are called "provinces," rather than "states") explains the fact that before the National Constitution was amended in 1996, a number of provinces/states had recognized the human right to the environment and the mirror public obligation to respect it, while in Uruguay – with a unitary form of government – rules of similar hierarchy could not be sanctioned in the local realm at the level of the *departamentos*, the latter being administrative semiautonomous territorial units under the single Constitution of the country.[3]

Such difference, in turn, determines the relative heterogeneity of the provincial constitutional and legal environment provisions in the Argentinean Republic, compared with the uniqueness of the constitutional terms of the Uruguayan Constitution. Thus, the situation, for the Argentinean legal operators (judges, attorneys, government agencies, including environmental/natural resource agencies, domestic and foreign investors), is somewhat confusing and creates ample opportunity for unwanted sources of litigation. Paragraph 3 of the above-quoted Section 41 of the National Constitution of Argentina opens the door to the possibility that "the Nation" defines the "minimum protection standards" in terms inconsistent with what seems to be the "pro-province" spirit of the Constitution, namely, reducing – rather than expanding or respecting – the powers of the provinces. Also controversial, in our opinion, is the National Constitution provision protecting the "local" (i.e., sub-provincial) autonomies, without including equivalent caveats to safeguard both local and provincial environmental jurisdictions from abuses – under the guise of extra protection – from the National Government.

The sub-constitutional levels of Argentina's legal system clearly reflect the above-referred contradiction stemming from a federal system the constitutional text has not fully succeeded in homogenizing.

Most provinces have positively addressed the environmental problem by incorporating regulations in such areas, which, obviously, are only enforceable within the territorial jurisdiction of each province. To be remembered is that only

---

[2] The full texts of the Federal Constitution of Argentina and the Constitution of Uruguay are available at http://leyes-ar.com/constitucion_nacional.htm and http://www.presidencia.gub.uy/normativa/constitucion-de-la-republica, respectively.

[3] Under Sección 262 of the Uruguayan Constitution, except for the public security services, which are national in scope, the government and management of each *Departamento* is entrusted to its Mayor and its Departmental Council (the executive and legislative powers within the limits of such territorial units, respectively).

in 1994 the Constitution (Section 41) mandated the Nation to regulate the provinces' "minimum protection standards" and required "the provinces [to sanction such standards as] necessary to reinforce them, without altering their local jurisdictions." In such context, a dense body of judicial decisions had seen the light[4] in the environmental realm before 1994, and even now "the Nation" has not yet fulfilled its constitutional mandate of setting minimum protection standards.

Progress seems to be mostly spearheaded by the Argentinean Judiciary, as remarked by Peyrano (1997), who highlights the implicit power of the courts to mandate the *restoration* of the environment when illegally impaired (rather than just awarding damages to the victim), and by foreign scholars (Duaygües 2014). The significant degree of "subjective interpretation" of court decisions remarked by Sánchez-Sáenz (2011) certainly makes case law a valuable instrument for progress in a relatively novel area such as that of Environmental and Natural Resources Protection Law. Instruments such as the so-called collective *amparo* (Sabsay 1996) are clear signs of the court-driven progress in the procedural law sphere.

Such nonlegislative help is further needed given the practical impossibility for any single province to regulate environmental issues as solely concerning its specific territory, since environmental problems seldom respect political borders. Such interprovincial legal "treaty" area is, to say the least, underdeveloped in Argentina.

A few examples for the environmental legislation in the largest and most industrialized provinces of Argentina (Cordoba, Buenos Aires, etc.) and the growing array of national laws give an idea of the inevitable interpretation challenges. The range of legal instruments that follow is but an example of the large spectrum of choices law enforcement entities are faced to in their legal harmonization efforts.

- In 1985, the *Province of Cordoba* passed its Environmental Preservation, Conservation, Defense, and Improvement Act.[5]
- Law 11.459 for the *Province of Buenos Aires*, on localization of industrial concerns, specifically mandates environmental assessments and other requirements for sustainable development.[6]

---

[4] In Argentina, a civil law country, case law is known as an "informal source of law," meaning courts' judgments (particularly those of the higher courts) do have bearing on future courts' decisions, even if the latter are not required to mimic legal precedents.

[5] Law No. 7343 *Principios Rectores para la Preservación, Conservación, Defensa y Mejoramiento del Ambiente* (Guiding Principles on Preservation, Conservation, Defense, and Improvement of the Environment). http://www.ambiente.gov.ar/?aplicacion=normativa&IdNorma=799&IdSeccion=0

[6] http://www.gob.gba.gov.ar/dijl/DIJL_buscaid.php?var=245

- Environmental-related federal laws include[7]
  - The General Environmental Law No. 25.675 on minimum budgets for sustainable and adequate management of the environment and biodiversity preservation and protection
  - The Federal Air Resource Preservation Law No. 20.284[8]
  - The Law No. 24.051 on Dangerous Residues
  - A range of (federal) civil and penal environmental law provisions, e.g., Art. 1113 of the Civil Code on civil liability; Law 14.051 (Art. 55) on Hazardous Waste Pollutants and personal liability of directors of corporations for environmental damages caused by their concerns. As remarked by:
  - The Law 25.612 on industrial and service activity residues
  - The Law 25.670 on management and disposal of PCBs
  - The Law 25.688 on the "Water Environmental Management Regime," including water-basin interprovincial committees
  - The Law 25.831 on "Free Access to Public Environmental Information" for the National Government and all provincial, municipal jurisdictions and those of the Autonomous City of Buenos Aires, as well as autonomous entities and private, public, or public-private partnerships providing public services
  - The Law 26.093 establishing the Regulation and Promotion Regime for Sustainable Production and use of "Biofuels"
  - The Law 2631 on Minimum Budgets for the Environmental Protection of the Native Forests

Interesting traits of the environmental legislation in Argentina determine that certain key pieces of federal legislation are open to replication by the provinces (voluntary adhesion by the provinces), and the federal civil and penal code environmental provisions are automatically enforceable at provincial level (mandatory implementation) (Alfaro 2014). Such mechanisms represent a measure of flexibility of the federal-vs.-provincial jurisdictions and reflect the importance given to the environmental matters, in what cannot but suggest new bases for the structure of the State that counter the process of centralization (Bidart Campos 1972). Certain jurisdictions (e.g., the provinces of Cordoba and Buenos Aires, the Autonomous City of Buenos Aires) have introduced more detailed/advanced environmental legislation than those of their territorial counterparts (Cámara de Comercio de los Estados Unidos en la República Argentina 2012).

[7] The complete texts of the Federal Argentinean laws the following excerpts belong to are available *in the Government of Argentina website:* http://www.infoleg.gov.ar/

[8] Regulations pending.

In addition, a significant number of international treaties and conventions on environmental matters have been adhered to by Argentina, e.g.,

Legal instrument[9]	Approval
UN Framework Convention on Climate Change	Law 24.295
Kyoto Protocol	Law 25.438
UN Convention on Protection of the World Cultural and Natural Heritage	Law 21.836
Montreal Protocol on Ozone Layer Depleting Substances	Law 25.389
MERCOSUR Framework Agreement on the Environment	Law 35.841
United Nations Convention to Combat Desertification	Law 24.701
Basel Convention	Law 23.922
Convention on Biodiversity	Law 24.375
Protocol to the Antarctic Treaty on Environment Protection	Law 24.261
Convention on Wetlands of International Importance	Law 23.919
Vienna Convention of Protection of the Ozone Layer	Law 23.724

Such combination of domestic legal instruments and international environmental treaties enforceable in Argentina does not contribute to easing the difficulties involved in implementing the environmental instruments in specific cases.

Furthermore, constitutional texts such as the one which forbids "the entry into the national territory of present or potential dangerous wastes" (Art. 41) add to the hardships of having to construe specific terms in concrete circumstances and determine, for instance, how "dangerous" need wastes to be for them to be deemed "dangerous" *from the constitutional standpoint*, thus forbidden, or what the exact time frame is to be considered for wastes no longer to be *constitutionally* presumed "potentially" dangerous (and allowed to enter the Argentinean territory).

Uruguay has also adhered to most international conventions and treaties on environment protection, as well as to multilateral regional environmental instruments sponsored by the Common Market of the South/Mercosur, such as with the European Union (European Commission 2007), as well as to bilateral treaties with a number of States, including several non-regional ones, such as Bulgaria, Russia, and Switzerland.

[9] Such international instruments texts/other references are available at the site of the Ministry of Foreign Affairs of Argentina at http://tratados.cancilleria.gob.ar/busqueda.php

Convention/treaty	Legal instrument of approval by Uruguay[10]
United Nations Frame Convention on Climate Change	Law No. 16.517
Vienna Convention for the Protection of the Ozone Layer	Law No. 15.986
Convention on Biodiversity	Law No. 16.408
United Nations Kyoto Protocol of 1997 on Climatic Change	Law No. 17.279
Basel Convention on the Control of Transboundary Movements of Hazardous Wastes and their Disposal	Law No. 16.221
MERCOSUR Frame agreement on the Environment – Decision N° 02/01.	Law No. 17.712
United Nations Convention to Combat Desertification	Law No. 17.026
Montreal Protocol on the Ozone Layer – Amendment	Law No. 16.744
Protocol Antarctic Treaty on Protection of the Environment	Law No. 16.518
Additional Protocol on the Environment of the American Convention on Human Rights	Law No. 16.519
United Nations Convention on the Prohibition of Military or Any Other Hostile Use of Environmental Modification Techniques	Law No. 16.374

The unitary structure of the Uruguayan State makes things easier for the interpreter; on the other hand, the Uruguayan environmental law system is less developed and includes more programmatic or aspirational provisions, which in our view does not represent a sound legal drafting technique.

22.3.2 *Differences Related to the Status of International Treaties*

Section 22 of the Argentinean Constitution states that certain international treaties on human rights (with the list being susceptible to amplification) are to be considered part of "the supreme law of the Nation," and on equal footing with the Constitution proper. In terms of human rights, they might even be above the Constitution, as can be seen in section 22, which reads:

> Treaties and concordats have a higher hierarchy than laws.
>
> The American Declaration of the Rights and Duties of Man; the Universal Declaration of Human Rights; the American Convention on Human Rights; the International Pact on Economic, Social and Cultural Rights; the International Pact on Civil and Political Rights and its empowering Protocol; the Convention on the Prevention and Punishment of Genocide; the International Convention on the Elimination of all Forms of Racial

[10] The full texts of Uruguayan laws are available at its national Government website: http://www.parlamento.gub.uy/indexdb/leyes/ConsultaLeyesSIPXXI.asp, with supplementary information to be found at the country's Ministry of Foreign Affairs site (http://www.mrree.gub.uy/frontend/page?1,inicio,tratados-busquedas,O,es,0) Given the importance ascribed in this paper to the Environmental Law/Reality connection, see also Organization of American States, 1992, particularly Item 4.2.6 and Annex 7.1.

> Discrimination; the Convention on the Elimination of all Forms of Discrimination against Woman; the Convention against Torture and other Cruel, Inhuman or Degrading Treatments or Punishments; the Convention on the Rights of the Child; in the full force of their provisions, they have constitutional hierarchy, do no repeal any section of the First Part of this Constitution and are to be understood as complementing the rights and guarantees recognized herein. They shall only be denounced, in such event, by the National Executive Power after the approval of two-thirds of all the members of each House.

No equivalent provisions exist in the Uruguayan Law, which makes international treaties enforceable only to the extent they had been ratified by Uruguay. In other words, international treaties only become obligatory upon ratification by the Congress followed by their promulgation (publication) by the executive branch of government, which turns them into regular domestic laws, in spite of their international nature and origin. Section 85(7) of the Uruguayan Constitution refers to this in conferring the General Assembly (namely, the Chamber of Deputates and the Senate acting together) the exclusive power to:

> ...approve, or refuse to approve, by an absolute majority of the total membership of each chamber, such peace, alliance, or trade treaties, as well as such conventions or contracts of any sort as agreed on by the Executive Power with a foreign power.

Such difference makes the scope of environmental sources of law wider in Argentina than in Uruguay, perhaps not in practical terms, since Uruguay has ratified most major environmental global treaties and conventions, but undoubtedly in potential terms, since international treaties which the country has not ratified as stipulated in the above-quoted constitutional subsection are not enforceable in its territory.

22.3.3 Differences Originated in the Constitutional Approach to the Nature of the Right-Holder or the Obligor in the Environmental Protection Equation

Argentina addresses the statute of the environmental values by identifying "the inhabitants" (of its territory) as those parties who enjoy the *right* to an environment having the attributes specified in the Argentinean Constitution. Uruguay, however, uses a different outlook and approaches the matter from the perspective of the *duty* of natural and legal "persons" to refrain from seriously depredating, destroying, or polluting the environment.

Such approach in the constitutional text of Uruguay has been criticized for failing to make the necessary emphasis in the *right* to a sound and safe environment. We consider such objection to be excessive, as illustrated by local scholars (Caumont 1997) who describe the environment "as the axis for a subjective right." If so, the "right" notion would be implicit in all mentions of the environment. Moreover, we note that no obligation (duty) makes sense (or is legitimately – rather than arbitrarily or despotically – enforceable) unless someone has *locus standi*[11] to demand such obligation be fulfilled under the law. Such right-holder is clearly identified in Articles 7 and 72 of the Uruguayan Constitution cited below (stress added):

> Article 7. The inhabitants of the Republic have the right of being protected in the enjoyment of life, honor, liberty, security, labor, and property. No one may be deprived of these rights except in conformity with laws which may be enacted for reasons of general interest.

> Article 72. The enumeration of rights, duties, and guarantees made in this Constitution does not exclude others which are *inherent in human beings* or which are derived from a republican form of government.

The Argentinean text, however, in our opinion, is more explicit than its Uruguayan counterpart (Art. 47, *supra*), in that the latter sows confusion between two mandates: active protection vs. obligation to refrain from acts that cause depredation, etc. Quoting again the first portion of such article:

> Environmental protection is a matter of common interest. All persons should abstain from committing any act that may cause serious degradation, destruction or contamination of the environment,

where the interpreter can contend (i) the second sentence has been included simply to clarify the idea of "protection," i.e., to convey the notion that the constitutionally mandated protection consists of (and only of) the prohibition for any "person" to incur any of the three types of acts mentioned in the article (degrade, destruct, contaminate) or contend (ii) such three types of behavior are the only unacceptable ones (taxative interpretation). Option (ii), if accepted, would seriously and unjustifiably narrow the scope of the "protection" desired by the constituents, in that if would leave aside active means of safeguarding and enhancing the environment.

Another issue with this Uruguayan text is that it fails to specifically enumerate the types of "persons" that are prohibited to act in detriment of the environment. Legal persons, including public bodies, should in our opinion have been clearly mentioned, no matter how obvious the hermeneutic result of such mention can be. Leaving the question to scholarly opinions rather than resolving it beforehand through transparent constitutional provisions ignores the reality that large financial interests can be involved in actual environmental disputes and make expert opinions less aseptic than they ought to be. Such scenario would be less likely if the constitutional text had dispelled such specifically mentioned both person categories.

[11] Namely, "[a] right to appearing in a court of justice, or before a legislative body, on a given question." Black's Law Dictionary. West Publishing Co., 4th Edition.

22.3.4 *Differences Connected with the Features of the Environment the Inhabitants Are Entitled to Enjoy*

The Argentinean Constitution specifies a series of qualities of the environment its Uruguayan counterpart remains silent about. The environment, as contemplated by Argentina, must be "healthy; balanced after any natural or man-made changes experienced, and fit for sustainable intergenerational human development."

However, the second paragraph of Article 41 of the Argentinean Constitution needs to be harmonized with both the first paragraph thereto and with the fact that it involves an *environment* notion whereby such concept includes:

1. The *natural and cultural heritage*
2. The *rational use of natural resources*
3. The *preservation of the biodiversity for future generations*
4. The obligation of the authorities to implement *environmental information and education* activities
5. The prohibition to introduce actually or potentially dangerous *residues* and radioactive *residues* in the nation's territory

The detailed definition cited above marks a fundamental difference with its Uruguayan counterpart, where such broad concept is completely absent.

Such methodological difference between the texts being considered in this paper leaves open the well-known question of the advantages or disadvantages of detailed constitutional provisions. Without going too deep into the discussion of this lateral matter, both texts can be criticized: the Argentinean one for being too meticulous and the Uruguayan one for being excessively sketchy. Both alternatives, in our opinion, create interpretation risks, particularly because the legal doctrine has not yet produced any universally accepted definition of "environment." Argentineans adopt a series of elements that seem to stem from the opinion than the human being himself/herself is an element of his environment[12] (an approach logically questionable, since "environment," ontologically, is "something that surrounds," and it is hard to conceive that something or somebody – here, human beings – can be simultaneously surrounded by something and part of such surrounding matter).[13]

---

[12] See, e.g., G. Bidart Campos, *Elementary Treaty of the Argentinean Constitutional Law*, Ediar, Buenos Aires, 1995, p. 298.

[13] "Environment:...*something that surrounds*...the conditions, circumstances, and influences surrounding and affecting the development, of an organism or group of organisms." Webster New World Dictionary of the American Language, 2nd College Edition. (stress added).

22.3.4.1 Procedural Differences

The Argentinean Constitution includes in its Article 43 a specific reference to the right of:

> [any person to] file an expeditious and swift action of 'amparo' whenever no other more appropriate judicial means exist, in order to challenge any act or omission by public authorities or private individuals, that presently or imminently harms, restricts, alters or threatens, in an arbitrary or openly illegal manner, the rights and guarantees recognized by this Constitution, a treaty, or a law.

The Uruguayan law, however, regulates the "amparo" procedure in an ordinary Act of Parliament (Law No. 16011),[14] does not mention treaties, and, in what is most relevant, does not modify, in regard to "amparo," the general rules set by the Civil Code, which exclude the possibility that "any person" files such action: even in the case that the petitioner is a legal representative of an environmentalist entity (rather than "any" individual, like in Argentina). The legal standing for "amparo" of such legal person will be evaluated case by case by the Uruguayan judge having jurisdiction, who can recognize or reject his/her standing.

22.3.4.2 Similarities

As described above, both Constitutions share a number of important traits. None of the texts being compared includes a formal definition of the environment, although the Argentinean formulae come much closer to a definition in incorporating a list of characteristics of the "environment" deserving protection.

Both constitutions invoke and incorporate natural law tenets. This is crucially important, since such feature tacitly operates as a limitation for the ability of the government to revoke or substantially reduce the scope of fundamental rights, including, people's water rights. The Uruguayan constitutional text does it in the following terms:

> Article 7. The inhabitants of the Republic have the right of protection in the enjoyment of life, honor, liberty, security, labor, and property. No one may be deprived of these rights except in conformity with laws which may be enacted for reasons of general interest.

> Article 72. The enumeration of rights, duties, and guarantees made in this Constitution does not exclude others which are inherent to the human nature or are derived from a republican form of government.

The above text is, therefore, more explicit than that of Argentina, which reads:

> Article 33. The declarations, rights and guarantees enumerated by the Constitution shall not be construed as a denial of other rights and guarantees not enumerated therein, but which issue from the principle of the sovereignty of the people and from the republican form of government.

[14] *Diario Oficial* of Uruguay, 29th Dec., 1988 – N° 22776.

As to the right holder, we think both Constitutions coincide, despite the fact that, as mentioned before, the Uruguayan text addresses the matter from the standpoint of duties, whereas the Argentinean approach consists of taking the rights of the beneficiaries as the point of departure.

Public entities, including the State as the major one, are, in both Constitutions, obligors in the environmental duties realm. Their obligations arise both from their potential to be violators of environmental rights of others and as hypothetical transgressors of their constitutional duty to protect the environment.

Such similarities are being enhanced – and the differences assuaged – by the fact that both countries are members of the MERCOSUR (under the March 1991 Treaty of Asuncion, Paraguay). The Common Market Council of MERCOSUR has sanctioned, among other instruments, the MERCOSUR Environmental Framework, 2001. A Specialized Environment Committee has also been created, as formed by the environmental public authorities of the member countries and responsible for providing advice to the Common Market Council. As mentioned before, both countries have also ratified treaties on the Uruguay River and the La Plata River limits (7 April 1961 and 19 November 1973, respectively) and accepted (along with the Latin American States at large) the "ecological good neighboring" principle.[15]

In what could be termed a "negative similarity," neither country has so far followed the example of Chile in creating a public autonomous entity responsible for the coordination of all environmental matters, the National Commission for the Environment (CONAMA, acronym for its name in Spanish) [Section 69 of Law 19.300 of 2004 on General Environmental Bases[16]]. CONAMA is a decentralized agency supervised by the President of the Republic.[17] However, water use is granted by a different agency, namely, the Dirección General de Aguas (DGA), based on availability of water resources. While exclusions apply when water is used for environmental conservation, water concessions for productive use may be lost when recipients do not consume the allocated volume (Donoso and Bosch 2014).

22.4 Final Remarks

A conclusion as to whether optimism is justified on the perspectives of progress in coordination and unification of environmental rules should not be confined, however, to the study of the relevant constitutional texts. Throughout the text of this chapter, we have presented and analyzed cases where additional efforts are

---

[15] Resolution 334, adopted during the XVth Regular Session of ECLAC. (Quito, Ecuador, 1973). http://www.cepal.org/cgi-bin/getprod.asp?xml=/noticias/paginas/1/21491/P21491.xml&xsl=/tpl-i/p18f-st.xsl&base=/tpl-i/top-bottom.xsl

[16] http://www.leychile.cl/Navegar?idNorma=30667

[17] Notwithstanding the mandate of CONAMA, the Chilean structure has been criticized as insufficient (Fernández Bitterlich 1998).

reasonably to be expected toward better implementation of constitutional technique tools, particularly in order to dispel interpretation issues, and we believe what we described at the beginning as the historic "legal cross-fertilization" of both countries can be conceived as a tool for both sister nations to advance together in so crucial constitutional matters as those of environmental (and particularly water) legal regimes.

Such efforts seem particularly necessary in the context of the protracted conflict between Uruguay and Argentina on the Uruguay River paper mills. The controversy gave rise to a 4-year-long blockage of the Fray Bentos-Gualeguaychú international bridge on the Uruguay River by Argentinean activists and ended with a ruling by the International Court of Justice in April 2010 (American Society of International Law 2010) stating that (1) Argentina had not proven contamination by the mills, but (2) Uruguay had nonetheless violated its Uruguay River Treaty with Argentina by failing to provide its neighbor prior clear information on the scope and nature of the activities intended to be implemented by the Finnish *Botnia* corporation on Uruguayan territory[18] (*Arbitration News* 2010).

Nonetheless, progress is observed in the area of the environmental regulations in Uruguay in what *The UNESCO Courier* described as "A World First," namely, the constitutionalization of mandatory state-managed ownership of water resources, a reform approved by 64.58% of the Uruguayan voters in the 2004 constitutional referendum (please see the above-transcribed Article 47 of the Constitution of Uruguay). Such constitutional amendment reflects an ongoing process started by the Uruguayan Constitution of 1967 (Cousillas 1998) and has been described as a worldwide initial step in a *policy of inevitability* (of the privatization of water public services) with specific reference to the Uruguayan constitutional 2004 reform (Hall et al. 2010).

Admittedly, the process of *legal unification* seems to face limits, but we firmly believe that if such process is to succeed, nations having so much in common as Argentina and Uruguay will be among the hemispheric forerunners. After all, proliferation of disparate legislations on crucial matters, such as water rights, is proven breeding grounds for long-lasting disputes that should be prevented.

An interesting side point in this respect is that Kenya – rather than any Latin American country – has demonstrated a degree of interest in taking the Uruguayan innovation as a model for its water resources law. Such paradox should spur LAC nations to further contacts and joint efforts toward solutions that cannot but be easier to achieve among countries sharing so many common challenges and values, born out of a common heritage.

Last but not least, *progress* in the environmental legal realm cannot be achieved unless three elements simultaneously coalesce: "theoretical definition and rationale,

[18] Walter Raymond contends the outcome was the worst scenario to be expected for both countries, since it condemns them to the "punishment of reaching an agreement" (on monitoring of the Finnish paper mill and the future compliance of agreements regarding the Uruguay River). Source: *Global Voices*, 2010 http://globalvoicesonline.org/2010/04/21/argentina-international-court-rules-in-paper-mill-conflict-with-uruguay/

international and national standard sanctioning, *and the required effectiveness*"[19] (Blengio 2003). We have underscored the third prong of the triad since it often appears to be the weakest one in our hemisphere.

The importance of the synergies among those three factors is also highlighted by the eminent Uruguayan scholars Magariños and Gorosito in *Medio Ambiente y Sociedad* (2005), who eloquently refer to the notion that environmental policy and law in general need to be based on reality (praxis), thus convincingly supporting and stressing the indispensable nexus between law and reality. In this same vein, Sagües (2004) ironically described as *constitutional alchemy* the preconcept that legal instruments are independent from human realities and enjoy unlimited power to model realities, which, as he further proclaims, need to be overcome by a legal and policy approach based upon and deferential to such realities.

Bibliography

Alfaro E (2014) The environmental laws in Argentina. http://www.alfarolaw.com/tapa/The%20Environmental%20Laws%20in%20Argentina.pdf

American Society of International Law (2010) Pulp mills on the River Uruguay: The International Court of Justice recognizes environmental impact assessment as a duty under international law. http://www.asil.org/insights/volume/14/issue/9/pulp-mills-river-uruguay-international-court-justice-recognizes

Arbitration News (2010) ICJ makes ruling on environmental protection – Pulp mills on the River Uruguay (Argentina v Uruguay). http://hsf-arbitrationnews.com/2010/05/07/icj-makes-ruling-on-environmental-protection-pulp-mills-on-the-river-uruguay-argentina-v-uruguay/, April

Bidart Campos G (1972) Manual de Derecho Constitucional Argentino. Ediar, Bs.As

Blengio M (2003) Derecho Humano a un Ambiente Sano. Revista de Derecho, Universidad de Montevideo. Year II, No. 4. http://revistaderecho.um.edu.uy/wp-content/uploads/2012/12/Revista-de-derecho-A%C3%83%E2%80%98O-II-2003-N4.pdf

Cámara de Comercio de los Estados Unidos en la República Argentina (2012) Environmental handbook. http://www.agendasocialweb.com.ar/media/uploads/pdf/amcham-environmental.pdf

Caumont A (1997) The environment as the axis for a subjective right and its co-respective duty. Uruguayan Civil Law Yearbook, FCU, Montevideo Uruguay

Constitución de la Nación Argentina. http://leyes-ar.com/constitucion_nacional.htm

Constitución de la República Oriental del Uruguay. http://www.presidencia.gub.uy/normativa/constitucion-de-la-republica

Cousillas M (1998) Constitutional protection of the environment – reflections on the 1996 Constitutional Reform. FCU, Montevideo

Donoso MC, Bosch MC (2014) Integrated water resource management in Latin America and the Caribbean. In: Setegn S, Donoso M (eds) Sustainability of Integrated Water Resources Management (IWRM) in the face of climate variability and change. Springer, Cham

[19] "En ese aspecto [del Progreso en la reconocimiento del medio ambiente como derecho humano] se destaca la búqueda de su conceptualización y fundamentación teórica, la consagración mnormativa a nivel internacional e interno *y su necesaria efectiividad*" [stress added].

Duaygües M (2014) Dos fallos judiciales representativos en la jurisprudencia ambiental argentina: recomposición y daño residual. Secretaría Nacional de Medio Ambiente y Recursos Naturales. Mexico. http://www2.inecc.gob.mx/publicaciones/libros/398/duaygues.html

ECLAC – Economic Commission for Latin America (1973) Resolution 334, adopted during the XV Regular Session of ECLAC. Quito, Ecuador. http://www.cepal.org/cgi-bin/getprod.asp?xml=/noticias/paginas/1/21491/P21491.xml&xsl=/tpl-i/p18f-st.xsl&base=/tpl-i/top-bottom.xsl

European Commission (2007) MERCOSUR Regional Strategy Paper (2007–2013). http://eeas.europa.eu/mercosur/rsp/07_13_en.pdf

Fernández Bitterlich P (1998) Institucionalidad Ambiental en Chile. Revista de Derecho, Special Issue, 35–42 (Valdivia). ISSN 0718-0950. http://mingaonline.uach.cl/scielo.php?pid=S0718-09501998000100005&script=sci_arttext

Global Voices (2010) Argentina: International Court Rules in Paper Mill Conflict with Uruguay. http://globalvoicesonline.org/2010/04/21/argentina-international-court-rules-in-paper-mill-conflict-with-uruguay/

Gobierno de Argentina. Federal Legislation of Argentina, full texts. http://www.infoleg.gov.ar/

Gobierno de Paraguay (1991) Tratado de Asunción. Asuncion

Gobierno de Uruguay. National Legislation of Uruguay, full texts. Website: http://www.parlamento.gub.uy/indexdb/leyes/ConsultaLeyesSIPXXI.asp, with supplementary information at the country's Ministry of Foreign Affairs site: (http://www.mrree.gub.uy/frontend/page?1,inicio,tratados-busquedas,O,es,0). See also Estudio impacto ambiental – Uruguay. pshconsultora.com.ar (2014) Todo el Derecho http://www.todoelderecho.com/Uruguay/ambiental.htm

Hall D, Lobina E, De la Motte R (2010) Making water privatization illegal: – new laws in Netherlands and Uruguay. Public Services International. http://PSIRU.ORG/REPORTS/2004-11-w-CRIM.doc

Magariños MJ, Gorosito R (2005) Medio Ambiente y Sociedad. Editorial F.C.U, Montevideo

MERCOSUR (2001) Framework Agreement on the Environment of MERCOSUR. Montevideo. http://www.ecolex.org/server2.php/libcat/docs/TRE/Full/En/TRE-153663.pdf

Ministerio de Relaciones Exteriores y Culto de la Republica Argentina – Digital Library. http://tratados.cancilleria.gob.ar/busqueda.php

Organization of the American States (1992) National Environmental Study, Item 4.2.6 – Main Limitations of the Legal Framework, and Annex 7.1, Environmental Protection Tenets. Uruguayan Government. OAS, IDB, Washington, DC. http://www.cicplata.org/pdf_oea/p9.pdf

Peyrano G (1997) Effective execution of environmental judgments. Jurisprudencia Argentina N° 6068 31-37. Buenos Aires

Sabsay DA (1996) Environmental protection through the so-called collective Amparo. El Derecho 167(4/16/96), Buenos Aires, Argentina, 61

Sagüés NP (2004) Interpretación constitucional y alquimia constitucional. Revista Iberoamericana de Derecho Procesal Constitucional, ISSN 1870-8390, N°. 1, pp 151–170

Sánchez Sáez AJ (2011) La "restitutio in pristinum" como mecanismo deseable para la reparación de los daños causados al medio ambiente. http://personal.us.es/patroclo/publicaciones_pdf/la_restituio_in_pristinum.pdf

Schulz A, Muriá Tuñón A, Villanueva Meza R (2005) Les instruments américains du droit international privé – Une note sur leurs rapports avec une future convention de La Haye sur les accords exclusifs d'election de for. Document préliminaire No. 31 de juin 2005 à l'intention de la Vingtième session de juin 2005, Conférence de la Haye du Droit International Privé. http://www.hcch.net/upload/wop/jdgm_pd31e.pdf

UNESCO (2006) When water becomes a political challenge. The UNESCO Courier, May 2007. ISSN 1993-8616. http://unesdoc.unesco.org/images/0019/001915/191576e.pdf

Part V
Climate Change Resiliency Actions Related to Water Resources Management Sustainability

Chapter 23
The Importance of Water-Energy Nexus for Sustainable Development: A South America Perspective

Janaina Camile Pasqual and Shimelis Gebriye Setegn

Abstract Water and energy are indispensable to the social and economic development of a country. The rising pressure on resource demands, new production, and consumption models requires a better understanding about the connections between water and energy. In a world of growing population and urbanization, cities are becoming the focus of international efforts of sustainability. According to the United Nations, over 100 years ago, only 10 % of the world population lived in the cities. Nowadays, this rate is of 15 %, and the tendency is for that number to rise over the next years, reaching 75 % in 2050. In order to accommodate this increasing population pressure, cities need to become more intelligent, well prepared, and organized, aiming to reduce poverty, providing education and health, managing and optimizing natural resources, protecting the environment, and facing climate change.

This chapter gives a better understanding about the water-energy nexus. It shows the importance of an integrated engagement in future interdisciplinary research and development to target water-use efficiency in the energy sector and energy efficiency.

Keywords Water and energy nexus • Sustainability • Water resources • Energy • Urban development

J.C. Pasqual (✉)
International Center of Hydroinformatic, Foz do Iguassu, Brazil

International Center of Renewable Energy-Biogas, Foz do Iguassu, Brazil

Pontifical Catholic University of Parana, Curitiba, Brazil
e-mail: janaina@maxc.com.br

S.G. Setegn
Department of Environmental and Occupational Health and Global Water for Sustainability Program (GLOWS), Florida International University, 11200 SW 8th Street, Miami, FL 33199, USA
e-mail: ssetegn@fiu.edu

S.G. Setegn, M.C. Donoso (eds.), *Sustainability of Integrated Water Resources Management*, DOI 10.1007/978-3-319-12194-9_23

23.1 Introduction

In a world of growing population and urbanization, cities are becoming the focus of international efforts of sustainability. According to the United Nations (2012a), over 100 years ago, only 10 % of the world population lived in the cities. Nowadays, this rate is of 15 %, and the tendency is for that number to rise over the next years, reaching 75 % in 2050. Urban population growth went from 750 million in 1950 to 3.6 billion in 2011, and it is also projected to gain 2.6 billion, passing from 3.6 billion in 2011 to 6.3 billion in 2050.

Predictions from the Organization for Economic Co-operation and Development (OECD 2011) depict that three-quarters of the world's population will begin to occupy areas that are currently covered by vegetation, possibly using more water and other natural resources without restriction. This scenario leads to a dramatic growth in demand for new infrastructure and energy and water consumption, resulting in various challenges for the public and private spheres.

In order to accommodate this increasing population pressure, cities need to become more intelligent, well prepared, and organized, aiming to reduce poverty, providing education and health, managing and optimizing natural resources, protecting the environment, and facing climate change (UN 2012a). According to UNEP (2012), currently "urban areas contributes up to 75 % of the total waste generation, producing from 60 to 80 % of all the greenhouse gas emission, consuming 75 % of the natural resources and corresponding to two-thirds of the energy consumption.

The most recent United Nations World Water Development Report, published in 2014, depicts that water resources on the planet are under pressure due to rapid growth of population and the demands for water, and, if there is not an implantation of public policies focusing on water resources and renewable sources of energy, this scenario tends to become serious.

Considering that water and energy are inseparably linked, the present chapter aims to analyze the scenario of the water and energy nexus and sustainable management highlighting South America. The chapter briefly addresses public policies which are being implemented in order to avoid the shortage of water and energy resources.

23.2 Theoretical Framework

23.2.1 Overview of Sustainable Development and Water Concerns

Sustainable development integrates economic development, social development, and environmental protection. According to the United Nations' current conception, sustainable development has three objectives and essential requirements:

(1) poverty reduction, (2) changing unsustainable patterns of production and consumption, and (3) protecting and managing the natural resource base of economic and social development (UNDESA 2008). From the 1970s, water started to appear as one of the most important chapters in the international agenda. After the United Nations Conference on the Human Environment, held in Stockholm (1972), the Mar del Plata conference (1977) was the first genuinely global multilateral event to focus on the issue of water, under the auspices of the United Nations. The decade of the 1980s was declared the "International Drinking Water Supply and Sanitation Decade," with the premise that "all people, whatever their stage of development and their social and economic conditions, had the right to have access to drinking water in quantity and of a quality equal to their basic needs" (UN 1977). The World Commission on Environment and Development was created in 1983, after an evaluation of 10 years since the Stockholm conference, aiming to promote hearings around the world and produce a formal result of the discussions. The resulting document defined that "sustainable development is the development that meets the needs of the present without compromising the ability of future generations to meet their own needs" (UN 2010). The report made direct reference to the water concern, declaring that:

> Many of us live beyond the world's ecological means, for instance in our patterns of energy use.... At a minimum, sustainable development must not endanger the natural systems that support life on Earth: the atmosphere, the waters, the soils, and the living beings. (UN 2012b)

The concept of sustainable development was definitively incorporated as a principle during the United Nations Conference on Environment and Development (UNCED) and The Earth Summit in 1992. It was emphasized that the sustainable development seeks the equilibrium between environmental protection and economic development, and it was the basis for the formulation of the Agenda 21, with which more than 170 countries compromised in the occasion of the Conference.

The Agenda 21 represented a comprehensive set of goals for the creation of a more sustainable and balanced world, and the theme of water resources was a relevant chapter of the document, not only because of its length but for concerning a point of interest that is strategic to all countries. The "multi-sectoral character of the development of water resources in the context of socioeconomic development" was recognized, "as well as the multiple interests in the use of these resources." Chapter 19 included seven programmatic areas: (a) integrated water resources development and management; (b) water resources assessment; (c) protection of water resources, water quality, and aquatic ecosystems; (d) drinking water supply and sanitation; (e) water and sustainable urban development; (f) water for sustainable food production and rural development; and (g) impacts of climate change on water resources (UN, Agenda 21, 1992).

The crucial importance of water to the various aspects of human health, development, and well-being has led to specific objectives concerning water and the support to each of the eight Millennium Development Goals (MDG), established by the UN in the year 2000.

In 2003, the Executive Director Council created the "UN Water" – an interagency mechanism to coordinate the actions of the United Nations System and reaching the goals related to water that were on the Millennium Declaration of the World Summit of Sustainable Development of 2002, held in Johannesburg (South Africa). This Declaration affirmed that the Sustainable Development is construed over three interdependent and mutually sustained pillars – social and economic development and environmental protection. According to the UN (2010), these concepts recognized "the complexity and interrelation of critical question such as poverty, waste, environmental degradation, urban decay, population growth, gender equality, health, conflict and violence to human rights." Franco (2001) corroborates with the theme, affirming that the term sustainable development "presents, thus, a dynamic character that moves far from the idea of static balance, referring to a sustainable evolutionary process of continuous change."

In order to further reinforce the global action for attending to the targets of the MDG concerning water, the General Assembly proclaimed the International Decade for Action, "Water for Life" (2005–2015). The Decade started in March 22, 2005, a date that is celebrated annually as the World Water Day (UN 2012a).

From then on, UN Water publishes every 3 years the World Water Development Report (WWDR) that provides detailed analysis of water scenario in the world and evaluates the implications for the world's pursuit of sustainable development and the Millennium Development Goals. It is the fruit of the collaboration by UN Water, the UN interagency coordination mechanism dedicated to all freshwater-related issues. The most recent report, published in 2014, depicts that the water resources of the planet are under pressure of the rapid growth of the demands for water and climate change. The report also affirms that "all the aspects of the development have a component based on water: it is the only resource that connects different sector and through which the main world crisis can be jointly managed" (UN Water 2014).

23.2.2 Urban Growth and Sustainable Development

The accelerated process of urbanization observed in the last two decades has caused a worldwide concern, mainly due to the high demand and consumption of water and energy resources. According to the OECD (2011), 90 % of the urban growth expected until 2030 should occur in cities of developing countries, which are deficient in adequate infrastructure, water supply, sewage system, waste collection, and energy efficiency, making it even more difficult to guarantee health, education, and conservation of the environment. The UN expects that until 2030, more people from all regions of the globe will leave rural zones and go to urban areas, even in Africa and Asia, which are currently among the less urbanized regions of the globe.

According to the report of UN-HABITAT, named "State of Latin America and Caribbean Cities 2012: Towards a new urban transition," until 2030, Latin America will continue to be the most urbanized region of the world, since 91.4 % of its population will be living in urban areas, followed by Europe (90.7 %) and North America (90.2 %). The less urbanized regions will continue to be Asia (66.2 %) and Africa (61.8 %). Nowadays, the region of Latin America and the Caribbean is predominantly urban, although there are areas that have little population.

In a context of new urban realities, Latin America and the Caribbean are molding the conditions for a new urban transition, as much as in terms of resources as of capacity, creativity, and political initiatives from local and national governments. To achieve a model of sustainable cities, it is necessary to reaffirm public commitment in urban planning, working on policies for social and territorial cohesion, in urban national policies and revising legal and institutional framework, since the cities of this region, overall, are and remain the most unequal in the world (UN-HABITAT 2012).

Brazil is the third most urbanized country in Latin America, with about 87 % of its population residing in urban areas, behind Argentina and Venezuela. In Mexico and the countries of the Andean-equatorial region, the urban rate is no more than 85 %; in the Caribbean and Central America, urban growth is constant and should reach 83 % and 75 % in 2050, respectively.

According to the World Water Development Report (WWRD) (2014), "population growth leads to increased water demand, reflecting growing need for drinking water, health and sanitation, as well as for energy, food and other goods and services that require water for their production and delivery." Besides that, intense urbanization will increase the demand for water supply, sanitation, and electricity for domestic use, making it necessary to rethink the utilization of water resources.

Facing this scenario, OECD (2011) affirms that sustainable growth is necessary, because challenges are about to increase as growth continues to deplete natural capital. If that is not taken into account, it will mean an increase on water scarcity, the strangulation of natural resources, more pollution, climate changes, and a loss of irretrievable biodiversity. According to Leff (2013), "the environmental crisis expresses the limits to growth, the unsustainability of economic rationality and technological reason."

23.3 Scenario of Water and Energy in South America

23.3.1 Availability of Water Resources in South America

Approximately, 70 % of the Earth surface is covered by water. Of that total, 97.5 % is saltwater and only 2.5 % is freshwater. Of the total of freshwater (34.6 million km^3), only 30.2 % can be utilized for human life, and 69.8 % is found in polar ice

Table 23.1 World water resources by region

Region	% of world resources
Southern America	28.3
Southern and Eastern Asia	26.8
Europe	15.2
Northern America	15.2
Africa	9.0
Oceania and Pacific	2.1
Central America and Caribbean	1.8
Near East	1.1
Central Asia	0.6

Source: FAO (2003)

caps, glaciers, and frozen ground. Considering the volume of freshwater, about 98.7 % (10.34 million km^3) corresponds to groundwater, and only 0.9 % (92.2 thousand km^3) is surface water, directly available to human demands, which corresponds to only 0.008 % of the total water in the world (UN 2012b).

According to the WWDR 2014, "global water demand (in terms of water withdrawals) is projected to increase about 55 % by 2050, mainly because of growing demands from manufacturing (400 %), thermal electricity generation (140 %) and domestic use (130 %)." As a result, more than 40 % of the global population is projected to be living in areas of severe water stress through 2050, and freshwater availability will be increasingly strained over this period.

Analyzing the availability of water resources on the planet (Table 23.1), it is possible to verify that South America and Asia are the regions that stand out the most, representing 28.3 % and 26.8 % ,respectively, followed by Europe and North America (15.2 %).

South America is the most privileged region in terms of availability of water resources, because it has some of the most important watershed in the world, such as Amazonas and São Francisco (Brazil), Orinoco (Venezuela and Colombia), and the Platine basin (Brazil, Uruguay, Bolivia, Paraguay, and Argentina).

According to the FAO report (2003), titled "Review of World Water Resources by Country," South America can be subdivided into the following subregions:

- Guyana: Guyana, French Guiana, and Suriname
- Andean: Bolivia, Colombia, Ecuador, Peru, and Venezuela
- Brazil
- Southern: Argentina, Chile, Paraguay, and Uruguay

In general, this region is relatively well endowed with water resources, "receiving 26 % of the world's precipitation and generating 28 % of its water resources. The region is home to 5.7 % of the world's population and the water resources per person in the region are about 35.000 m^3/year, well above the world average." (FAO 2003). Figure 23.1 shows the distribution of water resources among the

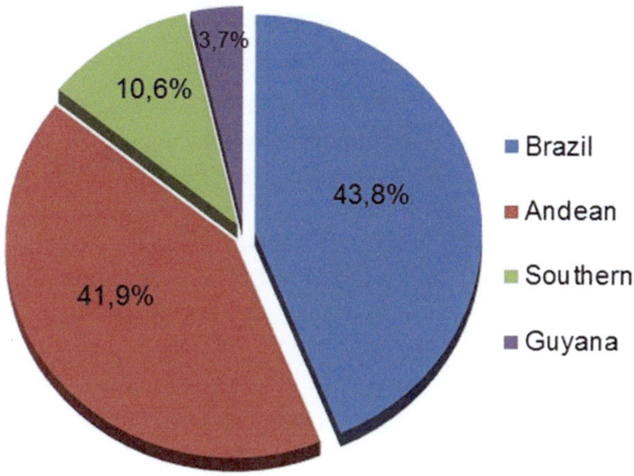

Fig. 23.1 South America water resources by subregions (Source: Author, data from FAO 2003)

subregions, but these averages hide the local scarcity conditions that tend to coincide with the most populated areas of the region and some desert or semidesert areas, in climatic terms. This is the case of Valle Central in Chile, the regions of Cuyo and the South in Argentina, the Peruvian and south Ecuadorian coast, the Cauca and Magdalena valleys in Colombia, the Bolivian Altiplano, the Gran Chaco (shared by Bolivia, Argentina, and Paraguay), and northeast Brazil. None of the countries in the South America region is in a situation of water scarcity, but it's necessary to consider internal variations.

Among the countries that compose South America, Brazil is the one that stands out the most in availability of water resources, with approximately 12 % of the reserves of freshwater worldwide (UN 2012a, b). Among the biggest watershed in the world, two are located, in the most part, in Brazil (Amazonas[1] and Platine). With this water potential, the Brazilian energy matrix remains strongly dependent of hydroelectricity, which represents a current and future dependence of this sector in relation to the water use available in the country.

Scott and Pasqualetti (2010) affirm that managing water and energy separately is "shortsighted and inefficient." Treating them together, on the other hand, can contribute to identify current and future problems and propose solutions for sustainability challenges. Considering that water and energy are strongly linked, the energetic scenario of South American countries will be presented in the next topic, highlighting the renewability index.

[1] The Amazonas basin has 7 million km^2 and is located in seven South American countries (Brazil 63 %, Peru 17 %, Bolivia 11 %, Colombia 5.8 %, Ecuador 2.2 %, Venezuela 0.7 %, and Guyana 0.2 %).

23.3.2 *Energetic Matrix of South America and Energy Policies*

Energy plays an important role in water supply. Stillwell 2009 indicated that half of a city's energy demand is caused by water supply and waste water treatment. Moreover, 75 % of municipal water processing and distribution costs come from electricity consumption (Glassman et al. 2011). Water and energy are inextricably linked. According to the WWDR (2014), water is essential for the production, distribution, and use of energy, and energy is an indispensable resource in order for life to exist in the planet and all human activities can be developed. Odum (apud Franco 2001) states that "the city is an ecosystem and as such contains a community of living organisms – where men predominate, a physical environment that transforms itself, product of an internal activity and a functioning based on the exchange of matter, energy and information." The author reinforces that one of the special characteristics of these ecosystems is "the volume of energy that walks out of living organisms, the energy that makes the system work and the enormous horizontal mobility, which permits to explore other ecosystems on longer or shorter distances." This represents the great need and dependence that the human being has in relation to energy consumption. According to the Electrical Energy Atlas of Brazil 2012 (ANEEL 2012):

> Consumption of energy is one of the main indicators of economic development and the level of life quality in any society. It reflects both the rhythm of activity of industrial, commercial and services sectors and the capacity of the population to acquire goods and services technologically more advanced, like automobiles (which demand fuel), household appliances and electronic goods (which demand access to electric network and put pressure on the consumption of electrical energy).

This interrelation was the main reason for the elevated increase in world energy consumption observers in the last years, as it is demonstrated in the following chart (Fig. 23.2).

According to a data from the Institute of Applied Economic Research (PEA), world economy has lived a cycle of vigorous expansion from 2001 to 2007, reflected in the growing variation of GDP: 3.6 % in 2003, 4.9 % in 2004, 4.4 % in 2005, 5 % in 2006, and 4.9 % in 2007. In the same period, the cumulative variation of the energy consumption was of 13 %, going from 9.828 millions of tons equivalent of oil (toe) in 2003 to 11.099 million toe in 2007. In other words, the energy consumption growth is usually associated with the GDP growth tendency.

Besides economic development, other variable that determines the energy consumption is population growth.[2] WWDR (2014) precede a demographic growth estimated between two and three billion people in the next 40 years, which implies in food and energy increase demands and consequently, in water use.

---

[2] This index is obtained as much from the comparison between birth and death rates as from measuring migration flows (ANEEL 2012).

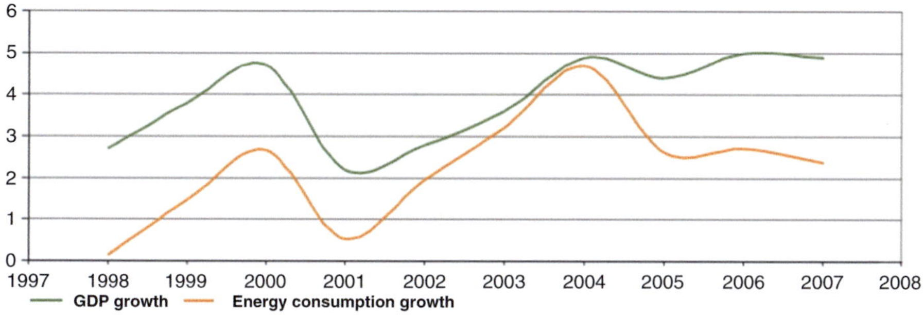

Fig. 23.2 GDP growth and change in the world's energy consumption (1998–2007) (Source: Institute of Applied Economic Research 2008)

According to the International Energy Agency (2012), approximately 90 % of the energy consumed in the world comes from nonrenewable fossil resources, whose depletion is constantly higher, which is why the countries of the world are seeking alternative energy. The world total primary energy supply in 2012 was composed of 32,4 % oil, 27,3 % coal/peat, 21,4 % natural gas, 5,7 % nuclear, 10 % biofuels and waste, 2,3 % hydro, and 0,9 % other.

Regarding the renewability index that presents the degree of utilization of renewable sources in relation to final energy consumption, South America has one of the best indexes in the world (33 %), as compared to other regions of the world, which average is 14 % and OECD countries 6 %. Even so, in 2009 nearly three-quarters of its structure corresponded to fossil fuels (OLADE 2011) (Fig. 23.3).

Paraguay, Brazil, and Uruguay have the largest index, mainly due to the hydropower potential. It is noted that in the region, only Argentina has renewability index below the world average (8 %). The majority of the energy consumed in the country in 2013 was not renewable (90.9 %) and that the main sources were natural gas (51.6 %), oil (35 %), and nuclear (2.8 %) (ARGENTINA 2013).

In order to increase the share of renewable energy in the energy matrix of countries, it is necessary to implement effective public policies. The following table shows briefly the major policies implemented by some countries in South America related to renewable energies (Table 23.2).

Although South America has the highest water availability, compared to other regions, and presents one of the best indexes in renewable energy, it's necessary to implement clear policies and strategies in each country to motivate and encourage the production of clean energy.

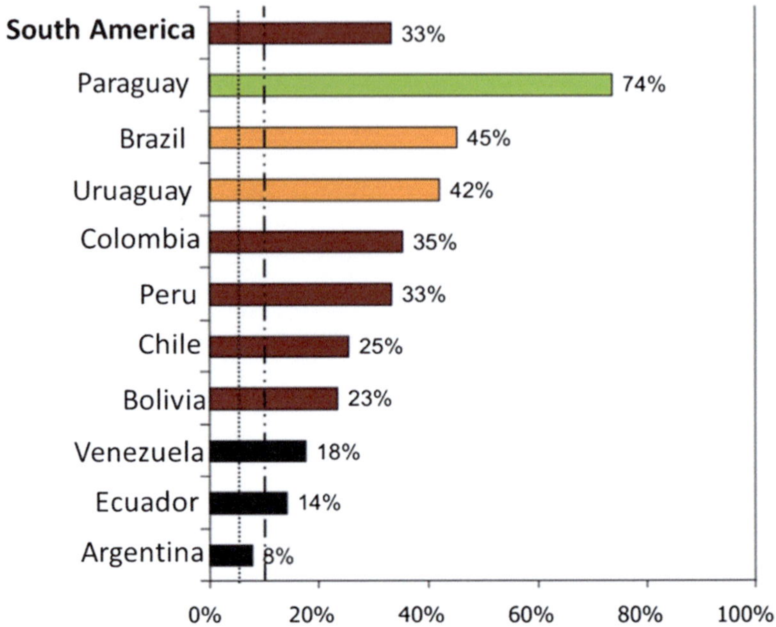

Fig. 23.3 South America renewability index (Source: Federation of Industries of the State of São Paulo, Brazil 2010)

23.4 Conclusion

Water and energy are indispensable to the social and economic development of a country. The rising pressure on resource demands, new production, and consumption models requires a better understanding about the connections between water and energy, because choices made in one area impact – positively or negatively – the other. The production model in energy has consequences on the quantity and quality of available water. This chapter assesses the water-use intensities and associated water quality and wider environmental impacts, across the extraction, processing, storage, and transport of the array of energy sources including hydropower, natural gas, oil, and transportation biofuels.

The United Nations WWDR depicts that water resources on the planet are under pressure of the rapid growth of the demands for water, and if there is not an implantation of public policies focusing on water resources and renewable sources of energy, this condition tends to become serious. As we could analyze, despite South America having the highest water availability worldwide and one of the best indexes in renewable energy, there are actions and strategies being implemented by the governments (Chile's National Energy Strategy, Good Living Plan in Ecuador, PROINFA (Incentive Program for Alternative Sources of Electric Energy) and National Agro-energy Plan in Brazil, Energy Strategy Guidelines and Energy Policy Plan 2005–2030 in Uruguay, etc.) to diminish the use of petroleum and the

Table 23.2 Some public policies on renewable energy in South American countries

Country	Some public policies on renewable energy
Argentina	*National Law 26.190* (2006) includes wind, solar, geothermal, tidal, hydro (up to 30 MW), biomass, and biogas from various sources. The law sets a goal: by 2016, 8 % of local power consumption should be supplied by renewable energy sources
	National Law 26.093 (2006): national initiative for biofuels
Bolivia	*Alternative Energy Policy for the Electricity Sector in the Plurinational State of Bolivia* (2011): development and promotion of research and use of new forms of alternative energy production, consistent with the environment conservation
Brazil	National Law 10.438 (2002): *Incentive Program for Alternative Sources of Energy* (PROINFA), in order to increase the share of electricity produced by wind, biomass, and small hydroelectric plant
	Decree 5.163 (2004) allows distributed generation
	National Law 11.097 (2005): *National Program for Production and Use of Biodiesel* aims to implement sustainable way, both technically and economically, the production and use of biodiesel
	National Agro-energy Plan (2006) aims to ensure sustainability, competitiveness, and greater equity (equal rights) between agents of agro-energy chains, in accordance with the expectations of society and public policy from the energy sector and social, environmental, and agricultural supply
	National Water Resources Plan (2006) establishes guidelines and public policies aimed to improve the water supply, in quantity and quality
Chile	National Law 20.257 (2008): *Law of Energy Generation with Renewable Nonconventional Energies* which establishes that in 2014 electrical companies must credit 5 % of the yearly energy sold and must come from renewable sources; this amount will increase 0,5 % annually starting in 2015 until it reaches 10 % in 2024
	National Energy Strategy has 6 pillars: growth with energy efficiency, nonconventional renewable energy resources, higher importance of the hydro sector, new focus in electrical transmission, more competitive energy market, and a sustainable advance in the options of the regional electrical interconnection
Colombia	*National Laws 69* 3 (2001), 788 (2002), 693 (2004), and 1.083 and 1.111 (2006): allow the government to force the mixture of ethanol with gasoline
	National Law 697 (2001): *Rational Energy Use* and incentives, the use of renewable sources, encourage wind, solar, geothermal, and biomass energy use
Peru	National Law 2.7345 (2000): *Promoting Energy Efficiency*, in order to ensure the supply of energy, protect consumers, promote the competitiveness of the national economy, and reduce the negative environmental impact of the use of energy
	Peru National Energy Policy 2010–2040 : its first objective is to have a diversified energy matrix, with emphasis on renewable energy and energy efficiency
Paraguay	Law proposal efficiency and rational use of energy
Uruguay	*Energy Strategy Guidelines and Energy Policy Plan 2005–2030* (2006) aims to create a consistent system of power supply with low cost and significant progress in incorporating renewable sources of energy
	Decree 354 (2009): granting specific tax incentives for renewable energy sector, National Law 18.585 (2009)
	Promotion of Solar Thermal Energy Decree 173 (2010) authorizes subscribers connected to the low voltage distribution network to generate energy from wind, solar, biomass, and mini-hydro sources. This decree is part of the National Energy Policy 2030, which promotes the use of renewable energy

(continued)

Table 23.2 (continued)

Country	Some public policies on renewable energy
	In July 2010, *Microgeneration* with renewable energy was allowed, and it's possible to sell this energy to the grid
Venezuela	*Development Plan of the National Electrical Service* (2005–2024): future energy requirements and electrical power are established, as well as the requirements of gas for electricity production until 2024
Ecuador	*Plan for Good Living* (2008) promotes policies, strategies, goals, and objectives for the diversification of the energy matrix. The master electrification plan impulses the immediate execution of hydroelectric plants as well as other renewable energy sources

promotion of renewable energy. This chapter on water and energy nexus and its contribution to sustainable development gives a better understanding about the water-energy nexus. It shows the importance of an integrated engagement in future interdisciplinary research and development to target water-use efficiency in the energy sector and energy efficiency. It also emphasizes reductions in energy intensities, in the water, and wastewater sectors.

References

ARGENTINA (2013) Matriz energética argentina: Situación actual y posibilidades de diversificación. http://www.eeaoc.org.ar/upload/contenido/pdf/20120228122933000000.pdf. Accessed 3 Aug 2014

BRASIL, ANEEL (National Electricity Agency) (2012) Electrical energy atlas of Brazil 2012. http://www.aneel.gov.br/visualizar_texto.cfm?idtxt=1689 Accessed 1 Aug 2014

Federation of Industries of the State of São Paulo (2010) Energy security in South America: a Brazilian Panorama. http://cebri.org/midia/documentos/seguranca_energetica.pdf. Accessed 1 Aug 2014

Food and Agriculture Organization of the United Nations FAO (2003) Review of world water resources by country. ftp://ftp.fao.org/agl/aglw/docs/wr23e.pdf. Accessed 1 Aug 2014

Franco MAR (2001) Planejamento Ambiental para a Cidade Sustentável. Annablume, São Paulo

Glassman D, Wucker M, Isaacman T, Champilou C (2011) The water – energy nexus: a world policy paper. World Policy Institute & EBG Capital, New York

Institute of Applied Economic Research (2008) Sustentabilidade Ambiental no Brasil. http://www.ipea.gov.br/agencia/images/stories/PDFs/comunicado/110215_comunicadoipea77.pdf. Accessed 8 July 2014

International Energy Agency (2012) Key world statistics. http://www.iea.org/statistics/. Accessed July 2014

Leff E (2013) A Latin America Perspective. Revista Desenvolvimento e Meio Ambiente, vol 27. UFPR, Curitiba

OLADE (2011) Manual de estadísticas energéticas año 2011. http://biblioteca.olade.org/iah/fulltext/Bjmbr/v32_2/old0179.pdf. Accessed August 2014

Organization for Economic Co-operation and Development OECD (2011) Green growth. http://www.oecd.org/greengrowth/48536946.pdf. Accessed 10 July 2014

Scott CA, Pasqualetti MJ (2010) Energy and water resources scarcity: critical infrastructure for growth and economic development in Arizona and Sonora. Nat Resour J 50(3):645–682

Stillwell AS (2009) Energy -water nexus in texas. A presentation by Ashlynn Stillwell at Water Wise Panel 7, EPRI & CEC
UNDESA (2008) Climate change and indicators of sustainable development, power point presentation by Bruckner M to conference on climate change, development and official statistics, Seoul, Dec 2008
United Nations (1977) Report of the United Nations water conference, Mar del Plata. UNESCO, Paris
United Nations (1992) Agenda 21. http://sustainabledevelopment.un.org/content/documents/Agenda21.pdf. Accessed 12 June 2014
United Nations (2010) World economic and social survey 2010: retooling global development. Sales no. E.10.II.C.1
United Nations (2012a) Sustainable development: from Brundtland to Rio 2012. http://www.un.org/wcm/webdav/site/climatechange.pdf. Accessed 10 July 2014
United Nations (2012b) Our common future. http://www.un-documents.net/our-common-future.pdf. Accessed 15 June 2014
United Nations – Water (2014) World water development report 2014: water and energy. UNESCO, Paris
United Nations Environment Programme – UNEP (2012) Global initiative for resource efficient cities. http://www.unep.org/pdf/GI-REC_4pager.pdf. Accessed 15 June 2014
United Nations-HABITAT (2012) State of Latin America and Caribbean cities 2012: towards a new urban transition. UN-Habitat, Nairobi

Chapter 24
Climate Change Mitigation and Adaptation: The Role of International Ocean and Freshwater Agreements

Ryan B. Stoa

Abstract Climate change presents the international community with an environmental process that is both challenging to monitor and foresee and requires a complex legal and regulatory framework capable of promoting mitigation and adaptation. In the absence of comprehensive and targeted international climate change legislation, however, some mitigation and adaptation measures are being adopted and implemented through indirect policymaking and regulation. International environmental treaties and customary international environmental laws and principles not specifically focused on climate change may nonetheless indirectly or unintentionally contribute to climate change mitigation and adaptation efforts through administration or enforcement of the legal regime. The crucial role that the water cycle plays in climatic processes, however, makes international freshwater and ocean laws and policies a particularly rich source of indirect climate change adaptation and mitigation measures. This chapter analyzes the international legal regimes regulating freshwater resources and ocean and marine resources with an eye toward mechanisms that contribute to – or detract from – climate change mitigation and adaptation efforts. I find that potential for regulation of climate change is greatest when treaties are focused on discrete environmental issues such as wetland conservation and pollution from ships, while comprehensive treaties like the Watercourses Convention and the Convention on the Law of the Sea make less tangible contributions to indirect climate change regulation by reinforcing principles of international law that require states to take collective action on international environmental challenges such as climate change.

Keywords IWRM • Climate change • Mitigation • Adaptation • International ocean and marine resources law • International freshwater resources law

---

R.B. Stoa (✉)
Florida International University, College of Law and Global Water for Sustainability Program – GLOWS, Miami, FL 33199, USA
e-mail: rstoa@fiu.edu

S.G. Setegn, M.C. Donoso (eds.), *Sustainability of Integrated Water Resources Management*, DOI 10.1007/978-3-319-12194-9_24

24.1 Introduction

Climate change presents the international community with a monumental regulatory problem that transcends generations, sectors, and political boundaries. The full effect of increased greenhouse gas (GHG) emission levels in the atmosphere is delayed by years or decades, meaning that the actions of one generation are felt most poignantly by an entirely different generation. Spatially, the negative consequences of emissions released from one actor or group of actors are shared by the entire international community and felt disproportionately by poor and disadvantaged communities. Economically, climate change is a cross-sectoral issue, the regulation of which may curtail the profitability of certain industries (such as fossil fuel production, animal agriculture, and transportation), while ensuring the sustainability of others (such as renewable energy production and tourism). The divergence of these challenges has been called a "perfect moral storm" that pushes us toward moral corruption or inaction (Gardiner 2006). Whether or not that is an adequate description of the hurdles facing humanity, it is clear that climate change presents the international community with an extraordinarily complex regulatory challenge.

It should come as no surprise, then, that our collective track record of climate change regulation has been unremarkable. Efforts directed at curbing greenhouse gas emissions through a collective regulatory scheme have struggled to gain widespread buy-in from key states and industries. On the international level, the United Nations Framework Convention on Climate Change (UNFCCC) and associated mechanisms (e.g., the Kyoto Protocol and Conference of the Parties) have established a platform for the international community to share information and negotiate emission reductions but have largely failed to create a binding treaty instrument capable of reducing the effects of climate change in a meaningful sense. Domestically, some countries have been successful in directly addressing their contributions to climate change (e.g., the European Union Emissions Trading System), but on the whole countries appear unwilling to alter the course or nature of their economic development without reciprocal commitments from the rest of the international community. Perhaps the most successful climate change initiatives have occurred at the subnational level, where decision makers and strong leaders enjoy greater influence over legislative efforts, cooperation between political actors is more harmonious, and/or the benefits of showing leadership by tackling a collective problem outweigh the perceived loss in competitive edge. Innovative and comprehensive climate change legislation emerging from cities and individual states in the United States, for example, is in stark contrast with the failure of the federal government to directly address climate change.

In the absence of comprehensive climate change legislation on the international and national level, a multitude of legal mechanisms that indirectly regulate climate change have emerged. Whether intentional or unintentional, statutory schemes whose objectives are environmental in nature may – and are even likely to – contribute to climate change mitigation and adaptation. Much attention has been paid to the United States Supreme Court's decision in Massachusetts v. Environmental Protection Agency (EPA), for example, in which the Court ruled that carbon emissions are a pollutant as defined in the federal Clean Air Act. The determination required the EPA to regulate carbon emissions by setting standards limiting the amount of carbon released into the atmosphere (Massachusetts v. EPA 2007).

Since most environmental lawmaking takes place at the national level of governance, research on indirect climate change mitigation and adaptation mechanisms have understandably ignored the international arena. However, international environmental law has emerged as a particularly robust field of international law. The nature of natural resources and environmental processes forces states to either cooperate or suffer the consequences of a tragedy of the commons management paradigm. A cursory review of environmental treaties in force reveals that many share a common focus: the water cycle. As water plays a central role in most sectors of human import – including agriculture, energy, human health, industrial production, transportation, tourism, and recreation – as well as a critical player in natural and environmental processes, it follows that many international environmental agreements will attempt to regulate the management of freshwater and ocean resources.

One of the environmental processes affected by and affecting water resources is climate change, and the relationship between the water cycle and climate change cannot be overstated. The water cycle incorporates the world's oceans, lakes, rivers, groundwater, glaciers, and polar ice caps, all of which shape the earth's climate through evaporation, precipitation, and weather patterns. The world's oceans can hold 1,000 times the heat energy of the atmosphere, while redistributing heat energy from the tropics to the poles. Even the greenhouse effect is predominantly caused by water vapor and clouds (Goreham 2013).

For these reasons, it is likely that regulation of water resources – whether freshwater or saltwater – will contribute to or detract from climate change mitigation and adaptation. This chapter reviews the major treaties and customary international environmental laws relating to freshwater and ocean resources in order to identify legal provisions or devices that have an indirect climate change impact. The result may offer the international community with potentially innovative mechanisms to address climate change in other seemingly unrelated agreements.

24.2 Climate Change and International Freshwater Resources Law

While climate change effects and subsequent negotiations gained widespread recognition in the 1990s, the importance of protecting freshwater resources has been recognized by the earliest human civilizations. The world's major religions – including Christianity, Judaism, and Islam – consider water to be a vitally protected resource and even sacred (Salzman 2005). Nonetheless, development of international water law frameworks did not occur until the mid-twentieth century, when the international law community recognized the need to avoid messy transboundary resource conflicts. From there, international water law has developed mainly along customary international law lines. In other words, what international water law exists today has been developed more by state practice and custom than robust multilateral treaties. Accordingly, few treaty mechanisms will prove instructive when searching for indirect climate change regulatory instruments. However, international water law has developed several principles of international law that may spur cooperation on climate change action.

Starting in 1966, international water law gained significance when the International Law Association (ILA) convened in Helsinki, Finland, to create the Helsinki Rules on the Uses of the Waters of International Rivers. The goal was to codify customary legal norms and principles, in addition to setting in motion further development of international water law. Given the preliminary nature of the endeavor, the Helsinki Rules were appropriately modest in their ambition, establishing the groundwork for future action and establishing principles of water law that reflected prevailing notions of water resources management.

The most salient feature of the Helsinki Rules was the establishment of the principle of equitable use. The principle of equitable use stipulates that a watercourse state has the right to a reasonable and equitable share of an international watercourse. The right is tempered by factors determining reasonableness, the most significant of which is the presence of significant harm incurred by a downstream state. Subsequent international water law instruments would reveal an inherent tension between the principles of equitable use and "no significant harm," but the Helsinki Rules award prominence to the principle of equitable use, to be defined by balancing a set of factors with undetermined weight. Interestingly, even in 1966 the Helsinki Rules created certain limited implications for climate change, as the "climate affecting the basin" is listed as a factor to be considered when making a reasonable use determination (Helsinki Rules, Article V(II)(3)). While the original intent of the clause was unlikely to impose climate change-induced obligations on watercourse states, the Helsinki Rules notably acknowledge that environmental processes affecting water resources must be taken into consideration when allocating resources and resolving conflicts.

Thirty years later, the international community attempted to further solidify and codify international water laws and principles in the form of a binding treaty instrument. The 1997 United Nations Convention on the Law of Non-Navigational Uses of International Watercourses contains 37 articles laying down basic norms of international water law. The cornerstone of the Convention, however, is Article 5, Equitable and Reasonable Utilization and Participation. Echoing the Helsinki Rules, Article 5 lays down the equitable use principle and affirms the concept of equitable participation to encourage states to resolve issues of equitable use jointly and cooperatively.

However, the Watercourses Convention departs from the Helsinki Rules in one important respect: Article 7's obligation not to cause significant harm stands in opposition to the equitable use principle. In contrast to equitable use, the principle of "no significant harm" imposes a higher standard on basin states by requiring them to refrain from taking actions that would cause substantial damage to another state's water resources. The no significant harm principle prevents upstream states from using water resources – even if their use is reasonable and beneficial – if downstream states are adversely affected. This can be problematic in cases where, for example, an upstream state decides to make reasonable use of a transboundary river to develop clean energy such as hydropower in order to meet emission reduction targets, to the detriment of a downstream state whose prior flow rates are diminished. Not surprisingly, the no significant harm principle is favored by downstream states.

Presented in tandem, the relationship between the principles of equitable use and no significant harm is unclear without a declarative interpretation of how the two should or can coexist. Nonetheless, the Watercourses Convention was adopted by an overwhelming number of states, with 106 votes in favor to only three against. This initial triumph is juxtaposed by the fact that without sufficient ratifications, the Convention has yet to enter into force and therefore remains a nonbinding treaty. As of January 2014, only 33 states have ratified the Convention, with 35 needed for entry into force (Stoa 2014).

As a result, the ILA reconvened in 2004 to synthesize customary international water law in light of the Watercourses Convention and the development of international environmental laws since the adoption of the 1966 Helsinki Rules. The 2004 Berlin Rules on Water Resources contributed several layers to the development of international water law. First, the Berlin Rules extended the applicability of international water laws to waters that were purely national. The right of public participation, the obligation to use best efforts to achieve both conjunctive and integrated management of waters, and duties to achieve sustainability and the minimization of environmental harm are principles either new or modified vis–àvis the Helsinki Rules and the Watercourses Convention, both of which restrict their scope to purely international watercourses.

Importantly, the Berlin Rules maintained the equitable use – no significant harm dichotomy, but acknowledged the apparent tension between the two principles by incorporating one into the other: "Basin States shall in their respective territories manage the waters of an international drainage basin in an equitable and reasonable manner having due regard for the obligation not to cause significant harm to other Basin States." Reconciling the two principles requires a case-by-case balancing test; while "vital human needs" are given priority, no other use is per se more preferable than another. In the absence of a sovereign body to oversee transboundary water resource conflicts, states are left to cooperate in good faith.

This third requirement – the duty to cooperate – has emerged over 40 years as the third leg of the international water law tripod. In the absence of clear standards for resolving conflicts between upstream and downstream states and equitable use and no significant harm as their supporting principles, respectively, international water law has adopted and reinforced the duty to cooperate as a customary principle of international water law. The duty to cooperate finds support in general international law instruments as well, such as the UN Charter (Chapter 1(3)), the 1970 UN

Declaration on Principles of International Law, the 1972 Stockholm Declaration, the 1992 Rio Declaration, and the UN Convention on the Law of the Sea (discussed further below). However, the nature of transboundary watercourses – being a necessarily shared and indivisible physical unit – has made international water law a catalyst for promoting the principle. However, the duty to cooperate may just as well be applied to global climate change for the very same reason: the earth's climate and regulating processes are a necessarily shared and indivisible physical unit, the regulation of which requires cooperation and good faith negotiation among states.

Indeed, the duty to cooperate is increasingly viewed as a legal obligation of states with respect to climate change. A 2009 report of the UN High Commissioner on Human Rights found that, while climate change does not violate human rights per se, human rights law does establish a duty to cooperate on climate change matters (Knox 2009). It is not yet clear if states are generally meeting that obligation; while treaties like the UNFCCC and the Kyoto Protocol have been adopted, the pace of negotiations has been slow and emission reduction commitments scarce. What is clear is that the basis for the duty to cooperate – a shared and indivisible natural resource – is shared by international water law and climate change negotiations. By extension, the continued application of the principle in international water law instruments should reinforce the principle as an international legal obligation on states that extends and applies to climate change.

Since 2004, other international instruments have played a role in developing international water law. Lacking a binding treaty instrument governing all types of groundwater, the UN's International Law Commission (ILC) produced the Draft Articles on the Law of Transboundary Aquifers in 2008. The Draft Articles elucidate some relatively uncontroversial and forward-thinking principles governing transboundary aquifers while reinforcing the dual principles of equitable use and no significant harm. However, the Draft Articles depart from previous understandings of international water law in Article 3, which provides that each aquifer state has sovereignty over the portion of a transboundary aquifer or aquifer system located within its territory, in accordance with international law. The Special Rapporteur to the ILC Chusei Yamada indicated that the inclusion of this principle (which does not appear in the Helsinki Rules, Watercourses Convention, or Berlin Rules) was a necessary concession to aquifer states that hold the view that aquifers are analogous to mineral resources and are governed by the principle of territorial sovereignty (Yamada 2011). The United Nations' Sixth Legal Committee convened in 2011 to determine if the Draft Articles were ripe for a binding convention. The Committee declined to move forward, calling instead for further study and exploration of the topic.

The Draft Articles represent a recent and significant attempt to move forward with development of international water law and may have implications for climate change policy if the sovereignty principle causes the international water law regime to move away from viewing transboundary watercourses as shared resources and toward a notion of territorial sovereignty. While each international environmental process can be viewed as part of a distinct legal regime, in reality international environmental law instruments often convey and reinforce similar sets of principles. Accordingly, while the inclusion of the sovereignty principle in an as-yet unratified treaty governing groundwater may seem innocuous, it may represent a shift away from viewing collective action as the appropriate mechanism for addressing shared resources. Such a shift would have dilatory effects on the development of a robust climate change regime.

Regional treaties and legal instruments may also indirectly regulate climate change. The United Nations Economic Commission for Europe (UNECE)'s Convention on the Protection and Use of Transboundary Watercourses and International Lakes, for example, is in force and ratified by 36 countries and the European Union. In 2003, the Convention was amended to allow ratification and participation from states outside the UNECE region, thus expanding the potential scope of the Convention. The UNECE Convention's legal foundation rests on the equitable use and no significant harm principles (as well as the principle of state cooperation) in a manner comparable to the 1997 Watercourses Convention and the 2004 Berlin Rules. Since the UNECE Convention has entered into force, however, it is binding on party states and significantly persuasive in the region. Therefore, the duty to cooperate and consider climate change as an element of equitable use may be providing support to climate change regulatory efforts. It remains to be seen how much broader its jurisdiction will become with the 2003 amendment, but as of writing, the composition of the regime is predictably European. States of the European Union have generally been more willing to adopt binding climate change regulatory policies, so the impact of the UNECE Convention on climate change mitigation and adaptation may remain limited unless and until the Convention incorporates party states that would not otherwise adopt meaningful climate change policies.

Other international environmental treaties regulating water resources may have an impact on climate change or induce climate change mitigation or adaptation efforts indirectly. The Ramsar Convention on Wetlands of International Importance especially as Waterfowl Habitat plays a particularly crucial role in encouraging contracting states to set aside wetlands for preservation as habitat and ecosystem service delivery. The Convention requires contracting states to designate at least one area as a Ramsar Wetland of International Importance, in which the wise use of the wetland must be promoted in order to maintain its ecological character. With 168 contracting parties, the Ramsar Convention enjoys widespread adoption of its text and endorsement of its principles. To date 2,168 sites have been listed as Ramsar Wetlands, covering a total area of over 797,811 square miles (Ramsar Secretariat 2014).

The impact of wetlands conservation on climate change mitigation and adaptation is a function of the crucial relationship between wetlands and climate change. Wetlands cover over 6 % of the Earth's surface and play a key role as sinks for organic carbon. Already up to 30–90 % of the world's wetlands have been destroyed or significantly impaired, reducing the Earth's ability to counteract the effects of increases in atmospheric carbon dioxide (Junk et al. 2013). The Ramsar Convention's ability to mobilize international support for wetlands conservation and wise utilization is therefore a key – and often overlooked – component of the international community's mitigation and adaptation approach to climate change. Whether or not wetlands are listed for this purpose, the Ramsar Convention's understated impact on climate change mitigation and adaptation is an innovative and indirect response to climate change.

Similarly, the 1994 United Nations Convention to Combat Desertification (UNCCD) was created to combat desertification and mitigate the effects of drought through international cooperation and partnership arrangements. As with the Ramsar Convention, the UNCCD enjoys widespread adoption, with all United Nation member states – excepting newly formed South Sudan – party to the Convention.

The UNCCD explicitly recognizes the contribution "that combating desertification can make to achieving the objectives of the UNFCCC," presumably because the challenges of combating desertification and mitigating the effects of drought are so intricately linked with climate change. Not only does climate change exacerbate desertification by making precipitation patterns more irregular, but also more direct forms of desertification – such as unsustainable agricultural practices and deforestation – eliminate another barrier ecosystem capable of absorbing atmospheric carbon dioxide. The UNCCD's ability to mobilize support for combating desertification therefore has a significant impact on climate change mitigation and adaptation for this causal connection between desertification and climate change. In addition, the UNCCD's unique integration with the UNFCCC provides a model for future international environmental agreements to fit their objectives into a climate change framework.

Due to their subject matter's important role in climate change mitigation and adaptation, as well as their widespread adoption, the Ramsar Convention and the UNCCD may be the two most influential freshwater-related international agreements with indirect impacts on climate change regulation. At the very least, their impact on climate change is more quantifiable than the influence that customary international water law principles have on climate change regulation and ongoing negotiations. The contribution that the UN Watercourses Convention and the Berlin Rules are making toward indirect climate change regulation may be significant however, especially as the duty to cooperate becomes more entrenched as a general principle of international law.

Nonetheless, the differences between the Ramsar Convention and UNCCD on the one hand and the Watercourses Convention on the other provide a potential framework for understanding and addressing indirect climate change regulation in future international environmental agreements. Most apparent is the target subject matter of the agreements. While the Watercourses Convention attempts to codify customary principles of management over a vital natural resource (inviting debate over sovereignty, rights, and obligations) (Stoa 2014), the Ramsar Convention and UNCCD provide a framework for states to protect certain ecosystem types, in this case wetlands and desert-prone areas. While water rights are highly contentious, wetlands and deserts rarely invoke sovereignty concerns for states, instead encouraging the preservation and sustainable management of ecosystems that are vital to combating climate change. This suggests that international environmental agreements seeking to contribute to indirect climate change regulation may find an easier path toward entry into force and continued support by focusing on the protection and preservation of ecosystems and not transboundary natural resources. Nonetheless, it is equally clear that the principles supporting the international water law regime – equitable use, no significant harm, and cooperation – have broad application to other fields of international law and climate change frameworks in particular. While the nascent state of the Watercourses Convention cautions against interpreting the principles as absolute legal rules of allocation, their broad acceptance provides justification for continued development of the climate change regulatory regime.

24.3 Climate Change and International Ocean and Marine Resources Law

Over 96 % of water on earth is located in the world's oceans (Gleick 1996). Exactly how the oceans regulate the earth's climate is complex and has been a point of some debate, but research increasingly suggests that oceans are the earth's primary heat sink, and therefore its first line of defense against climate change. The Intergovernmental Panel on Climate Change reported in 2013 that 90 % of the energy increase in the earth's climate system from 1971 to 2010 was absorbed in the ocean, including 60 % that was absorbed by the upper ocean (0–700 m), (IPCC 2013). Polar ice melts, rising sea levels, and ocean acidification are all a result of the climate change-induced alterations in oceanic processes.

Accordingly, laws regulating oceans, coasts, and marine resources may have significant potential to indirectly affect climate change adaptation and mitigation. Relative to international freshwater agreements, ocean and marine agreements are both more plentiful and more robust in nature, providing the international community with several mechanisms capable of indirectly regulating climate change.

Perhaps the most sweeping and comprehensive international environmental agreement in human history, the United Nations Convention on the Law of the Sea contains 320 articles regulating jurisdiction, seabed mining, fisheries, dispute resolution, pollution, military activity, and navigation, among others. The Convention was the result of a nearly 10-year negotiation process that aimed to achieve consensus from member states on a wide variety of issues. Although climate change was not a primary issue at the time – negotiations took place from 1973 to 1982, before climate change had come to the international community's attention – several of the Convention's provisions and enumerated principles have set in motion significant implications for global climate change regulation.

Foremost among these is the Convention's governing principle on areas beyond national jurisdictions, such as the deep seabed. Article 136 declares these areas to be the "common heritage of mankind." While a seemingly innocuous passage, the common heritage of mankind principle implicates collective action on matters or areas of collective concern, thereby declaring management of "the commons" the responsibility of all nations. The common heritage of mankind principle finds itself in the texts of other international environmental agreements – such as the Outer Space Treaty of 1967 – but derives considerable legitimacy from its prominent role in the Convention on the Law of the Sea.

The common heritage of mankind principle has several legal implications for international climate change regulation. As a principle of customary international law, states may be bound to take affirmative action to protect areas held in common between states. Where this concerns the Arctic or Antarctic, for example, collective action to preserve the marine environment and marine resources is likely to have a mitigating or adaptive effect on climate change. The Antarctic Treaty System, for example, provides a framework for states to cooperatively manage, observe, and protect the Antarctic region – the Antarctic Treaty obliges states to share information and scientific research while limiting sovereignty claims and nuclear activity, while the Protocol on Environmental Protection to the Antarctic Treaty bans all mineral exploitation in the region and requires environmental assessments of all activities. While polar ice melt may be primarily caused by warmer atmospheric temperatures, any legal mechanisms that purport to limit activity that would have deleterious impacts on Arctic or Antarctic ecosystems would likely contribute to climate change mitigation and adaptation. That the Antarctic Treaty expressly aligns its mission in "the interests of all mankind" may suggest that areas held in common are increasingly viewed as requiring collective action. Not only does that heighten the impetus for collective action on climate change, but it may also develop more regional environmental agreements that indirectly contribute to climate change mitigation and adaptation.

Second, while the common heritage of mankind principle originally sought to protect the interests of living human beings anywhere on earth, a significant outgrowth of the principle incorporates future generations as well. Intergenerational equity aims to curtail unsustainable human activity on the grounds that environmental or cultural degradation violates the rights of future generations. Article 8 of the UNESCO Declaration on the Responsibilities of the Present Generations Towards Future Generations, for example, permits the use of the common heritage of mankind by present generations only if that use does not compromise our heritage irreversibly. Intergenerational inequity is a fundamental challenge for climate change regulation frameworks. Requiring that present generations make carbon footprint sacrifices in the interest of future generations they may not know or care about is a hard sell for contemporary policymakers. Since the Convention on the Law of the Sea significantly reinforces the common heritage of mankind principle – itself a buttress for intergenerational equity – the Convention itself acts as a mandate for climate change regulation on an intergenerational basis.

The Law of the Sea Convention may also indirectly contribute to climate change mitigation and adaptation by providing an incentive for states to prevent further increases in sea-level rise. One of the most significant achievements of the Convention is the way in which it defines jurisdictional boundaries. All states are entitled to certain privileges that extend outward from their nautical baseline. For example, states can set laws and regulate all resource use within 12 miles of the baseline or establish exploitation rights over natural resources within 200 miles of the baseline. However, the baseline is defined as the low-water line along the coast (UNCLOS Article 5), and sea-level rise would naturally submerge lands that are unlikely to be productive underwater while diminishing the reach of a state's various jurisdictional zones. Even a withdrawal of a few feet can have tremendous consequences for a state's legal rights, especially in areas where jurisdictional zones are contiguous or overlapping. Ironically then, the Law of the Sea Convention's silence on climate change and sea-level rise is likely to incentivize climate change mitigation and adaptation.

Finally, the Law of the Sea Convention establishes a framework for the protection and conservation of the ocean and marine resources by creating over 50 treaty articles related to marine conservation. Of most relevance to climate change mitigation and adaptation is Article 212, obliging states to establish national and international laws to prevent marine pollution occurring from or through the atmosphere. It is well established that atmospheric carbon dioxide can act – and be defined – as a pollutant at high enough levels (see, e.g., Massachusetts v. EPA). Moreover, the Law of the Sea Convention's International Tribunal for the Law of the Sea provides a mechanism for states to bring complaints of violations of the Convention. Using the international customary law principle of no significant harm as a buttress, for example, a small-island state could claim that the industrialized nations' failure to regulate greenhouse gas emissions violates Article 212 of the Law of the Sea Convention since the resultant sea-level rise threatens to significantly reduce – or even eliminate – the state's land area. While the success of such a lawsuit would seem unlikely due to the difficulty of proving causation of climate change impacts, small-island states have demonstrated their willingness to pursue these and similar claims for several years (Seneviratne 2002). Accordingly, either through tribunal mandate or preemptive regulation, the Law of the Sea Convention's prohibition on harmful air pollution may lead to significant climate change regulation in the near future.

The Law of the Sea Convention is not the only marine-related international agreement to indirectly regulate climate change. The International Convention for the Prevention of Pollution from Ships (MARPOL) aims to eliminate oil pollution and other harmful substances purposefully or accidentally released by ships. Annex VI of MARPOL establishes caps on the type and quantity of air pollutants contained in ships' exhaust gasses, including nitrous oxides and sulfur oxides. Annex VI also prohibits the deliberate release of ozone-depleting gasses and establishes a mechanism for creating emission control areas where emissions are further restricted. While regulations on pollution from ships may not seem terribly significant to global climate change processes, consider that in 2012 alone, the world fleet was approximately 80,000 ships with a total tonnage of over 1.05 billion tons, carrying an overwhelming majority of the world's freight transport (Equasis 2012). In this context, regulating air pollutant emissions from merchant vessels has the potential to significantly reduce global greenhouse gas emissions. Further amendments to

Annex VI may yield even more promising climate change mitigation measures, as recently adopted energy efficiency standards for ships are being touted as the first legally binding international treaty since the Kyoto Protocol to directly regulate climate change (IMO Technical and operational measures 2014). With international shipping estimated to contribute approximately 2.7 % of global emissions in 2007, MARPOL's emission reduction schemes and widespread political support (counting 152 countries party to the Convention) are a promising source of climate change mitigation and adaptation (IMO Overview of greenhouse gas emission from ships 2012).

The 1992 Basel Convention on the Control of Transboundary Movements of Hazardous Wastes and Their Disposal may also play a significant role in climate change mitigation and adaptation. The Basel Convention aims to limit the transportation of hazardous wastes – especially from developed to less-developed countries – while encouraging states to limit the amount of hazardous waste produced. The Convention enjoys widespread support from the international community, with 180 countries party to the Convention and 73 countries supporting a total ban on shipping of hazardous wastes to less-developed countries (Stoa 2014). This framework may indirectly regulate climate change in three ways. First, the Convention enthusiastically adopts the duty to cooperate (Article 10), particularly when developing countries require assistance to carry out terms of the Convention. Reinforcing a duty for developed countries to provide assistance to developing countries on matters of collective environmental concern has parallels in the climate change regime, where developing countries typically insist that industrialized nations bear the burden of global emission reductions. Second, by limiting transboundary shipping and encouraging states to dispose hazardous wastes domestically, the Convention indirectly regulates climate change by reducing the overall cumulative impacts of international shipping.

Finally, the Convention may provide an opportunity to regulate greenhouse gas emissions by treating them as hazardous wastes. The Convention provides two mechanisms for determining if a waste product is hazardous and therefore subject to the Convention: the waste can be listed in Annex I and display a hazardous characteristic described in Annex III, or it can be deemed hazardous by the country of export, transit, or import. Greenhouse gasses are not listed in Annex I, but there may be a sufficient basis for a country to independently determine that they are a hazardous waste. The Convention itself describes ecotoxicity as a hazardous characteristic, defined as "substances or wastes which if released present or may present immediate or delayed adverse impacts to the environment by means of bioaccumulation" (Annex III H12), and Annex IV's description of disposal operations include several provisions that resemble contemporary carbon capture and storage (CCS) methods. CCS is still in its infancy, but should it eventually facilitate transboundary waste disposal of carbon dioxide, the Basel Convention may force exporting states to store carbon themselves and incentivize emission reductions therewith.

As with freshwater agreements, the effect of international marine agreements on climate change mitigation and adaptation can be divided between a comprehensive catchall agreement that aims to incorporate a majority of international ocean issues on the one hand and agreements that limit their scope to discrete issues. The Law of the Sea Convention is remarkable in its breadth and is a shining example of international consensus building. Jurisdictional boundaries provide incentives to combat sea-level rise, and provisions on transboundary air pollution may provide a basis for climate change claims under the Tribunal for the Law of the Sea. But the Convention's most significant contribution to climate change regulation may be its expansion of the common heritage of mankind principle. On the other hand, while MARPOL and the Basel Convention are less comprehensive, they enjoy broad support on provisions that significantly limit the cumulative impacts of international shipping on the environment. It may be impossible to quantify their impacts, but international ocean and marine agreements are already playing an indirect role in climate change regulation, with the potential to further contribute to climate change mitigation and adaptation.

24.4 Conclusion: Climate Change, the Water Cycle, and International Agreements

Until the international community can develop a legal regime that imposes firm emission reductions on countries and their constituents in a way that substantially curbs the global effects of climate change, mitigation and adaptation efforts will be necessarily piecemeal. That being said, even if such a legal regime was enacted, other international environmental agreements would still have a role to play in combating climate change by complementing emission reductions programs with their own indirect regulatory impacts. Because of the water cycle's crucial role in climatic processes, international agreements regulating various aspects of the water cycle are particularly likely to indirectly contribute to climate change mitigation and adaptation.

This chapter has analyzed the various mechanisms by which international freshwater and marine legal instruments indirectly regulate climate change. From the analysis, several patterns emerge that prove instructive. First, comprehensive agreements that attempt to regulate or establish laws for a resource as a whole are difficult to negotiate and enter into force, limiting their potential climate change mitigation and adaptation impact. This is particularly true of the UN Watercourses Convention, which has yet to enter into force despite its vague provisions balancing equitable use, no significant harm, and the duty to cooperate. Until it does, allocation and management of transboundary watercourses is governed more by interpretations of customary international laws and principles than binding legal obligations. Similarly, while the Law of the Sea Convention enjoys broad support today, it took almost 40 years for the Convention to be negotiated and entered into force. What is apparent from these experiences is that comprehensive resource-based treaties take time to negotiate and implement. For parties keen to move quickly on climate change regulation, these treaties are unlikely to provide immediate relief.

However, the comprehensive nature of the Watercourses Convention and the Law of the Sea Convention lends their regimes an air of legitimacy with respect to customary principles. Both treaty instruments reinforce the duty to cooperate on matters of collective environmental concern, for example, a principle that is critical to states' understandings of their own duty to negotiate a robust climate change agreement. Similarly, broad-based agreements can establish the foundational principles of their regime. The Law of the Sea Convention establishes the common heritage of mankind principle over areas beyond national jurisdictions; this principle may contribute to an understanding that the Earth's climate system is a commons requiring collective action, while notions of intergenerational equity may eventually impose legal obligations on states to preserve the environment for future generations. While it is difficult to quantify the impact that evolving customary legal principles are having on the international community's willingness to combat climate change, it is nonetheless encouraging that both the Watercourses Convention and the Law of the Sea Convention codify principles that provide an impetus for climate change regulation.

In opposition to the comprehensive agreements are treaties that focus on discrete environmental problems. The Basel Convention on Wetlands, the Convention to Combat Desertification, MARPOL, and the Ramsar Convention are relatively limited in scope and unlikely to reverse global climate change impacts on their own. However, by exercising restraint and respecting sovereignty concerns, these agreements have made significant cumulative contributions to indirect climate change mitigation and adaptation efforts. The conservation and protection of wetlands and desert-prone areas limits the carbon footprint of potential development in these areas while simultaneously providing carbon sinks. Both objectives are enhanced by the Conventions' widespread support from party states.

Similarly, MARPOL and the Ramsar Convention establish rules for international transportation that have significant cumulative impacts. MARPOL's Annexes targeting fuel emissions and energy efficiency, for example, are likely to significantly reduce the marine shipping industry's cumulative impacts on climate change. The Ramsar Convention makes a similar contribution by reducing global shipping activity and may eventually encourage states to develop their own carbon capture and storage programs. While international shipping and hazardous waste disposal is only a fraction of the global climate change equation, MARPOL and the Ramsar Convention provide a model for other industries to develop their own indirect climate change regulatory frameworks.

Despite the inherent limitations of international law, it is apparent that the water cycle's crucial role in regulating the Earth's climate will make even modest international freshwater or marine agreements sources of indirect climate change regulation. Whether treaties contribute to the development of customary principles that increasingly mandate climate change action, or simply regulate particular agents of the climate change process, freshwater and marine agreements are playing and will continue to play an important role in climate change mitigation and adaptation efforts.

References

Equasis (2012) The world merchant fleet in 2012. Equasis Statistics

Gardiner SM (2006) A perfect moral storm: climate change intergenerational ethics and the problem of moral corruption. Environ Value 15:397–413

Gleick PH (1996) Water resources. In: Schneider SH (ed) Encyclopedia of climate and weather, vol 2. Oxford University Press, New York, pp 817–823

Goreham S (2013) Climate change is dominated by the water cycle, not carbon dioxide. The Washington Times. http://communities.washingtontimes.com/neighborhood/climatism-watching-climate-science/2013/oct/7/climate-change-dominated-water-cycle-not-carbon-di/. Accessed 23 Jan 2014

International Maritime Organization (2012) Overview of greenhouse gas emissions from ships. In: International shipping facts and figures – information resources on trade, safety, security, environment. Maritime Knowledge Center, IMO

International Maritime Organization (2014) Technical and operational measures. IMO

IPCC (2013) Climate Change 2013: the physical science basis, summary for policy makers. Stocker TF et al (eds). IPCC, Switzerland

Junk WJ et al (2013) Current state of knowledge regarding the world's wetlands and their future under global climate change: a synthesis. Aquat Sci 75(1):151–167

Knox J (2009) Linking human rights and climate change at the United Nations. 33 Harv Envtl L Rev 477:492

Massachusetts et al. v. EPA et al, 549 U.S. 497, (2007)

Ramsar list of wetlands of international importance. The Secretariat of the convention on wetlands. http://www.ramsar.org/pdf/sitelist.pdf. Accessed 23 Jan 2014

Salzman J (2005) Thirst: a short history of drinking water. Yale J L Human 18(3): Article 6

Seneviratne K (2002) Tiny Tuvalu steps up threat to sue Australia. U.S. IPS

Stoa R (2014) The United Nations watercourses convention on the dawn of entry into force. Vanderbilt J Trans Law 47(5):1321–1370

United Nations (1992) Basel convention on the control of transboundary movements of hazardous wastes and their disposal

United Nations (1997) Convention on the law of non-navigational uses of international watercourses

Yamada C (2011) Codification of the Law of transboundary aquifers (groundwater) by the United Nations. Water Int 36(5):557–565

Chapter 25
International Perspective on the Basin-Scale Water-Energy Nexus

Luis Metzger, Belize Lane, Shimelis Gebriye Setegn, Jenna Kromann, Mathew Kilanski, and David MacPhee

Abstract This chapter addresses the current state of water and energy resources management in different regions of the world using basin-scale case studies from North America, Latin America, and Africa. It focuses on the characterization of the current state and future projections of water and energy resources available in each basin as well as management of information gaps and potential links for integrating water and energy management. Agriculture demands large amounts of water in each basin and tends to be a priority when water distribution decisions are being made. Overall, this chapter provides a worldwide view of the state of water and energy in semiarid regions, showing cases of water management strategies that are being carried out in the case study basins considered in this chapter.

Keywords Arid regions • Basin scale • Water-energy nexus

L. Metzger (✉)
National Service of Meteorology and Hydrology, Lima, Peru
e-mail: lmetzger@senamhi.gob.pe

B. Lane
Department of Land, Air and Water Resources, University of California, Davis, CA, USA
e-mail: baalane@ucdavis.edu

S.G. Setegn
Department of Environmental and Occupational Health and Global Water for Sustainability Program (GLOWS), Florida International University, 11200 SW 8th Street, Miami, FL 33199, USA
e-mail: ssetegn@fiu.edu

J. Kromann • M. Kilanski
Department of Geological Sciences, Earth and Energy Resources, University of Texas, Austin, TX, USA
e-mail: jennakromann@gmail.com; kilanskimatthew@gmail.com

D. MacPhee
Department of Mechanical Engineering, San Diego State University, San Diego, CA, USA
e-mail: davidwmacphee@gmail.com

S.G. Setegn, M.C. Donoso (eds.), *Sustainability of Integrated Water Resources Management*, DOI 10.1007/978-3-319-12194-9_25

25.1 Introduction

Increasing demands on water and energy resources from different sectors (public and private), coupled with a growing global population and the over exploitation of fossil fuels, are pushing governments to adjust management practices to mitigate the impact on their economies in the near future. Some parts of the world are moving toward cleaner, more renewable energy sources which are the most promising alternatives in terms of investment, generation, and operational costs.

Another driver in the shifting water-energy nexus is climate change. Impacts of climate change are often exacerbated in semiarid and arid regions where water resources are naturally scarce and extreme events are more prevalent. Based on trends from 1900 to 2005, precipitation has increased significantly in eastern regions of North and South America, northern Europe, and north and central Asia and declined in the Sahel, the Mediterranean, southern Africa, and parts of southern Asia. Globally, the area affected by drought has likely increased since the 1970s (IPCC 2007).

According to the United Nations, an area is said to be experiencing water stress when annual water supplies fall below 1,700 m^3/s, water scarcity below 1,000 m^3/s, and absolute water scarcity below 500 m^3/s. Currently, physical water scarcity is affecting 1.2 billion people or almost one-fifth of the world's population (UN-Water n.d.). Furthermore, a report by the United Nations Department of Economic and Social Affairs found that 1.8 billion people will live in countries or regions with absolute water scarcity by 2025 (UNESCO-WWAP 2006).

Throughout history, water conflicts have occurred for different reasons, including development disputes, religious and military goals, and political objectives. More recently, however, the frequency and severity of water conflicts have increased through time; this trend is expected to continue into the future (Pacific Institute 2012). Conflicts over water are becoming an increasing concern, emerging on the international, national, and local level.

Clearly, we are now living in a changing climate. The alterations in climate have vast implications for water management, including changes to hydrologic processes and regimes and the quantity, quality, and timing of freshwater resources (Bates et al. 2008). The emerging challenge for this generation is to develop a framework for synergistic water-energy management capable of meeting the needs of our shifting paradigm by increasing the use of renewable energy, improving water and energy efficiency, and reducing CO_2 emissions.

This review paper provides basin-scale case studies of water and energy systems in three countries: the USA, Ethiopia, and Peru. The existing and potential links between water and energy under different climatic, demographic, and economic contexts are presented. For a better understanding of the nexus investigated here, the authors have based their analysis on quantitative indicators that are applicable to the water and energy sectors supported by observed and projected climatic and hydrologic information in order to characterize and compare in quantitative terms the performance of the watershed as a whole system.

The focus of this review paper is to describe experiences from different countries with different socioeconomic realities facing a common problem: the ability to supply water and energy to meet increasing population demands within the context of climate change. These are the challenges and opportunities that the future climate scenarios will pose to these countries and humankind at large.

25.2 Background

Throughout human history, food, water, and energy have been essential elements for all human beings, and competition has always existed to obtain these elements. Climate has played an important role in resource development and management. It was clear in ancient civilizations like Mesopotamia, Egypt, Indus Valley, and China that they would have prevailed above others if they had access to water in the right quantity and quality. Today, that philosophy remains unchanged, and in the new global economy, there are additional elements that are frequently considered as important as water: maintaining sustainable economic growth, food security, and clean energy generation.

The global population has increased from two to seven billion people in the last 85 years (Fig. 25.1). Of this total, three billion people live in rural areas, where 2.5 billion people derive their livelihoods from agriculture. Focusing on these statistics, it seems that there is a balance between urban and rural populations in all countries, but this is not true. The proportion of populations that are urban or rural is not well balanced across the globe. In Northern America, Latin America, the Caribbean, Europe, Oceania, Africa, and Asia, urban inhabitants make up 82.2 %, 79.1 %, 73 %, 70.7 %, 60.5 %, and 44.9 % of the population, respectively, although these values may differ from country to country. Further 1.6 billion people, about 25 % of the world's population, do not have access to electricity (ElBaradei 2007).

According to the Food and Agriculture Organization (FAO) Statistical Yearbook, from 1990 to 2010, there has been a slight increase in agricultural populations in Africa and Asia, but it has decreased in the Americas and Europe. Out of a global agricultural population of 2.5 billion people, almost two billion or 80 % are from Asia, but Asia's production is dedicated only to domestic consumption. If growth rates in high-fertility countries continue to increase, as projected by the United Nations Population Division, by 2050 there will be an additional two billion people, most of whom will live in urban areas where they will not be able to sustainably produce food for themselves (FAO 2013).

Consequently, there is enormous pressure on all countries, not only to food consumers but food producers, to keep their agricultural economies growing. According to FAO, the net food production by region ranks Latin America as number one, with an increase in 50 % from 2000 to 2012, followed by sub-Saharan Africa with 40 %; Eastern Europe, Asia, Middle East, and North Africa with 36 %; and North America with only a 20 % increase. As for net exports of food (crops and livestock), evaluated by region from 2004 to 2006 with constant

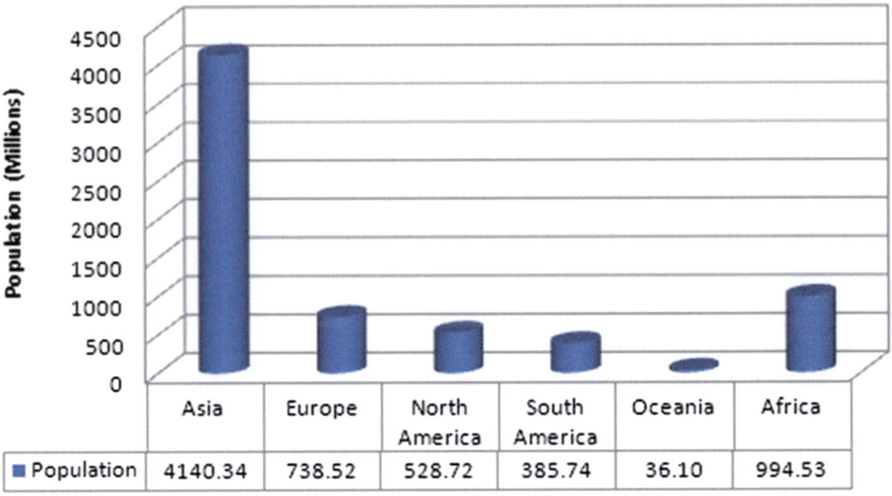

Fig. 25.1 World population by continents (Source: US Census Bureau (March 2012))

reference international prices, North America and Latin America are the leaders with about $35 billion in 2011 exports, followed by Oceania with $15 billion, and Eastern Europe and Central Asia with $4 billion (FAO 2012).

There is a need to understand the links and feedbacks between water and energy and their interactions with the economic sectors. Incomplete information regarding these interactions may produce policies designed to increase efficiency in one sector and creating additional demands in the other sector (Hussey and Pittock 2012). Advanced and useful tools like the Water Evaluation And Planning (WEAP) and Long-range Energy Alternatives Planning (LEAP) models are available in the scientific community to assess the water and energy nexus in order to provide decision-makers with more robust and science-based tools (SEI 2012).

25.3 Climate

Climate change is threatening many regions in the world, and several of the most vulnerable areas are naturally dry regions with high-population growth rates and projections for decreasing precipitation.

The Palmer Drought Severity Index (PDSI) trends estimated using historical data around the world show that, from 1900 to 1949, the Guinea Coast, southern Africa, parts of Canada, and southern and central Europe became drier, while it became wetter in most of Asia, Alaska, and parts of South America (Fig. 25.2). From 1950 to 2002, an evident increase in precipitation over Argentina, the southern USA, and most of Western Australia resulted in wetter conditions (i.e., higher PDSI) in these regions. However, the eastern part of Eurasia, central Africa,

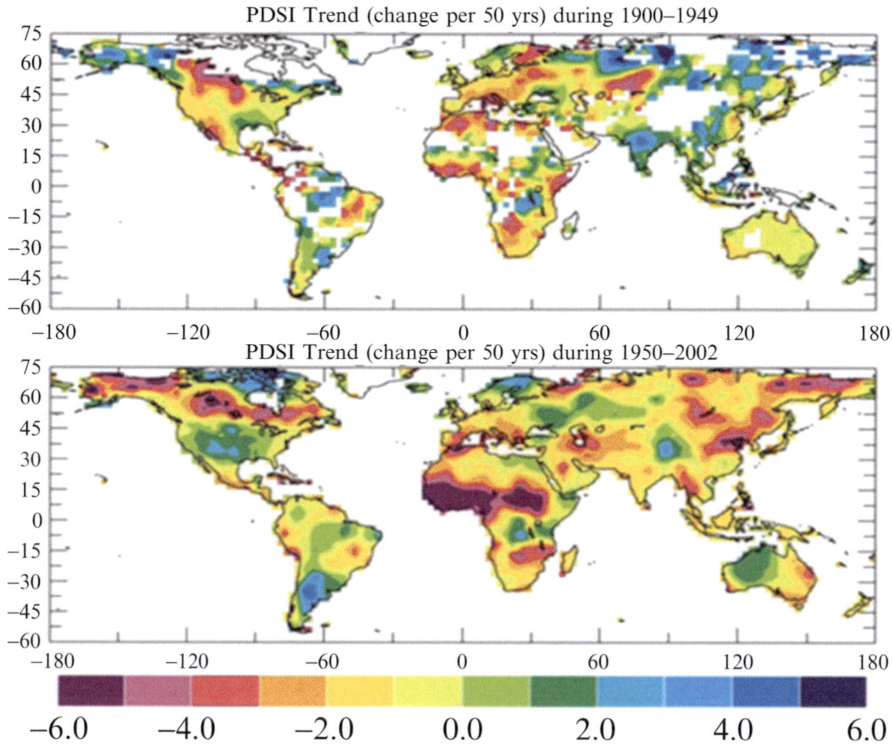

Fig. 25.2 Maps of linear trends of PDSI changes calculated with both precipitation and temperature changes (Source: Dai et al.)

Canada, Alaska, and eastern Australia became drier in the same period, partly because of large surface warming since 1950 over these regions (Dai et al. 2004).

It is a fact that in the last decades, changes in climate have increased considerably, causing negative impacts on natural and human systems on all continents.

According to a recent document by the (IPCC 2014), there is a very high confidence that impacts from recent climate-related extremes, such as heat waves, droughts, floods, cyclones, and wildfires, reveal significant vulnerability and exposure of some ecosystems and many human systems to current climate variability. Likewise, there is high confidence that glaciers continue to shrink almost worldwide due to climate change as well as causing permafrost warming and thawing in high-latitude regions and in high-elevation regions. Finally, the report mentions that there is medium confidence that changing precipitation or melting snow and ice are altering hydrological systems, affecting water resources in terms of surface runoff, quantity, and quality.

In that sense, the climate change summit has provided a space for governments to discuss important issues such as measures to reduce the emission of gases and adaptation measures against climate change, among others. Thus, during the 2009

Copenhagen Climate Change Summit, President Barack Obama committed to reduce carbon dioxide emissions in the range of 17 % below 2005 levels by 2020. The Energy Information Administration showed on its April 2013 report that the USA had a 12 % reduction during the 2005–2012 period.

On the eighteenth Conference of the Parties (COP 18) on climate change carried out in Doha, Qatar, the Kyoto Protocol was ratified for its second period from January 1, 2013 until December 31, 2020. The most important goal in the summit was to establish a climate agreement to ensure that the global temperature increase does not exceed 2 °C.

25.4 Energy

Water and energy are closely related; together they are used by humans from all around the world as main elements to satisfy their basic needs. Energy generation requires water as well as water treatment, and distribution requires energy. More than 7 % of the world's energy supply is used to move water (James et al. 2002).

The four main demands for the energy of the world are the population, economic growth, oil and gas resources, and technic-economic characteristics of energy technologies (European Commission 2006).

According to the Latin American Energy Organization (OLADE), in 2007, Latin America produced 1,223,092 GWh of electricity, of which 65 % came from hydropower, 27.5 % thermoelectric, 3.4 % nuclear power, and 4.1 % geothermal plants. Brazil and Mexico combined generated 56 % of the region's energy. Of this energy production, Brazil generated 54 % of the total hydropower produced while Mexico provided 44 % of the total generation through thermal power plants (Fig. 25.3).

It is important to highlight that, in a period of 30 years (1977–2007), the total energy consumption doubled in Latin America, with a large increase in the transport and industry sectors, which represented 72.5 % of the energy demand in 2007.

Trends on energy consumption and CO_2 emissions from 1990 to 2006 indicate that electricity generation increased by 60 % while associated with CO_2 emissions increased by 57 % and represented 38 % of global fuel combustion-related CO_2 emissions (IEA 2009).

On the international scale, energy consumption has experienced an increase around the world. According to regional energy use records for 1990–2008 (IEA/OECD), the USA, EU-27, Middle East, China, Latin America, Africa, and India have experienced population growth, but of these, the Middle East and Africa have increased their population by 51 % and 55 %, respectively, while EU-27 has increased by only 5 %. In energy consumption (kWh/capita), the USA and EU-27 experienced a growth of −2 % and 1 %, respectively, while China increased in 111 %.

The International Energy Agency (IEA 2011) indicated that in 2009, there was a global investment of about $9.1 billion in energy infrastructure. As a result,

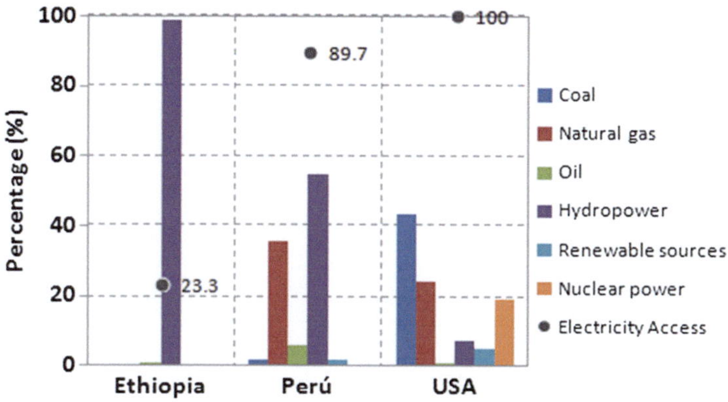

Fig. 25.3 Sources of electricity production and electricity access (Source: World Bank 2014)

20 million people gained access to electricity. In the New Policies Scenario, it is expected an investment of $296 billion between 2010 and 2030; this is $14 billion per year or 54 % more than in 2009. With this new scenario, 550 million people would gain access to electricity. However, an average of $48 billion per year is required to provide universal access to energy by 2030.

According to this scenario, the number of people without access to electricity in Latin America would decrease from 26 to 8 million people in rural area, while in urban area, it would decrease from 4 to 2 million people (Fig. 25.4).

25.5 Case Studies

Water and energy management as an integrated system that is continuously interacting with the different users is not an easy task, especially if we are located in regions where aridity is part of the system.

On the basin scale, there are internal driving factors such as population, land planning, water and energy management, and water and energy demands. The external factor and main driving force in human decisions is climate, which is an uncertain variable that humans do not have control of. Below is shown on a map the location of basins of study, and then there is a description of each basin regarding the water-energy nexus.

25.5.1 Lower Feather River Watershed, California, USA

As California's demand for water increases with population growth, climate change, and socioeconomic activities, so does water-related energy consumption.

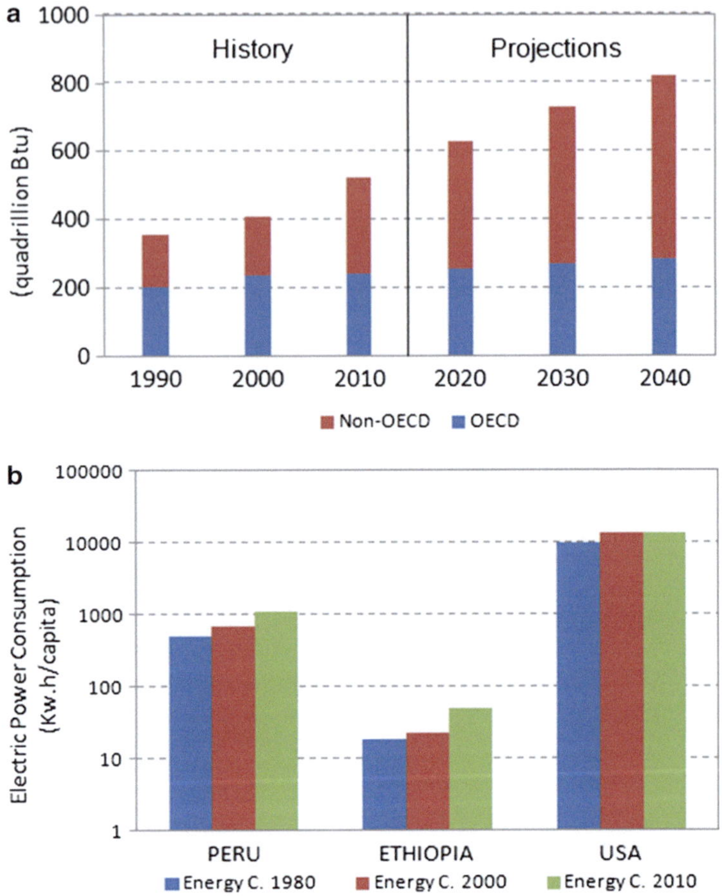

Fig. 25.4 (**a**) World energy consumption 1990–2040, (**b**) electric power consumption (Source: (**a**) US Energy Information Administration (http://www.eia.gov/forecasts/ieo/); (**b**) World Bank (data.worldbank.org/indicator/EG.USE.ELEC.KH.PC)

The water sector is the largest energy user in California, consuming about 19 % of all electricity in the State as well as 30 % of all non-power plant-related natural gas consumption (CEC 2005). This vast energy consumption is used to transport water from places of abundance with lower population densities (e.g., Northern California) to more arid regions with higher population densities (e.g., Southern California). Further complicating the issue is the fact that almost all types of electricity generation require significant water use. By the year 2050, the population in California could be as high as 65 million, with a per capita water use of 221 gallons per day, making California by far the largest water user in the USA (DWR 1998). Given current trends, it is clear that rising energy and water demands will likely require system expansion at the cost of additional financial burdens and environmental impacts. Basin-scale integrated water resources management (IWRM) has the potential to greatly reduce the demand for water and consequently save energy in California without drastically impacting financial obligations or environmental integrity.

25.5.1.1 Background

The Feather River drains a total of 16,000 km^2 of Northern California's Sierra Nevada, southern Cascades, and Sacramento Valley. The Lower Feather River Watershed (LFRW) (2,080 km^2, 13 % of total watershed area) begins at the outlet of Oroville reservoir where the forks of the upper Feather meet, receiving the Yuba and Bear Rivers from the east before becoming the main tributary to the Sacramento River. From here, the water is routed through the Sacramento-San Joaquin Delta as the largest input to the California State Water Project (SWP). The Feather River provides water to central and southern California as well as hydropower for Sutter, Yuba, and Butte Counties.

The current population of ~100,000 in the tri-county area is expected to double over the next 40 years, and much of that growth will likely occur in the unincorporated areas of the counties (http://www.sacriver.org/aboutwatershed/roadmap/watersheds/feather/lower-feather-river-watershed). Major challenges to watershed planning will involve providing water and energy to these developing areas and mitigating development impacts on agriculture, habitat, water quality, and hydrology. Furthermore, the projected effects of climate change on the region have major implications for water management in California and the LFRW. Increases in temperature, changing precipitation patterns, and reduced snowpack are already being observed in the state (Moser et al. 2009). These changes are resulting in a suite of threats to regional human water supply and quality. In addition, earlier runoff and reduced spring and summer streamflows are leading to severe consequences for aquatic and riparian ecosystems adapted to the natural flow regime (Poff et al. 1997).

25.5.1.2 Hydrology

The LFRW's semiarid Mediterranean climate is typical of northern California, with hot, dry summers and cool, wet winters. The watershed contains elevations ranging from 20 to 3,778 ft above sea level and has an average annual precipitation of 17–60 in. Prior to the completion of Oroville Dam in 1968, the maximum observed discharge at the mainstream Feather confluence was 230,000 cubic feet per second (cfs) (Mar. 1907), and the average daily discharge fluctuated between 500 and 170,000 cfs; since completion (1969–2009), the maximum flow has been reduced to 161,000 cfs (Jan. 1997), and the average discharge from the dam has been stabilized at around 300 cfs. The Lower Feather River is on the Clean Water Act Section 303 (d) list of impaired water bodies for temperature, chlorpyrifos, diazinon, mercury, and unknown toxicity. In addition to surface water inputs, the watershed contains portions of four major groundwater aquifers: East Butte, North Yuba, South Yuba, and Sutter aquifers. Water quality constituents of concern for groundwater include total dissolved solids, nitrate, and several other chemicals (CA 2006).

25.5.1.3 Main Water Users

Oroville Dam, the tallest dam in the USA, provides flood control for the Sacramento Valley and water storage for downstream agricultural users. The Sacramento area, for which the Feather River is the main tributary, currently claims the highest flood risk of any major metropolitan city in the USA (over $20 billion in projected damages in a single flood event). Water supply in the LFRW is allocated mainly to municipal (111 million cubic meters/year) and agricultural uses (2,714 million cubic meters/year), with agriculture making up the vast majority (>95 %) of demand.

Directly downstream of Oroville Dam is the Oroville-Thermalito Complex, the most important facility in the CA Department of Water Resources' SWP; the SWP is the largest publicly built and operated water and power development and conveyance system in the world. At maximum capacity, the system can produce 762 MW (HLPCO 2014) and store around 4.47 km^3. Moreover, this system has the ability to essentially store electricity in the form of pumped hydrostorage during off-peak electricity demand, further demonstrating the inseparable links between energy and water in the LFRW.

25.5.1.4 Main Energy Users

Butte, Yuba, and Sutter Counties completely contain the LFRW, and energy needs are met almost entirely by natural gas and nearby hydroelectricity. In 2011, the three counties consumed 2.5 GWh of electricity and 84 million therms of natural gas (ECDMS 2014). The per capita electricity consumption in the area is roughly 6.5 MWh/year, well below the national average of 12.2 (Energy Almanac 2010); this is primarily due to relatively mild weather, reducing the need for space heating and cooling.

Most of the natural gas consumption is used either directly for electricity generation (40–45 %) or in the gas mining process (~10 %). The remainder of the gas consumption end use is residential and is comprised mainly of space and water heating (CNGDS 2014).

In general, it is difficult to pinpoint electricity end use without expensive surveying or smart grid technology. However, it is quite easy to discern the percentages of electricity used in the commercial and residential sectors through regular billing practices; an analysis of this data gives an almost equal split between residential and nonresidential electricity use (ECDMS). This fact, coupled with the region's relatively low-population density, confirms this area as a substantial exporter of energy and water, making integrated water resources management a key area of focus – especially so due to the heavily populated regions to the south relying on its water and energy exports.

25.5.1.5 Water-Intensive Energy Production

In general, the region only uses natural gas and hydroelectric sources. Technically speaking, hydroelectricity is very water intensive – but only when filling reservoirs. Once reservoirs are filled, there is not much "use" of water, except due to increased drainage/evaporation from the increased surface areas of the reservoirs themselves. A non-negligible amount of water is likely consumed in cooling towers for many industrial applications as well; however, this is likely less per capita than many other urban areas due to the low demand for electricity. In essence, this region is a relatively low-water-extraction region for the purposes of energy production; this is mainly due to the hydroelectric and gas turbine power production, both of which do not necessarily consume water in their core processes.

25.5.1.6 Energy-Intensive Water Production/Transmission

No significant pumping projects exist in this region as part of the California State Water Project for purposes of elevation change. This is due to the high elevation of the region in general, allowing for natural (gravity-induced) transportation of water to the lower, more densely populated regions of Central and Southern California.

25.5.1.7 Water and Energy Governance/Environmental Regulations

The management of California's water system is highly decentralized with over 3,000 water districts and agencies governing water management and policy at various scales and with varying objectives. Federal agencies include the US Bureau of Reclamation and the Army Corps of Engineers. State and regional water agencies such as the State Water Project and the Metropolitan Water District of Southern California provide water to retailers and end users, and water rights and water quality are regulated by the State Water Resources Control Board. However, most water agencies and the majority of funding and expertise reside in the thousands of local water agencies (Lund et al. 2009).

25.5.2 Lower Colorado River Basin, Texas, USA

The Lower Colorado River Basin is located in eastern central Texas and is predominantly the Colorado River and a few other tributaries. There are a number of different demands on water that the basin supplies primarily irrigation, steam-electric, and municipal use. The basin is located in a semiarid region that frequently experiences drought. Most future climatic scenarios predict that the region will have reduced precipitation and higher temperatures. Population in the Lower Colorado River Basin is expected to increase significantly. With future projections for drought and an increasing population, the demand on water and energy resources will only increase. The State of Texas manages water resources on a regional level including planning regions for surface water and groundwater that establish a regime for allocating water each basin. The planning regions implement a variety of strategies to meet demand such as conservation programs, new infrastructure facilities (such as reservoirs), and reuse technologies.

25.5.2.1 Background

The Lower Colorado River Basin starts in central Texas and extends to the Gulf of Mexico; most of the basin is part of the Colorado River (Fig. 25.5). The total basin area is estimated to be 34,652 km^2. The basin is composed of all or a portion of 14 counties in Texas. The biggest city in the basin is Austin, the capital of Texas. The Lower Colorado River Basin is one of the 16 planning districts in the state. Approximately 6 % of the total population of Texas resides in the Lower Colorado River Basin, and it is projected that the population will increase by 100 % to 2,831,937 people by 2060 (TWDB 2012a, b).

25.5.2.2 Hydrology

The Lower Colorado River Basin is located in a semiarid region, prone to droughts and high temperatures. This region is characterized by a Post Oak Savanna with subhumid mixed prairie, savanna, and woodlands (TWDB 2012a, b). Precipitation in the region mostly comes in the form of occasional storms, where average annual precipitation is 24–48 in per year. Projections made with climate models say that temperatures will rise and precipitation will decrease in most of Texas (Nielsen-Gammon 2011). There are both surface water and groundwater drinking water sources in this area; 77 % of the population relies on surface water, while 23 % relies on groundwater. There are nine surface reservoirs in the region that supply drinking water. Groundwater is sourced from 11 major in minor aquifers in this area; the aquifers that supply the most water are the Edwards-Trinity (Plateau), Trinity, Edwards (Balcones Fault Zone), Carrizo-Wilcox, and the Gulf Coast. Total water supply in the Lower Colorado River Basin was estimated to be 1,162,884 acre-feet in 2010; it is expected to increase minimally to 1,169,071 acre-feet in 2060 due to an increase in smaller local water suppliers. There is a need for more water under drought conditions of approximately 367,671 acre-feet by 2060 (TWDB 2012a, b). Streamflow in the river ranges from 0 to 60 cfs in the upper basin and 0 to 919 cfs in the lower basin, the streamflow is variable depending on location within the basin, weather conditions, and time of year (LCRA).

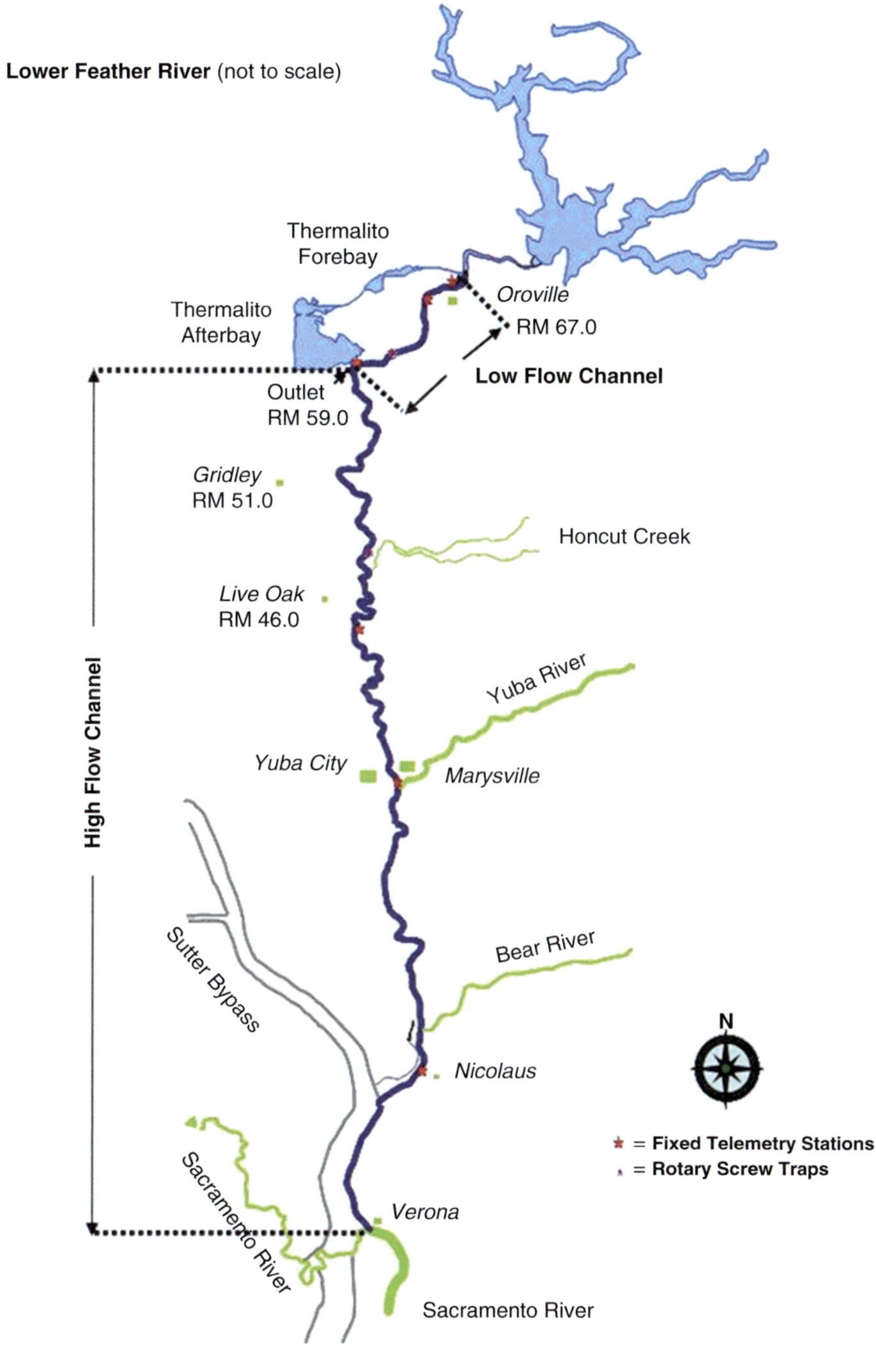

Fig. 25.5 Lower Feather river and tributaries

25.5.2.3 Primary Water Uses

Irrigation, municipal use, and steam-electric generation account for 89.7 % of water use in the region. The Texas Commission on Environmental Quality has actively permitted over four million acre-feet of water use in Lower Colorado River Basin annually, even though it is estimated that there is under 900,000 acre-feet of supply available. There are 11 dams along the Lower Colorado River, six of them are used for hydroelectric power, four are used to hold supply for thermoelectric cooling, and one is used to store supply for crop irrigation (rice). In the early to mid-twentieth century, a series of six dams was built on the Highland Lakes to control flooding and provide electricity to central Texas. Together, these man-made lakes hold nearly 650,000 acre-feet of water and generate 230,000 kW of power. The largest of the Highland Lakes is the Mansfield Dam which forms Lake Travis, the municipal water supply for Austin, Texas.

The South Texas Project has the largest dammed lake on the Lower Colorado River. The lake was built to cool water for the South Texas Nuclear Generating Station. The plant produces 2,700 MW of power and is permitted to use 204,000 acre-feet of water each year, but only uses about 75,000 acre-feet per year. The South Texas Nuclear Generating Station has two 102,000 acre-feet water right permits, one of which is the third most senior active water right on the entire river, meaning in times of extreme drought, it could divert supply from everyone on the river except for the City of San Angelo (whom has a permit for 3,000 acre-feet of water) and the Southwest Graphite Company (whom has a water right for 400 acre-feet per year).

25.5.2.4 Water and Energy Governance/Environmental Regulations

The Lower Colorado River Authority (LCRA) is a key water right holder and energy provider. It owns five of the six Highland Lake dams as well as 2,747 MW of conventional power in the basin. LCRA sells wholesale power to 43 city-owned utilities in the region.

The Texas Commission on Environmental Quality (TCEQ) permits the use of surface water for the entire State of Texas. It enforces permits and resolves conflicts. The Texas Water Development Board develops the State's water plan and issues bonds to fund water development projects throughout Texas.

Texas recognizes two basic doctrines of surface water rights: the riparian doctrine and the prior appropriation doctrine. While the State of Texas "owns" the surface water, the riparian doctrine allows entities with property adjacent to a body of water to use it for useful purposes. The State permits use to the water through the prior appropriations doctrine, generally known as "first in time, first in right," meaning the first entity to claim the water has the priority to use it (some of the water rights in Texas date back hundreds of years). Surface water rights can be bought and sold; however, if water is transferred outside of its original basin, it

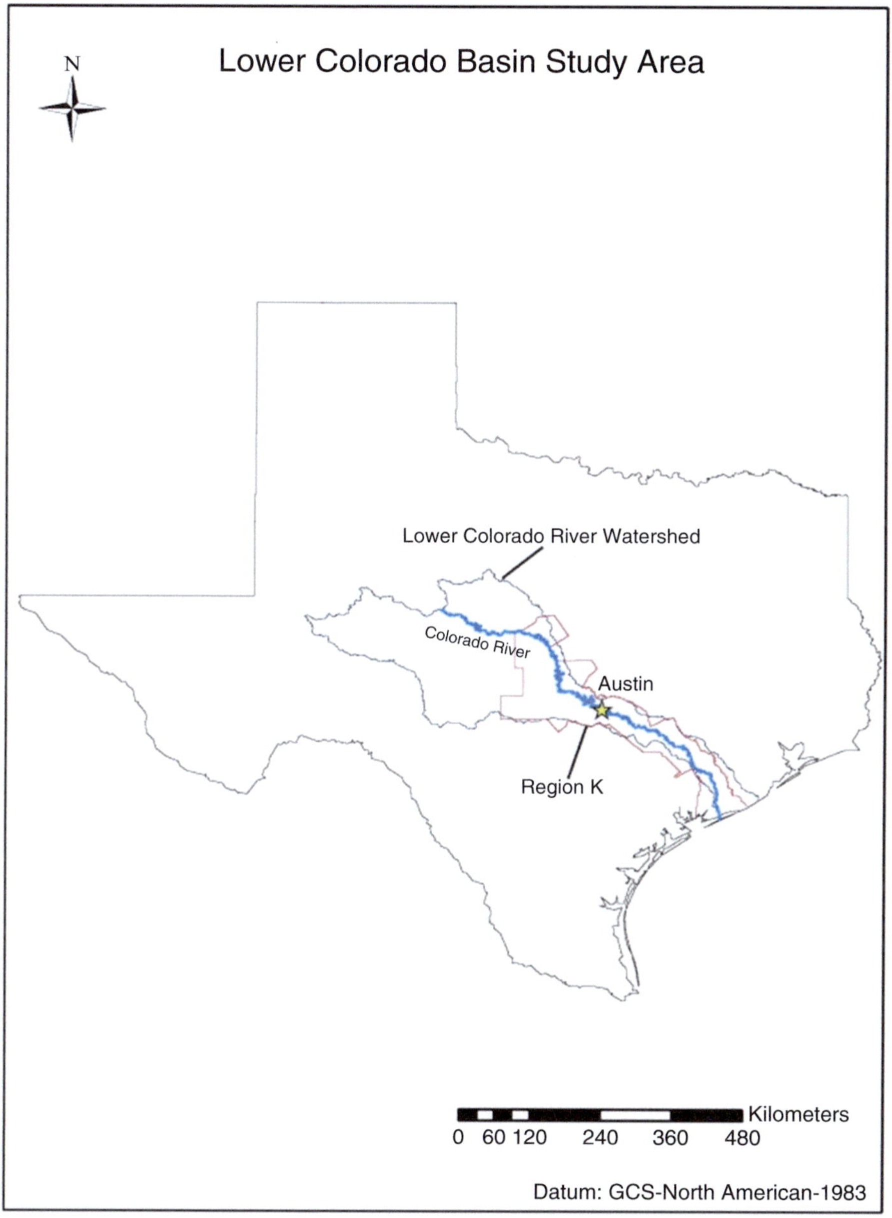

Fig. 25.6 Map showing the Lower Colorado River Watershed

becomes the lowest priority in the entire basin. This makes it difficult for parts of the state that need water to attain it from other parts of the state that have excess (Fig. 25.6).

Groundwater is a property right governed on the principles of the English common law "rule of capture," meaning users can pump as much groundwater as their pump will allow as long as it is not wasteful. Groundwater conservation districts are established across most of the state to permit groundwater pumping; however, groundwater conservation districts authority to enforce permits is limited.

The South Texas Project nuclear facility's senior water rights can be a driving force in surface water right allocation in the Lower Colorado River Basin. Behind that the Lower Colorado River Authority has the authority, and has used it as recently as 2012, to withhold water from downstream farmers. This shows that while agriculture is the primary user of water, it is not the most important. If future projections hold, irrigators may need to seek alternative water sources. The region has significant groundwater resources, but has been reluctant to increase reliance upon them for a variety of reasons.

In the Lower Colorado River Basin and Texas more generally, energy production tends to be prioritized for water rights, followed by municipal use, and then irrigation and industry.[1] As the population of the state swells, Texas municipal water and energy use will increase as well. While municipalities will continue to implement conservation measures, the resource pinch will most likely be felt by irrigators (Fig. 25.7).

25.5.3 Blue Nile River Basin, Ethiopia

25.5.3.1 Background

The Nile River drains an area of 2.9 million km^2 that covers 10 % of the African continent. It is an "international" river as its water resources are shared by 11 countries, namely, Ethiopia, Egypt, Sudan, South Sudan, Eritrea, Uganda, Tanzania, Kenya, Burundi, Rwanda, and Democratic Republic of Congo. With a course of 6,695 km, it is the longest river in the world. The two major tributaries are the Blue Nile, stemming from Lake Tana in Ethiopia and flowing to Sudan, and the White Nile, from Lake Victoria in the East African Community. Lake Victoria is fed by several tributaries: Kagera, Yala, Sondu, Nyando, Mara, Mbalageti, Simiyu, and Konga rivers (van Griensven et al. 2012). The Blue Nile is the source of most of the water and fertile soil. The Blue Nile and the White Nile join each other in Khartoum to form the Nile River that flows Northeast (Fig. 25.8).

Ethiopia has embarked on a development and transformation pathway with the vision of attaining middle-income country status by 2025. This is currently guided

[1] There is a major exception to the prioritization, the Texas panhandle, the part of the state that uses the most water, prioritizes irrigation above all else. The region relies on the vast Ogallala aquifer for water. The water-energy nexus is a completely different equation there. In the panhandle, the question is "How much energy does it take to pump water from increasing depths?" and "When does it become too expensive to pump from that depth?"

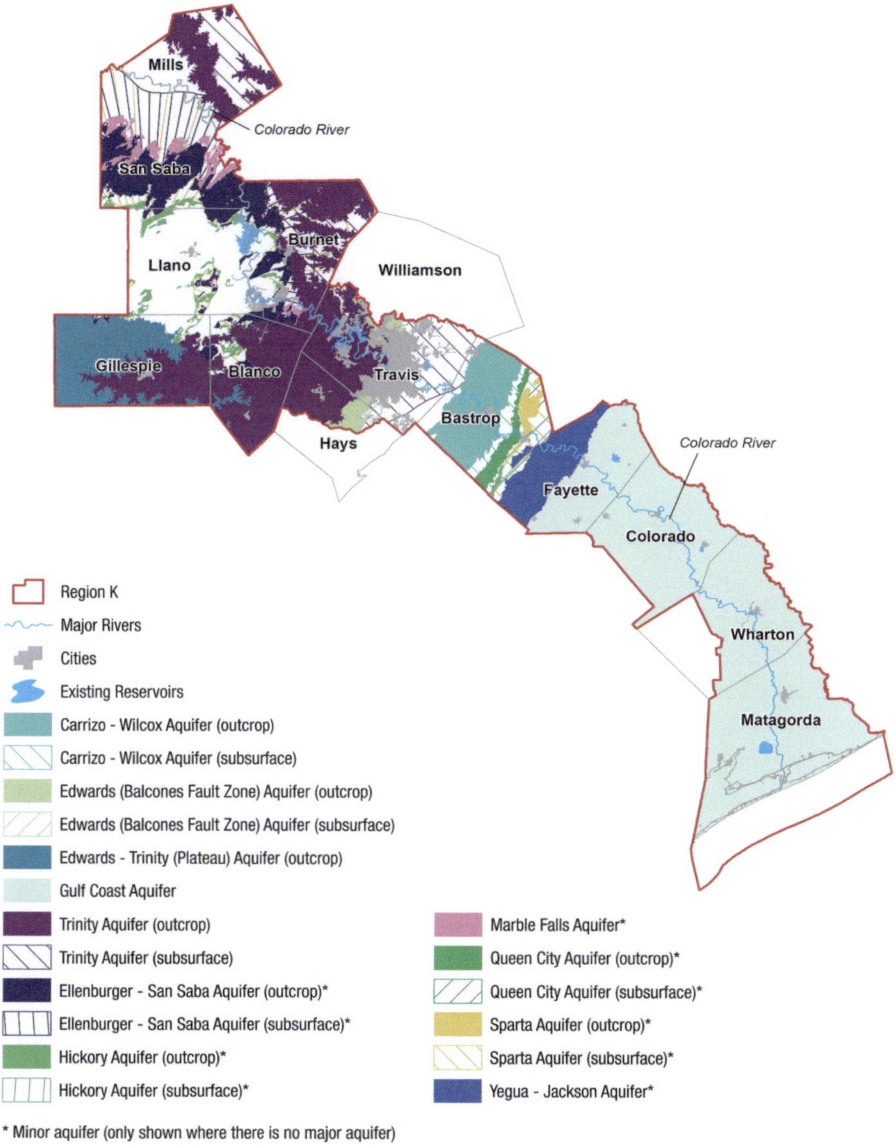

Fig. 25.7 Lower Colorado River Basin overview (planning region K) (TWDB 2012c)

by the national Growth and Transformation Plan (GTP) for 2011–2015 and the 2011 Climate-Resilient Green Economy (CRGE) strategy. Renewable energy generation is to be rapidly expanded, improving local access to modern energy and providing excess power for export (SEI 2013). Many organizations have been implementing projects in the Upper Blue Nile River (e.g., Lake Tana and Beles River basins). The projects use the nexus approach to compare the outcomes of

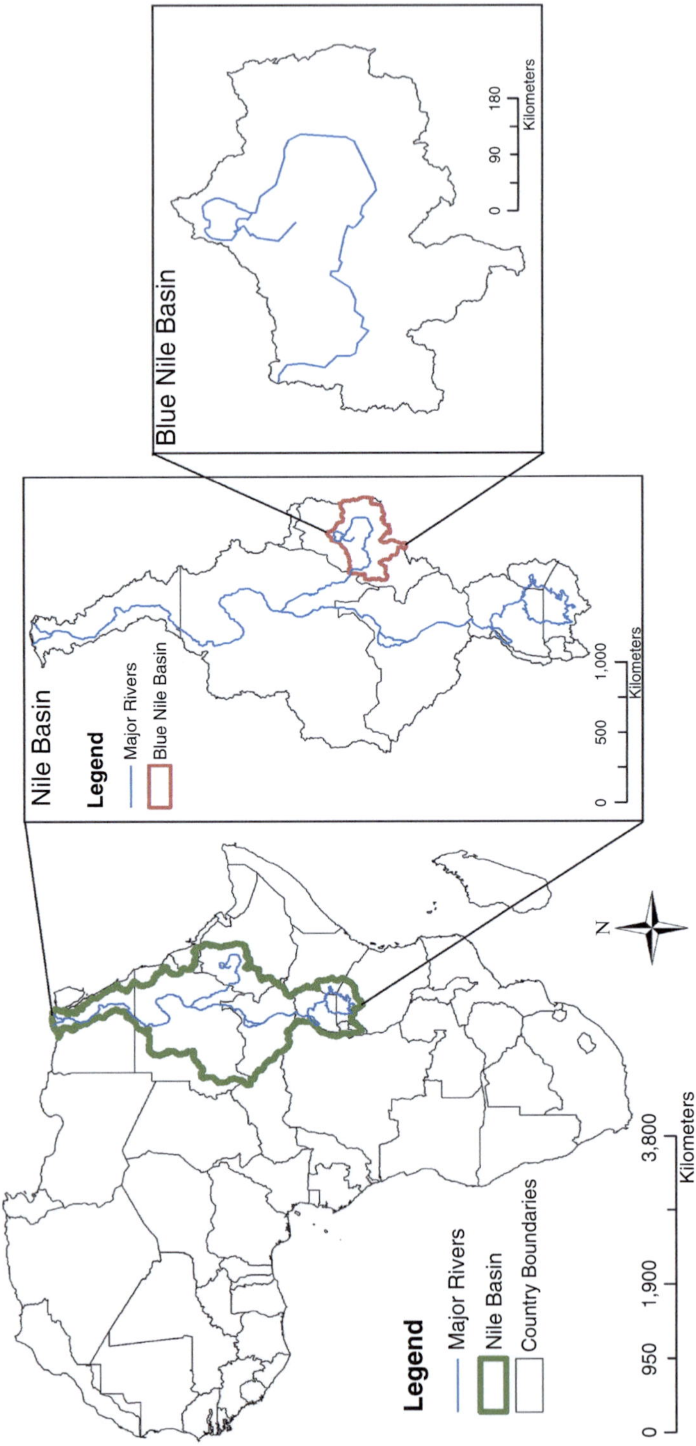

Fig. 25.8 Location map of Nile River basin

different development pathways and interventions, at scales from national to local. The purpose of this approach is to help decision- and policy makers and planners to identify opportunities to improve food and energy security and promote economic development. According to Ethiopia's 25-year Power Master Plan, the country aims to become the "battery" of Eastern Africa through harnessing its huge hydropower potential by erecting huge dams on many of its rivers mainly the Rivers Omo and the Blue Nile and their tributaries. According to the Plan, the country aims to hit the gauge of 37 gigawatts by 2037.

25.5.3.2 Hydrology

Nile River basin is a complex hydrological system which consists of distinct hydrological regimes at the different reaches of the river system. The runoff response of the system to the climate change is greatly associated with the variation in the behavior of the hydrological regimes of the river system.

The Blue Nile originating from the Lake Tana, Ethiopia. The Blue Nile eventually joins the White Nile at Khartoum, Sudan, and the Nile continuous through Egypt to the Mediterranean Sea at Alexandria. The Blue Nile River has a drainage area of 199,812 km^2 and supplies nearly 84 % of the water of the Nile River during high-flow season. Between the months of June and September, flow increases dramatically due to seasonal heavy rains occurring between June and September.

Moges and Gebremichael (2014) divided the Nile River basin into four hydrological regimes: the water source or the high rainfall regime, the energy source regime or the transition regime, the water-losing ecosystem regime, and the water use regime. According to Moges and Gebremichael (2014), the outflow from the water source and energy source regimes are highly sensitive to climate change, while the outflow from the water-losing ecosystems are less sensitive to climate change effects. The energy source regime is located in the lower reaches of the water sources area between the highland (high rainfall regime) and lowland (low-rainfall regime). It is a transitional region located largely in the lower reaches of the main tributaries of the Nile. The lower reaches of Blue Nile, Tekeze (Atbara in Sudan), Baro, Kagera in Rwanda, and other tributaries flowing into Lake Victoria. Like the water source regime, the energy source regime is highly sensitive to climate change.

Development and management of hydropower-based multipurpose large-scale storage schemes in these reaches offer multipurpose benefits as well as buffers the effect of climate change and climate variability. It produces large amount of stable green energy pool to the regional market as well as the upstream countries of the water source. It also provides the capacity to reduce the sensitivity of flow and buffer the impacts of climate change to the lower riparian countries. Developing appropriate basin-wide adaptation strategies that promote water development in the basin requires an understanding of the sensitivity of the different hydrological regimes of the basin to climate change.

25.5.3.3 Water Resources and Energy in the Nile Basin

Africa is blessed with water and other natural resources, but food insecurity continues to prevail, with millions of people struggling to survive hunger and poverty over the last several decades. The increasing population and the associated water and energy demands exacerbated by climate variability and change contribute to food insecurity. Water is a vital strategic natural resource for all economies – mainly in food production, domestic use, and in the production of renewable energy. Attaining food security by raising agricultural productivity will inevitably involve an increase in energy inputs for water supply and management, agricultural production, agro-processing, community lighting, and drinking water.

The Nile represents a crucial resource for the economy of eastern and northeastern Africa. Water is a critical resource for all countries that share the basin. Water will be even more critical in the future as these countries face larger populations and therefore an even greater demand for water. The major water users in the Nile River basin are irrigation and hydropower.

The Blue Nile is very rich in water resources. The annual runoff amount is more than 48 billion m^3 per year that flows to the neighboring countries, contributing to 62 % of the Nile River water flow. The water resources are however hardly used in Ethiopia for economic development and poverty alleviation in the basin.

The Blue Nile possesses more than 15,000 MW of energy (Desalegn et al. 2011). Blackmore and Wittington (2008) indicated that construction of more cascaded large storage dams in the Upper Blue Nile will provide more economic benefit, enhance water availability in the system, and increase adaptive capacity toward climate change impacts to lower riparian countries. One of the projects that has been undertaken by the Ethiopian government is the implementation of the 6,000 MW Grand Ethiopian Renaissance Dam Project (GERDP) located on the Blue Nile just upstream of the Ethiopian-Sudan border as well as other hydropower projects in the upper part of the GERD.

25.5.3.4 Nile River and Hydropower Development

Hydropower is an important source of modern energy. Hydroelectric power supplies 32 % of Africa's energy (UNEP 2012). Although endowed with considerable hydropower potential, African countries have developed only a small fraction of it. The key hubs, sometimes called "water towers," for potential hydropower generation are in the Congo River basin, the Fouta Djallon highlands in West Africa, and the Ethiopian highlands in East Africa. According to World Water Assessment Programme (WWAP) (2014), 60 % of the region's hydropower potential is in the Congo and Ethiopia. In light of the lack of exploitation to date, hydropower remains the main energy option to promote sustainable development and to power trade, regional integration, and poverty eradication in Africa. Given the increasing demand for clean, reliable, and affordable energy, the role of hydropower is gaining importance, particularly as a means to reduce poverty and attain sustainable development (WWAP 2014). Hydropower provides the opportunity for long-term and trans-generational investment in clean energy for the growth of Africa, economically, socially, and environmentally.

The Ethiopian government's policy is to significantly increase large-reservoir water storage in the Blue Nile basin in order to support national development. The planned increases in water storage will facilitate significant increases in hydropower generation and irrigation in the basin. If all planned development occurs, large-reservoir water storage will exceed 160 km^3 (approximately 14 times present levels and 3 times the current mean annual flow at the Ethiopia-Sudan border), irrigation will exceed 360,000 ha (23 times current levels), and installed hydropower generating capacity will be in excess of 10,000 MW (47 times current levels) (McCartney and Girma 2012).

25.5.3.5 The Implications of Climate Change for Water Resource Development in the Blue Nile River

Climate change projection studies in the Nile basin offer a glimpse of uncertainty and high level of sensitivity of flows to changes in the climate. Climate change adaption recommendation must combine the projection uncertainty as well as the growing demand.

There is great uncertainty about the impacts of climate change in the basin, and there have been few systematic studies of the possible implications for water resource development. Results indicate that changes in climate will affect both water availability and demand (Setegn et al. 2011; Moges and Gebremichael 2014). The anticipated changes in climate will have significant impacts on both hydropower generation and water supply to irrigation schemes.

Different studies evaluated the impact of climate change on the performance of existing and planned irrigation and hydropower schemes in the Blue Nile River basin. They indicated that climate change will constrain the technical performance of large reservoirs with knock-on effects for agriculture and electricity production. This requires much more systematic planning of water storage and greater cooperation between the riparian states.

Water resource development in the basin requires interventions that bolster resilience and water security. To moderate the negative impacts of climate change requires much better planning and management of water storage. In the Nile basin, this requires much greater cooperation between riparian states. Careful consideration needs to be given to integrated, possibly transnational, storage "systems" that maximize the benefits to be obtained from the complementarities of different storage options (e.g., surface water used conjunctively with groundwater). However, planning for climate change requires going beyond water alone to consider other sectors in the water-energy-food nexus.

Although there remains great uncertainty about how climate change will impact the water resources of the basin, McCartney and Girma (2012) indicated that performance of planned irrigation and hydropower schemes is likely to be severely constrained by the end of the twenty-first century. Hence, adaptation to climate change, effective water resources management, and development are needed to be considered together. Nile basin needs development-based adaptation approach that builds the resilience capacity and the need for economic growth of the Nile basin society.

25.5.4 *Tambo River Watershed, Moquegua, Peru*

Peru is located in the western part of South America. Its geography is diverse, including coastal desert, the Andes mountain range, and the Amazon jungle. The total population of the country is about 30 million, of which 8.5 million live in the capital city of Lima (INEI 2012). A recent study developed by the International Hydrologic Programme – UNESCO – in 2010 ranks Peru as the country with the sixth highest percentage of arid zones in Latin America and the Caribbean. Moquegua is located in southern Peru, with an area of 15,734 km^2 and a contribution of 1.42 % to the Gross National Product. The electricity and water generated in Moquegua represent 0.12 % of Peru's Gross National Product.

25.5.4.1 Background

The Tambo river watershed covers regions of Arequipa (lower part), Moquegua (medium and upper part), and Puno (upper part). The watershed covers an area of 12,718 km^2, of which 8,149 km^2 are humid. According to the 2007 national census by Instituto Nacional de Estadística e Informática (INEI), the population in the watershed was 88,790 people, of which 52,519 were from Arequipa, 33,867 from Moquegua, and 2,404 from Puno. Also, in a special bulletin issued by INEI (2010), the population in Arequipa, Moquegua, and Puno is expected to increase in average 8.4 % by 2015 and 19.5 % by 2025 compared to the population in 2007.

In November 2003, the Tambo river watershed was the center of a social conflict caused by political disputes regarding the management of the Pasto Grande dam that is located in the upper watershed. The Pasto Grande Special Project, which expanded the agricultural and power generation frontiers, was created in November 1987 with the purpose of providing water for municipal demands, agriculture, and industrial uses in the provinces of Mariscal Nieto and Ilo and expanding agricultural frontier and power generation.

25.5.4.2 Hydrology

According to Köppen climate classification, the Tambo river watershed has three climate types: desertic or subtropical arid from 0 to 2,000 m.a.s.l., steppe and low valleys from 1,000 to 3,000 m.a.s.l., and valleys mesoandinos with cold climate above 3,000 m.a.s.l. Precipitation ranges from 150 to 500 mm per year. The Tambo river is formed by the confluence of four rivers: Carumas, Coralaque, Ichuña, and Paltiture. The headwater of this watershed begins in the Andes at about 5,000 m.a.s. l., and the flow goes from east to west up to the Pacific Ocean.

Based on the aridity index, the Tambo river watershed is characterized as hyperarid in its lower watershed, arid in the medium part, and semiarid in the upper watershed. The water resources are mainly located in the upper watershed; these resources can be found in the form of lakes, springs, rivers, and groundwater.

The Tambo river has an irregular regime with maximum discharge from January to March and minimum discharge from October to December. The average surface water registered by "La Pascana" station from 1956 to 1999 is 993.4 MCM/year equivalent to 1,112.4 cfs. The groundwater coming from the Tambo aquifer has been estimated in 450 MCM/year and the one from the Chilota-Huachunta aquifer located in the sub-catchment of the Vizcachas river in 47 MCM/year (BARRIENTOS 2011).

An analysis made on the discharge of the Tambo river based on records of "La Pascana" station indicates that from 1989 to 1999 (after Pasto Grande dam) the average discharge without naturalization was 784.3 MCM, and considering the naturalization, the discharge was 874.8 MCM (INADE 2001) (Fig. 25.9).

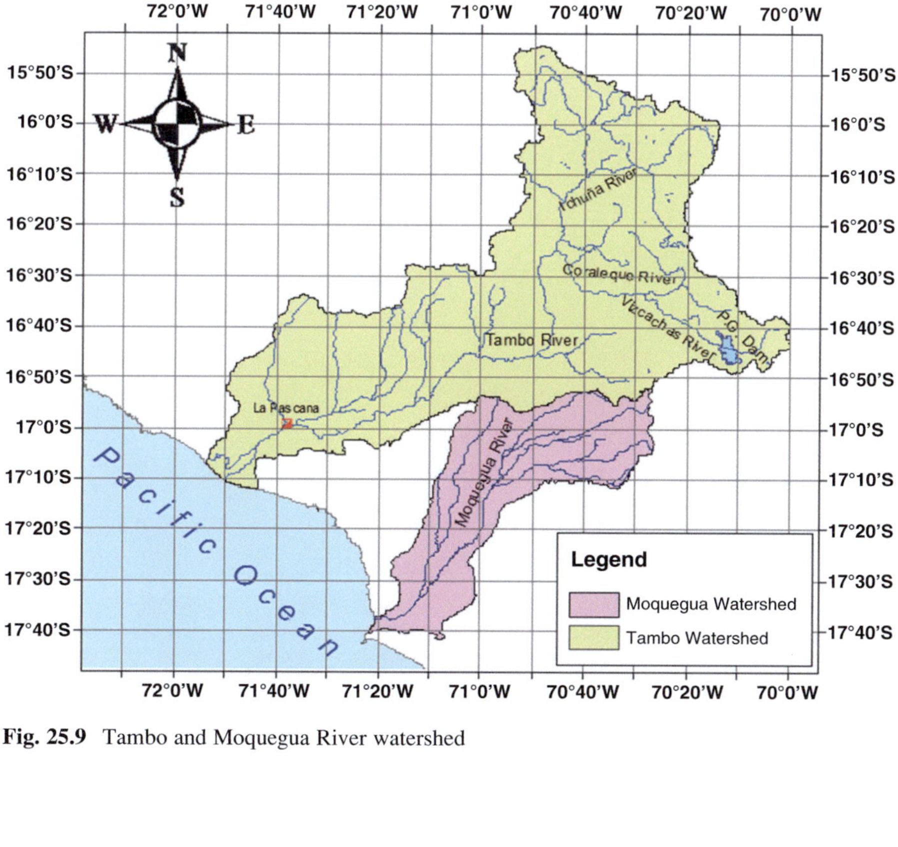

Fig. 25.9 Tambo and Moquegua River watershed

Main Water Users Agricultural, Domestic, Mining, Energy, and Industrial

The Pasto Grande dam located at 4,520 m.a.s.l. in the Tambo watershed was built in 1989; it has a maximum storage capacity of 210 MCM and an active storage of 190 MCM. The purpose of this dam was to export an annual average of 70 MCM to the Moquegua river in order to improve the irrigation of Moquegua-Ilo (3,450 Ha) and Torata (300 Ha) valleys and to expand in 2,688 Ha the agricultural frontier in Pampas Estuquiña, San Antonio, and Jahuay-Rinconada. Water users for irrigation of the Tambo valley say that the derivation of water to the Moquegua watershed has caused a decline in the quantity and quality of water available for irrigation in the Tambo valley, especially during the low-flow season from October to December (MINAG 2005).

Between 2001 and 2010, there has been a gradual increase in the harvested areas of the Tambo and Moquegua watersheds of 47 % and 30 %, respectively (Regional Office of Agriculture of Arequipa and Regional Agricultural Directorate of Moquegua).

Water supply for agricultural use in the Tambo watershed is 440.3 MCM/year followed by domestic use with 6.6 MCM/year. The total demand of water for the Tambo river watershed is 451.4 MCM; this means that agriculture has the highest demand in the watershed with 97.5 % of the total water demand (BARRIENTOS 2011).

Water users for mining activities are Minera Pampa de Cobre S.A. and Aruntani SAC that are located in the upper part of the Tambo watershed with 1.07 MCM/year (Fig. 25.10).

25.5.4.3 Main Energy Users and Sources

In the period 1990–2010, the production of electrical energy in Peru has increased in almost 4 times, from 9,547.7 GWh in 1990 up to 35,908 GWh in 2010. During

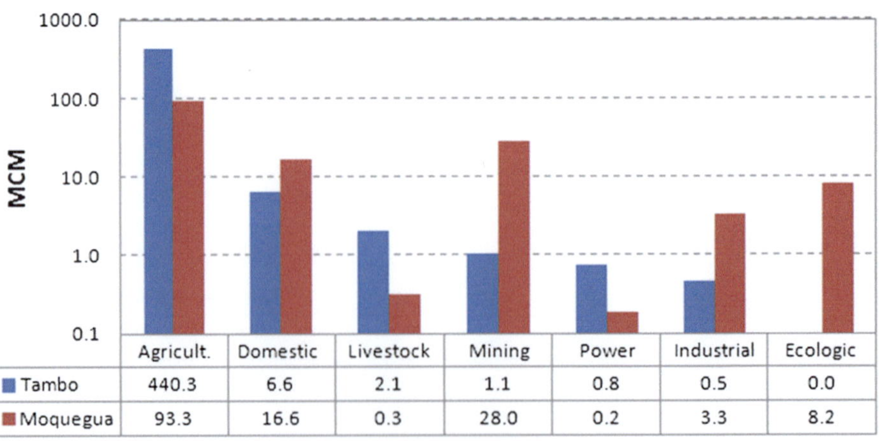

	Agricult.	Domestic	Livestock	Mining	Power	Industrial	Ecologic
Tambo	440.3	6.6	2.1	1.1	0.8	0.5	0.0
Moquegua	93.3	16.6	0.3	28.0	0.2	3.3	8.2

Fig. 25.10 Water uses for the Tambo and Moquegua watersheds

this time, the hydraulic generation has doubled its production from 10,170 GWh up to 20,052 GWh; it is important to outline that thermoelectric generation has increased 5 times from 2,992 GWh up to 15,855 GWh. Eolic generation starts in 1996 with 0.4 GWh and has increased up to 1.2 GWh in 2010.

The economic growth that Peru has experienced in recent years generates at the same time a higher demand for electric power. For instance, the maximum demand on electricity has doubled since year 2000 from 2,621 MW up to 5,291 MW in 2012. Also, in May 2013, a new record of maximum demand was reached with 5,389 MW to meet this demand: hydro-energy participated with 49 %, natural gas with 48.9 %, diesel and residual with 1.8 %, and biomass and biogas with only 0.3 %.

According to the 2010 electrical statistics, Moquegua generated 1,623 GWh of electrical energy of which 48.7 GWh were generated by hydropower centrals and 1,574.3 GWh by thermoelectric power plants. Also, the electrical energy consumption was 1,825.6 GWh. The energy sources used in thermoelectric generation are coal, diesel, natural gas, residual 6, residual 500, vapor, and biogas. Another important indicator is the per capita electricity consumption; the average at country level in 2010 was 1.08 MWh/year. The highest electricity consumer by region was Moquegua with 10.66 MWh/year, followed by Ica with 2.58 MWh/year (MINEM 2010).

In 2010, the billing of energy consumption in Moquegua was $150,966,000 of which mining was the higher consumer with $140,615,000 followed by residential, commerce, community activities, street lighting, and manufacture with $5,529,000, $912,000, $727,000, $638,000, and $564,000, respectively (Electrical statistics by region 2010).

25.5.4.4 Water and Energy Governance/Environmental Regulations

The Peruvian Water Law N° 29338 regulates the use and management of water resources. It includes surface water, groundwater, continental, and assets associated with this. It extends to the seawater and atmospheric as applicable.

The government of Peru through its Ministry of Energy and Mining encourages investment in renewable energy to diversify its energy matrix. The legislative decree N° 1002 was given in May 2008 to promote generation of electricity with renewable energy.

The National Water Authority establishes criteria for the delimitation of areas of water management authorities such as considering the hydrographic watershed as a unit of water management, integrated hydrological systems, and homogeneity of climatic characteristics, among others. The Tambo watershed belongs to the Administrative Water Authority of Caplina – Ocoña.

By law, water is patrimony of the nation, and there is no private ownership of water. Also, the National Water Authority (ANA) is the highest level technical authority of the national system of water resources management. ANA is in charge of authorizing and canceling water use rights and water use licenses.

25.6 Conclusions

Recent analyses made by IEA confirm that the current trends (1990–2006) in water and energy demands and management are leading to an unsustainable global energy future. Energy consumption has increased in IEA member countries and IEA nonmember countries by 19 % and 35 %, respectively. CO_2 emissions increased in IEA member countries and IEA nonmember countries by 14 % and 51 %, respectively.

Agriculture is a major driver of these unsustainable conditions. Agriculture represents 95 % of water use in the LFRW in California and 97.5 % in Peru's Tambo watershed. In Texas, agriculture is a significant and historically important industry. However, recently the Lower Colorado River Basin prioritized municipal and energy interests above agriculture. These basins contribute to the massive water demands of agricultural production in part through the influence of agribusiness subsidies and lobbying, which often extends the availability of cheap water for agriculture past what is socioeconomically sustainable.

Our case studies illustrate a number of methods being employed to manage dwindling water and energy resources. In the case of Peru, large irrigation projects are planned in the Tambo and Moquegua watersheds with the goal of substantially increasing local water storage and electricity production. In contrast with the infrastructure-based solutions being applied in much of South American countries, the main method being applied in the Lower Colorado River Basin is water conservation. New technology and political reform are driving conservation measures in central Texas in the hopes of helping to create a more sustainable water supply for central Texas. California and Ethiopia are attempting to harness the water-energy nexus to sustain water and energy resources, mainly through hydroelectricity. The ability to harness the water-energy nexus is expected to be a crucial component of moving toward a more sustainable resource future.

An increase in energy consumption has been observed in the last 20 years, the projections for the next 25 years indicate that this consumption will double the consumption registered in year 2000. This increase is basically due to the fact that non-OECD countries are consuming more than OECD countries. It is expected that by 2040 the energy consumption of non-OECD countries would double the consumption of OECD countries.

References

Barrientos J (2011) Integrated water resources management model of the Moquegua and Tambo river watersheds. Thesis to obtain the master degree in Management and environmental auditing, Piura, Perú, 135 pp

Bates BC, Kundzewicz ZW, Wu S, Palutikof JP (eds) (2008) Climate change and water. Technical paper of the Intergovernmental Panel on Climate Change, IPCC Secretariat, Geneva, 210 pp

Blackmore D, Whittington D (2008) Opportunities for cooperative water resources development on the Eastern Nile: risks and rewards. An independent report of the scoping study team to the Eastern Nile Council of Ministers. NBI, Addis Ababa
CEC (2005) Integrated energy policy report. http://www.energy.ca.gov
CNGDS (2014) California natural gas data and statistics. http://energyalmanac.ca.gov/naturalgas/. Accessed May 2014
Dai A, Trenberth K, Qian T (2004) A global dataset of palmer drought severity index for 1870–2002: relationship with soil moisture and effects of surface warming. NCAR, Boulder
Desalegn DT, Awlachew SB, Moges SA (2011) Blue Nile (Abbay) hydropower potential, prioritization, and trade-offs on priority. In: Melesse AM (ed) Nile River basin hydrology, climate and water use. Springer, New York/Dordrecht, pp 321–334
DWR (Department of Water Resources) (1998) California water plan update. State of California, Department of Water Resources, Sacramento
ECDMS (2014) California's energy consumption data management system. http://energyalmanac.ca.gov/electricity/ecdms.html. Accessed May 2014
ElBaradei M (2007) Security today and tomorrow. IAEA Bull 48(2), 5, Vienna, Austria
Energy Almanac (2010) U.S. Per Capita electricity use by state in 2010. Retrieved from http://energyalmanac.ca.gov
European Commission (2006) Energy futures – the role of research and technological development, Italy. ISBN 92-79-01639-3
FAO (2012) The state of food and agriculture 2012. Agricultural Development Economics Division (ESA), Rome. ISBN 978-92-5-107317-9
FAO (2013) FAO statistical year book 2013, world food and agriculture. Statistics Division (ESS), Rome. ISBN 978-92-5-107396-4
HLPCO (2014) Hydropower License Planning and Compliance Office: Oroville facilities, project No. 2100. Retrieved from http://www.water.ca.gov
Hussey K, Pittock J (2012) The energy–water Nexus: managing the links between energy and water for a sustainable future. Ecol Soc 17(1):31
IEA (2009) Towards a more energy efficient future. IEA/OECD, Paris
IEA (2011) World energy outlook – energy for all. IEA/OECD, Paris
INADE (National Institute of Development) (2001) Water budgets of the Tambo, Moquegua and Ilo river watersheds, Lima, Perú. http://www.scribd.com/doc/215752633/Simulacion-Tambo-Moquegua
INEI (National Institute of Statistics and Informatics of Perú) (2010) Perú: estimations and projections of population by departments, calendar years and ages 1995–2025. Technical direction of demography and social indicators. Special Bulletin N° 22, 209 pp, Lima, Perú
INEI (National Institute of Statistics and Informatics of Perú) (2012) Lima would have 8.5 million people. Press Note, N° 008-17/01/2012, 1 pp, Lima, Perú
IPCC (2014) Summary for policymakers. Phase I Report Launch, Yokohama, Japan, 44 pp
IPCC Fourth Assessment Report (AR4) (2007) Climate change 2007 synthesis report, Valencia, Spain, 52 pp
James K, Campbell SL, Godlove CE (2002) Watergy: taking advantage of untapped energy and water efficiency opportunities in municipal water systems. Alliance to Save Energy
Lund JR, Howitt RE, Medellín-Azuara J, Jenkins MW (2009) Water management lessons for California from statewide hydro-economic modeling. A report for the California Department of Water Resources, Center for Watershed Sciences Department of Civil and Environmental Engineering Department of Agricultural and Resource Economics University of California, Davis, June 2009
McCartney MP, Girma MM (2012) Evaluating the downstream implications of planned water resource development in the Ethiopian portion of the Blue Nile River. Water Int 37(4):362–379
MINAG (2005) Water resources's strengthening in the Tambo valley. Feasibility study, vol. III, Annex IV – Project Engineering, Lima, Perú, 241 pp
MINEM (2010) Electrical statistics by region. General Direction of Electricity, Lima

Moges SA, Gebremichael G (2014) Climate change impacts and development-based adaptation pathway to the Nile River Basin. In: Melesse AM et al (ed) Nile River Basin: ecohydrological challenges, climate change and hydropolitics, vol XV. Springer, 718 p. 233 illus; pp 339–362

Moser S et al (2009) The future is now: an update on climate change science impacts and response options for California. California Energy Commission Public Interest Energy Research Program CEC-500-2008-071

Nielsen-Gammon (2011) August 2011 Climatic Bulletin. http://climatexas.tamu.edu/climatic-bulletins/august-2011

Pacific Institute (2012) Water conflict chronology list. http://www.worldwater.org/conflict/list/

Poff NL, Allan JD, Bain MB, Karr JR, Prestegaard KL, Richter BD, Sparks RE, Stromberg JC (1997) The natural flow regime: a paradigm for river conservation and restoration. BioScience 47:769–784

Setegn SG, Rayner D, Melesse AM, Dargahi B, Srinivasan R (2011) Impact of climate change on the hydroclimatology of lake Tana Basin, Ethiopia. Water Resour Res 47:W04511. doi:10.1029/2010WR009248

SEI – Stockholm Environment Institute (2012) SEI links powerful water, energy and climate change mitigation planning software tools to help policy-makers grapple with 'nexus'. http://www.sei-international.org/press/press-releases/2472-sei-links-powerful-water-energy-and-climate-change-mitigation-planning-software-tools-to-help-policy-makers-grapple-with-nexus

SEI – Stockholm Environment Institute (2013) Using a Nexus approach to support development and environmental planning in Ethiopia. http://www.sei-international.org/mediamanager/documents/Publications/SEI-DB-2013-Nexus-Blue-Nile-Ethiopia.pdf

TWDB (Texas Water Development Board) (2012a) Texas state water plan. https://www.twdb.state.tx.us/waterplanning/swp/2012/

TWDB (2012b) https://www.twdb.state.tx.us/publications/state_water_plan/2012/02.pdf

TWDB (Texas Water Development Board) (2012c) Water for Texas 2012 state water plan, 2012 water for Texas. http://www.twdb.texas.gov/publications/state_water_plan/2012/2012_SWP.pdf

UNEP (United Nations Environment Programme) (2012) The UN-Water status report on the application of integrated approaches to water resources management. UNEP, Nairobi. http://www.un.org/waterforlifedecade/pdf/un_water_status_report_2012.pdf

UN-Water (n.d.). International decade for action: water for life, 2005–2015. Website. Zaragoza, Spain, UNWater/UN-IDfA. http://www.un.org/waterforlifedecade/scarcity.shtml

van Griensven A, Ndomba P, Yalew S, Kilonzo F (2012) Critical review of SWAT applications in the upper Nile basin countries, Hydrol. Earth Syst Sci 16:3371–3381

World Bank (2014) World development indicators: electricity production, sources and access. http://wdi.worldbank.org/table/3.7

WWAP (United Nations World Water Assessment Programme) (2006) World water development report 2: water: a shared responsibility. UNESCO, Paris/New York. http://www.sacriver.org/aboutwatershed/roadmap/watersheds/feather/lower-feather-river-watershed

WWAP (United Nations World Water Assessment Programme) (2014) The United Nations world water development report 2014: water and energy. UNESCO, Paris

Chapter 26
Efficient Use of Water Resources for Sustainability

Cecilia Lartigue

Abstract Strategies to achieve sustainable water management throughout the world are urgently needed as different regions are already experiencing high water stress with serious negative consequences on human health and ecosystems. Acknowledging the difference between efficiency and sustainability in water management, this paper considers the need of implementing actions that take into account future demand as well as the integrity of ecosystems. Consequently, recommendations are made regarding the investments that should be made on (1) access to clean water for human consumption, (2) ecosystem conservation and restoration, (3) watershed management, (4) collection and dissemination of data to the general public and decision makers, (5) finding ways to prevent and solve conflicts over water, and (6) encouraging community participation through local leadership, control over water resources, a fair allocation of benefits, and effective funding mechanisms.

Keywords Efficient water management • Sustainability • Community participation • Water observatories

26.1 The Urgent Need for Sustainability of Water Resources Management

26.1.1 Water Demand and Scarcity

According to the World Bank (2013), the main development challenges in this century, such as food and energy security, urban expansion control, human development, and adaptation to climate change, will only be met if water resources are managed adequately. Water is abundantly available globally, but it is unevenly distributed across the world and likely to become one of the most critical resources. For instance, Europe is densely populated and heavily industrialized, and it receives

C. Lartigue (✉)
Programme for Management, Use and Reuse of Water (PUMAGUA), National Autonomous University of Mexico, Mexico City, Mexico
e-mail: cecilia.lartigue@gmail.com

S.G. Setegn, M.C. Donoso (eds.), *Sustainability of Integrated Water Resources Management*, DOI 10.1007/978-3-319-12194-9_26

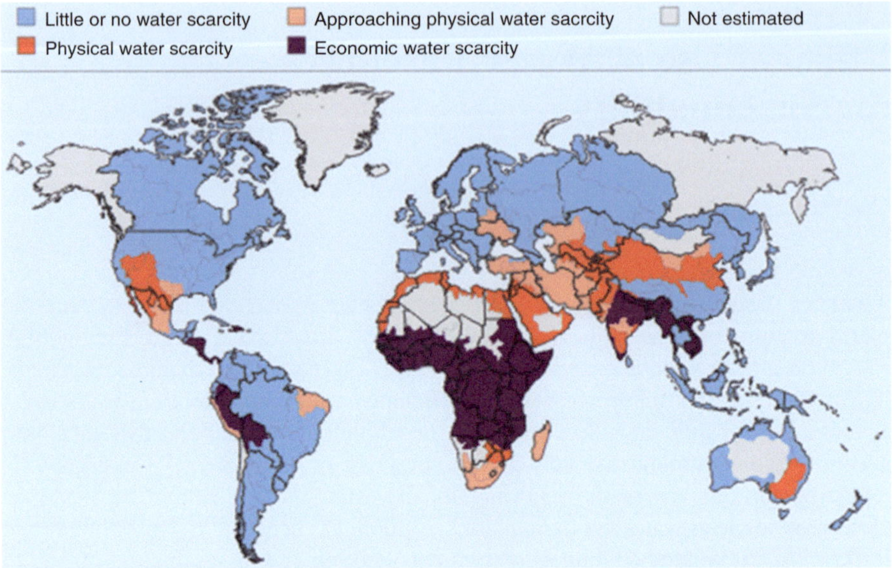

Fig. 26.1 Areas of physic and economic scarcity in the world (*Source*: Molden 2007)

only 7 % of global rainfall, and Asia accounts for 50 % of total world population and has only 30 % of rainfall. In contrast, North America with a share of less than 6 % of population receives nearly 14 % of rain (Gross 1986). In Mexico, for example, 77 % of the population lives in regions with only 31 % of the country's water availability. The northern region, which occupies more than 50 % of the territory, gets about 25 % of rainfall, while the southern region (28 % of the territory) receives nearly 50 % of rainfall (Comisión Nacional del Agua 2005).

Many regions are already experiencing moderate to high water stress (consumption level exceeding 20 % of availability supply). The United Nations assessment in 1997 determined that already one third of the world population was living in regions under this condition, and by 2025, it is likely that the share of the population undergoing this situation will be two thirds (Raskin et al. 1997). Figure 26.1 shows areas of the world approaching economic or physical scarcity. Economic scarcity is caused by a lack of investment in water or lack of human capacity to satisfy the demand for water, while physical scarcity occurs when there is not enough water to meet all demands, including environmental flows (Molden 2007).

While population triplicated between 1900 and 1995, water demand rose sixfold because of an increased demand for agriculture, industry, and domestic use (World Resources Institute 1999). Total global freshwater withdrawals are estimated at 3,800 km^3 of which 70 % are used by agriculture, followed by industry (20 %) and domestic use (10 %). Industrial and domestic uses are increasing relative to that for agriculture (Molden 2007). Industrial water use, for example, is predicted to double by 2025 (World Resources Institute 1999). Urban populations will expand from a current estimate of 3.5 billion (50 %) to 6.3 billion (69 %) by 2050 (United Nations 2010), while domestic and industrial demand for electric power will increase by about 50 %, of which hydropower is expected to supply about one third (Cook et al. 2011).

Water use in agriculture is expected to increase as world food demand rises. Agriculture already accounts for about 70 % of water consumption worldwide, and the United Nations projects a 50–100 % increase in irrigation water demand by 2025 (World Resources Institute 1999). Moreover, due to growing urban populations, demand for animal products will most likely increase by 74 %. This raise in food production is predicted to be achieved through intensification of production systems, but about 15 % is expected from expansion of agricultural lands (Cook et al. 2011).

Much of the projected increase in water demand will occur in developing countries, where population growth and industrial and agricultural expansion will be greatest. However, per capita consumption continues to augment in the industrialized world as well (World Resources Institute 1999).

In summary, current water stress is likely to increase in the near future due to water demands of a growing human population with increasing consumption needs. Additionally, climate change is already increasing the pressure on water resources. In consequence, strategies for sustainable management of water resources are urgently needed. The best way to do this is by implementing processes, institutions, and technologies that consider both efficiency and ecosystem conservation (Brooks and Brandes 2011).

26.1.2 Water Quality Degradation

Water availability is strongly related to its quality, because pollution is one of the main causes of its scarcity for uses such as human consumption, agriculture, industry, or biodiversity conservation (Peters and Meybech 2000). Studies show that human health is highly affected due to scarcity of water and pollution of water supplies, especially in rapidly urbanizing areas. Particularly, many developing countries are now facing pollution problems such as eutrophication, heavy metal deposition, acidification, and persistent organic pollutants while still trying to sort out typical problems of poor water supply and lack of sanitation services (World Health Organization 1997). The quality of both surface and groundwater in a catchment is a result of a combined effect of different processes affecting water in its hydrologic cycle. Its chemical composition depends on the solids, liquids, and gases generated internally or with which the water interacts. According to Peters and Meybech (2000), human beings have changed water quality by altering the lands in which the hydrologic cycle takes place with activities, such as agriculture, urbanization, dam construction, and deforestation, and also through the direct addition of substances (sewage discharges, application of pesticides and fertilizers, etc.) . Pollution is a very serious matter when it affects groundwater supplies, where contamination is slow to dilute, and it is quite expensive to apply purification measures (World Resources Institute 1999).

Some alarming facts regarding water quality degradation worldwide are the following (United Nations 2010):

- Every day, two million tons of sewage and industrial and agricultural waste are discharged into the water.
- Annually, 1, 500 km^3 of wastewater are generated.
- Worldwide, 2.5 billion people live without adequate sanitation, of which 70 % live in Asia (see Fig. 26.2).
- 18 % of the world's population experience open defecation.
- Infectious diseases such as waterborne diseases are the main cause of death of children under 5 years old (World Health Organization 2012).
- Unsafe or inadequate water, sanitation, and hygiene cause approximately 3.1 % of all deaths worldwide (World Health Organization 2012).
- 24 % of mammals and 12 % of birds connected to inland waters are considered threatened.
- Freshwater species have faced an estimated extinction rate five times greater than that of terrestrial species.
- Additionally, climate change is already putting more pressure on water resources. It is expected to increase hydrologic variability, resulting in extreme weather events such as droughts, floods, and major storms. As it is the case of most natural disasters, the poorest people will suffer most.
- But the other way around is also true: changes in water resources availability, water quality, and the destructive potential of storms and floods will determine, to a large extent, how climate change will affect human beings and the ecosystems (Miller 2008).
- Also, poor water management of reservoirs can also increase the release of greenhouse gases (GHG). It is estimated that emissions emanated from rotting vegetation and carbon inflow in watersheds account for between 1 and 28 % of the global warming potential of GHG emissions (Scanlon et al. 2004).

26.2 Efficient Water Management Is Not the Same as Sustainable Management

Kjellén and Mcgranahan (1997) pointed out that "The current burden of water-related diseases in urban areas is not, by and large, the outcome of the city-wide water supply and pollution problems that threaten sustainability. Low income urban neighbourhoods and households are more likely to lack water because they cannot access the city's water supplies due to limited availability. A comparatively healthy overall water balance can be accompanied by extremely unhealthy conditions in disadvantaged neighborhoods (...) Urban health essentially involves furthering the interests of today's poor, while ecological sustainability entails protecting the rights of future generations. When narrowly pursued, health and sustainability goals can conflict and create trade-offs between the interests of future generations and today's poor."

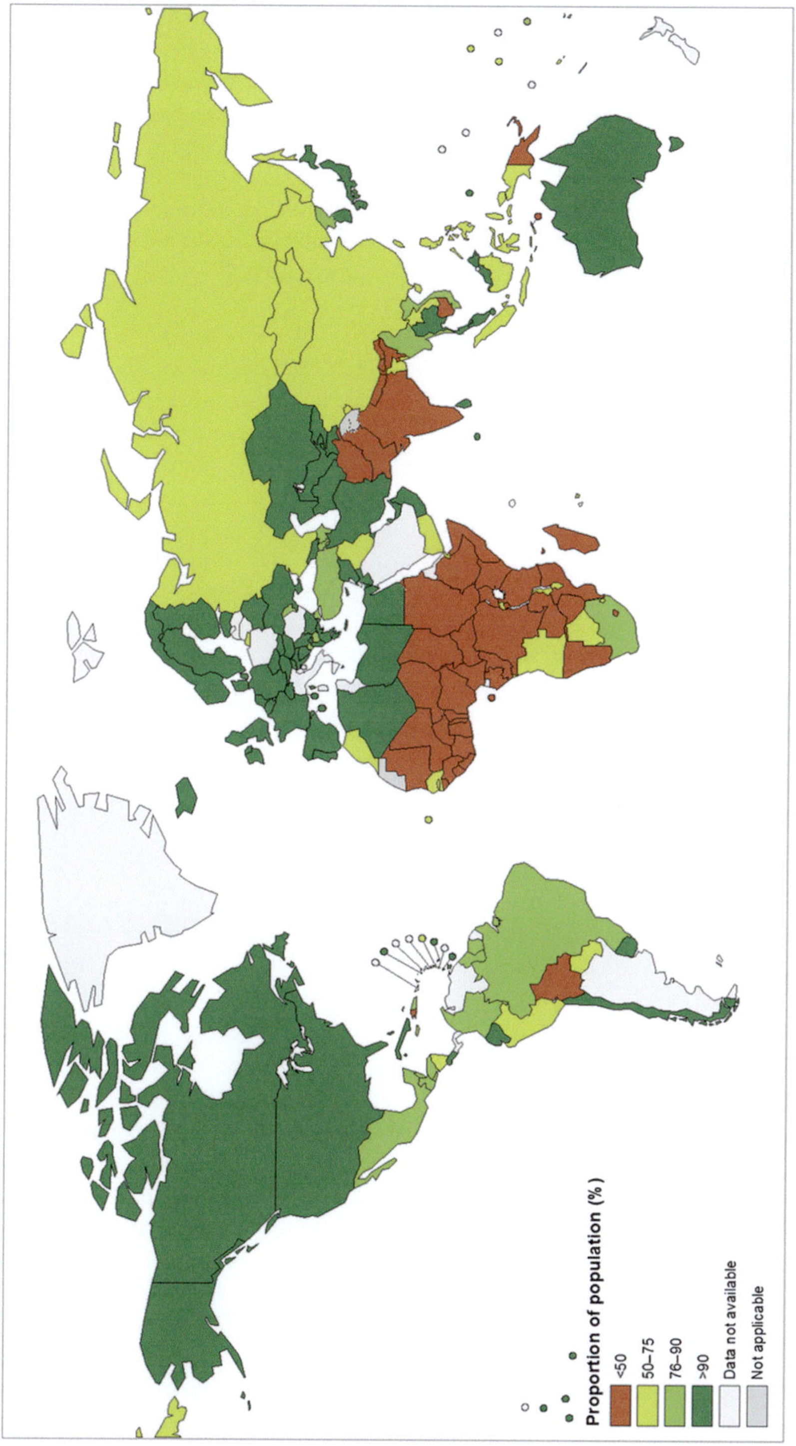

Fig. 26.2 Percentage of population with access to improved sanitation facilities, 2010 (*Source*: World Health Organization 2012)

Water used for agriculture is no longer available for wetlands, streams, deltas, and plants and animals. And as aquatic and terrestrial ecosystems are damaged, ecosystems change (Rosegrant 1997). According to the definition of the Brundtland Commission (World Commission on Environment and Development 1987), a system is sustainable if today's needs are met without compromising future generation to meet their own needs. Sustainability implies taking into account society, economy, and the environment, because people, habitats, and economic systems are interrelated. This interdependence may not be obvious in the short term, but sooner or later we are reminded of its existence by some alarm of crisis (Strange and Bayley 2008).

As the United States Environmental Protection Agency states: "Sustainability creates and maintains the conditions under which humans and nature can exist in productive harmony, that permit fulfilling the social, economic and other requirements of present and future" (Environmental Protection Agency 2013).

When referring specifically to sustainable water management, Gleick's definition (1998) includes the importance of the hydrological cycle, as follows: "Sustainable water use is the use of water that supports the ability of human society to endure and flourish into the indefinite future without undermining the integrity of the hydrological cycle or the ecological systems that depend on it." A similar definition is that of Mays (2006), "…the ability to use water in sufficient quantities and quality from the local to the global scale to meet the needs of humans and ecosystems for the present and the future to sustain life, and to protect humans from the damages brought about by natural and human-caused disasters that affect sustaining life."

26.3 Toward Sustainable Water Management

26.3.1 A Basic Water Requirement Will Be Guaranteed to All Humans to Maintain Human Health

In order to make decisions on how to allocate and use water resources, several criteria and goals can be established (based on Gleick 1998):

Goal number 7, target 10, from the Millennium Development Goals (MDG) formulated in 2000 by the United Nations, states as follows: "To halve, by 2015, the proportion of people without sustainable access to safe water and basic sanitation." According to the MDG Report 2013, although there has been huge progress in terms of access to drinking water, this issue is still a matter of serious concern, as 768 million people still drew water from an unimproved source in 2011 (United Nations 2013). Likewise, from 1990 to 2011, 1.9 billion people gained access to a latrine, flush toilet, or other improved sanitation facility. In order to meet the MDG sanitation target, this number must increase by another 1 billion people by 2015. As in 1990 only 49 % of the global population had improved sanitation; coverage must extend to 75 % to meet the target, up from the current level of 64 % (United Nations 2013).

According to the World Health Organization (2010), it is estimated that it would cost about US$ 23 billion per year to achieve the target by the year 2015. Taking into account that governments currently spend US$ 16 billion a year in building new infrastructure, only an additional US$ 7 billion a year would be needed to supply good water and sanitation. This amount is less than one tenth of what Europe spends on alcoholic drinks each year and half of what the United States spends each year on pet food.

The World Health Organization (2010) suggests concentrating efforts on the following issues:

- *Getting health back into the water agenda*, which means putting an emphasis on safe water supply, adequate sanitation, and environmental management, instead of relying only on strictly medical interventions to enhance health.
- *Better planning for water and health*, which includes not transferring costs of development to health sector and giving priority to well-being of people.
- *Reaping the benefits of science*. An adequate budget must be assigned to research on health and technological innovation. Also, what has already been investigated by researchers from different parts of the world must be integrated and disseminated.
- *Taking advantage of globalization*, i.e., the integration of the world could help to achieve a safer and cleaner environment.

26.3.2 A Basic Water Requirement Will Be Guaranteed to Restore and Maintain the Health of Ecosystems

In order to achieve sustainability in water management, it is essential that water storage and diversion for human purposes are programmed in a manner that does not affect ecosystems. This necessarily implies that there is a limit to the amount of water that may be withdrawn without compromising the ecological integrity of the affected ecosystems, resulting in the loss of native species and valuable ecosystem products and services for society (Richter et al. 2003). To determine this limit, a fundamental concept is environmental flow: the "quantity, timing and quality of water flows required to sustain freshwater and estuarine ecosystems and the human livelihood and well-being that depend on these ecosystems" (Brisbane Declaration, River Foundation 2013). In order to sustain freshwater ecosystems, not only is a minimum low flow required but also a naturally variable regime of flow (Poff et al. 2010).

Poff et al. (2010) developed ELOHA, a framework to develop environmental flow standards based on both a scientific and social process. Hydrologic analysis and classification are developed in parallel with flow alteration–ecological response relationships, which provide scientific input into a social process that balances this information with societal values and goals to set environmental flow standards (Fig. 26.3).

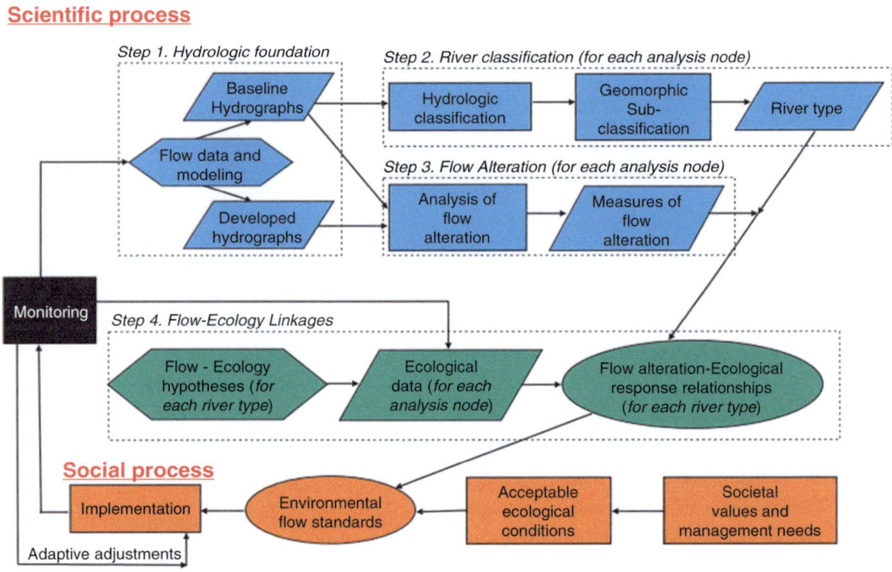

Fig. 26.3 The ecological limits of hydrological alteration framework. It comprises both a scientific and social process. Hydrologic analysis and classification (*blue*) are developed in parallel with flow alteration–ecological response relationships (*green*), which provide scientific input into a social process (*orange*) that balances this information with societal values and goals to set environmental flow standards. This paper describes the hydrologic and ecological processes in detail and outlines the scientist's role in the social process (*Source*: Poff et al. 2010)

In order to measure ecosystem health, indicators have been developed worldwide. An indicator is a measure, either qualitative or quantitative, of facts or conditions of particular issue. Environmental indicators should represent the key elements of a complex ecosystem or environmental issue, in this case, water. If the indicators are observed regularly, they can analyze changes during the observation period (Juwana et al. 2012). One example is the set of indicators established for the North American Great Lakes basin (Shear et al. 2005). First, the issues of the basin were determined according to its particular geographical, geological, and human settlement condition such as land use, climate change, toxic pollutants, nutrients, invasive species, ecosystem ecology, habitat status, and data collection methods. The indicators were then classified into state indicators, such as biodiversity; status of native species; pressure indicators, such as phosphorus concentration and fecal coliform levels; and response indicators, like human stewardship activities (Shear et al. 2005).

Another example is the set of indicators used in Queensland, Australia, by the Department of Environment and Health Protection (2013)and classified in physicochemical (including pH, nutrients, oxygen, temperature, and salinity), biological (or fish diversity, benthic algal growth and benthic oxygen demand, chlorophyll a), habitat indicators (width, continuity, extent of shading and species composition, bank erosion, presence of woody debris, among others), and flow indicators

Table 26.1 Components, indicators, and sub-indicators used by the West Java Water Sustainability Index

Component	Indicator	Sub-indicator
Conservation	Water availability	
	Land use changes	
	Water quality	
Water use	Water demand	Coverage
	Water services provision	Water loss
Policy and governance	Information disclosure	Education
	Governance structure	Poverty
	Public participation	Health impact
	Law enforcement	Sanitation

(changes in base flows, peak flows, no flow periods, seasonality of flows). Also, the Government of Canada developed a series of water indicators (Table 26.1).

Some indicators might be combined together to form an index or composite indicator. One of them is the Water Poverty Index (WPI) (Lawrence et al. 2002). Other examples are the Canadian Water Sustainability Index (CWSI), Watershed Sustainability Index (WSI), and West Java Water Sustainability Index (WJWSI) (Table 26.2). All these indices seek to measure sustainability, can assist decision makers and other stakeholders in achieving sustainability, and can be used to communicate the progress of sustainability to society (Juwana et al. 2012).

26.3.3 The Watershed: An Adequate Physical Unit to Manage Water Resources

Because water and soil are closely related, they must be managed jointly (Wang 2001). The most logical unit in which to carry out these management processes is the watershed, as it forms natural boundaries within a land mass and it is hydrologically defined. Using the watershed approach means focusing in water and other natural resources in a holistic way (Mylavarapu et al. 2012). A watershed approach is a flexible framework for managing water resource quantity and quality.

The watershed approach allows protecting water resources and water quality through proper land-use planning. It is of paramount importance to include all stakeholders as, for instance, water quality management and land-use planning are frequently implemented by different agencies with different objectives. Planners and policy makers at different levels should bring stakeholders together to implement a diagnosis of the watershed, identify sources of the problems, understand the relationship between the sources and consequences, and find out how to solve these problems. Some common solutions include implementing vegetation buffers, improving water quality of discharges to rivers, restricting certain land uses, and promoting others (Wang 2001).

Table 26.2 Components and water indicators used by Environment Canada

Component	Indicators
Water quantity	Water quantity in Canadian rivers
	Regional water quantity in Canadian rivers
	Local water quantity in Canadian rivers
	Canada's water quantity in a global context
	Water availability in Canada
	Residential water use
	Water withdrawal and consumption by sector
Water quality	Freshwater quality in Canadian rivers
	Regional freshwater quality in Canadian rivers
	Local freshwater quality in Canada
	Canada's freshwater quality in a global context
	Shellfish growing area quality indicator
	Polybrominated diphenyl ethers (PBDEs) in fish and sediment
	Drinking water advisories in Canada
Regional ecosystems	Restoring the Great Lakes areas of concern
	Phosphorus levels in the Great Lakes
	Phosphorus levels in the St. Lawrence River
	Reducing phosphorus loads to Lake Simcoe
	Phosphorus and nitrogen levels in Lake Winnipeg
Pressures on water quality	Land use impacts on freshwater quality
	Household use of chemical pesticides and fertilizers
	Municipal wastewater treatment
	Soil and water quality indicators for agriculture
	Release of toxic substances to water
	Managing disposal at sea

26.3.4 Access to Data on Water Resources Availability, Use, and Quality

According to the World Meteorological Organization (2012), in order to support sustainable economic and social development, there is a pressing need for accurate information on the condition and trend of a country's water resources. This information is used very frequently and in many different ways for planning, development, or operational purposes. As clean water becomes scarce and competition increases, water information grows in value. Several countries have demonstrated the benefits of hydrological information and analysis. For example, in Canada and Australia, a benefit-to-cost ratio for hydrological data collection of 9.3 and 6.4, respectively, has been found in studies.

The World Meteorological Organization (2012) also states that many countries have undertaken assessments of water resources at local, regional, or national scales, but in many cases there is a lack of good-quality data. Some of the necessary information are for instance, long-term records of climate, river flows, reservoir levels, and groundwater levels. However, in many parts of the world, governments have overlooked the importance of maintaining long-term, good-quality environmental records. Additionally, because of climate change, there is a need for new scenario-based approaches to future resource modeling and continuing reliable long-term records to determine trends in water availability.

Water observatories play an important role in data collection and analysis, in dissemination of information, and in involving citizens in planning processes. They are also an excellent means of enhancing public participation through a co-responsibility approach, where citizens are informed about water conditions, and institutional efforts to improve these conditions, and consequently they are exhorted to also "do their part." Some interesting water observatories throughout the world are found in Europe (PEER-EURAQUA Network of Hydrological Observatories 2013) and specifically in France (Observatoire Départemental de Vendée (http://observatoire-eau.vendee.fr/), Observatoire de l'eau en Bretagne (http://www.observatoire-eau-bretagne.fr/)), in the United States (for instance, the Chicago Waterway Observatory of the United States Geological Survey (http://il.water.usgs.gov/data/cwo/)), and also in Latin America, in countries such as Argentina (Observatorio del Agua Universidad Nacional de Patagonia San Juan del Bosco (http://observatoriodelagua.org.ar/)) and Mexico (Observatorio Ciudadano del Agua (http://www.h2observa.net/)). The National Autonomous University of Mexico (UNAM) has recently launched its water observatory (http://www.agua.unam.mx/observatorio/), with six indicators, referred to percentage of leaks, compliance of water quality regulation, and participation in responsible use of water of UNAM's authorities and the community.

26.3.5 Mechanisms to Prevent and Resolve Conflicts Over Water

Water conflicts normally arise from the fact that the freshwater resources of the world are not partitioned to match the political borders, nor are they evenly distributed in space and time (Nandalal and Simonovic 2003). Almost half of the land area of the earth is part of an international river basin, and more than 220 nations share water with a neighboring country. Sharing a limited water resource by several stakeholders can create conflicts among them when their requirements exceed availability (Gleick 1998).

Additionally, because of both, the global communication and the democratic revolution, conflicts over water that used to be local have been brought into the international arena. They usually result from the building of large infrastructure projects, changes in community access to water, or impacts of critical socio-ecological systems (Conca 2006). It is likely that climate change will jeopardize water conflicts as it may alter the flow of many rivers due to changes in precipitation patterns and increase the demand for water, due to more frequent droughts and greater stress being placed on other sources of water (Tir 2012).

For solving conflicts over water, there exist some treaties among river basin nations allocating water, setting up management oversight, and developing acceptable standards for operations and water quality. Unfortunately, many international rivers do not have such treaties and in other cases do not address current problems adequately (Gleick 1998). As climate change puts more pressure on water systems, for treaties to adapt to this stress, there must be institutional design that includes provisions for joint monitoring, conflict resolution, treaty enforcement, and the delegation of authority to intergovernmental organizations. Treaties that contain more of these features are expected to better manage conflicts caused by water stress (Tir 2012).

In order to solve the technical aspect of conflicts, computer-based models can prove to be effective tools. One of the major types of models are simulation models, which addresses "what if" questions. Given the assumptions for a system design and operation, a simulation model can predict how well it will perform. The other major types are optimization models, which address "what should be" questions and what design and operating policy will best meet the specified objectives (Loucks 2008).

Models can be used for many different purposes, such as predicting runoff from watersheds or interactions between groundwater and surface water bodies. They can also be used to study reservoir operation; to forecast floods and plan flood-control programs; to predict storm surges, embankment erosion, and dam breaks; and to plan for ecosystem restoration (Loucks 2008).

They can help to further understand the problems, formalize performance objectives, develop and evaluate alternatives, and provide confidence in solutions, and they can also work as negotiation forums, by employing the logic of model studies and development to structure the negotiation process; however, they only are useful when objectives and functions are well defined (Lund and Palmer 1997). According to Nandalal and Simonovic 2003, there are several alternative solutions, and the best one is not obvious, the impacts resulting from decisions are not evident without modeling, and there is readily obtainable data to estimate parameters.

Water planning and decision making will be democratic, ensuring representation of all affected parties and fostering direct participation of affected interests.

To improve the quality, effectiveness, and sustainability of development actions, community participation is fundamental. When people are placed at the center of actions, there is a greater possibility of empowering stakeholders who have a sense of ownership of results and therefore a strong interest on positive results (Fisher and Urich 1999). By community involvement, we understand that users participate to a project cycle by assuming a responsibility and exercising an authority and a control over the setting up of the water services (Tandia 2006).

For participation to be transformative, information needs to be available for all stakeholders preferably during the early stages of the project cycle. This enables stakeholders to engage in a dialogue with the project designers to negotiate parameters for participation and consultation (Fisher and Urich 1999).

Among the benefits of social participation, Von Korff et al. (2010) point out the following:

- Improved legitimacy for decision makers as stakeholders feel that they are responding to their value and thus start trusting them
- Better and less expensive decisions because stakeholders share information and participate with their ideas
- More possibilities of implementing decisions because people are unlikely to oppose to a project that they contributed to build capacity building of stakeholders during participant interaction

However, it is possible that participation does not bring benefits because stakeholders become disappointed as their expectations are not fulfilled, objectives are unclear, or some stakeholders receive more benefits than others (Von Korff et al. 2010). This lack of equity was found by Hoffet et al., (2012), where the authors state that uneven benefits persist because of issues of inequity in allocation at the national, regional, and local scales. They suggest that participation should go beyond the watershed challenging power differences between stakeholders (Hoffet et al. 2012).

Also, it is likely that participation does not occur properly due to conflicts among stakeholders or lack of inclusion of all the involved parties (Von Korff et al. 2010). Agrawal and Gibson (1999) present a recommendation that could foster efficient community participation in natural resource conservation: redefining the concept or community and instead of considering it as a fixed small spatial unit, as a homogeneous social structure, and as shared norms, a more political approach should be considered, examining communities in the context of development and conservation by focusing on the multiple interests and actors within communities, on how these actors influence decision making, and on the internal and external institutions that shape the decision-making process.

From Agrawal and Gibson (1999) and Haddad et al. (2007), the following are some aspects that could improve community participation and promote sustainable systems:

- Founding projects on principles of check and balances among stakeholders.
- Involving all who are directly involved in water management and making them feel that they have ownership over the resource and/or ownership for the way it is used and managed.
- Understanding the different roles of men and women to target action appropriately.
- Making strenuous efforts to empower local groups.
- Focusing on building capacities of local representatives of governments to intermediate between the local community and the government institutions.
- Strengthening local leadership to make changes.
- Developing management abilities of the different participating agencies.
- Seeking to implement reasonable processes of decision making instead of trying to ensure successful outcomes. Reasonable processes mean that all stakeholders' interests (especially those that are normally excluded) are represented in

decision making, that there exist mechanisms to guarantee that the outcomes of current decision processes are going to form part of the data on which future decisions will be based, and that decision makers are constantly evaluated by those affected by decisions.

- Ensuring that local groups have access to adequate funds for implementing the rules they create. It is preferable that funds are local instead of granted by central governments. This could allow communities to demand control over resources.

26.4 Conclusions

Water has become a pressing natural resource in large areas of the world. Its scarcity and poor quality have serious consequences on human populations and on natural ecosystems as well. There is therefore an urgent need for efficient water management in terms of both quantity and quality. For water management to be sustainable, future generations' welfare must be acknowledged, and environmental, economic, and social aspects must also be considered. Based on Gleick (1998), several goals and criteria for sustainable water management can be established:

- Meeting basic human requirements through sufficient clean water, by investing on infrastructure but also on health research and technological innovation.
- Meeting environmental water needs that promote conservation and restoration of ecosystems. For this purpose, it is helpful to develop indicators that represent the key elements of complex ecosystems.
- Using a watershed approach to manage water resources through proper land-use planning and focusing on adequate management of other resources in a holistic way.
- Collecting data and making it available for planning, development, or operational purposes. Observatories can prove a useful tool for these aims.
- Finding mechanisms to prevent and solve conflicts over water, which can include treaties that take into account provisions for joint monitoring, conflict resolution, treaty enforcement, and the delegation of authority to intergovernmental organizations.
- Involving communities in planning and decision making, strengthening local leadership, and making strenuous efforts on promoting equity among stakeholders.

References

Agrawal A, Gibson CC (1999) Enchantment and disenchantment: the role of community in natural resource conservation. World Dev 27(4):629–649

Brooks DB, Brandes OM (2011) Why a water soft path, why now and what then? Int J Water Resour Dev 27(2):315–344

Comisión Nacional del Agua (2005) Estadísticas del agua en México. Síntesis. Comisión Nacional del Agua, México
Conca K (2006) The new face of water conflict. Navigating peace. Woodrow Wilson Int Center Scholars 3:1–3
Cook S, Fisher M, Tiemann T, Vidal A (2011) Water, food and poverty: global- and basin-scale analysis. Water Int 36(1):1–16
Department of Environment and Health Protection (2013) http://www.ehp.qld.gov.au/water/monitoring/assessment/water_quality_indicators.html. Accessed 1 Aug 2013
Environmental Protection Agency (2013) http://www.epa.gov/sustainability/basicinfo.htm. Accessed 1 Aug 2013
Fisher KT, Urich PB (1999) Information dissemination and communication in stakeholder participation: the Bohol-Cebu water supply project. Asia Pacific Viewpoint 40(3):251–269
Gleick PH (1998) Water in crisis: paths to sustainable water use. Ecol Appl 8(3):571–579
Gross AC (1986) Water quality management worldwide. Environ Manage 10(1):25–39
Haddad HF, Al Zoubi R, Alaween M, Shraideh F (2007) Local community participation for sustainable water resource management. MEDA WATER international conference on Sustainable Water Management, Tunis, 21–24 Mar 2007
Hoffet N, Daoud I, Tourrand JF, Alay V, Moselhy N (2012). Participation, power and sustainable water resource management. A case study of the Rainfed Desert Region of Matruh, Egypt. Communication IFSA/International Farming Systems Association, Denmark, July 2012
Juwana I, Muttil N, Perera BJC (2012) Indicator-based water sustainability assessment – a review. Sci Tot Environ 438:357–371
Kjellén M, Mcgranahan G (1997) Comprehensive assessment of freshwater resources of the world. Urban water: towards health and sustainability. Stockholm Environmental Institute, Stockholm
Lawrence P, Meigh J, Sullivan C (2002) The water poverty index: an international comparison. Keele Economics Research Papers. Centre for Economic Research, Research Institute for Public Policy and Management, Keele University, Staffordshire
Loucks D (2008) Water resource management models. Bridge 38(3):24–30
Lund JR, Palmer RN (1997) Water resources system modeling for conflict resolution. Water Resour Update 3(108):70–82
Mays LW (2006) Water resources sustainability. McGraw-Hill Professional
Mylavarapu R, Hines K, Obreza T, Means G (2012) Watersheds of Florida: understanding a watershed. http://edis.ifas.ufl.edu. Accessed 28 July 2013
Miller M (2008) Climate change and water resources: the challenges ahead. J Int Aff 61(2):35–50
Molden D (2007) Water for food, water for life. A comprehensive assessment of water management in agriculture. International Water Management Institute, Sterling
Nandalal KDW, Simonovic S (2003) Conflict resolution support system: a software for the resolution of conflicts in water resource management. Division of Water Sciences, UNESCO, Paris, 134 p
Observatoire Départemental de Vendée. http://observatoire-eau.vendee.fr. Accessed 6 Aug 2013
Observatoire de l'eau en Bretagne. http://www.observatoire-eau-bretagne.fr. Accessed 6 Aug 2013
PEER EurAqua Network of Hydrological Observatories. http://www.euraqua.org/content/activities/hydrologicalobservatories.4.3cd20f1b1243376c116800033.html. Accessed 6 Aug 2013
Observatorio Ciudadano del Agua. http://www.h2observa.net/. Accessed 6 Aug 2013
Observatorio del Agua Universidad Nacional de Patagonia San Juan del Bosco. http://observatoriodelagua.org.ar. Accessed 6 Aug 2013
Peters NE, Meybech M (2000) Water quality degradation effects on freshwater. Water Int 25(2):185–193
Poff NL, Richter BD, Arthington AH, Bunn SE, Naiman RJ, Kendy E, Acreman M, Apse C, Bledsoe BP, Freeman MC, Henriksen J, Jacobson RB, Kennen JG, Merritt DB, O'Keeffe JK, Olden JK, Rogers J, Tharme RE, Warner A (2010) The Ecological Limits of Hydrologic Alteration (ELOHA): a new framework for developing regional environmental flow standards. Freshw Biol 55:147–170

Raskin P, Gleick P, Kirshen P, Pontius RG, Strzepek K (1997) Water futures: assessment of long-range patterns and problems, comprehensive assessment of the freshwater resources of the world. Stockholm Environment Institute, Stockholm

Richter BD, Mathews R, Harrison DL, Wigington R (2003) Ecologically sustainable water management: managing river flows for ecological integrity. Ecol Appl 13(1):206–224

River Foundation (2013) http://www.riverfoundation.org.au/images/stories.pdfs/bnedeclaration.pdf. Accessed 29 July 2013

Rosegrant MW (1997) Water resources in the twenty-first century: challenges and implications for action, food, agriculture, and the environment. Discussion Paper 20. International Food Policy Research Institute, Washington, DC

Scanlon J, Cassar A, Nemes N (2004) Water as a human right? IUCN, Gland

Shear H, Bertram P, Forst C, Horvatin P (2005) Development and application of ecosystem health indicators for the North American Great Lakes Basin. In: Jorgensen SE, Constanza R, Xu FL (eds) Handbook of ecological indicators for assessment of ecosystem health. CRC Press, Boca Raton

Strange T, Bayley A (2008) Sustainable development: linking economy, society, environment. OECD insights. OECD Publishing, Paris

Tandia CT (2006) Involvement/community participation in hygiene and water in Central and Western Africa. In: International conference on community health in African region. World Bank, World Health Organization, UNICEF. Addis Ababa, Ethiopia, 20–22 November 2006

Tir J (2012) Weathering climate change: can institutions mitigate international water conflict? J Peace Res 49(1):211–225

United Nations (2010) United Nations, Department of Economic and Social Affairs, Population Division: World urbanization prospects, the 2009 revision: highlights. United Nations, New York

United Nations (2013) Millennium development goals report. New York

United States Geological Survey. Chicago waterway observatory http://il.water.usgs.gov/data/cwo/. Accessed 6 Aug 2013

Universidad Nacional Autónoma de México. Observatorio del Agua de la UNAM. http://www.agua.unam.mx/observatorio/. Accessed 6 Aug 2013

Von Korff YP, d'Aquino P, Daniell KA, Bijlsma R (2010) Designing participation processes for water management and beyond. Ecol Soc 15(3):1

Wang X (2001) Integrating water-quality management and land-use planning in a watershed context. J Environ Manage 65:25–26

World Bank (2013) Water resources management overview. http://www.worldbank.org/en/topic/waterresourcesmanagement/overview. Accessed 28 July 2013

World Commission on Environment and Development (1987) Our common future. Oxford University Press, New York

World Health Organization (1997) Health and environment in sustainable development: five years after the earth summit. World Health Organization, Geneva, pp 54–55

World Health Organization (2012) Public health information and geographical information systems. http://gamapserver.who.int/mapLibrary/Files/Maps/phe_Global_sanitation_2010.png

World Health Organization (2010) World water day report. http://www.who.int/water_sanitation_health/takingcharge.html. Accessed 30 July 2013

World Meteorological Organization (2012) Technical material for water resources assessment. World Meteorological Organization, Génève

World Resources Institute (1999) Water: critical shortages ahead? World Resources 1998–99. Washington

Chapter 27
Land Use and Climate Change Impact on the Coastal Zones of Northern Honduras

Arie Sanders, Denisse McLean, and Alexandra Manueles

Abstract In this paper, we examine the effect of land use change on climate change sensitivity on the north coast of Honduras. We ran simulations to analyze the spatial and temporal variations on sensitivity derived from land use dynamics and their implications on the use of land use policy as a tool for climate change adaptation in integrated coastal zone management. We developed two scenarios (trend and normative scenarios) for different spatial development trends for the 2010–2050 period. The biggest change in the trend scenario would be a decrease in pasture (19.4 %) and forestry (8.1 %) as a result from an increase in palm oil plantations. There would be more fragmentation, and the region will become more vulnerable to climate change. In the case of the normative scenario, we expect a 50.2 % decrease in pasture and an 18.3 % increase of the broad-leaved forest area, making the region less vulnerable to climate change. The national and local governments have a decisive role in assuring the implementation of their land use policies (normative scenario) to protect the region against climate change impact.

Keywords Climate change impact • Land use change • Conversion of Land Use and its Effects (CLUE)

27.1 Introduction

The north coast of Honduras is one of the most vulnerable areas in the Caribbean Basin to storms and hurricanes (IPCC 2007). However, climate change (CC) has made extreme climate conditions more frequent and intense. Projected impacts from CC on the Caribbean region include rising sea levels, stronger tropical cyclones, altered rainfall, storm surges, and increasing sea temperatures. According to Schatan et al. (2010), the scenarios on climate change effects in northern Honduras forecast that climate conditions will become increasingly extreme. Additionally, the region's vulnerability is made more severe by the lack of effective

A. Sanders (✉) • D. McLean • A. Manueles
Department of Environment and Development, Zamorano University, Francisco Morazan, Honduras
e-mail: asanders@zamorano.edu

S.G. Setegn, M.C. Donoso (eds.), *Sustainability of Integrated Water Resources Management*, DOI 10.1007/978-3-319-12194-9_27

governance structures, high population growth rates, and urbanization, as well as poor land use planning, which results in environmental degradation and habitat destruction.

The rapid development in northern Honduras during the last decades has drastically changed the geographical landscape. The lack of spatial planning has not only created competition and conflict between the different uses, such as residential, industrial, recreational, and agricultural activities, but has also deteriorated existing ecosystems' capacity for resilience in the face of CC. Despite ongoing conservation efforts by the national government, deforestation levels of broad-leaved forests and mangroves as a result of agricultural activities and urbanization remain high. Further ecosystem degradation may result in changes to the north coast, including reduced precipitation and increased droughts, which will make the region vulnerable to CC.

In recent years, there have been growing calls for a more integrated management of the Honduran north coast as a fundamental prerequisite for sustainable development. A good example is the Biological Corridor Project (PROCORREDOR) managed by the Secretariat of Natural Resources and Environment (SERNA) to establish and implement a comprehensive management plan for the protected areas on the north coast. However, little attention has been paid to land use conversion driven by human activities and climate change, which leads to serious forms of environmental degradation and habitat destruction on the north coast.

Even with the recognized importance of CC effects on the vulnerability of human populations, climate-related assessments in Honduras have mostly addressed sensitivity to natural phenomena under their normal rates of occurrence (Argeñal 2010; COPECO 2010). Assessments carried out by the Permanent Contingency Commission (COPECO) explore sensitivity to river floods, landslides, droughts, and forest fires at the municipal level for some of the country's municipalities. Only modest progress has been achieved on exploring climate change extreme weather patterns and their implications for human settlements.

The few CC assessments that have been undertaken in Honduras concentrate on the impacts on water resources and on watershed adaptation strategies (MIRA 2005; SERNA 2006a, b, 2007). Nevertheless, the determinants of human response to weather anomalies are related not only to the response of natural systems but to the economic, social, and cultural characteristics that shape human populations (Klein and Nicholls 1999). Those characteristics are reflected in the ways communities make use of their available resources. One evident manifestation of this is a region's land use patterns and land use changes over time.

The land use characteristics of a population in given geographical area directly reveal the main activities undertaken, their intensity level, their relative importance, and the competition and potential conflict between them. Indirectly, land use patterns reveal a population's priorities, habits, and associated opportunity cost. Moreover, given each location's natural suitability for certain land uses, alternative arrangements of land use distributions can influence the state and productivity of the interrelated natural, economic, and social systems. Thus, land use provides a suitable framework to analyze a population's characteristics and how they shape the population's ability to cope with extreme weather events.

Therefore, the objective of this paper is to examine the effect of land use change on climate change sensitivity on the north coast of Honduras. To reach this objective, we ran simulations to analyze the spatial and temporal variations on sensitivity derived from land use differences and their implications on the use of land use policy as a tool for climate change adaptation in integrated coastal zone management (ICZM). By combining land use change scenarios with land use sensitivity data, this study will analyze how climate change could affect the region in the near future.

27.2 Methodology

27.2.1 Study Area

The study area compromises the department of Atlántida in northern Honduras which fronts the Caribbean Sea (Fig. 27.1). With 344,000 inhabitants, Atlántida is characterized by its economic and environmental importance at a national level. It includes part of the Mesoamerican Biological Corridor and the city of La Ceiba, the largest coastal city in Honduras characterized by its unregulated expansion (Rubio 2012). The department has three contrasting natural regions: the coastal plain, the mountain zone, and an intermediate hillside zone (Buckles 1999). The flat terrain of the coastal plain, a narrow strip along the coast that is less than 100 m above sea level, has the best agricultural land in the region. Slopes are typically less than 10 %

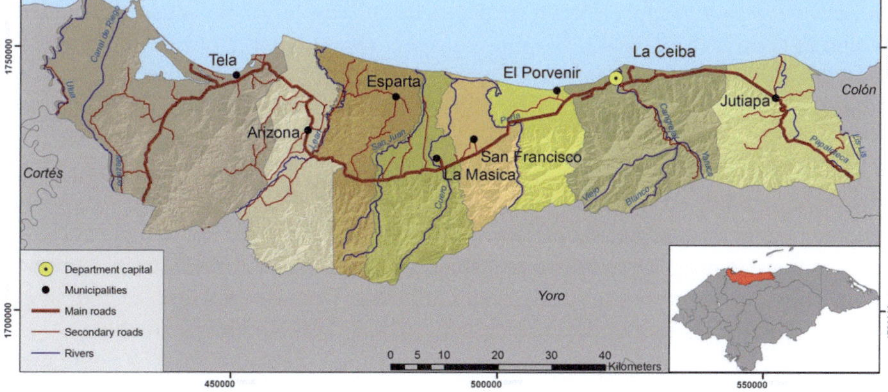

Fig. 27.1 Map of the municipalities of the department of Atlántida, Honduras

and never exceed 20 % throughout the zone. This is where the area's urban growth has taken place, alongside the development of large agro-exporting industries like banana, pineapple, and oil palm. The hillside zone – below 600-m asl – is less suitable for agriculture than the coastal plain because of the very steep slopes, yet corn and bean production (subsistence crops) is concentrated in this hillside zone. The mountain zone is generally unsuitable for agriculture as it has very steep slopes and thin undeveloped soils. Very humid subtropical mountain forest is the primary vegetation type at 800–1,800-m asl, and cloud forest still predominates at higher elevations where access is difficult and where the slopes are mostly steep.

The department has been subject to major land cover changes as a result of various social and economic processes. The economic conditions are influenced by a long tradition of banana plantation agriculture, which has dominated the Honduran northern landscape and commerce throughout the first decades of the twentieth century. In the 1960s, commercial agriculture replaced the banana plantations in importance. The growing demand for beef in the United States increased extensive livestock production. According to the latest 1993 agriculture census, more than 50 % of the farmland was composed of natural and cultivated pasture (Baumeister and Wattel 1996; Sunderland and Rodríguez 1996). In her study about the "cattle boom" in northern Honduras, Humphries (1998) analyzed the rapid growth of ranching in the region from the 1950s to the 1990s. During this period, the cattle population in Atlántida expanded from 27,583 to 147,233 heads, representing an increase of 434 %. This growth rate was possible with loans and special projects from the World Bank and USAID (Sunderland and Rodríguez 1996). At the same time, forests in this area declined from 945,200 ha in 1962 to 258,700 ha in 1990, representing a 72.6 % decrease over the period (*ibid*).

Owing to the availability of plantation employment alternatives, favorable agroclimatic conditions, and relatively good infrastructure, the Atlantic Coast has been an attractive migrant destination for farmers from less-favored zones in the country, strongly influencing agricultural development (Jansen et al. 2006). As a result of fluctuations in the labor needs of the agricultural plantation sector in combination with large-scale migration in the country beginning in the 1970s, farmers were increasingly forced to resort to cultivation of the steep hillsides of Atlántida (Porch et al. 2007).

In the 1980s, to restrict the use of some of the important and vulnerable areas in the department, the Honduran government established reserves and national parks. In the Atlántida area, there are seven terrestrial reserves and parks with a total area of 187.8 thousand hectares (43 % of the total land area). The reserves and national parks are comanaged by local nongovernmental organizations, which in general lack the financial resources for effective protection of the areas. As in other national parks in the country, due to the lack of financial resources and institutional support, the effectiveness of the protected areas is limited. In most of the protected areas, you will find agriculture and livestock activities, as well as illegal dwellings.

The north coast is an area of high demographic pressure with a population growth rate of 2.3 % (UNDP 2009) and increasing built-up areas. As consequence of national migration, there has been rapid urbanization and accelerated urban sprawl around the two main cities of La Ceiba and Tela and next to the highway between those cities. Most of the urbanization is less than 20 km from the coast, and more than 50 % of the population of Atlántida is living in this area.

The climate of northern Honduras is classified as humid tropical with temperatures between 25 and 28 °C with high annual precipitation in a bimodal distribution (2,000–3,000 mm per year). However, Honduras lies within the hurricane belt, and the Caribbean coast is particularly vulnerable to hurricanes and tropical storms that travel inland from the Caribbean. Although the climatic conditions in northern Honduras are always wet during the last decennials, precipitation has varied periodically.

Porch et al. (2007) found an increase of 0.022 °C/year for minimum temperature and 0.018 °C/year for maximum temperature for Honduras. They also analyzed the average yearly rainfall, and although a negative slope was found, changes in precipitation were not statistically significant. Warmer weather will cause a series of other events, such as a greater number of warm spells/heat waves over most land areas, more frequent intense tropical cyclones, a greater number of extreme high sea level events, and severe drought (CEPAL 2010). According to Schatan et al. (2010), the scenarios on climate change effects in northern Honduras agree that climate conditions will become increasingly extreme.

27.2.2 Climate Change Sensitivity

For this study, we have chosen to limit our focus to simulating the expected climate change scenarios developed by the Intergovernmental Panel on Climate Change (IPCC) for coastal zones and determining the implications for the region's vulnerability. We considered three kinds of plausible climate change derived impacts for the region: sea level rise (SLR), mean temperature change (MTC), and mean precipitation change (MPC). Approximately, 85 % of the disasters in Central America are related to floods and droughts (CEPAL 2010). The climate change scenarios for Central America suggest that depending whether it is rainy or dry season, the possibility of a higher frequency of floods and droughts, respectively, is very likely (CEPAL 2010). The National Strategy for Climate Change (SERNA 2010) estimates sea level increases of up to 0.6 m for Honduras by the year 2100. The A2 and B2 IPCC scenarios adapted for Honduras by Argeñal (2010) explored the effects of climate change on mean monthly temperature and mean monthly precipitation. Results for the department of Atlántida showed that by the year 2050, mean monthly temperature is expected to increase by between 1.4 and 1.9 °C and mean monthly precipitation is expected to decrease by between 2 and 25 %, depending on the month of the year.

The latest land use map for the area, elaborated by the National School of Forest Sciences (ESNACIFOR) for the PROCORREDOR Project in 2010, was used as the basis for the study. Three hundred ground control points were determined using a high-precision GPS along the main road of the study area for field verification. Different land use intensities in areas with similar cover were disaggregated into additional land use categories to allow for a better estimate of climate change sensitivities. The road and administrative maps were also validated.

Based on emergent recommendations from local experts during introductory meetings, an improved land use/land cover map was developed through photographic interpretation of the 0.3-m resolution aerial photographs provided by PROCORREDOR. The land use classification employed in the interpretation was based on the CORINE land cover methodology. The improved land use map was developed at a 1:1,500 scale, with a smallest mapping unit of 225 m^2.

Sea level rise exposure was simulated using a flood model projecting a sea level increase of 0.5 m by 2050. Using the elevation points obtained from the PROCORREDOR photogrammetric flight from a 5-m resolution, digital elevation model (DEM) was developed. From this DEM, all pixels with 0 or 1 m of elevation were selected as floodable areas for the year 2050. Following Yoo et al. (2011), sea level rise sensitivity was estimated as the percentage of flooded area within each land use category, municipality, and village in the department in order to identify the most vulnerable entities at different scales, so that:

$$\mathrm{SSLR}_i = \mathrm{FA}_i / A_i^* - 1 \tag{27.1}$$

in which SSLR_i is the sensitivity to sea level rise of land use, municipality, or village i; FA_i is the flooded area in land use, municipality, or village i; and A_i is total area of land use, municipality, or village i in landscape.

MTC and MPC sensitivity were estimated through expert judgment elicitation and an extended review of literature. Expert judgment is a qualitative research technique used in decision making and risk analysis to predict the occurrence of future events and the consequences of decisions based on experts' opinions, knowledge, and experiences (Lannoy and Procaccia 2001; Martin et al. 2012). The methodology is usually employed in areas where information about model parameters is uncertain or incomplete and where empirical evidence is unavailable.

Guidelines provided by Burgman et al. (2007), Speirs-Bridge et al. (2008), and EPA (2011) were used to develop the questionnaire, select the expert panel, design and conduct the elicitation, and aggregate the results from the study. Explicit considerations to minimize overconfidence, anchoring, motivational, and accessibility biases were included in the questionnaire and elicitation protocol. Given the diversity of disciplines needed in the expert panel and the location and availability of the selected experts, an electronic single-phase elicitation protocol was employed to better capture the breadth and depth of information regarding the different land use categories.

27.2.3 CLUE Model Description

The CLUE (Changes in Land Use and its Effects) methodology was used to allocate the estimated land use changes from the scenarios, subject to spatial rules and restrictions (Verburg et al. 1999; Verburg and Overmars 2009). Starting with the current land use map and distribution, the CLUE model executes a dynamic allocation process based on the competition between the future demands for the different land uses (established by a land use demand table) and the spatial constraints originated by the factors that determine the suitability of each land unit for a particular land use. These factors include the geographic, climatic, demographic, and accessibility characteristics of an area (established by a series of binary logistic regression models), the spatial policies that restrict or encourage land use patterns in a certain area (established by the use of land use restriction maps and rules), the natural sequences of transition among land uses (established by a transition allowance matrix), and the relative mobility of the different land use categories (established by an indicator of conversion elasticities). The model combines a nonspatial module where the qualitative and quantitative scenario descriptions and the aggregate demands for each land use and other model parameters are specified, with a spatial module that simulates the dynamic competition and distributes the land use demands according to the defined rules (Fig. 27.2).

To characterize the occurrence of each land use category, the location factors were plotted as independent variables against the current distribution of each land use in the binary logistic regression models. The regressions identified which are characteristics of an area were significant for determining the occurrence of each land use type (assuming that the current land use pattern represented the allocation for the highest suitability). The model represented the suitability of the area by estimating the probability of occurrence of each land use category based on the logistic regression coefficients for every surface grid cell, using the following equation:

$$\log(P_{ku}/1 - P_{ku}) = \beta_0 + \beta_1 X_{1,k} + \beta_2 X_{2,k} \ldots + \beta_n X_{n,k} \qquad (27.2)$$

in which P_{ku} is the probability of occurrence of land use k on grid cell u; X is the location geographic, climatic, demographic, and accessibility factors included in the regression; and β is the binary logistic regression coefficient for the corresponding factors.

This determined which land use was more likely to be established on the surface grid cells according to their characteristics. To allocate the land use changes specified as future demands in the scenarios for each time step, the model first identified which cells were allowed to change according to the restriction rules and the transition matrix. Then, the total probability for the occurrence of each land use class for every grid cell was estimated by combining:

$$\mathrm{TP}_{ku} = P_{ku} + \mathrm{ELAS}_k + \mathrm{ITER}_k \qquad (27.3)$$

where TP_{ku} is the total probability of occurrence of land use k on grid cell u, P_{ku} is the probability of occurrence of land use k on grid cell u estimated based on the logit model, ELAS_k is the conversion elasticity for land use k added *only* when grid cell u is already under land use k, and ITER_k is the iteration variable for land use k.

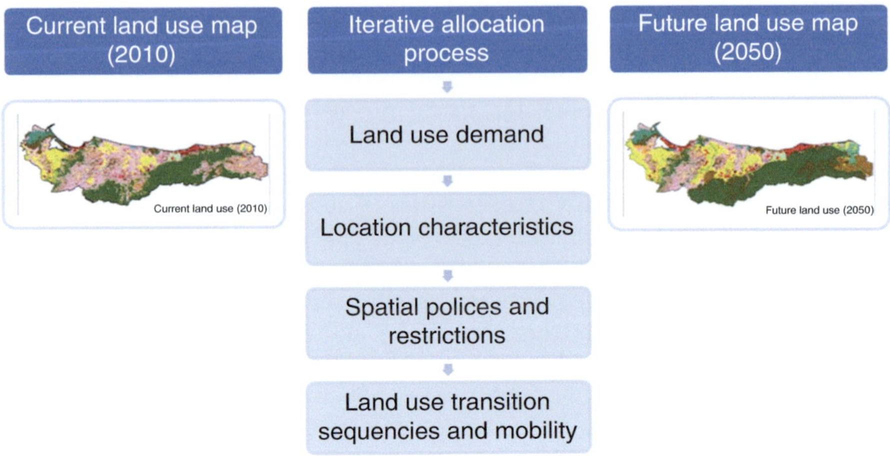

Fig. 27.2 Methodological framework for future land use modeling with CLUE

The $ELAS_k$ conversion elasticity parameter for each land use varies between 0 and 1, where 0 indicates that the conversion is readily reversible (the land use can be easily relocated from one location to another) and 1 indicates that the conversion is difficult to reverse (the use is harder to relocate). It was used as a measure of the tendency of an area to maintain its current land use. The $ITER_k$ iteration parameter represents the competitive strength of each land use. It started with the same value for all land use types, when the model allocated the land use with the highest preference on every grid cell. The model then compared the allocated area with the specified demand and modified the iteration parameter to assign the highest preferences for the uses where the former was lower than the latter and vice versa. The procedure was repeated in successive iterations until the allocated area matched the specified demand for every time step in the simulation.

Future land use maps obtained from the CLUE model for each scenario were used to reassess climate change sensitivity using the methodology described in the section above. By comparing the current state with the future scenarios, the main trends in changes in climate change sensitivity were identified. By comparing the sensitivities among scenarios, the effect of alternative land-use-based adaptive strategies on the magnitude and distribution of sensitivity changes were explored.

27.2.4 Fragmentation

To assess the landscape fragmentation in the current state and future land use scenarios, a set of fragmentation metrics were estimated for the natural land use categories using the FRAGSTATS Spatial Pattern Analysis Program, version 4 (McGarigal et al. 2012). These metrics included total number of patches, patch density (number of patches per unit of area), mean patch area, proportion of like adjacencies and the proximity, connectance, and cohesion indices.

Based on McGarigal et al. (2012), percentage of like adjacencies was estimated as:

$$\mathrm{PLADJ} = \left(g_{kk} \middle| \sum_{k=1}^{n} g_{ko} \right)(100) \tag{27.4}$$

in which g_{kk} is the number of like adjacencies or joints between grid cells of the same land use k and g_{ko} is the number of adjacencies between grid cells of land use k and all different land uses o.

The index ranges between 0 and 100 from maximum disaggregation (no like adjacencies, i.e., one-celled patches surrounded by patches of different land use categories) to maximum aggregation (like adjacencies only, i.e., landscape comprises a single land use allocated in one patch).

The proximity index was estimated as:

$$\mathrm{PROX} = \sum_{a=1}^{n} A_{ab} / d_{ab}^{2} \tag{27.5}$$

in which A_{ab} is the area of patch a within specified neighborhood of patch b, a and b being patches of the same land use, and d_{ab}^2 is the squared distance between patch a and b computed from cell center to cell center.

The index ranges from 0 to 100, where 0 represents a patch with no neighbors of the same land use class within the specified neighborhood. The upper limit of index depends on the neighborhood radius, which is user specified (200 m for the purpose of this study).

The connectance index was estimated as:

$$\mathrm{CONNECT} = \left(\sum_{a=b}^{n} C_{abk} \middle| [n_k(n_k - 1)]/2) \right)(100) \tag{27.6}$$

where C_{abk} are the joinings between patch a and b (1 if joined, 0 otherwise) of the corresponding land use k and n_i the number of patches of land use k in landscape.

The index ranges between 0 and 100 and represents the percentage of functional joinings of patches of the same type from the maximum possible connectance given the number of patches in the land use class. The threshold distance that defines a functional joining was user specified as 200 m.

Finally, the cohesion index was estimated as:

$$\left[1 - \left(\sum_{a=1}^{n} p_a \middle| \sum_{a=1}^{n} p_a \sqrt{A_a} \right) \right] / \left[\left(1 - 1 \middle| \sqrt{Z}\right)^{-1} \right](100) \tag{27.7}$$

in which p_a is the perimeter of patch a (number of surface grid cell), A_a the area of patch a (number of surface grid cell), and Z the total number of grid cells in landscape.

The index ranges between 0 and 100, where 0 represents land use entirely subdivided and physically disconnected and increases as the landscape becomes more aggregated.

27.3 Land Use Changes in Northern Honduras

27.3.1 Land Use Categories

The disaggregated land use map contained 18 land use categories, and their definitions are given in Table 27.1. For the forest area, we distinguished the categories broad-leaved and pine forests, which include all kinds of classes, including secondary, open, and fragmented forest. The permanently flooded forest areas at the shore line are classified as mangroves. Livestock activities can be classified as extensive cattle raising and dairy production systems. The former system can be

Table 27.1 Land use categories for the department of Atlántida (2010)

Land use categories	Brief description	Land use categories	Brief description
Natural vegetation		*Agriculture*	
Mangroves	Forestry in flooded areas	Shrub lands	Land with potential for agriculture but not under use
Broadleaf forest	Primary and secondary forest (open and closed)	Extensive agriculture	Arable land used for growing annual crops (horticulture)
Pine forest	Secondary pine forest	Intensive agriculture	Arable land used for growing crops, both annual and perennial (corn, beans, sugar cane, etc.)
Water bodies	Surface water	Banana/ Plantain	Intensive monoculture plantations
Beaches		Fruit crops	Including oranges and other citruses
		Pineapple	Intensive monoculture plantations
Human settlement		Coffee	Monoculture and shaded plantations
Discontinuous human settlement	Isolated dwellings and small settlements in the rural areas	African oil palm	Monoculture plantations
Continuous human settlement	Urban and small towns	*Livestock (pastures)*	
Commercial and industrial areas	Area in use for commercial and productive activities	Extensive livestock	Natural grazed grassland in the more isolated areas in use for cattle raising
		Intensive livestock	Cultivated pasture and natural grazed pastures in less isolated areas

characterized as large natural grassland areas that are extensively used for livestock grazing, while the latter refers to natural grasslands with a higher grazing density, generally used by dual-purpose cattle systems (milk and beef production) and located in the more urbanized areas. In total, eight agricultural land use categories were established, including shrub land. Shrub land habitats are agriculture land or pasture with natural or seminatural woody vegetation. Generally, those areas are left fallow for 1–5 years to recover soil fertility. The category intensive agriculture refers mainly to horticulture, including pineapple, while the category extensive agriculture refers to annual crops like corn and beans (subsistence crops) and sugar cane.

Our study assesses the changes of a relatively large group of land use classes because these allow us to evaluate the potential geographical spread of increasing agriculture and livestock production and expanding urbanization throughout the department of Atlántida. However, the land use categories included in the CLUE model for the study are fewer than those in the original land use map. Certain categories had to be aggregated since the model is unable to adequately allocate land use categories and land use changes of relatively small magnitude (less than 1 % of the total area). Fruit crops, pineapple, and coffee plantations were aggregated into a single category. The intensive agriculture and banana land uses were also grouped since the area covered by banana was too small for the model to run, but its changing rate was similar to that of intensive agriculture. A category including beaches and water bodies was established and remained constant through the modeling period. After running the allocation model, these categories were disaggregated into their original components.

Table 27.2 shows the land use categories with their factor coefficients integrated into the study's regression model at a 95 % confidence level. To test the robustness of the inference of the regression from the available data, a relative operating characteristic (ROC) curve analysis of the predicted probabilities was performed. The ROC values for the models range from 0.68 to 0.99, suggesting that the models are capable of explaining the spatial variation of land use patterns.

The logistics models estimate the spatial variation in occurrence of the different land use types. The model results for all land uses indicate that land locations are jointly determined by biophysical parameters (elevation, slope, and soil type), climate (precipitation and temperature), and socio-geographic characteristics, such as distance to major roads, rivers, and urban areas and population density. The models demonstrate that the remaining mangroves are influenced by all of these factors with the exception of population density. Mangroves are found in low flat areas where the temperature is generally higher than in the other areas. The location and distribution of broad-leaved forests is highly related to the biophysical parameters. They are found in elevated and sloped areas with poor soils and lower temperatures. These areas are distant from main roads and human settlements. Pine forests have similar characteristics, but are found in areas with lower elevations and higher temperatures.

The analytical results showed that agricultural and livestock activities are multifaceted and influenced by both physical and socio-geographic characteristics,

Table 27.2 Regression analysis results

	Mangroves	Broadleaf forest	Pine forest	Scrublands	Extensive agriculture	Intensive agriculture/ Banana	Fruit crops/ Pineapple/ Coffee	African oil palm	Extensive livestock	Intensive livestock	Discontinuous human settlement	Continuous human settlement	Commercial and industrial areas	Water bodies/ Beaches
Distance to roads	−.000053		−.002046	−.000052	−.000096	−.000179	−.000427	−.000240		−.001046	−.000078	−.000721	−.000604	−.000247
Distance to mayor cities	−.000031	−.000047	−.000146	−.000043	−.000035	−.000036	−.000134	−.000034	−.000011		−.000073	−.000142	−.000109	−.000041
Distance to rivers	.000323	−.000495	.	−.000266	−.000392	−.000477	−.000388	−.000385	−.000275		−.000317			
Elevation	−.013711	.000399	.007545	−.000568				−.000756	−.000215		−.000607		−.001284	−.000729
Slope	.012346	−.004662	.011473				.009079	.003177	.003662		.006475	.008802	.014842	.005349
Precipitation	.002263	.000368		−.000799	−.000342	−.001516	−.000825	−.001056	−.001186	−.002026	−.001002	−.001738	−.002002	−.000764
Temperature	2.548431	−.027884	−.671127	−.191195	−.153103	.	−.453168	−.515114	−.211647	−.223245	−.178367	−.287265	−.541462	−.226972
Soil type	−.352582	.415313		.021789	.141678		−.164085	−.195475	.076650			−.235422	−.334024	−.047801
Population density		−.000253		.000017	−.000157		−.000390		−.000025		−.000107			
Constant	−69.619151	−3.819141	6.572128	5.150401	.788105	−.353826	11.954605	15.062477	7.097964	5.179017	4.016541	10.285466	16.850190	4.081739
ROC	.967524	.876632	.996252	.687000	.758620	.796243	.846005	.766662	.784001	.904329	.734988	.823204	.838086	.675377

particularly by temperature, slope, and soil conditions. Extensive agriculture, the production of corn and beans by small-scale farmers, is concentrated in areas with poor soils distant from main roads and cities. The locations of extensive livestock are influenced by all of the factors. The regression results indicate that extensive livestock activities are influenced by natural conditions (slope and soil conditions) and human activities (distance from human settlements). Intensive agriculture activities, including palm plantations, fruit crops, and pineapple, are located in the flat lowlands with favorable soil conditions. The regression results confirm that each location possesses specific soil characteristics and slopes that influence the potential for natural and agricultural vegetation.

Continuous human settlement is constrained by slope, soil type, and socio-geographic conditions. Originally, human settlements were based in the more favorable areas of fertile soil and low elevations and slopes. New settlements are influenced by all factors except for slope. Soil type, road distance, and population pressure have negative coefficients, implying that population pressure caused the expansion of new settlements to more remote and less fertile areas.

27.3.2 Land Use Change Scenarios

Based on the binary logistic regression models, land use demand, and land use conversion rules, the CLUE model was applied to simulate two scenarios to account for different spatial development trends for the 2010–2050 period. The first scenario assumes the continuation of present trends of land use change determining demand (trend scenario) with no implementation of spatial policies with respect to the allocation of agricultural, livestock, and urban expansion. The second scenario is based on the same overall expansion rate for agriculture, livestock, and urbanization. However, we have assumed a spatial policy of land use as defined by the national and local government to ensure the protection of the declared protected areas and the implementation of reforestation projects and planned urbanization (normative scenario).

In the case of the trend scenario (E1), a decrease in pasture (19.4 %) and forestry (8.1 %) would result from an increase in palm oil plantations (see Fig. 27.3). A change from pasture to palm oil plantation would create pressure on the other land use categories, creating more pressure on the protected and/or state-owned areas. Cultivated land losses in the La Ceiba and Tela area are mainly caused by competition between cultivated land and urbanization. This type of land use conversion is indirectly related to the productive capacity of the land because both cities are situated in the lower and more fertile areas. The total urban area, including industrial and commercial areas, will increase by 1.6 %. Increases in horticultural and fruit areas are expected in the more fertile areas between Tela and La Ceiba. The results of this scenario forecast that nearly a fifth (19.2 %) of the current land use will be changed by 2050, especially in areas with relatively good soils and low elevations.

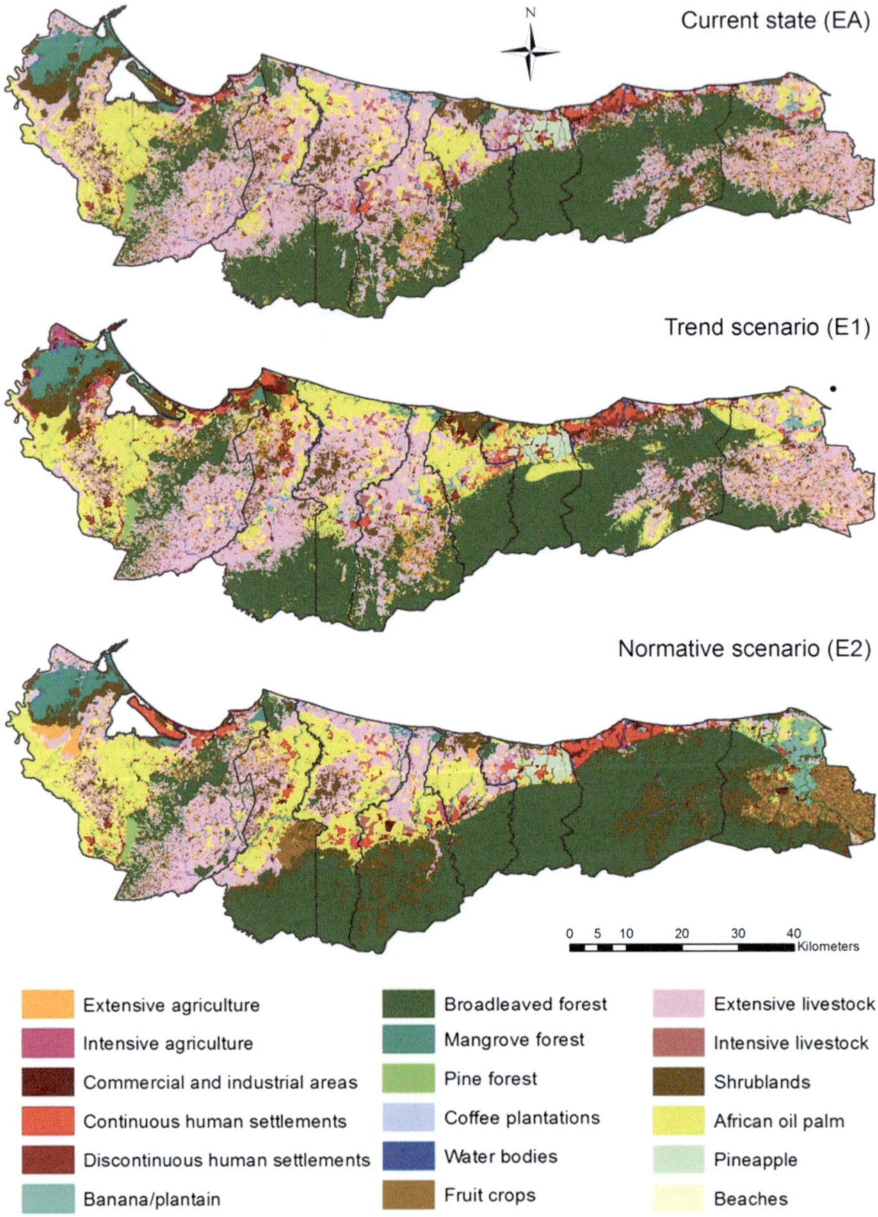

Fig. 27.3 Observed and simulated land use for scenarios E1 and E2

Compared to the trend scenario, the E2 scenario shows some important differences. With local and national government implementation of their land use management plans, broad-leaved forest in the department of Atlántida would increase by 18.3 %. Palm oil would register the same increase as in the trend scenario, but the

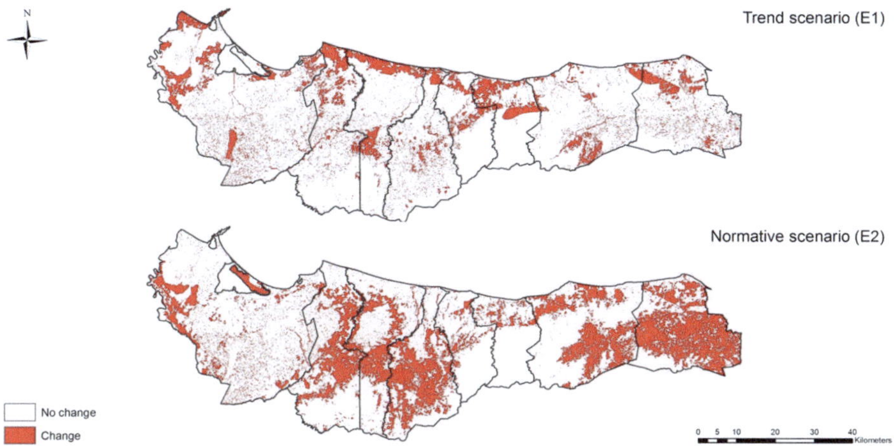

Fig. 27.4 Land use change between 2010 and 2050; trend scenario E1 and normative scenario E2

expansion would basically involve replacement of grazing areas. The most drastic change would be a 50.2 % decrease in extensive livestock activities. Because of increasing land pressure and the inability to open "new land" in the protected areas, land use intensification with more profitable crops per hectare would be the expected result. A good example is the case of Jutiapa, the most eastern municipality where extensive livestock would lose its importance and would be replaced by banana and citrus plantations and to some extent by reforestation.

Figure 27.4 shows the resulting maps of each land use change between 2010 and 2050 for each of the two scenarios. The land use changes between 2010 and 2050 demonstrate that the most extensively changed areas are located in the central and eastern parts of the department of Atlántida, especially in areas with extensive livestock and low elevation.

27.3.3 Sensitivity of Land Use Categories to CC

Because of their complexity, it is difficult to estimate the expected climate change effects for each land use pattern in quantitative terms. These effects are often diverse and not always direct and are therefore also hard to measure. As mentioned in the methodology section, to estimate land use sensitivity to climate change (temperature and precipitation changes), we used expert opinions. The opinions were verified with an extended literature review on land use and CC. Based on these results, we constructed a sensitivity value for each of the land use patterns per category (Table 27.3). The categories are as follows: natural land cover (forest and wetlands), nonnatural land cover (human intervention), human settlements, and livestock. In the short term, the experts do not expect a direct climate change impact for each of the land use patterns. However, heavy rainfall and increasing storm surges during the last decade could be linked to climate change (World Bank 2009).

Table 27.3 MTC and MPC sensitivities estimated per land use category

	MTC sensitivities		MPC sensitivities	
Less < Level of sensitivity > More	*Natural vegetation*			
	Beaches	−0.80	Beaches	−0.76
	Water bodies	−0.73	Water bodies	−0.75
	Mangroves	−0.58	Broadleaved forest	−0.54
	Broadleaved forest	−0.52	Mangroves	−0.49
	Pine forest	−0.49	Pine forest	−0.33
	Agriculture			
	Coffee plantations	−0.54	Extensive agriculture	−0.54
	Shrublands	−0.54	Fruit crops	−0.53
	Fruit crops	−0.51	African oil palm	−0.52
	Intensive agriculture	−0.50	Intensive agriculture	−0.47
	Extensive agriculture	−0.48	Banana/Plantain	−0.44
	Banana/Plantain	−0.45	Pineapple	−0.41
	African oil palm	−0.39	Shrub lands	−0.26
	Pineapple	−0.32	Coffee plantations	−0.19
	Livestock			
	Extensive livestock	−0.50	Extensive livestock	−0.44
	Intensive livestock	−0.45	Intensive livestock	−0.43
	Human settlements and built up areas			
	Continuous human settlement	−0.60	Continuous human settlement	−0.54
	Discontinuous human settlement	−0.54	Commercial and industrial areas	−0.45
	Commercial and industrial areas	−0.50	Discontinuous human settlement	−0.38

In the long term, the impact of CC is expected to be substantial. The extent of CC effect is related to the kind of land use.

In the category of natural vegetation, beaches and water bodies are most sensitive to CC. Climate change, as a result of rising sea levels and temperature raise and drought, will lead to coastal erosion and tend to degrade or remove natural protective features (e.g., mangroves and sand). That in turn will increase extreme water levels and hence the risk of coastal flooding. However, as with other considerations about coastal degradation, it is difficult to separate the effects of human-induced forces from those that result directly from climate change (Nicholls et al. 2009). The immediate effect of CC for water bodies in coastal areas is the infiltration of saltwater and increased flooding, as well as saltwater intrusion into surface waters and wetland loss.

Many aspects of projected climate change will likely affect forest growth. However, land use change is the greatest threat to species diversity of tropical forests, but climatic change alone could decrease the diversity of plant types at the boundaries of biomes, particularly in the tropics (Kirschbaum and Fischlin 1995).

The experts classified the sensitivity of broad-leaved forests as high. Species in moist tropical forests, including hardwoods, are the least drought adapted in the tropics, and their survival (with the attendant loss in diversity) in some areas must be considered at risk from climatic change (Kirschbaum and Fischlin 1995). Although many trees are resilient to some degree of drought, increased temperatures could make future droughts more damaging than those experienced in the past. In addition, drought increases the risk of forest fires, which is a widespread problem in most of the Honduran forest areas.

Nonnatural land use, mainly agriculture, is highly dependent on specific climate conditions. Trying to understand the overall effect of climate change is complicated. Increases in temperature can be beneficial for some crops in some places; crops tend to grow faster under warmer conditions (Parry et al. 2004). The crops most vulnerable to higher temperatures are coffee, fruits, and vegetables, while crops like fruits and African palm are more sensitive to droughts. The general effects of CC also need to be considered along with other factors that affect agricultural production, such as farming practices, property rights, and technology.

The experts classified livestock activities as not very sensitive to CC. However, changes in climate could affect livestock activities both directly and indirectly. Direct effects are heat stress, which increases animal vulnerability to disease, but will also have significant effects on milk production and reproduction in dairy cows (Reilly 1995). Longer and more intense droughts, resulting from higher dry season temperatures and reduced precipitation, could reduce the amount of forage available to grazing livestock. At the same time, CC may increase the prevalence of parasites and diseases that affect livestock.

The sensitivity for human settlements and built-up areas are evaluated at an intermediate level compared with other land uses. Both main cities, Tela and La Ceiba, are located near the coastline and are vulnerable to rising sea levels, extreme weather events, and flooding. In recent years, both cities have suffered substantial damage to physical infrastructure such as buildings, roads, drainage, and energy systems – which in turn has impacted the welfare and livelihoods of its inhabitants. In both cities, the population continues to grow in the absence of effective urban planning, and more than 65 % of the population is poor and vulnerable. Their neighborhoods are located in the low-lying areas which are most exposed to the effects of CC.

Rural Honduras is characterized by high poverty levels; 70 % of the rural population lives in poverty, and people rely on rainfed agriculture for their livelihoods (World Bank 2006). The impact of higher temperatures and drought can be linked to decreasing drinking and irrigation water sources in wells and springs, especially in the more northern part of the department.

27.3.4 Fragmentation

Land use changes not only reduce or increase the land cover of a specific category but can also lead to habitat fragmentation in which a large patch of natural habitat is divided into smaller patches. Habitat fragmentation might increase CC impacts because it not only blocks the ability of species to expand their range as a response

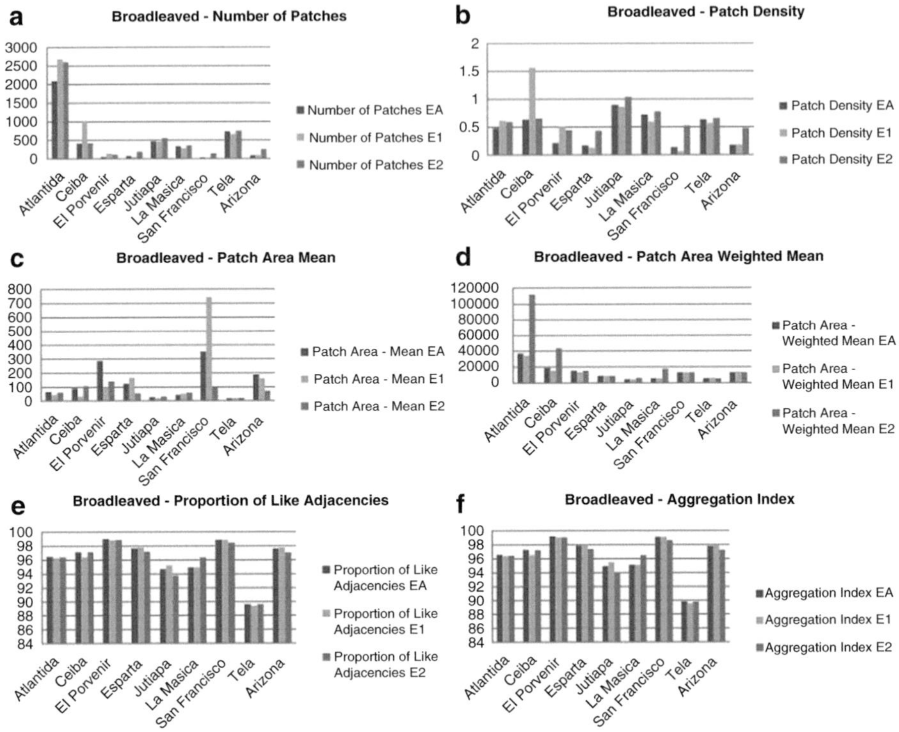

Fig. 27.5 Landscape metrics for land use change patches: (**a**) number of patches, (**b**) patch density; (**c**) patch area mean; (**d**) patch area weighted mean; (**e**) proportion of like adjacencies; (**f**) aggregation index

to shifting suitable climate zones but also reduces the resilience capacity of the local region to climate change impact. We analyzed the fragmentation of the broadleaved forest area under different land use scenarios. We used five fragmentation indices including number of patches, patch density (number of patches per unit of area), mean patch area, proportion of adjacencies and the proximity, connectance, and cohesion indices.

The number of patches for a particular habitat may affect a variety of ecological processes; under both scenarios the number of patches will increase, but the impact is less severe for the normative scenario (Fig. 27.5a). The patch density measures the spatial heterogeneity. Landscape with greater patch density is considered more fragmented than a landscape with a lower patch density of a given patch type (Fig. 27.5b). At the regional level, no differences were found, but the trend scenario forecasts an increase in the patch density for the La Ceiba area, which indicates more spatial heterogeneity in the specific municipality. A landscape with a smaller mean patch size for the target patch type than another landscape might be considered more fragmented (Fig. 27.5c). In the case of the municipality of San Francisco, the forest patch has a greater mean patch size than the other patch types in the municipality so it might be considered that the forest area in this specific

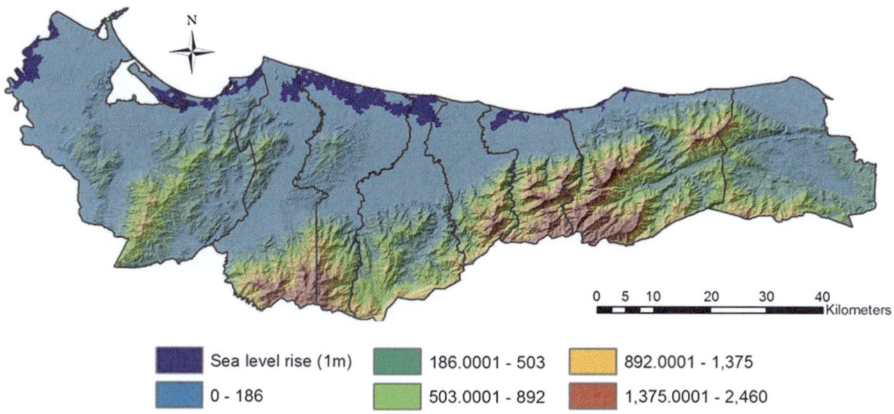

Fig. 27.6 Digital elevation and sea level rise model of the department of Atlántida, 2050

municipality is less fragmented. By weighting patches according to their size, larger patches are weighted more heavily than smaller patches in calculating the average patch mean, which is useful when characterizing the landscape structure. The normative scenario favors the creation of greater forest areas and will reduce the fragmentation of current forest areas (Fig. 27.5d).

The proportion of like adjacencies is the percentage of cell adjacencies involving the corresponding patch type that are like the adjacent patch type. An increasing percentage of like adjacencies implies greater aggregation of the patch type (Fig. 27.5e). The results for both scenarios indicate that the expected changes will be small. The municipality with the lowest index is Tela, which implies a smaller aggregation of the forest areas. The aggregation index (AI) is directly related to the fragmentation index, and it is expected to decrease with the increasing number of patches. This is because increasing a class on a landscape, e.g., forestry, increases the probability of forming larger (more aggregated) patches, and AI decreases if patch size remains unchanged. The results were as expected; Tela reports the lowest level of aggregation and has the most fragmented forestry habitat (Fig. 27.6f). Moreover, the AI of broad-leaved forest in scenario E2 increases slowly in La Masica and La Ceiba, but decreases in the other municipalities.

27.3.5 Sea Level Rise

The municipalities most sensitive to sea level rise by 1 m are shown in Fig. 27.6. According to the elevation contour, 11,608 ha or 2.6 % of the department is estimated to be flooded by the year 2050. This flooding would mostly occur in villages from the municipalities of Esparta, La Masica, and Tela. Those municipalities have many low areas, resulting in higher proportion of flooded area with sea level rise.

The maximum elevation obtained from the DEM corresponded to 2,460 m above sea level. The proportion of the department of Atlántida located between 0 and 20 m above sea level was estimated at 26 %. This surface is mostly occupied by productive land uses and human settlements.

The municipalities in Esparta and La Masica are not densely populated, and their land use involves a very large proportion of agricultural lands, especially oil palm cultivation, which suggests that there is an urgent need to prevent and manage flooding in agricultural fields in this area. Both main cities of the department of Atlántida, La Ceiba, and Tela are also affected by sea level rise; the area affected in those municipalities was about 25 % of the total flooded area. In summary, a higher sensitivity to sea level rise was observed in the lower areas where land use is dominated by oil palm cultivation, and the population density is relatively low.

27.3.6 Climate Change Sensitivity

We present two kinds of maps to analyze the CC impact for each land use cover. Figure 27.7 presents the current state and the results for both scenarios based on

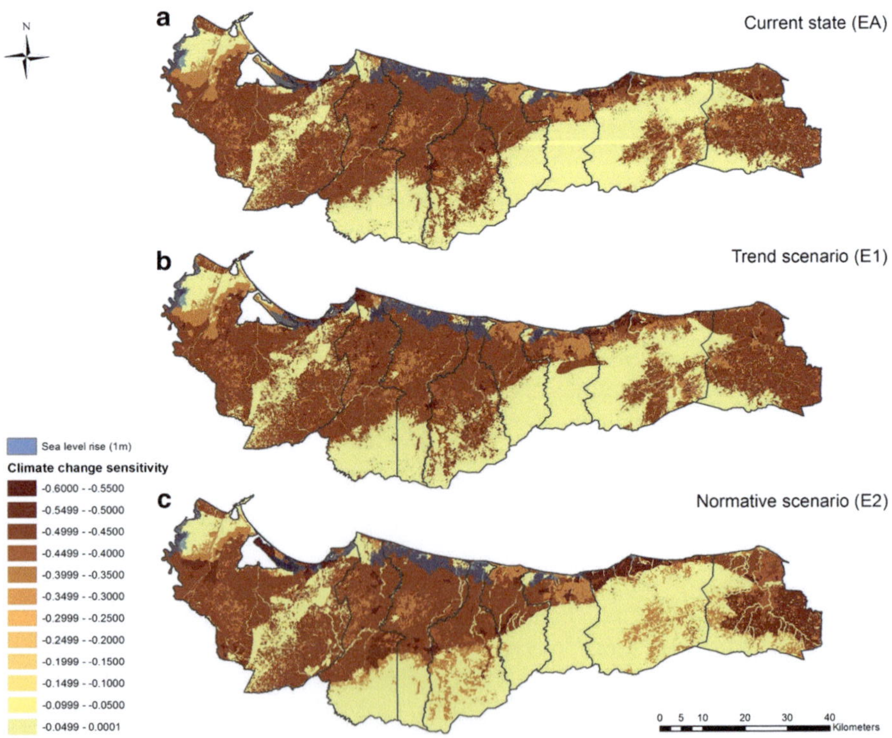

Fig. 27.7 Climate change sensitivity at land cover level: (**a**) current state, (**b**) trend scenario, (**c**) normative scenario

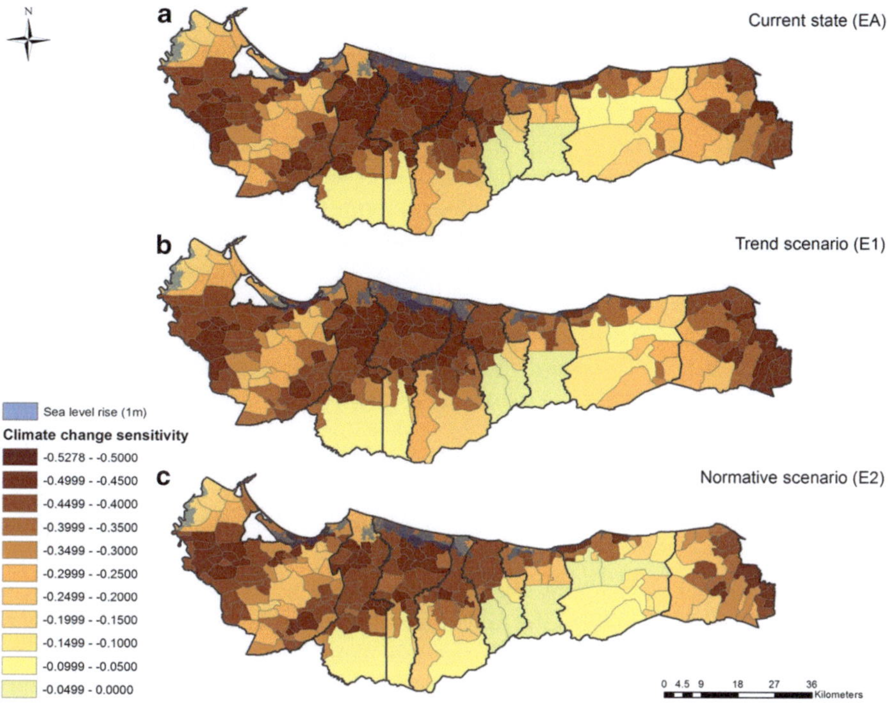

Fig. 27.8 Climate change sensitivity at village level: (**a**) current state, (**b**) trend scenario, (**c**) normative scenario

land cover with changes in land use for the scenarios analyzed at patch level. In Fig. 27.8, we calculated a weighted average for all land use categories at village level. Villages with more sensitive land cover will present a higher sensitivity to climate change. In both figures, sea level rise is presented as an independent variable and not linked with the CC variables of temperature and precipitation.

Figure 27.7a shows that under the current situation, 60 % of all land cover is considered as highly sensitive to CC. The most sensitive area is in the western part of the department where extensive livestock and agriculture activities dominate the area. On average, the villages in the municipality of Esparta are the most sensitive to CC, including flooding.

For the E1 scenario, the "business as usual approach," we assumed that land use change would continue as in the past. During the 2010–2050 period, we estimate that 18.9 % of the total land use will change. According to Fig. 27.6b, this change will occur on the sea coast and in areas of the municipalities of Jutiapa, Tela, and El Porvenir. The expected changes for 2050 are mainly a further expansion of oil palm replacing pasture and other agriculture crops. The total impact of the land use change on sensitivity to climate change on a municipality level depends on whether the replaced land cover was more or less sensitive to CC than the new land cover. According to our analysis, nearly 80 % of the land cover change for 2050 involves conversion to oil palm cultivation. As can be seen in Fig. 27.7b, the municipalities where oil palm is expanding will be less sensitive to CC in 2050 than in 2010 because oil palm is not very sensitive to temperature rise and is relatively resistant to droughts.

The normative scenario, Fig. 27.7c and 27.8c, presents a more optimistic outlook. The land use plans developed by the national and local governments would control the expansion of agriculture activities to the protected areas located in the higher part of the department. The total land cover change will be 38 % by 2050. The natural areas, especially broad-leaved forests, will expand by 18 %. The model forecasts a dramatic reduction in extensive livestock and agriculture activities (52 % and 25 %, respectively). African oil palm will grow at the same rate (15.7 %) as in the trend scenario, but the replacement impact (crowding out the replaced agriculture activities to protected areas) will be limited.

27.4 Discussion

By combining land use patterns with climate change sensitivity for each land use cover, it is possible to analyze the department of Atlántida's sensitivity to climate change over the next 40 years. In this context, important climate changes impact include a change in the length of the growing season. Rainfall and temperature changes are major factors determining agricultural productivity. A number of climate models agree that in much of Honduras, the growing season is projected to decrease by between 5 and 15 % by 2050, with particularly negative effects in the dryer areas of south Honduras (IPCC 2007). Over the last 30 years, Honduras has already shown a gradual increase in both mean maximum and mean minimum temperatures. Although the average temperature shows an increasing trend, the data is highly variable (Argeñal 2010). In the case of precipitation patterns, more variation and decreasing mean annual rainfall is expected; this will have a significant impact on crop yields. Porch et al. (2007) found that shorter growing seasons and higher temperatures represent important production constraints for bean farmers in the department.

The results of our logistic models for all land use categories explain that the location and expansion of agriculture and livestock activities and urbanization are determined by biophysical parameters (altitude, slope, and soil type); demographic characteristics, such as distance to major roads, cities, and rivers and population density; and climate variables (precipitation and temperature). Those driving factors shaped the current land cover. They are the main factors that will shape future land use.

The CLUE methodology was applied to associate land use demand and possible trends to simulate land use scenarios for the 2010–2050 period. As we could observe, land use change is directly linked to biophysical, climate, and demographic variables (E1 scenario), but can be influenced by effective policy measures to protect specific areas (E2 scenario). Simulation results indicated that land use changes for the E1 scenario occur more frequently on the coast and in areas with low elevations. More dramatic changes were found for the E2 scenario; the effect of replacing extensive livestock with African oil palm was less harmful for the protected areas. However, it is plausible to expect that the replacement of natural forest areas with pastures will occur in other areas with less control than the protected areas (departments of Cortes and Gracias a Díos). The reduction of natural areas will make the region more vulnerable to climate change. However, the replacement of different kinds of crops and livestock activities with African oil palm will make the region less vulnerable to climate change.

By using landscape metrics, it was possible to evaluate and compare before and after conditions in a landscape plan for a particular landscape, in our case the forestry area. At the landscape level, the E1 scenario would lead to a more fragmented forestry area during the simulation period than the E2 scenario. Fragmentation would not only reduce the critical threshold for species' survival, but would also affect the forest's resilience to recover from severe disturbances and to maintain the rates of supply of goods and services to the region (Thompson et al. 2009). To reduce further fragmentation of the forest area, conservation policies could have an important impact on the forest area patch patterns, making the region less vulnerable to CC.

The 1-m projection of sea level rise could increase flooding, particularly in the lower areas near the coast line. Sea level rise will probably increase the biophysical and demographic vulnerability of the area. However, it is in this area where we expect increasing population growth and further expansion of the main cities in the region, Tela and La Ceiba. In recent years, flooding has affected especially the poorest populations because they are concentrated in the more hazardous areas.

The combination of flooding with a decrease in rainfall could increase the inland penetration of salt water, affecting the quality and availability of fresh water. Also it could impact the natural wetlands and lagoons (*Laguna de los Micos* and *Cuero y Salado*); those areas are protected by low barrier beaches, and sea level rise could induce overtopping and affect the ecosystems. African oil palm plantations and other agricultural activities could be at risk of flooding and soil salinization. In general, the results stress the biophysical and demographic vulnerability of settlements and agricultural activities located in the very low areas near the coast line.

Without active land use planning for the coming years, the increasing development pressures from a growing population and expanding agricultural activities could impact the capacity of ecosystems to adapt to climate changes, making the region more vulnerable. The essence of land use analysis is to show policy makers how possible land use cover changes could impact future land use. The current land use plans of local and national governments will have a decisive role in determining the most appropriate land use for the upcoming 40 years. It is important to find a balance between conservation of crucial areas and economic and agricultural development. Although trends of oil palm expansion will continue, it is important to limit further agriculture expansion in the protected areas. Reforestation and the adoption of Reducing Emissions from Deforestation and Forest Degradation (REDD) projects could be important measures to protect the region against climate change.

27.5 Conclusions

The coastal zone of northern Honduras is highly exposed to the potential impacts of climate change. In particular, the region is vulnerable to sea level rise and to changes in temperature and precipitation. The potential effects of CC, which can include significant socioeconomic implications for the population, and the relation with land cover change are complex. Changes in land use due to such activities as agriculture, urban sprawl, and transportation infrastructure are generally recognized in the literature as major causes of increasing vulnerability to climate change. Assessing land use sensitivity is important as it enables the identification of at-risk areas and the threats posed by a decrease in or loss of such resources that could threaten future sustainable development in the region (Berry et al. 2006).

The aim of this study was to analyze sensitivity variations derived from spatial and temporal land use differences and their implications on the use of land policy as a climate change adaptation tool in integrated coastal zone management. In the case of northern Honduras, agricultural expansion is the most important proximate cause of land use change, followed by deforestation and infrastructure development. Growing export demand for palm oil and increasing numbers of subsistence farmers on hillsides have been the primary drivers for converting natural areas (directly or indirectly) into land for agricultural and extensive livestock use. The total land cover change for the next 40 years will be about 18.9 % of the total area. Those land cover changes will not only affect the biophysical landscape but will also influence the agrarian structure – more land will become concentrated in fewer hands.

This study shows that some specific land use patterns are more sensitive to climate change than others. Specifically, beach areas, water bodies, and broad-leaved forests are considered as highly vulnerable to higher temperatures and drought. In general, crops are vulnerable to CC, which will likely combine to reduce yields and increase production risks in the region. However, the economic drivers that are causing the ongoing land use changes from agriculture and livestock to primarily oil palm production seem to be effectively mitigating climate change. Oil palm is less sensitive to CC than other crops and adapts relatively well in the region. The main problem of the expansion of oil palm is the pressure on other crops and natural areas. While oil palm is less sensitive to CC and we found no direct linkage between the expansion of oil palm cultivation and deforestation, indirectly oil palm production leads to the conversion of natural areas into cropland or pasture because of the increasing land pressure on lower-lying areas.

These results have important implications for future land use policies. For instance, future conversion from cropland to other land types could cause increased sensitivity (particularly through urbanization and deforestation) while future expansion of cropland could also cause improved CC adaptation, particularly through the expansion of less sensitive crops like oil palm into marginal areas. However, the increasing land pressure caused by palm oil production has a potential negative impact on natural areas, causing deforestation and making the whole region more vulnerable to climate change over the next 40 years. National and local governments have a decisive role in assuring the implementation of their land use policies to protect the region against climate change impact.

Acknowledgments We would like to acknowledge the support of the "Programa de gestión de los recursos naturales y cuencas del Corredor Biológico Mesoamericano en el Atlántico Hondureño (PROCORREDOR)" who gave us access to their geographic database. We are also grateful for the advice and financial support provided by the Lincoln Institute of Land Policy for the implementation of this study.

References

Argeñal FJ (2010) Variabilidad Climática y Cambio Climático en Honduras. SERNA and UNDP. Tegucigalpa, Honduras

Baumeister E, Wattel C (1996) El Agro Hondureño y su Futuro. CDR-ULA, San José

Berry PM, Rounsevell MDA, Harrison PA, Audsley E (2006) Assessing the vulnerability of agricultural land use and species to climate change and the role of policy in facilitating adaptation. Environ Sci Pol 9:189–204

Buckles D (ed) (1999) Cultivating peace: conflict and collaboration in natural resource management. International Development Research Center, Ottawa

Burgman M, Fidler F, McBride M, Walshe T, Wintle B (2007) Eliciting expert judgments: literature review. Australian Centre for Excellence in Risk Analysis (ACERA), University of Melbourne, Melbourne

COPECO (Permanent Contingency Commission) (2010) Proyecto Mitigación y Desastres Naturales 2010: Plan Municipal de Gestión de Riesgos. La Ceiba, Atlántida

CEPAL (2010) Efectos del cambio climático en la Costa de América Latina y el Caribe Dinámicas: Tendencias y Variabilidad Climática. CEPAL, Santiago de Chile

EPA (2011) Expert elicitation task force white paper. http://www.epa.gov/stpc/pdfs/ee-white-paper-final.pdf. Accessed 8 Apr 2012

Humphries S (1998) Milk cows, migrants, and land markets: unraveling the complexities of forest-to-pasture conversion in Northern Honduras. Econ Dev Cult Chang 47(1):95–124

IPCC (Intergovernmental Panel on Climate Change) (2007) Fourth Assessment Report 2007: climate change: impact, adaptation and vulnerability. Working group II report impacts, adaptation and vulnerability. Cambridge University Press, Cambridge

Jansen HJ, Pender J, Damon A, Schipper R (2006) Land use in the hillside areas of Honduras: a quantitative livelihoods approach. IFPRI, Washington, DC

Kirschbaum M, Fischlin A (1995) Climate change impacts on forests. In: Watson RT, Zinyowera MC, Moss RH (eds) Working group II report "Impacts, adaptations and mitigation of climate change: scientific-technical analyses". IPCC, Cambridge University Press, Cambridge

Klein RJT, Nicholls RJ (1999) Assessment of coastal vulnerability to climate change. Ambio 28(2):182–187, Springer on behalf of Royal Swedish Academy of Sciences

Lannoy A, Procaccia A (2001) L'utilisation du jugement d'experts en sûreté de fonctionnement. Editions TEC & DOC. Dec 2001

Martin TG, Burgman MA, Fidler F, Kuhnert PM, Low-Choy S, Mcbride M, Mengersen K (2012) Eliciting expert knowledge in conservation science. Conserv Biol 26:29–38

McGarigal K, Cushman SA, Ene E (2012) FRAGSTATS v4: spatial pattern analysis program for categorical and continuous maps. Computer software program produced by the authors at the University of Massachusetts, Amherst. http://www.umass.edu/landeco/research/fragstats/fragstats.html. Accessed 27 Jan 2012

MIRA-USAID (Environmental Resource Integrated Management Programme) (2005) Efectos del cambio climático en el desarrollo costero de La Ceiba. www.cambioclimaticohn.org. Accessed 27 Jan 2012

Nicholls RJ, Woodroffe C, Burket V (2009) Coastline degradation as an indicator of global change. In: Letcher TM (ed) Climate change: observed impacts on planet earth. Elsevier Science, Oxford

Parry ML, Rosenzweig C, Iglesias A, Livermored M, Fischer G (2004) Effects of climate change on global food production under SRES emissions and socio-economic scenarios. Glob Environ Chang 14:53–67

Porch T, Bernsten RR, Rosas JC, Jahn M (2007) Climate change and the potential economic benefits of heat-tolerant bean varieties for farmers in Atlántida, Honduras. J Agric U Puerto Rico 91(3–4):133–148

Reilly J (1995) Agriculture in a changing climate: impacts and adaptation. In: Watson RT, Zinyowera MC, Moss RH (eds) Working group II report "Impacts, adaptations and mitigation of climate change: scientific-technical analyses". IPCC, Cambridge University Press, Cambridge

Rubio D (2012) Evolución del crecimiento urbano del departamento de Atlántida, Honduras. INYPSA, Honduras

Schatan C, Montiel M, Romero I (2010) Cambio climático y retos para el sector turismo de Centroamérica. Naciones Unidades – CEPAL, México

SERNA (Ministry of Natural Resources and the Environment) (2006a) Estrategia de Adaptación al Cambio Climático y Plan de Acción para la Cuenca del Río Aguan en Honduras. SERNA, Tegucigalpa

SERNA (Ministry of Natural Resources and the Environment) (2006b) Vulnerabilidad actual de la cuenca del río Aguan en Honduras. SERNA, Tegucigalpa

SERNA (Ministry of Natural Resources and the Environment) (2007) Vulnerabilidad de Honduras ante los efectos del cambio climático. SERNA-Dirección de Cambio Climático, Tegucigalpa

SERNA (Ministry of Natural Resources and the Environment) (2010) Estrategia Nacional de Cambio Climático. SERNA-Dirección de Cambio Climático, Tegucigalpa

Speirs-Bridge A, Fidler F, McBride M, Flander L, Cumming G, Burgman K (2008) Eliciting reliable expert judgments for ecological models. Australian Centre for Excellence in Risk Analysis (ACERA), University of Melbourne, Melbourne

Sunderland W, Rodríguez JA (1996) Cattle, broadleaf forests and the agricultural modernization law of Honduras: the case of Olancho. CIFOR, Tegucigalpa

Thompson I, Mackey B, McNulty S, Mosseler A (2009) Forest resilience, biodiversity, and climate change: a synthesis of the biodiversity/resilience/stability relationship in forest ecosystems. Secretariat of the Convention on Biological Diversity, Montreal

UNDP (Programa de las Naciones Unidas para el Desarrollo) (2009) Informe sobre Desarrollo Humano. Honduras: De la exclusión social a la ciudadanía juvenil. UNDP, Tegucigalpa

Verburg PH, Overmars KP (2009) Combining top-down and bottom-up dynamics in land use modeling: exploring the future of abandoned farmlands in Europe with the Dyna-CLUE model. Landsc Ecol 24(9):1167–1181

Verburg PH, Koning H, DeKok GHJ, Veldkamp K, Bourma J (1999) A spatial explicit allocation procedure for modeling the pattern of land use change based upon actual land use. Ecol Model 116:45–61

World Bank (2006) Honduras poverty assessment: attaining poverty reduction, vols 1 and 2. World Bank, Washington, DC

World Bank (2009) Honduras country note on climate change aspects in agriculture. World Bank, Washington, DC

Yoo G, Hwang JH, Choongik C (2011) Development and application of a methodology for vulnerability assessment of climate change in coastal cities. ROYAUME-UNI, Elsevier, Kidlington

Part VI
Tools in Support of Sustainability for IWRM

Chapter 28
Understanding the Spatiotemporal Variability of Hydrological Processes for Integrating Watershed Management and Environmental Public Health in the Great River Basin, Jamaica

Shimelis Gebriye Setegn, Assefa M. Melesse, Orville Grey, and Dale Webber

Abstract The demand for adequate and safe supplies of water is becoming crucial especially in the overpopulated urban centers of the Caribbean islands. Moreover, population growth coupled with environmental degradation and possible adverse impacts of land use and climate change are major factors limiting freshwater resource availability. The main objective of this study is to develop a hydrological model and analyze the spatiotemporal variability of hydrological processes in the Great River basin, Jamaica. Physically based hydrological model, Soil and Water Assessment Tool (SWAT), was calibrated and validated in the basin. Spatial distribution of annual hydrological processes, water balance components for wet and dry years, and annual hydrological water balance of the Great River basin are discussed. The basin water balance analysis indicated that surface runoff contributes more than 28 %, whereas the groundwater contributes more than 18 % of the stream flow. The water balance components differ spatially between each subbasin. The actual evapotranspiration varies between subbasins which range from 887 to 1,034 mm. The variation in evapotranspiration between subbasins is mainly due to variations in land cover. The model can be used to predict watershed responses to climate and land use changes. Hydrological water balance analysis can be used to predict the existing water resource component that can help manage water

S.G. Setegn (✉) • A.M. Melesse
Department of Environmental and Occupational Health and Global Water for Sustainability Program (GLOWS), Florida International University, 11200 SW 8th Street, Miami, FL 33199, USA
e-mail: ssetegn@fiu.edu

O. Grey
Department of Earth and Environment, Florida International University, 11200 SW 8th Street, Miami, FL 33199, USA

D. Webber
Department of Life Sciences, Faculty of Pure and Applied Sciences, The University of West Indies, Kingston, Jamaica

S.G. Setegn, M.C. Donoso (eds.), *Sustainability of Integrated Water Resources Management*, DOI 10.1007/978-3-319-12194-9_28

availability and predict where and when there will be water shortages. The output of water balance study can be used in irrigation potential assessment, runoff assessment, flood control, and pollution control.

Keywords SWAT • Hydrological modeling • Water balance • Stream flow • Great River • Jamaica

28.1 Introduction

Availability of freshwater resources to meet all the consumptive and nonconsumptive needs is a global challenge facing many nations. This challenge is further complicated at a time when demands for freshwater are increasing owing to population increase and the need for more food production in the face of uncertainty of the rainfall emanating from a changing climate. The severity of this problem is even further exacerbated in the smaller islands of the Caribbean where freshwater resources and healthy coastal lines are the mainstay of the economy for tourist attractions. The vulnerability of these islands to natural- and human-induced disasters and their inability to cope with the problem necessitate the understanding of the hydrological processes and responses of the watersheds to various stressors.

There are a number of major environmental threats to the Great River basin of Jamaica which include water contamination and pollution mainly due to soil erosion, agrochemical runoff, and sedimentation causing threats to the coral reefs. Drained by small streams and flows to the Montego Bay, the river is considered as one of the most important rivers of the island. Hence, there is a need for hydrological research of the Great River basin that can support improved catchment management programs. This can better safeguard the alarming degradation of soil and water resources in the island. Understanding the effect of land use dynamics on the hydrology of the river and hence the level of runoff, sediment, and pollutant delivery to coastal waters will necessitate the use of models.

The availability of high-speed computers has resulted in a widespread use of computer models in the analysis and prediction of hydrological variables for research as well as for practical design and management purposes. In recent years, watershed models are increasingly used to implement alternative management strategies in the areas of water resource allocation, flood control, impact of land use change and climate change, and environmental pollution control. Many of these watershed models are applied for runoff and soil-loss prediction (e.g., Setegn et al. 2009, 2010; Grønsten and Lundekvam 2006; Morgan 2001; Srinivasan et al. 1998), water quality modeling (e.g., Debele et al. 2006; Santhi et al. 2006; Abbaspour et al. 2007), land use change effect assessment (e.g. Sheng et al. 2003; Claessens et al. 2006; Wu et al. 2007), and climate change impact assessment (e.g. Andersson et al. 2006; Huang et al. 2005; Zhang et al. 2007; Setegn et al. 2011). Among the available models, physically based SWAT model is a well-established model for analyzing the impact of land management practices on water, sediment, and agricultural chemical yields in large complex watersheds. A comprehensive review of the SWAT model applications is given by Gassman et al. (2007).

The resources of the Great River basin are critically important to the economic well-being of its inhabitants, and management of these resources determines improvement or degradation for both agricultural and urban areas. Planning and management are needed to ensure protection of resources and prevention of further degradation in the basin. Therefore, sustainable planning and management are essential for sustainability of resources in the watershed. The main objective of this study is to develop a hydrological model and analyze the spatiotemporal variability of hydrological processes in the Great River basin, Jamaica. The specific objectives of this study are to (1) analyze the hydrometeorological time series trend of the basin; (2) identify the most sensitive flow parameter in the watershed, calibrate the SWAT model for flow, and analyze the model prediction uncertainties; (3) determine the annual and seasonal hydrological water balance of Great River basin; and (4) determine the spatial distribution of hydrological processes in the Great River basin.

28.2 Description of Study Area

28.2.1 Location of Study Area

The Great River basin is located in northwestern Jamaica where it covers an area of approximately 327.27 km^2. The Great River is of major importance as it supplies water for domestic, agricultural, and industrial purposes. It is the major source of potable water to Montego Bay, a major tourist resort area. Figure 28.1 shows the location of the Great River basin in the island of Jamaica.

28.2.2 Geology and Hydro-stratigraphy

The geology of the Great River basin plays important roles in determining the occurrence of water resources and water availability. The watershed is predominantly a limestone area. The dominant rocks include the Montpelier Formation, the Gibraltar (Bonnygate formation), and the Troy/Clairemont/Sommerset formation.

The dominant hydro-stratigraphic units in the Great River basin include the basement aquiclude, limestone aquifer and aquiclude, and coastal aquiclude. These aquicludes are characterized by a dense network of streams, with peak flows in the rainy season and relatively low flows in the dry season.

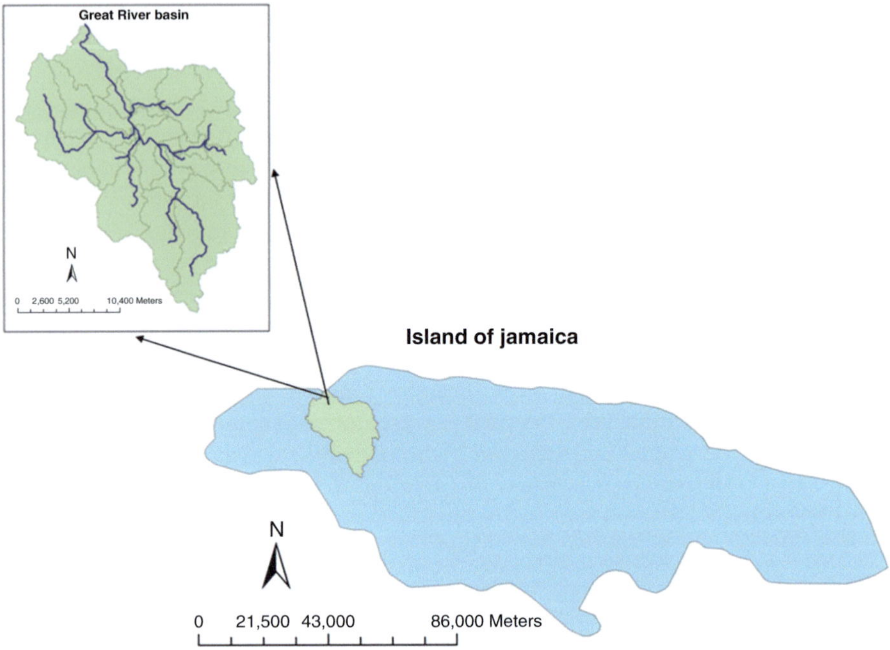

Fig. 28.1 Island of Jamaica showing the location of the Great River basin

28.2.3 Basin Topography

The terrain in the Great River basin is relatively rugged and has some steep escarpments. The area is hilly and undulating, with moderate rocky slopes. Slopes are generally less than seven degrees (68 %). Ninety-nine percent of the watershed area is comprised of slopes below 20° (Table 28.1).

Elevation in the watershed ranges from 762 m near Berkshire to less than 15.24 m on the coast. There is marked variation in elevation, which ranges from 0 at sea level where the Great River enters the sea to 762 m at Berkshire.

28.2.4 Land Use and Soil

Major agricultural crops include vegetables such as cabbage, lettuce, tomatoes, banana, citrus, coffee, shrubs, and forests. Forest vegetation is more widespread in the lower reaches of the watershed. Deforestation is more evident in the upper watershed area, especially on the Westmoreland side. The soils of the watershed are mainly clays formed from yellow limestone and Cretaceous clastic rocks. These soils tend to weather quickly. In the upper watershed, clays are susceptible to slope failures, especially along road cut. Terra rossa soils of the watershed are associated

Table 28.1 Slope classification and distribution in the Great River basin

Class	Slope	Area (ha)	% of area
1	<7°	22,508.10	68.81
2	7–15°	9,713.97	29.70
3	15–20°	404.55	1.24
4	>20°	81.54	0.24

Adapted from NRCA 1998 as sited by Alicia Hayman – Rapid Rural Appraisal of the Great River Watershed, 2001

with hard limestone and are usually linked to sinkholes, cracks, and voids in the rocks. This soil is often mixed with pieces of limestone and is generally very shallow.

28.2.5 *Hydro-climatology*

The annual rainfall has a predominantly bimodal distribution pattern with a dominant peak from September to October and a secondary peak from May to June. Figure 28.2 shows the seasonal characteristic of the monthly rainfall and average air temperature (1992–2000) measured at Norman Manley International Airport in Montego Bay.

The watershed is characterized by a high variability of stream flows owing to rainfall variability and also a rapid movement of runoff across the basin due to the steep topography resulting in less recharge.

28.3 Modeling of Watershed Systems

The study focuses on the characterization of the temporal and spatial variability of hydrological processes in the Great River basin, Jamaica. The physically based watershed model, Soil and Water Assessment Tool (SWAT), was applied for prediction of stream flow and other hydrological components in the Great River basin, Jamaica. The seasonal and annual basin hydrological water balance of the Great River basin was determined. The application of the SWAT models *involved calibration and validation* and *sensitivity and uncertainty analysis*. For this purpose, sequential uncertainty fitting (SUFI-2) calibration and uncertainty analysis algorithms were used.

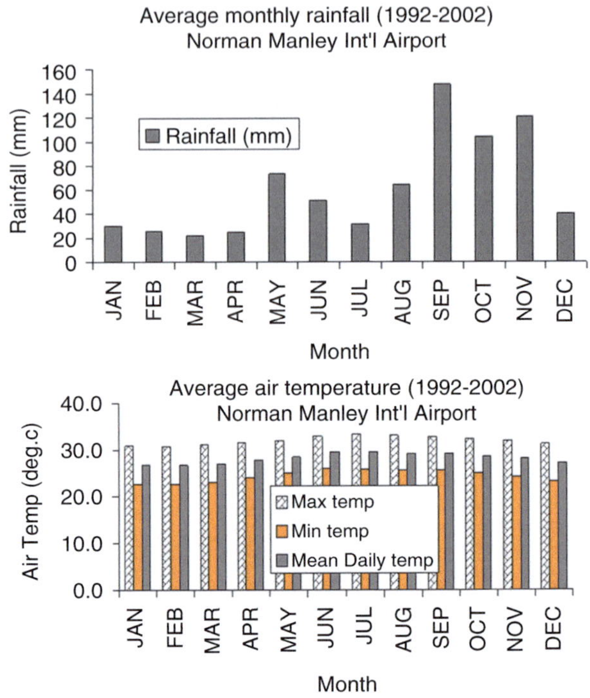

Fig. 28.2 The monthly distribution of rainfall and average air temperature (1992–2000) measured at Norman Manley International Airport

28.3.1 Description of the SWAT Model

SWAT is a river basin scale, a continuous time, long-term, distributed parameter model (Arnold et al. 1998) developed to predict the impact of land management practices on water, sediment, and agricultural chemical yields in large complex watersheds with varying soils, land use, and management conditions over long periods of time (Neitsch et al. 2005). SWAT is embedded in an ArcGIS interface called ArcSWAT. It is computationally efficient, uses readily available inputs, and enables users to study long-term impacts. SWAT divides a watershed into subbasins connected by a stream network and further delineates each subbasin into hydrological response units (HRUs), which consist of unique combinations of land cover, slope, and soil type.

28.3.1.1 Hydrological Component of SWAT

The simulation of the hydrology of a watershed is done in two separate divisions. One is the land phase of the hydrological cycle that controls the amount of water, sediment, nutrient, and pesticide loadings to the main channel in each subbasin. Hydrological components simulated in land phase of the hydrological cycle are canopy storage, infiltration, redistribution, evapotranspiration, lateral subsurface flow, surface runoff, ponds, tributary channels, and return flow. The second division is routing phase of the hydrologic cycle that can be defined as the movement of water, sediments, nutrients, and organic chemicals through the channel network of the watershed to the outlet.

Land Phase of Hydrological Cycle

In the land phase of hydrological cycle, SWAT simulates the hydrological cycle based on the water balance equation:

$$\mathrm{SW_t} = \mathrm{SW_0} + \sum_{i=1}^{t}\left(R_{\mathrm{day}} - Q_{\mathrm{surf}} - E_{\mathrm{a}} - w_{\mathrm{seep}} - Q_{\mathrm{gw}}\right)_i \tag{28.1}$$

in which $\mathrm{SW_t}$ is the final soil water content (mm), $\mathrm{SW_o}$ is the initial soil water content on day i (mm), t is the time (days), R_{day} is the amount of precipitation on day i (mm), Q_{surf} is the amount of surface runoff on day i (mm), E_{a} is the amount of evapotranspiration on day i (mm), W_{seep} is the amount of water entering the vadose zone from the soil profile on day i (mm), and Q_{gw} is the amount of return flow on day i (mm).

More detailed descriptions of the different model components are listed in Arnold et al. (1998)and Neitsch et al. (2005).

Surface runoff occurs whenever the rate of precipitation exceeds the rate of infiltration. SWAT offers two methods for estimating surface runoff: the Soil Conservation Service (SCS) curve number procedure (USDA-SCS 1972) and the Green and Ampt infiltration method (Green and Ampt 1911). Using daily or sub-daily rainfall, SWAT simulates surface runoff volumes and peak runoff rates for each HRU. In this study, the SCS curve number method was used to estimate surface runoff because of the unavailability of sub-daily data for Green and Ampt method.

The SCS curve number equation is (USDA-SCS 1972):

$$Q_{\mathrm{surf}} = \frac{\left(R_{day} - 0.2S\right)^2}{\left(R_{\mathrm{day}} + 0.8S\right)} \tag{28.2}$$

in which Q_{surf} is the accumulated runoff or rainfall excess (mm), R_{day} is the rainfall depth for the day (mm), and S is the retention parameter (mm). The retention parameter is defined by Eq. 28.3:

$$S = 25.4\left(\frac{100}{\mathrm{CN}} - 10\right) \tag{28.3}$$

SWAT version includes two methods for calculating the retention parameter; the first one is retention parameter that varies with soil profile water content and the second method is the retention parameter that varies with accumulated plant evapotranspiration. The soil moisture method (Eq. 28.4) overestimates runoff in shallow soils. But in calculating daily CN (curve number) as a function of plant evapotranspiration, the value is less dependent on soil storage and more dependent on antecedent climate:

$$S = S_{\max}\left(1 - \frac{\mathrm{SW}}{[\mathrm{SW} + \exp(w_1 - w_2 \cdot \mathrm{SW})]}\right) \tag{28.4}$$

in which S is the retention parameter for a given day (mm), $S_{\max}$ is the maximum value that the retention parameter can have on any given day (mm), SW is the soil water content of the entire profile excluding the amount of water held in the profile at wilting point (mm), and w_1 and w_2 are shape coefficients. The maximum retention parameter value, $S_{\max}$, is calculated by solving Eq. 28.3 using CN_1:

$$S_{\max} = 25.4\left(\frac{100}{\mathrm{CN}_1} - 10\right) \tag{28.5}$$

When the retention parameter varies with plant evapotranspiration, the following equation is used to update the retention parameter at the end of every day:

$$S = S_{\mathrm{prev}} + E_{\mathrm{o}}\exp\left(\frac{-\mathrm{cncoef} - S_{\mathrm{prev}}}{S_{\max}}\right) - R_{\mathrm{day}} - Q_{\mathrm{surf}} \tag{28.6}$$

in which S_{prev} is the retention parameter for the previous day (mm), E_o is the potential evapotranspiration for the day (mm/day), cncoef is the weighting coefficient used to calculate the retention coefficient for daily curve number calculations dependent on plant evapotranspiration, $S_{\max}$ is the maximum value the retention parameter can achieve on any given day (mm), R_{day} is the rainfall depth for the day (mm), and Q_{surf} is the surface runoff (mm). The initial value of the retention parameter is defined as $S = 0.9S_{\max}$.

The SCS curve number is a function of the soil's permeability, land use, and antecedent soil water conditions. SCS defines three antecedent moisture conditions: I, dry (wilting point); II, average moisture; and III, wet (field capacity). The moisture condition I curve number is the lowest value the daily curve number can assume in dry conditions. The curve numbers for moisture conditions I and III are calculated with Eqs. 28.7 and 28.8:

$$\mathrm{CN}_1 = \mathrm{CN}_2 - \frac{20 \cdot (100 - CN_2)}{(100 - \mathrm{CN}_2 + \exp[2.533 - 0.0636 \cdot (100 - \mathrm{CN}_2)])} \tag{28.7}$$

$$\mathrm{CN}_3 = \mathrm{CN}_2\exp[0.00673\ (100 - \mathrm{CN}_2)] \tag{28.8}$$

Typical curve numbers for moisture condition II are listed in various tables

(Neitsch et al. 2005). The values are appropriate for a 5 % slope. Williams (1995) developed an equation to adjust the curve number to a different slope:

$$\mathrm{CN}_{2S} = \frac{(\mathrm{CN}_3 - \mathrm{CN}_2)}{3}\,[1 - 2\exp(-13.86 \cdot \mathrm{sl\,p})] + \mathrm{CN}_2 \tag{28.9}$$

in which CN_1 is the moisture condition I curve number, CN_2 is the moisture condition II curve number for the default 5 % slope, CN_3 is the moisture condition III curve number for the default 5 % slope, CN_{2S} is the moisture condition II curve number adjusted for slope, and slp is the average percent slope of the subbasin.

SWAT calculates the peak runoff rate with a modified rational method. There are many methods that are developed to estimate potential evapotranspiration (PET). Three methods are incorporated into SWAT: the Penman-Monteith method (Monteith 1965), the Priestley-Taylor method (Priestley and Taylor 1972), and the Hargreaves method (Hargreaves et al. 1985). For this study, Hargreaves method was used due to limitation of weather data such as wind speed, humidity, and sunshine hours.

The simulation of groundwater is partitioned into two aquifer systems, i.e., an unconfined aquifer (shallow) and a deep-confined aquifer in each subbasin. The unconfined aquifer contributes to flow in the main channel or reach of the subbasin. Water that enters the deep aquifer is assumed to contribute to stream flow outside the watershed (Arnold et al. 1993). In SWAT, the water balance for a shallow aquifer is calculated with Eq. 28.10:

$$\mathrm{aq}_{\mathrm{sh},i} = \mathrm{aq}_{\mathrm{sh},\mathrm{i}-1} + w_{\mathrm{rchrg}} - Q_{\mathrm{gw}} - w_{\mathrm{revap}} - w_{\mathrm{deep}} - w_{\mathrm{pump,sh}} \tag{28.10}$$

in which $\mathrm{aq}_{\mathrm{sh},i}$ is the amount of water stored in the shallow aquifer on day i (mm); $\mathrm{aq}_{\mathrm{sh},i-1}$ is the amount of water stored in the shallow aquifer on day $i-1$ (mm); w_{rchrg} is the amount of recharge entering the aquifer on day i (mm); Q_{gw} is the groundwater flow, or base flow, into the main channel on day i (mm); w_{revap} is the amount of water moving into the soil zone in response to water deficiencies on day i (mm); w_{deep} is the amount of water percolating from the shallow aquifer into the deep aquifer on day i (mm); and $w_{\mathrm{pump,sh}}$ is the amount of water removed from the shallow aquifer by pumping on day i (mm). The steady-state response of groundwater flow to recharge is estimated by Eq. 28.11 (Hooghoudt 1940):

$$Q_{\mathrm{gw}} = \frac{800K_{\mathrm{sat}}}{L_{\mathrm{gw}}} h_{\mathrm{wtbl}} \tag{28.11}$$

in which K_{sat} is the hydraulic conductivity of the aquifer (mm/day), L_{gw} is the distance from the ridge or subbasin divide for the groundwater system to the main channel (m), and h_{wtbl} is the water table height (m). A water table fluctuation due to non-steady-state response of groundwater flow to periodic recharge is calculated by Eq. 28.12 (Smedema and Rycroft 1983):

$$\frac{dh_{\mathrm{wtbl}}}{dt} = \frac{w_{\mathrm{rchrg,sh}} - Q_{\mathrm{gw}}}{800\mu} \tag{28.12}$$

in which $\frac{dh_{\mathrm{wtbl}}}{dt}$ is the change in water table height with time (mm/day), $w_{\mathrm{rchrg,sh}}$ is the amount of recharge entering the aquifer on day i (mm), and μ is the specific yield of the shallow aquifer (m/m). Assuming that variation in groundwater flow is linearly related to the rate of change in water table height, Eqs. 28.11 and 28.12 can be combined to obtain

$$\frac{dQ_{\mathrm{gw}}}{dt} = 10 \cdot \frac{K_{\mathrm{sat}}}{\mu \cdot L_{\mathrm{gw}}^2}\left(w_{\mathrm{rchrg,sh}} - Q_{\mathrm{gw}}\right) = \alpha_{\mathrm{gw}}\left(w_{\mathrm{rchrg,sh}} - Q_{\mathrm{gw}}\right) \tag{28.13}$$

in which α_{gw} is the base flow recession constant or constant of proportionality. The base flow recession constant, α_{gw}, is a direct index of groundwater flow response to changes in recharge (Smedema and Rycroft 1983). α_{gw} varies from 0.1–0.3 for land with slow response to recharge to 0.9–1.0 for land with a rapid response. Although the base flow recession constant may be calculated, the best estimates are obtained by analyzing measured stream flows during periods of no recharge in the watershed.

Routing Phase of the Hydrological Cycle

In SWAT, water is routed though the channels network using either the variable storage-routing or Muskingum River-routing methods. The details of the water-routing methods are discussed in Neitsch et al. (2005).

28.3.2 *Model Inputs*

The spatially distributed data (Geographic Information System (GIS) input) needed for the ArcSWAT interface include the digital elevation model (DEM), soil data, land use, and stream network layers. Data on weather and river discharge were also used for prediction of stream flow and calibration purposes.

28.3.2.1 Digital Elevation Model

A 90-m by 90-m resolution DEM was downloaded from SRTM (Shuttle Radar Topography Mission) website (Jarvis et al. 2006). Topography is defined by a DEM that describes the elevation of any point in a given area at a specific spatial resolution. The DEM was used to delineate the watershed and to analyze the drainage patterns of the land surface terrain. Subbasin parameters such as slope gradient, slope length of the terrain, and the stream network characteristics such as channel slope, length, and width were derived from the DEM.

28.3.2.2 Soil Data

SWAT model requires different soil textural and physicochemical properties such as soil texture, available water content, hydraulic conductivity, bulk density and organic carbon content for different layers of each soil type.

The dominant factors responsible for formation of soils in the Great River basin are the parent material, topography, climate, and the rate of weathering. The major soil types in the Great River basin are described in Table 28.2. More than half of the watershed carries Bonnygate soil (a stony loam classified according to the USDA Soil Taxonomy as an association of Lithic Troporthents and Lithic Ustorthents). These are highly eroded soils with a uniform profile of the AC type, in which the A horizon shows no significant organic matter contents (ochric horizon). Moreover, no diagnostic layers originating from pedogenetic phenomena can be identified in the C Horizon. There is some controversy on the formation of Bonnygate and similar soils types, as they do not exhibit the chemical features of the limestone on which they rest. They have a stony-loam surface layer and bedrock at 5–30-cm depths. They exhibit rapid internal drainage, very low moisture-holding capacity, low natural fertility, and a high erosion hazard, when cultivated.

28.3.2.3 Land Use and Vegetation

Land use is one of the most important factors that affect runoff, evapotranspiration, and surface erosion in a watershed. Land use in the Great River basin is extremely diverse, with a mixture of agriculture, forestry, and small urban and rural settlements. Table 28.3 outlines major land uses in the Great River basin. This information was taken from a map prepared at a 1:250,000 scale.

28.3.2.4 Weather Data

The hydrological model SWAT requires daily meteorological data that could either be read from a measured data set or be generated by a weather-generator model. In this study, the weather variables used for driving the hydrological balance are daily precipitation and minimum and maximum air temperature for the period of 1997–2006. These weather variables were collected from the National Meteorological Service for the period of 1997–2006 for Cacoon Castle, Catadupa, and Montpelier gauge stations.

Table 28.2 Soils of the Great River basin

Soil	Property	Area (ha)	% of area
Valley	Clay	72	0.22
Fontabelle	Clay loam	201	0.62
Sunbury	Clay	7	0.02
Shoothill	Clay	49	0.15
Wirefence	Clay	789	2.41
Pennants	Clay loam	21	0.06
Halifax	Clay loam	20	0.06
Donnington	Clay	124	0.38
Agualta	Gravelly Clay Loam	64	0.20
Cave Belfield	Clay	184	0.56
Highgate	Clay	534	1.63
Marymount	Clay	84	0.26
Hall's Delight	Channery	2	0.01
Llandewey	Clay loam	386	1.18
Bloxburgh	Silt	2	0.01
Newell	Loam	24	0.07
Churdleigh	Clay	2,637	8.06
Lucky Hill	Clay loam	1,378	4.21
Union Hill	Stony loam	1,901	5.81
Bonnygate	Stony loam	17,320	52.93
St. Ann	Clay loam	63	0.19
Bundo	Clay	363	1.11
Seawell	Stony loam	42	0.13
Windsor	Stony loam	721	2.20
Killancholly	Clay	113	0.34
Nonsuch	Clay	10	0.03
Carron Hall	Clay	2,425	7.41
Wait A Bit	Clay	2,193	6.70
Wild Cane	Sandy	80	0.24
Deepdene	Stony loam	130	0.04
St. Toolies	Clay	11	0.03
Donnington	Gravely clay loam	2	0.01
Total		31,952	100

Source: Rapid Rural Appraisal of the Great River Watershed, 2001 prepared by Alicia Hayman

28.3.2.5 River Discharge

The Great River is a well-defined second-order drainage system, with a trunk stream and a number of tributaries. The Great River rises at Pisgah at approximately 381 m above sea level. A number of sub-catchments can be found along the river. The tributaries include Quashies River, Lambs River, Seven River, Brown's River, and Roaring River. These rivers are extremely important, as they are a source of

Table 28.3 Land use in the Great River basin

Land use type	Area (ha)	Area (%)
Banana, coconut, mixed coconut, and improved pastures	548	1.7
Food forest: a mix of fruit trees including banana, pimento, coconut, ackee, breadfruit, coffee, etc., with no clear dominance	2,222	6.8
Pimento and pastures. Areas where pimento trees are used as shade trees and cover 20–50 % of the indicated area	297	0.9
Intensive mixed agriculture. Areas formed by many small fields of onions, cabbages, tomatoes, peppers, etc., in pure or mixed stand	2,911	8.9
Extensive mixed agriculture, crop production is dominant in 15–60 %	5,790	17.7
Improved pastures	7,793	23.8
Unimproved pastures	1,440	4.4
Grassland on steep slopes	63	0.1
Forest of conifers, broadleaf evergreens and various broadleaf species. These forests have a crown closure of 75 % or more	7,073	21.6
Bush	4,663	14.2
Total	*32,800*	*100*

Mercatti (1998), as sited by Alicia Hayman – Rapid Rural Appraisal of the Great River Watershed, 2001

water for many communities. There are a number of springs in the watershed that provide water for many communities including Shettlewood, Mafoota, and Blue Hole. Daily river discharge values for Lethe gauge station were obtained from the Water Resource Authority (WRA 1995).

These daily river discharges at Lethe gauge station were used for model calibration (1999–2002) and validation (2003–2006). Even if there are long years of measured stream flow data at Lethe gauge station, this study only utilized data from 1997 to 2006 due to missing data in the previous years.

28.3.2.6 Model Calibration and Evaluation of Predictions Performance

The calibration and uncertainty analysis were done using sequential uncertainty fitting (SUFI-2) (Abbaspour et al. 2004, 2007). SUFI-2 accounts for several sources of uncertainties including uncertainty in model input variables (e.g., rainfall, conceptual model, parameters, and measured data). The degree, to which uncertainties are accounted for, is quantified by a *p*-factor which is the percentage of measured data bracketed by the 95 % prediction uncertainty (95PPU). The 95PPU is calculated at the 2.5 and 97.5 % levels of the cumulative distribution of an output variable obtained through Latin hypercube sampling method (Abbaspour et al. 2007). Another measure quantifying the strength of a calibration or uncertainty analysis is the *r*-factor which is the average thickness of the 95PPU band divided by the standard deviation of the measured data. The goodness of calibration and prediction uncertainty is judged on the basis of the closeness of the *p*-factor to

100 % (i.e., all observations bracketed by the prediction uncertainty) and the *r*-factor to 1. The average thickness of the 95PPU band ($\overline{r}$) and the *r*-factor are calculated by Equations 14 and 15. The details of the SUFI-2 procedures are described in Yang et al. (2008).

$$\overline{r} = \frac{1}{n}\sum_{t_i}^{n}\left(y^M_{t_i,97.5\ \%} - y^M_{t_i,2.5\ \%}\right) \tag{28.14}$$

$$R\text{-factor} = \frac{\overline{r}}{\sigma_{obs}} \tag{28.15}$$

in which $y^M_{t_i,97.5\ \%}$ and $y^M_{t_i,2.5\ \%}$ represent the upper and lower boundaries of the 95PPU and σ_{obs} is the standard deviation of the measured data.

The other factor is the goodness of fit that can be quantified by the coefficient of determination (R^2) and Nash-Sutcliffe efficiency (NSE) (Nash and Sutcliffe 1970) between the observations and the final best simulations. Coefficient of determination (R^2) and Nash-Sutcliffe coefficient (NSE) are calculated by Eqs. 28.16 and 28.17:

$$R^2 = \frac{\left[\sum_i \left(Q_{\text{m},i} - \overline{Q_{\text{m}}}\right)\left(Q_{\text{s},j} - \overline{Q_{\text{s}}}\right)\right]^2}{\sum_i \left(Q_{\text{m},j} - \overline{Q_{\text{m}}}\right)^2 \sum_i \left(Q_{\text{s},i} - \overline{Q_{\text{s}}}\right)^2} \tag{28.16}$$

$$\text{NSE} = 1 - \frac{\sum_i \left(Q_{\text{m}} - Q_{\text{s}}\right)_i^2}{\sum_i \left(Q_{\text{m},i} - \overline{Q_{\text{m}}}\right)^2} \tag{28.17}$$

in which Q_{m} is the measured discharge, Q_{s} is the simulated discharge, $\overline{Q}_{\text{m}}$ is the average measured discharge, and $\overline{Q}_{\text{s}}$ is the average simulated discharge.

28.4 Results and Discussion

28.4.1 Hydrometeorological Time Series Trend Analysis

28.4.1.1 Climatic Characteristics

Meteorological variables such as rainfall and temperature were collected from the National Meteorological Service of Jamaica for the period of 1997–2006 for Cacoon Castle, Catadupa, and Montpelier gauge stations. Figure 28.3 depicts the

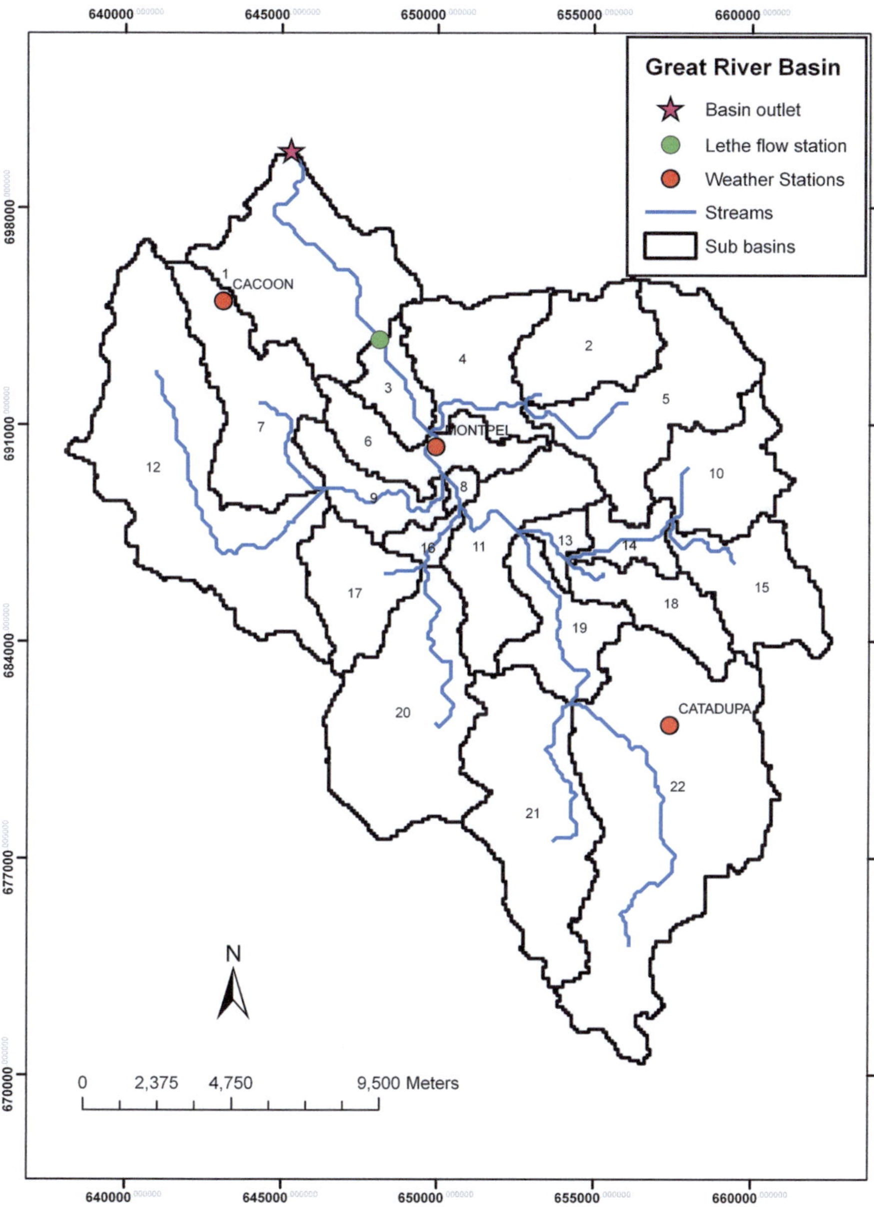

Fig. 28.3 Diagrammatic representation of the Great River basin showing the Great River monitoring station (Lethe), weather stations, stream network, and subbasins

Table 28.4 Average temperature, evaporation, and rainfall for the Great River basin

Climate variable	Jan	Feb	Mar	Apr	May	Jun	Jul	Aug	Sep	Oct	Nov	Dec
Rainfall (mm)	58	48	64	114	229	231	160	160	211	236	188	86
Temperature (°C)	22	22	22.6	23.3	24.2	24.7	25.1	24.6	24.9	24.6	24.0	23.1
Evaporation (mm)	93	95	118	135	133	138	136	115	117	96	93	93

subbasin boundaries, the location of the three weather stations, stream flow gauging station (Lethe), and stream network in Great River basin.

Table 28.4 shows the monthly average rainfall, temperature, and evaporation data for the Great River basin. The record indicated that the high-rainfall period is during May to November. Similarly, the temperature starts to rise from the month of May with the maximum value in July (25.1 C). The minimum rainfall and temperature record in the watershed is in January and February. The amount, intensity, and distribution of rainfall determine the soil moisture content and availability of water for surface runoff in the Great River basin.

The annual distribution of rainfall in the watershed was determined using data collected from the National Meteorological Service for the period of 1997–2000. Based on the analysis of the existing rainfall data in the area, the rainfall trend is noticeably seasonal. There are two rainy seasons in the basin. The smaller rain season is between September to November and the longer and higher rain season is from May to June. Due to the climate variability and other factors, these seasons are sometimes not uniform. Annual rainfall for the watershed ranged from 2,162 mm in Cacoon Castle to 1,032 mm at St. Leonard's.

28.4.1.2 Analysis of Long-Period River Discharge

The mean annual discharge was estimated based on the 52 years of (1955–2007) observed stream flow data at Lethe gauge station. The maximum daily discharge recorded within this 53-year span is 481.7 m^3/s that was in September 2002. The minimum daily discharge was less than 1 m^3/s that was observed in February 2006. The monthly average discharge was analyzed based on the data from 1997 to 2007. The result indicated that the highest and lowest average monthly discharge were 74.5 and 1.0 m^3/s for September 2004 and February 2006, respectively. Figure 28.4 shows the monthly mean stream flow for the Lethe gauge station of Great River flow. The maximum stream flow occurs in the months of September and October and the minimum flow is observed in the month of March. One monitoring station in the watershed does not allow accurate analysis of the temporal and spatial variations of stream flow in the watershed. There is a need for other flow stations.

28.4.2 Parameter Sensitivity Analysis

The ability of a watershed model to sufficiently predict stream flow in a river basin is evaluated through sensitivity analysis, model calibration, and model validation. Before the calibration process, sensitivity analysis was conducted to determine the most sensitive flow parameters on prediction of stream flow and other hydrological components. The method in the ArcSWAT interface combines the Latin hypercube (LH) and one-factor-at-a-time (OAT) sampling (Van Griensven and Meixner 2006). Sensitivity analysis evaluates how different parameters influence a predicted

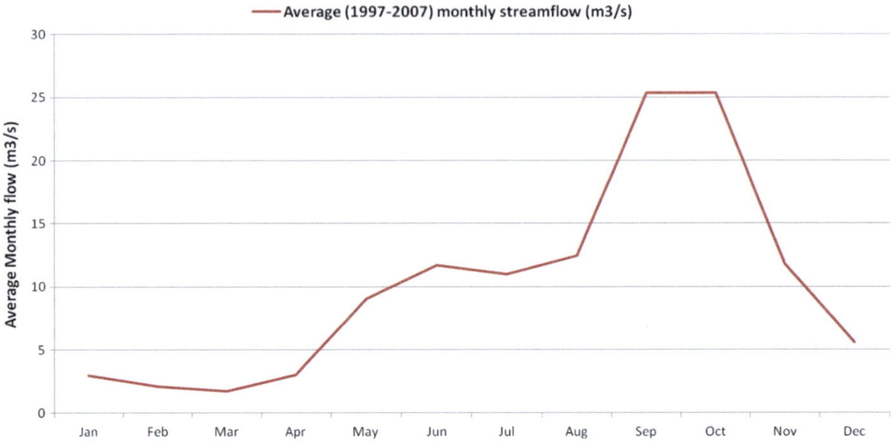

Fig. 28.4 Average monthly Great River flow at Lethe gauge station (1997–2007)

output. It is measured as the response of an output variable to a change in an input parameter. The greater the change in output response, the greater will be the sensitivity of a parameter. The parameters that significantly influence predicted outputs are used to calibrate the model. In this analysis the most sensitive parameters were curve number (CN2), base flow alpha factor (Alpha_Bf) [days], soil evaporation compensation factor (ESCO), available water capacity (Sol_Awc) [mm water/mm soil], threshold depth of water in the shallow aquifer for "revap" to occur (REVAPMN.gw) [mm H_2O] [days], groundwater "revap" coefficient (Gw_Revap), channel effective hydraulic conductivity (Ch_K2) [mm/h], and threshold depth of water in the shallow aquifer for return flow to occur (GWQMN.gw) [mm H_2O]. These sensitive parameters were considered for model calibration.

28.4.3 Model Calibration and Validation

In this study, the data from 1999 to 2002 were used for calibration and from 2003 to 2006 were used for validation of the model at Lethe gauge station. Periods 1997–1998 were used as "warm-up" periods for calibration purposes. The warm-up period allows the model to get the hydrological cycle fully operational. Calibration of hydrological models requires the alteration of parameter values, simulation, and comparison of predicted output of interest to measured data until a defined objective function is achieved.

The comparison between the observed and calibrated flow discharge values for calibration period indicated that there is a good agreement between the observed and simulated flows with values of coefficient of determination and Nash-Sutcliffe efficiency (NSE) greater than 0.5. Different researchers suggested that model

Table 28.5 Comparison of annual and monthly stream flow simulations at Lethe gauge station

		Goodness of fit measures Jamaica	
Simulation period	R^2	NSE	
Calibration (1997–2000)	Annual	0.85	0.75
	Monthly	0.75	0.68
Validation (2001–2004)	Annual	0.81	0.71
	Monthly	0.72	0.66

simulation can be judged as satisfactory if R^2 is greater than 0.6 and NSE is greater than 0.5. Hence, our results agree reasonably well with the recommended values. The model validation results indicated generally a similar relationship between measured flow and predicted output to calibration results. There is a good agreement between measured and predicted outputs on an annual and monthly scale. Calibrated and validated model predictive performances for the Great River basin are summarized in Table 28.5. The NSE and R^2 evaluation coefficients are relatively lower on a monthly temporal scale than an annual scale, which may be attributable to seasonal variation in evapotranspiration and soil moisture conditions.

A good agreement shows that the model parameters represent the processes occurring in the basin. They may also be used to predict watershed response for various outputs. Figure 28.5 shows the time series of observed and simulated monthly flow for calibration and validation period at Lethe gauge station.

The model prediction efficiency was also evaluated based on comparison of annual means and standard deviation for each year of simulation with the purpose of describing the similarity correlation of those two variables. Table 28.6 includes statistical comparisons of long-term means and standard deviation for each year of simulation. The annual mean of the daily discharge for each simulation year indicated that there is a good agreement between the simulated and observed stream flow. The overall 9-year annual averages for observed and simulated stream flow are 10.9 and 11.1 m^3/s, respectively.

The annual average observed flow volume in million cubic meters (MCM) was also compared with simulated flow. The annual average observed flow volume was 343.7 MCM and the predicted flow volume was 350 MCM. This analysis indicated that in most of the simulation years, the model reasonably predicted the stream flow.

28.4.4 Uncertainties in Modeling

An important issue to consider in the field of hydrology, sediment yield, and water quality is uncertainties in the predictions. The main sources of uncertainty are simplifications in the conceptual model; processes occurring in the watershed but not included in the model; processes that are included in the model, but their occurrences in the watershed are unknown to the modeler, for example, reservoirs,

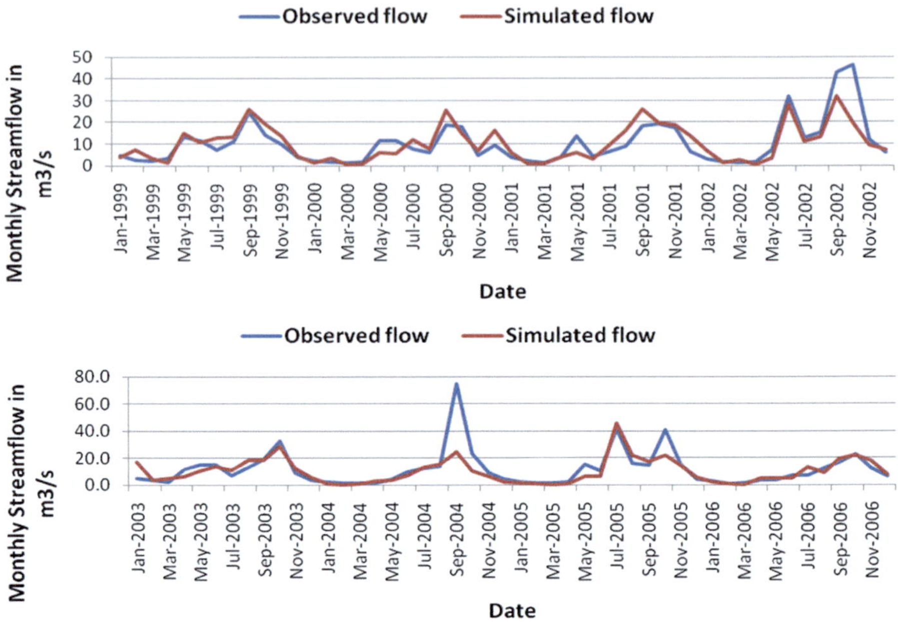

Fig. 28.5 Time series of observed and simulated monthly flow for calibration (*top*) and validation (*bottom*) period at Lethe gauge station

Table 28.6 Comparison of daily annual average observed and simulated stream flow at Lethe gauge station, Great River

Year	Annual total rainfall (mm)	Average daily stream flow (m^3/s)		Standard deviation (m^3/s)		Annual average flow volume in million cubic meter (MCM)	
		Observed	Simulated	Observed	Simulated	Observed discharge	Simulated discharge
1999	2,004	8.9	11.9	6.5	6.4	280.7	375.3
2000	1,827	7.8	9.2	6.1	7.0	246.0	290.1
2001	1,963	8.8	11.3	6.6	7.6	277.5	356.4
2002	1,892	15.2	12.3	16.2	9.4	479.3	387.9
2003	2,175	11.4	13.7	8.6	6.6	359.5	432.0
2004	1,592	13.0	7.8	20.4	6.5	410.0	246.0
2005	2,165	13.7	13.1	14.1	12.5	432.0	413.1
2006	1,805	7.9	9.6	6.7	6.6	249.1	302.7
Basin average	1,901	10.9	11.1	10.7	7.8	343.7	350.0

water diversions, or irrigation; processes that are not known to the modeler and not included in the model; or processes that may last for a number of years and drastically change the hydrology or water quality such as constructions of roads, bridges, tunnels and dams, and errors in the input variables such as rainfall and temperature.

In SUFI-2 the degree, to which uncertainties are accounted for, is quantified by a *p*-factor which is the percentage of measured data bracketed by the 95 % prediction uncertainty (95PPU). The 95PPU represents a combined model prediction uncertainty including parameter uncertainties resulting from the nonuniqueness of effective model parameters, conceptual model uncertainties, and input (i.e., rainfall) uncertainties. The combined effect of all uncertainties is depicted by the final estimates of parameter uncertainties. In this study, the *p*-factor brackets 73 % of the observation and *r*-factor equals 0.71 for the Great River basin. Figure 28.6 shows monthly calibration results for the Great River at Lethe gauge station showing the 95 % prediction uncertainty intervals along with the measured discharge. There are some uncertainties in the prediction due to errors in input data such as rainfall and temperature and/or other sources of uncertainties such as upstream water diversions and other unknown activities in the subbasins. There are also possibilities of model prediction uncertainties due to the methods used for estimation of the different hydrological components. For example, in this study we have used the Hargreaves method to calculate evapotranspiration that depends on minimum and maximum temperatures. Hargreaves method does not include the effect of other climate variables such as wind on evapotranspiration. In cases where other climate variables are predominating factors in a particular region, the method can introduce errors.

28.4.5 Hydrological Water Balance of Great River Basin

Hydrological water balance analysis can be used to predict the existing water resources component that can help manage water availability and predict where and when there may be water shortages. The main water balance components of the Great River basin includes the total amount of precipitation falling on the subbasin during the time step, actual evapotranspiration from the basin, and the net amount of water that leaves the basin and contributes to stream flow in the reach (water yield). The water yield includes surface runoff contribution to stream flow, lateral flow contribution to stream flow (water flowing laterally within the soil profile that enters the main channel), and groundwater contribution to stream flow (water from the shallow aquifer that returns to the reach) minus the transmission losses (water lost from tributary channels in the hydrological response units via transmission through the bed and becomes recharge for the shallow aquifer during the time step). Table 28.7 lists the simulated annual average water balance components for the Great River basin. The results indicated that above 50 % of the annual precipitation is lost by evapotranspiration in the basin during the model simulation period (including both calibration and validation period). Surface runoff contributes more than 28 %, whereas the groundwater contributes more than 18 %. The total groundwater recharge accounts about 20 %.

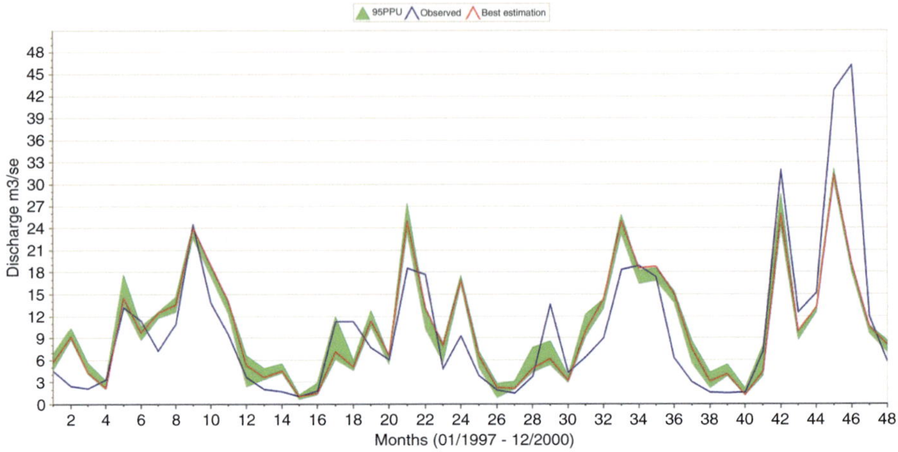

Fig. 28.6 Monthly calibration results for Great River at Lethe gauge station showing the 95 % prediction uncertainty intervals along with the measured discharge

Table 28.7 Annual average basin water balance of the Great River basin. This annual average is based on 9 years of simulation (1997–2006)

Water balance component	Annual average (mm)
Precipitation	1,906.2
Surface runoff	545.3
Lateral soil	45.61
Groundwater (shallow aquifer)	358.8
Revap (shallow aquifer $\geq$ soil/plants)	17.9
Deep aquifer recharge	19.80
Total aquifer recharge	396.1
Total water yield	948.2
Actual evapotranspiration	961.6
Potential evapotranspiration	1,512.0
Transmission losses	1.5

The monthly water balance of the Great River basin was analyzed based on the 9 years (1997–2006) of simulation. There is a high correlation between rainfall and surface runoff in the Great River basin. The highest rainfall month in the basin is September with an average value of 301 mm. This high rainfall results in high surface runoff in the basin which contributes for peak discharge in the river flow system.

Table 28.8 shows the annual monthly water balance of the Great River basin. The lowest rainfall month is February with a rainfall amount of 45 mm and surface runoff of only 10.4 mm. The highest rainfall is observed from May to October that ranges from 187 mm to a maximum of 305 mm. Similarly the highest surface runoff, which is the actual amount of surface water flowing into the river, was

Table 28.8 Annual monthly water balance of the Great River basin. This monthly average is based on 9 years of simulation (1997–2006)

Months	Rainfall (mm)	Surface runoff (mm)	Lateral flow (mm)	Water yield (mm)	AET (mm)	PET (mm)
Jan	65.5	22.8	1.7	41.2	40.7	94.3
Feb	45.1	10.4	1.3	18.7	55.1	106.3
Mar	67.7	9.0	1.4	12.9	116.5	131.7
Apr	155.3	17.2	2.4	20.9	129.8	137.3
May	249.3	40.6	4.9	53.2	114.3	148.5
Jun	186.5	48.0	4.6	73.7	79.1	151.4
Jul	245.0	83.9	5.2	122.8	87.2	155.0
Aug	213.8	60.2	5.4	112.7	87.0	151.5
Sep	305.5	116.8	6.8	181.9	86.0	135.2
Oct	214.0	75.8	6.4	154.1	76.0	120.2
Nov	114.0	33.7	3.5	93.6	50.7	95.1
Dec	85.8	27.1	2.1	63.0	40.5	88.2

observed from May to October with a peak flow in September. Low runoff occurs in February and March. During the wet seasons, runoff is estimated to be above 38 % of rainfall. The highest evapotranspiration is observed from March to May with a peak loss in April. In the wet months, evapotranspiration is considerably higher due to the fact that soil moisture available to plants is higher.

28.4.6 Water Balance Components for Wet and Dry Years

The water balance of the Great River basin was analyzed annually to understand the prediction performance of the SWAT model for different rainfall conditions. The study compared the annual average rainfall and other hydrological components for each year of the calibration and validation periods. Table 28.9 shows that year 2004 was a relatively "dry" year and 1999 was a wet year as compared to the 9 years of simulation period (calibration and validation period). The use of the term "dry" is relative as the annual total rainfall is greater than 1,500 mm. The wet years produce a larger water yield than the dry years. The water fluxes indicate that in a wet year surface runoff dominates water yield which is the total amount of water leaving the HRU and entering the main channel during the time step. However, in a dry year, lateral flow contribution makes up a larger part of the water yield.

Table 28.9 Water balance of the Great River for each year of simulation period (1999–2006)

Water balance component	1999	2000	2001	2002	2003	2004	2005	2006
Precipitation	2,036.2	1,816.1	1,978.9	1,968.1	2,196.1	1,600.4	2,129.3	1,845.7
Surface runoff	562.1	467.5	530.1	610.3	690.7	374.0	690.8	436.4
Lateral soil	48.8	37.9	44.3	54.9	53.0	35.1	41.7	43.7
Groundwater (shallow aquifer)	402.8	285.9	394.8	382.6	425.2	256.2	384.4	336.8
Revap (shallow aquifer $\geq$ soil/plants)	20.3	18.5	17.9	16.2	21.9	13.8	18.0	16.4
Deep aquifer recharge	21.9	16.4	22.2	20.6	23.1	14.3	21.3	18.8
Total aquifer recharge	437.0	327.2	444.0	411.8	461.1	285.2	425.2	375.7
Total water yield	1,012.1	789.9	967.6	1,046.5	1,167.3	664.1	1,115.3	815.5
Percolation out of soil	435.2	326.2	442.5	410.3	459.2	284.2	423.6	374.2
Actual ET	1,008.9	946.9	947.3	913.3	1,008.5	890.3	974.0	1,005.0
Potential ET	1,549.0	1,508.8	1,492.7	1,510.6	1,545.2	1,488.3	1,471.1	1,530.0
Transmission losses	1.6	1.3	1.6	1.3	1.6	1.2	1.6	1.4

28.4.7 Spatial Distribution of Annual Hydrological Processes in the Great River Basin

The hydrological water balance characteristics were analyzed for each subbasins aiming at analyzing the spatial distribution of hydrological processes in the Great River basin. Water balance components for each sub-watershed, which shows spatial distribution of annual hydrological processes in the Great River basin, are presented in Table 28.10. Great River basin was subdivided in to 22 subbasins using the 90-m resolution digital elevation model and stream network. The delineation of subbasins was done using ArcSWAT interface. The smallest is subbasin 8 (1 km^2) and the largest is subbasin 22 (56 km^2). According to the result presented in Table 28.10, the water balance components differ spatially between each subbasin. The actual evapotranspiration varies between subbasins which range from 887 mm to a maximum of 1,034 mm. Subbasin 1, 7, 12, and 14 are those subbasins which lost the highest amount of water (1,001, 1,027, 1,034, and 1,012 mm, respectively) through evapotranspiration. This is mainly due to the land cover type of those subbasins. In these subbasins, the majority of the land is covered by mixed herbs/shrubs, subsistence plantation, and grassland. The land use characteristics might play a great role for the loss of water through ET (evapotranspiration). Subbasins 16, 15, 4, 18, and 9 are covered by disturbed lowland seasonal evergreen forest. This type of vegetation usually needs more water for their consumptive use. Moreover, the canopy cover might contribute the extent of water loss through evapotranspiration. Slope of the land might also play a role for spatial variation of water balance comments. The groundwater flow varies from 190 mm in subbasin 8 to 921 mm in subbasin 15. The spatial variations in the groundwater flow may depend on the size of the area, the topographic condition, the vegetation cover, and the geology of the area.

28.5 Integrating Water Resource Management and Environmental Public Health

Understanding the spatiotemporal variability of a watershed system is an important step in understanding the human and ecological impact on the environment and public health. Degradation of the water quantity and quality significantly affects the public in different parts of the world. Especially in the Caribbean islands, availability and quality of freshwater highly affect the health status of communities in the coastal regions. The impact can affect the tourism and other industries in the countries. Intensification of the agricultural sector and industrialization play a major role in environmental degradation. Changes in land use and climate can significantly affect the ecosystem and livelihood. Nutrients and chemicals used in the agricultural areas can significantly affect population living in the surroundings. Therefore, proper watershed and water resource management will significantly

Table 28.10 Water balance components and spatial distribution of annual hydrological processes in the Great River basin

Subbasin	Area (km)^2	Rainfall (mm)	AET (mm)	SW (mm)	Percolation (mm)	Surface runoff (mm)	Groundwater (mm)	Water yield (mm)
1	28	2,084	1,001	279	325	737	294	1,053
2	13	1,595	924	275	241	424	216	646
3	6	1,595	927	275	245	412	219	643
4	15	1,595	887	253	212	491	190	686
5	23	1,595	924	275	239	424	214	646
6	11	1,595	919	274	227	445	203	653
7	20	2,084	1,027	249	335	715	302	1,026
8	1	1,595	921	275	235	439	210	650
9	8	1,595	917	274	223	453	200	655
10	13	2,104	990	260	524	535	478	1,067
11	16	1,595	925	275	242	421	216	645
12	46	2,084	1,034	232	324	720	292	1,019
13	3	1,595	931	276	255	395	228	638
14	5	2,104	1,012	278	438	627	397	1,049
15	14	2,104	870	164	993	14	921	1,162
16	2	1,595	847	228	187	560	168	729
17	12	1,595	922	275	236	432	212	649
18	9	2,104	912	202	634	330	584	1,140
19	11	2,104	979	261	390	722	353	1,086
20	29	2,104	991	276	398	701	361	1,072
21	29	2,104	988	276	388	710	351	1,076
22	56	2,104	937	227	611	408	561	1,116

contribute for a sustainable environmental health. Environmental health aspects should be considered in planning and designing an integrated water resource management plan.

28.6 Conclusion

The ability of SWAT model to sufficiently predict stream flow in the Great River basin was evaluated through sensitivity analysis, model calibration, and validation. The parameters that significantly influence predicted outputs are used to calibrate a model. The three most sensitive parameters that were given careful consideration during model calibration were curve number (CN2), soil evaporation compensation factor (ESCO), and base flow alpha factor (Alpha_Bf) [days]. The comparison between the observed and simulated discharge for calibration and validation period indicated that there is a good agreement between the observed and simulated flows with higher values of coefficient of determination and Nash-Sutcliffe efficiency (NSE). Finally, a good agreement in prediction shows that the model parameters represent the processes occurring in the basin. The calibrated model can be used to predict watershed response to climate and land use changes.

An important issue to consider in the prediction of hydrology, sediment yield, and water quality is uncertainties in the predictions. In SUFI-2, the combined effect of all uncertainties is depicted by the final estimates of parameter uncertainties. In this study, the p-factor brackets 73 % of the observation and the r-factor equals 0.71 for the Great River basin. The basin water balance analysis indicated that above 50 % of the annual precipitation is lost by evapotranspiration in the basin. Surface runoff contributes more than 28 %, whereas the groundwater contributes more than 18 %. The total groundwater recharge accounts about 20 %.

There is a high correlation between rainfall and surface runoff in the Great River basin. The higher rainfall in September resulted in high surface runoff that contributes for peak discharge in the river flow system. During the wet seasons, runoff is estimated to be above 38 % of rainfall.

The highest evapotranspiration is observed from March to May with a peak loss in April. In the wet months, evapotranspiration is considerably higher due to the fact that soil moisture available to plants is higher. In a wet year, surface runoff dominates water yield. However, in dry year, lateral flow contribution makes up a larger part of the water yield.

The spatial distribution of hydrological components indicated that the actual evapotranspiration varies between subbasins which range from 887 mm to a maximum of 1,034 mm. The variation in evapotranspiration between subbasins is mainly due to variations in land cover. Moreover the canopy cover might contribute to the extent of water loss through evapotranspiration. The spatial variations in the groundwater flow may depend on the size of the area, the topographic condition, the vegetation cover, and the geology of the area.

Hydrological water balance analysis can be used to predict the existing water resource component that can help manage water availability and predict where and when there will be water shortages. The output of water balance study can be used in irrigation potential assessment, runoff assessment, flood control, and pollution control.

One monitoring station in the watershed does not allow accurate analysis of the temporal and spatial variations of stream flow in the watershed. There is a need for other flow stations especially where water is diverted from the Great River.

Acknowledgment We are grateful to the Inter-American Institute for Global Change Research (IAI) for supporting the study, the Water Resources Authority of Jamaica for providing the hydrological data, and the National Meteorological Service of Jamaica for providing the necessary weather data.

References

Abbaspour KC, Johnson CA, van Genuchten MT (2004) Estimating uncertain flow and transport parameters using a sequential uncertainty fitting procedure. Vadose Zone J 3(4):1340–1352

Abbaspour KC, Yang J, Maximov I, Siber R, Bogner K, Mieleitner J, Zobrist J, Srinivasan R (2007) Modelling hydrology and water quality in the pre-alpine/alpine Thur watershed using SWAT. J Hydrol 333:413–430

Andersson L, Wilk J, Todd MC, Hughes DA, Earle A, Kniveton D, Layberry R, Savenije HHG (2006) Impact of climate change and development scenarios on flow patterns in the Okavango River. J Hydrol 331(1):43–57

Arnold JG, Allen PM, Bernhardt G (1993) A comprehensive surface groundwater flow model. J Hydrol 142:47–69

Arnold JG, Srinivasan R, Muttiah RR, Williams JR (1998) Large area hydrologic modeling and assessment part I: model development. J Am Water Resour Assoc 34(1):73–89

Claessens L, Hopkinson C, Rastetter E, Vallino J (2006) Effect of historical changes in land use and climate on the water budget of an urbanizing watershed. Water Resour Res 42(3)

Debele B, Srinivasan R, Yves Parlange J (2006) Coupling upland watershed and downstream water body hydrodynamic and water quality models (SWAT and CE-QUAL-2) for better water resources management in complex river basins. Environ Model Assess

Gassman PW, Reyes MR, Green CH, Arnold JG (2007) The soil and water assessment tool: historical development, applications, and future research directions. Trans ASABE 50(4): 1211–1250

Green WH, Ampt GA (1911) Studies on soil physics, 1. The flow of air and water through soils. J Agric Sci 4:11–24

Grønsten HA, Lundekvam H (2006) Prediction of surface runoff and soil loss in south eastern Norway using the WEPP Hillslope model. Soil Tillage Res 2006(85):186–199

Hargreaves GL, Hargreaves GH, Riley JP (1985) Agricultural benefits for Senegal River basin. J Irrig Drain Eng 111(2):113–124

Hayman A (2001) Rapid rural appraisal of the Great River Watershed, 2001. Prepared for the: Government of Jamaica's National Environment and Planning Agency and the United States Agency for International Development. Implemented by: Associates in Rural Development, Inc., unpublished document

Hooghoudt SB (1940) Bijdrage tot de kennis van enige natuurkundige grootheden van de grond. Versl Landbouwkd Onderz 46:515–707

Huang YF, Zou Y, Huang GH, Maqsood I, Chakma A (2005) Flood vulnerability to climate change through hydrological modeling – a case study of the swift current creek watershed in western Canada. Water Int 30(1):31–39

Jarvis A, Reuter HI, Nelson A, Guevara E (2006) Hole-filled seamless SRTM data V3. International Centre for Tropical Agriculture (CIAT). Available from http://srtm.csi.cgiar.org

Monteith JL (1965) Evaporation and the environment. In: The state and movement of water in living organisims, XIXth symposium. Soc For Exp Biol. Cambridge University Press, Swansea, pp 205–234

Morgan RPC (2001) A simple approach to soil loss prediction: a revised Morgan–Morgan–Finney model. Catena 44:305–322

Nash JE, Sutcliffe JV (1970) River flow forecasting through conceptual models. Part I – a discussion of principles. J Hydrol 10:282–290

Neitsch SL, Arnold JG, Kiniry JR, Williams JR (2005) Soil and water assessment tool, theoretical documentation: Version. USDA Agricultural Research Service and Texas A&M Blackland Research Center, Temple.

Priestley CHB, Taylor RJ (1972) On the assessment of surface heat flux and evaporation using large-scale parameters. Mon Weather Rev 100:81–92

Santhi C, Srinivasan R, Arnold JG, Williams JR (2006) A modeling approach to evaluate the impacts of water quality management plans implemented in a watershed in Texas. Environ Model Softw 2006(21):1141–1157

Setegn SG, Srinivasan R, Melesse AM, Dargahi B (2009) SWAT model application and prediction uncertainty analysis in the Lake Tana Basin, Ethiopia. Hydrol Process 24(3):357–367

Setegn SG, Dargahi B, Srinivasan R, Melesse AM (2010) Modeling of sediment yield from Anjeni-Gauged watershed, Ethiopia using SWAT model. J Am Water Resour Assoc 46(3): 514–526

Setegn SG, Rayner D, Melesse AM, Dargahi B, Srinivasan R (2011) Impact of climate change on the hydroclimatology of Lake Tana Basin, Ethiopia. Water Resour Res 47, W04511. doi:10.1029/2010WR009248

Sheng XB, Sun JZ, Liu YX (2003) Effect of land-use and land-cover change on nutrients in soil in Bashang area, China. J Environ Sci (China) 15(4):548–553

Smedema LK, Rycroft DW (1983) Land drainage-planning and design of agricultural drainage systems. Cornell University Press, Ithaca

Srinivasan R, Ramanarayanan TS, Arnold JG, Bednarz ST (1998) Large area hydrologic modeling and assessment. Part II: model application. J Am Water Resour Assoc 34(1):91–101

USDA Soil Conservation Service (SCS) (1972) USDA soil conservation service. National engineering handbook section 4 hydrology, Chapters 4–10

Van Griensven A, Meixner T (2006) Methods to quantify and identify the sources of uncertainty for river basin water quality models. Water Sci Technol 53(1):51–59

Water Resources Authority (1995) Jamaica: watershed management unit selection. Water Resources Authority, Kingston, 32p

Williams JR (1995) Chapter 25: The EPIC model. In: Singh VP (ed) Computer models of watershed hydrology. Water Resources Publications, Highlands Ranch, pp 909–1000

Wu W, Hall CAS, Scatena FN (2007) Modelling the impact of recent land-cover changes on the stream flows in northeastern Puerto Rico. Hydrol Process 21:2944–2956

Yang J, Reichert P, Abbaspour KC, Xia J, Yang H (2008) Comparing uncertainty analysis techniques for a SWAT application to the Chaohe Basin in China. J Hydrol 358(1–2):1–23

Zhang X, Srinivasan R, Hao F (2007) Predicting hydrologic response to climate change in the Luohe River basin using the SWAT model. Trans ASABE 50(3):901–910

Chapter 29
Rainfall-Runoff Modelling for Sustainable Water Resources Management: SWAT Model Review in Australia

Partha Pratim Saha and Ketema Zeleke

Abstract Water is considered as a vital resource for survival, development and ecological needs. In a world of increasing water demand, planning for a sustainable system satisfying the demands of present and future without degradation of the ecosystem is a major challenge. Being the driest inhabited continent, Australia is not without this challenge and uncertainty in future water availability in the scenario of climate change is creating more pressure. Hydrological models are a useful tool due to the capability of simulating the past and future scenarios of a water management system to assess the balance between human and environmental demands. The physically based semi-distributed hydrological model Soil and Water Assessment Tool (SWAT) is such a tool that is capable of simulating a wide range of hydrological processes with different management scenarios. Although the number of SWAT applications in Australia is limited compared to other regions of the world, there are some important and diverse applications, from simple water balance assessments to complex environmental policy and water market evaluation. The review of SWAT applications in Australia revealed that despite several limitations, the model has a promising scope to explain Australia's hydrology in the context of the sustainability of water resources management.

Keywords Australia • Hydrological modelling • Sustainability • SWAT • Water resources management

P.P. Saha
School of Environmental Sciences, Charles Sturt University, Wagga Wagga, NSW, Australia

K. Zeleke (✉)
Graham Centre for Agricultural Innovation, Charles Sturt University, Wagga Wagga, NSW, Australia

School of Agricultural and Wine Sciences, Charles Sturt University, Wagga Wagga, NSW, Australia
e-mail: kzeleke@csu.edu.au

S.G. Setegn, M.C. Donoso (eds.), *Sustainability of Integrated Water Resources Management*, DOI 10.1007/978-3-319-12194-9_29

29.1 Introduction

Water is the primary resource for life and its availability is essential for survival and socioeconomic development (Milburn 2006). Water scarcity is a fundamental barrier for both social and economic growth of many countries and regions, whereas one-third of the world's population is expected to live in significant water-stressed basins by 2030 (Addams et al. 2009). Less than 3 % of the available water is nonsaline, and 99 % of these freshwaters are not freely accessible by humans as they are either icecaps, glaciers or stored in aquifers as groundwater resources (Gleick 1996). River system stores significant amount (2,000 km^3) of the world's freshwater, and importantly, more than 20 times (45,000 km^3) of water of its storage flows annually through the rivers to the oceans which not only surpasses the annual withdrawal by humans (3,800 km^3) but also 10 % of it (4,500 km^3) replenishes the groundwater (Church 1996; Oelkers et al. 2011; Oki and Kanae 2006). Rainfall is another source of freshwater which supplies 80 % of the water for agricultural production worldwide (Oelkers et al. 2011). Water is a renewable resource and the world's total average available water is 10 times higher than the current per capita use, but still water is a scarce resource for many parts of the world due to unbalanced distribution of rainfall across the globe (Oelkers et al. 2011). Uneven distribution of water resources coupled with uneven population growth forced 35 % of the world's population to live in chronic water shortage ($<$1,000 m^3/capita/year) in 2005, whereas the percentages were 9 % and 2 % in 1960 and 1900, respectively (Kummu et al. 2010).

Water is one of the fundamental resources for development, but uneven distribution of available freshwater and uncontrolled use of both surface and groundwater are posing a threat not only to the future availability for development work but also to the sustainability of a balanced ecosystem. The global water use increased by four times from 1940 to 1990 which has been increasing at a rate more than twice of the population growth rate in the last century (FAO 2013). About 70 % of the world's freshwater is used by agriculture followed by industry (20 %) and domestic use (10 %) (UNWATER 2009). All levels of users have competitions regarding appropriate allocation, and this crisis is expected to increase with growing water demand in almost all countries (UNWATER 2012). Uncontrolled extraction of both the surface and groundwater already poses a threat to the future water availability, and water withdrawals are predicted to increase by 50 % in developing countries and 18 % in developed countries by 2025 (UNWATER 2012). Unless there is improvement in water use efficiency, development might be hampered by the shortage of available freshwater.

The combination of uneven distribution of population and freshwater availability forces many people to survive currently in water-stressed conditions, and the water demand is expected to increase with population growth unless everyone finds ways to conserve and recycle the available water resources (Foster and Chilton 2003). Ever-increasing demand poses more and more pressure on water managers, both in government and private sectors, in water allocation (UNDESA 2013), whereas predictions of uncertain water availability due to climate change will make the situation more critical. Water is one of the vital natural resources, and appropriate allocation of water required by different sectors is critical. Apart from different sectorial needs, there is also a wide recognition that rivers, lakes and wetlands have their own water needs, known as environmental flow, which is essential for the functioning ecosystem and world's biodiversity (Acreman and Dunbar 2004). The traditional fragmented approach to consider the water issues individually is no longer viable in the current situation, and a more holistic approach is essential to deal with the critical water allocation issues for better water management (UNDESA 2013). Under these circumstances, the concept of integrated water resources management (IWRM) was introduced in the World Summit on Sustainable Development (WSSD) in 1992 in Rio de Janeiro, Brazil. Global water partnership (GWP) defines IWRM as 'a process which promotes the coordinated development and management of water, land and related resources in order to maximize economic and social welfare in an equitable manner without compromising the sustainability of vital ecosystems and the environment' (Global Water Partnership 2013). Since its introduction, the IWRM concept has been adopted widely by water managers, decision-makers and politicians around the world, and out of the 95 countries examined in 4th World Water Forum in Mexico (2006), 74 % either had IWRM strategies in place or had initiated a process for the formulation of such strategies (Hassing et al. 2009). Along with IWRM, green growth concepts are also introduced to maintain a balance between economic development and environmental ecosystem which is essential for long-term sustainability of all developments in the water resources sector. Sustainable water resources can be defined as a 'water resources system designed and managed to fully contribute to the objectives of society, now and in the future, while maintaining their ecological, environmental and hydrological integrity' (Loucks 2000). Water resources should not only be dealt with in an integrated way but also its sustainability should be considered to ensure long-term effectiveness.

Special attention is essential to maintain the balance between different sectorial water demands including the ecological need which is the key to sustainable water resources management. The hydrological model is a suitable tool for modelling the water resources system to address such criteria. Continuous monitoring of a water resources system through a hydrological model provides necessary information that needs to be addressed for future sustainability. Among the available hydrological models, the Soil and Water Assessment Tool (SWAT), a physically based distributed hydrological model, has proved its efficiency to assess different hydrological problems for a broad range of scales and environmental conditions across the globe (Gassman et al. 2007). Like other regions of the world, achieving the goal of sustainable water management is a major challenge for Australia. About 70 % of Australia is semiarid to arid which has a unique ecosystem with low and erratic rainfall (Pavey and Nano 2006). Achieving a sustainable future for such environments requires adequate understanding on the complex interaction between arid zone ecosystems, biodiversity, land use and planning options (CSIRO 2011). These complex circumstances coupled with the climate change scenario make the situation more challenging to achieve the goal of sustainable water management. This paper reviews the limited SWAT applications in Australia and its scope as a hydrological modelling tool for sustainable water resources management.

29.2 Water Resources in Australia

Australia is the driest inhabited continent (DFAT 2008). More than 80 % of the country has an average annual rainfall below 600 mm, and 50 % of those receive less than 200 mm (ABS 2013a). About 90 % of this rainfall is lost through evapotranspiration (ET), and most of it disappears from the atmosphere leaving a small amount for the terrestrial phase of the water balance (Kollmorgen et al. 2007). With an average of only 45 mm, Australian runoff is the lowest among all the continents which is only one-fourth of Africa, one-seventh of Asia, Europe and North America and one-fourteenth of South America (Kollmorgen et al. 2007). Australian rivers show high variability in flows which is about double of those of the rest of the world (Chiew 2011; Kollmorgen et al. 2007).

The agriculture industry is the largest user of water in Australia. It consumed 59 % in 2011–2012, whereas water supply and sewerage, households, mining and manufacturing industry consumed 13 %, 11 %, 4 % and 3 %, respectively, during the same period (ABS 2013b). High variation in water availability always keeps Australian farmers in immense pressure and uncertainty. In 2004–2005, only 6 % of Australia's total runoff comes from the Murray-Darling basin which is the major basin for agricultural activity containing 70 % of Australia's irrigated crops and pastures and consumes more than 50 % of total water use (Kollmorgen et al. 2007). Irrigated crops and pastures use 90 % of the total water used for agriculture (Bureau of Meteorology 2012). Agricultural water use varies from year to year depending on the water availability. Records show that agricultural water use in NSW during 2011–2012 was double of the 2008–2009 period (ABS 2013b).

The Murray-Darling basin (MDB), one of the world's largest and driest river systems, is the largest river and most iconic river basin of Australia which covers 1.06 million km^2 or about 14 % of Australia (ABS 2008). The basin is a key resource and its sustainability is highly important due to its immense contribution to both the economy and biodiversity of Australia. MDB accounts for approximately 40 % of the gross value of Australian agricultural production (ABS 2008). MDB uses more than half of Australia's water consumption, but water use efficiency needs special attention as the total percentage of agricultural water use inside MDB is much higher than that of the gross agricultural production (ABS 2008). Long-term over-allocation of water for human use (mainly agriculture and domestic) along with occasional droughts set a dire impact on natural environments including river flow reduction and salinity intrusion (MDBA 2014).

Australia has an established water regulations system from local to national level. City councils have their own management plan for their locality, whereas most of the major rivers have their catchment management authorities to prepare plans suitable for the individual basins. Although each of the states has their own legislation to manage the water resources inside the states, an intergovernmental agreement was made in 2004 by all state and territory governments and the Australian government through the Council of Australian Governments (COAG) to synchronize the individual plans and manage the water resources in national scale. This is known as the National Water Initiative (NWI), and the main objective was to achieve a 'nationally compatible, market, regulatory and planning based system of managing surface water and groundwater resources for rural and urban use that optimizes economic, social and environmental outcomes' (Council of Australian Governments 2004). Since its beginning, the progress achieved by NWI was monitored continuously and the latest biennial report suggested that NWI has delivered significant and tangible benefits for Australia. Although the initiative is giving positive impact and this is not the appropriate time to assess the long-term efficiency, NWI is yet to achieve its intended primary goal of sustainable and efficient water management (National Water Commission 2011). Especially accountability for environmental outcomes remains weak including inadequate monitoring capacity and insufficient scientific research to link the environmental watering with ecological outcomes.

29.3 Modelling/Decision Support Tools and Sustainable Water Resources Management

Progress in computational capability during the last few decades allowed very fast development of technological tools. Access to these tools (e.g. hydrological models) allowed scientists to test their ideas and thoughts more effectively in a multidisciplinary way which would have not been possible without such tools. Hydrological modelling is a simplified and conceptual representation of a hydrological process. Complex processes are involved in hydrological systems and it is difficult to include all of those processes in a single model. Rainfall-runoff is the most important process among all the hydrological processes. The first attempt to model the rainfall-runoff system was made by an Irish engineer Thomas James Mulvaney nearly 150 years ago through a simple equation which became known as rational method (Beven 2012).

The availability of digital computers in the 1960s opened a new window to the hydrologists removing computational constraints. The Stanford watershed model is one of the successful models developed during that time and was widely used (Beven 2012). With the technological advancement, more and more components were added to the simple rainfall-runoff processes and more descriptive models became available. Currently, the progress of hydrological science is closely connected to modelling (Rosberg and Masdsen 2005).

Since the development of the first model, numerous types of hydrological models have been developed. Due to the abundance and diversity of the developed

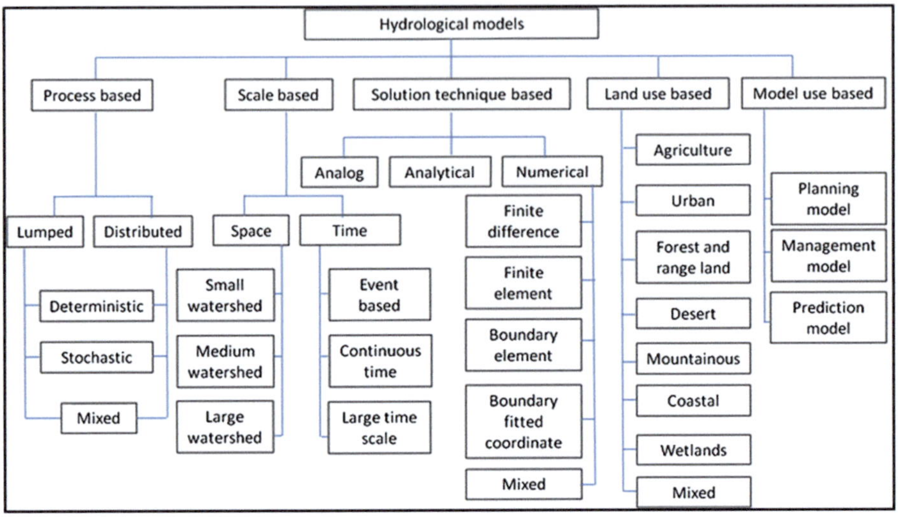

Fig. 29.1 Classification of hydrological models proposed by Sing (1995)

models, different schemes are available to classify hydrological models. A classification introduced by Sing (1995) is presented in Fig. 29.1.

Among the available models, physically based distributed hydrological models are widely used by researchers and water managers for their versatility in simulating different kinds of hydrological processes. They are based on our understanding of the physics of the hydrological processes. Currently, common physically based models are capable of simulating a wide range of hydrological processes using just a personal computer. The combination of the GIS interface with a modelling environment gives users more flexibility to prepare input data efficiently and view the outputs more realistically.

Refsgaard and Abbot (1996) assessed the need for a distributed hydrological modelling which can be applied for water resources assessment, irrigation, soil erosion, surface and groundwater pollution, land use and climate change scenario analysis, aquatic ecology and historical reconstructions of the impact of human activities. Advancement of technology and a wide range of research are now enabling distributed models to simulate more and more processes and management scenarios. The Soil and Water Assessment Tool (SWAT), one of the most widely used distributed models, is now capable of simulating plant growth, nutrients, pesticides, agricultural management, stream routing and pond/reservoir routing including bacteria, riparian zones, pothole topography, forest growth, channel downcutting and widening and input uncertainty analysis (Arnold and Fohrer 2005) in addition to the processes discussed by Refsgaard and Abbot.

As a water-scarce country, efficient water resources management is always in discussion among the scientists and policymakers of Australia. Numerous hydrological models have been developed and applied in Australia for various purposes in the past four decades. Water balance models developed and widely used in

Australia have been reviewed under the Water 2010 project of the Bureau of Rural Sciences, Australia (Ranatunga et al. 2007). The review presented an in-depth analysis of the models in terms of complexity, performance under various conditions, nationwide applicability and capability of using special data including data requirement. Although both physically and empirically based models are developed and used in Australia, most of these models are purpose specific and suitable for application at a given spatial scale (Ranatunga et al. 2007). Free availability of model and basic GIS data, user-friendly GIS interface linked with sensitivity, calibration and uncertainty analysis tool make SWAT a popular tool for both data-rich and data-scarce regions (van Griensven et al. 2012). The capability to simulate a wide range of hydrological processes as a single model and successful application results across the globe put SWAT in an adventurous position over other similar models.

29.4 SWAT and Its Application in Australia

The Soil and Water Assessment Tool (SWAT) is a watershed-scale semi-distributed, physically based hydrological model developed for the USDA Agricultural Research Service (Arnold et al. 1998; Neitsch et al. 2011). The purpose of SWAT development was to support the water resources managers for better assessment of water supplies including nonpoint pollution sources for large river basins and to achieve that emphasis was given on four key elements: (1) land management, (2) water quality loadings, (3) flexibility in basin discretization and (4) continuous time simulation during primary model development (Arnold et al. 1998). It runs on a daily time step and is capable of continuous simulation over a long duration (Gassman et al. 2007). The tool is suitable to assess the long-term impact of land management practices on water, sediment and agricultural chemical yields in large, complex watersheds with varying soils, land use and management conditions (Arnold et al. 1998; Neitsch et al. 2011). As a hydrological modelling tool, SWAT was subject to continuous improvement since its invention and is now capable of simulating most of the key hydrological processes at basin scale (Zhang et al. 2009). Although developed in United States, SWAT has been extensively applied and it has proved its capability for solving different types of hydrological problems in different types of catchments in other regions of the world including Asia, Europe and Africa (Gassman et al. 2007). SWAT's capability was successfully tested to simulate the hydrological processes of small to large watersheds (Lévesque et al. 2008; Tang et al. 2013) in arid to mountainous (Dong et al. 2013; Rahman et al. 2013) regions including both drought and flood flow conditions (Trambauer et al. 2013; Tzoraki et al. 2013). However, due to very limited application of SWAT, Australia was not included as a key country or region of SWAT application in the 2012 SWAT conference (Gassman et al. 2012).

Although SWAT has not been applied rigorously like other regions of the world, it has been successfully applied in several river catchments of Australia. The names and locations of the catchments are provided in Fig. 29.2 which indicates that

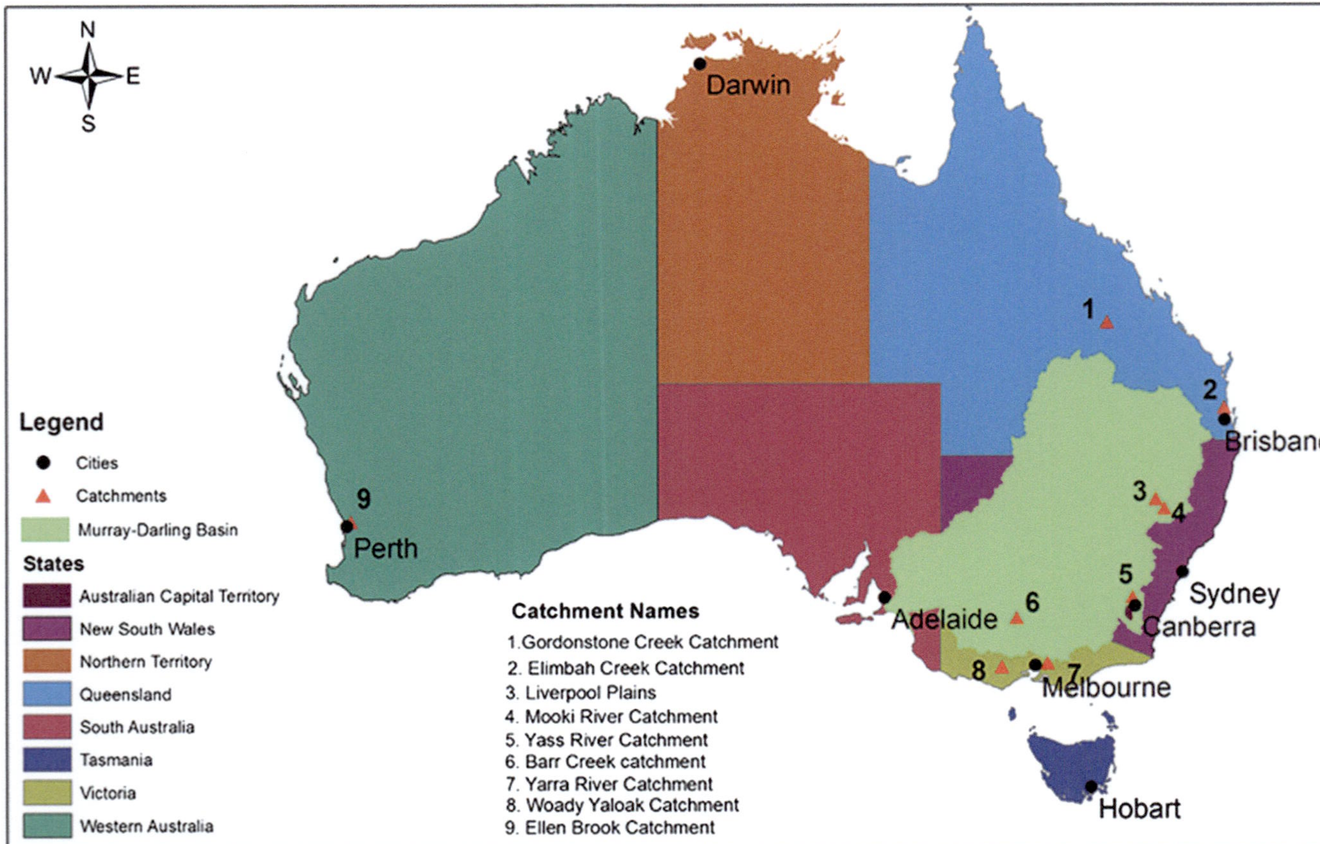

Fig. 29.2 Location of catchments where SWAT was applied in Australia including Murray-Darling basin

Table 29.1 List and description of SWAT applications in Australia

Catchment name	Catchment area (km^2)	State	Application	Reference
Gordonstone creek catchment	270	Queensland	Agricultural management practice on pollutant load	Dougalla et al. (2003)
Elimbah creek catchment	142	Queensland	Surface water and shallow groundwater interactions	Labadz et al. (2010)
Liverpool plains	437	New South Wales	Drainage and shallow groundwater recharge	Sun and Cornish (2005, 2006)
Moki river catchment	836	New South Wales	Environmental flow, integrated catchment management, water use efficiency and environmental policies targeting irrigated agriculture	Lee et al. (2007, 2012)
Yass river catchment	1,597	New South Wales	Streamflow simulation for high and low flow	Saha et al. (2013)
Barr creek catchment	600	Victoria	Recharge estimation using remotely sensed ET data	Githui et al. (2012)
Yarra river catchment	1,511	Victoria	Model setup, calibration, validation and sensitivity analysis	Das et al. (2013a, b)
Woady Yaloak river catchment	1,157	Victoria	Water balance	Watson et al. (2003a, b, 2005a, b)
Ellen Brook catchment	570	Western Australia	Predicting nitrogen concentration in river systems	Exbrayat et al. (2011)

SWAT applications in Australia have not been spread widely across the whole country but rather concentrated in the southeastern corner of the country. Although the number of applications is limited, SWAT studies reported inside Australia are diversely ranged between simple water balance to complex environmental and market policy analysis. Table 29.1 summarizes the SWAT application in Australia.

Table 29.1 indicates an increasing number of SWAT applications in Australia in recent times, but most of the studies are located in the eastern and southeastern part of the country with no study reported for the inland Australia which is the drier part of the country. Despite SWAT's capability in modelling the hydrology of large catchments, there is no reported application of SWAT in any of the large catchments of Australia. The number of studies inside the largest river basin in the country MDB is insignificant.

In general, as reported by the water balance studies, SWAT underpredicts the flow in wet years and overpredicts it in dry years for Australian catchments (Das et al. 2013a; Watson et al. 2003a). Despite of high variation of flow, SWAT's performance was found to be excellent for both annual and monthly time steps which indicates that the model is capable to capture the seasonal variations of Australian flow, whereas results were found to be satisfactory or average for daily time step (Saha et al. 2013; Watson et al. 2003a). Although SWAT was built for a long-term scenario analysis rather than a single-event flood simulation (Neitsch et al. 2011), good model performances in a daily time step were reported by several studies with daily rainfall input (Gassman et al. 2007). Significant improvement in runoff estimation, especially high flows, is possible if sub-hourly rainfall data is used (Jeong et al. 2010). Further improvement in predicting Australian runoff is possible using sub-daily or sub-hourly rainfall data with iterative use of SWAT in Australia.

SWAT was applied to model the shallow groundwater and surface water interactions of several catchments in Australia. In addition to conventional method of input data, a method of remotely sensed evapotranspiration data approach was used for recharge estimation. SWAT was found to be an effective tool for recharge estimation in irrigated semiarid regions like Australia (Githui et al. 2012) where better quantification of the 'revap' component of the model was suggested when conventional method was to be used (Sun and Cornish 2006).

SWAT was also successfully applied for sediment load and in-stream pollutant estimation adopting several predefined land management scenarios. The tool was found to be effective for achieving a certain goal of sediment load reduction by improving cropping practice examining the SWAT simulations (Dougalla et al. 2003). Applicability of SWAT was also tested to predict the N_2 concentration in a Western Australian catchment (Exbrayat et al. 2011).

An advanced study combining SWAT with economic model targeting irrigation efficiency improvement at both ecological and economic standpoints can provide a more precise estimation of the optimal choices between basin production activities on a site-specific basis (Lee et al. 2007). Case studies have shown that SWAT can provide valuable information to choose between technological options (drip or pivot) and source (ground- or surface water) of irrigation which is valuable to both farmers and policymakers for selecting a cost-effective system ensuring environmental flow requirement and water security (Lee et al. 2012).

29.5 Gaps in SWAT Model Application in Australia

SWAT needs several data as input to develop a model using the ArcSWAT, the ArcGIS interface for model development. Some data are compulsory to develop a model and the remaining optional as per required output. Digital elevation model (DEM), soil, land use and weather data are mandatory for a SWAT model, whereas reservoir, sediment, pesticide, fertilizer, chemical and water quality data is optional as per purpose of the model development. Although Australian catchments can be classified as data rich in terms of hydroclimatic data, scarcity of water quality and land use management data still remains a major problem for simulating different ecological and land use management scenarios (Das et al. 2013b). Detailed weather data for at least one weather station is essential for building the SWAT model. SWAT has a weather generator that is used for filling missing values of climatic data inputs. The database of 1,041 stations is available for the United States through the default ArcSWAT interface, but SWAT users in Australia need to create a weather generator database. The Bureau of Meteorology (BOM) and SILO (an enhanced climate data bank constructed from ground-based observed data hosted by the Science Delivery Division of the Department of Science, Information Technology, Innovation and the Arts) data drill was the main source of climate data where SILO contains weather data at 0.05° grids across Australia with missing or suspect values 'patched' with interpolated data in addition to BOM's numerous data stations. Parameter descriptions of common land use types are stored in the land use database of SWAT which does not contain some commonly available land cover types of Australia. During the early stages of SWAT application in Australia, unavailability of land use maps had forced to develop a land use map from a Landsat satellite image (Watson et al. 2003a), whereas nationwide 50 m raster grid land use data of the Australian Bureau of Agricultural and Resource Economics and Sciences (ABARES) was found to be the most consistent source of land use data for SWAT study. One study reported a better-resolution (30 m) local land use data developed by the Department of Primary Industries (DPI), Victoria (Githui et al. 2012), which might not be available for other regions of the country. Although the Digital Atlas of Australian Soils is available as a source of digital soil map, required soil parameters need to be calculated to create a user soil database from different interpretations and look-up tables available for the atlas (McKenzie et al. 2000; Western and McKenzie 2004), whereas a database for all soil types of the United States is available with a basic SWAT module. Although SWAT can incorporate ten layers of soil data, only two layers of data are available through the Digital Atlas of Australian Soils.

Apart from problems in model development, some inherent problems of the SWAT model were also reported by the limited studies in Australia. Model performance in simulating monthly streamflow was found to be better than daily flow which is common to other regions. Australia rainfall pattern varies from occasionally high values to continuous dry condition for long duration. It is always difficult to model daily hydrograph for such regions. Availability of sub-daily or hourly rainfall data might improve the results, but continuous rainfall data is only available for automatic weather stations (AWS) of the Bureau of Meteorology. The number of AWS is only about 550 (Bureau of Meteorology 2005) compared to simple rain gauge network of about 5,760 stations (Jones et al. 2007). Being a dry continent, low-flow hydrology is critical for Australia, but SWAT groundwater component was found to be simple to simulate base flow components with greater accuracy especially for summer months (Watson et al. 2003b). Eucalyptus trees cover significant portions of Australian land. The basic module of SWAT was found to be inadequate to simulate the LAI growth of Eucalyptus which can be improved by integrating 3-PG, a forest growth model, into SWAT (Watson et al. 2005a). Improvements of bypass flow simulation routine particularly for cracking vertosol soil and inclusion of a 'revap' component are required for better recharge estimation of dry catchments such as that of Australia (Sun and Cornish 2005).

Although limitations were found in inherent equations for simulating groundwater, drought and salinity including incapability of fire impact simulation, among the considered similar models, SWAT was found to be the most appropriate tool to model and simulate terrestrial nutrient cycle, transport and vegetation dynamics for catchment management in the Victorian region of Australia (Zyngier et al. 2010).

Although the limited number of SWAT applications in Australia covers a wide range of hydrological studies, there is scope to apply SWAT in other hydrological studies such as climate and land use change scenario analysis (Dile et al. 2013; Mango et al. 2011) and different crop growth management practices (Strauch and Volk 2013) with different in-stream processes. As SWAT has an option to modify its source code, different conceptual and computational studies can be initiated for better representation of Australian hydrology.

29.6 Conclusions and Recommendations

Sustainability of water resources management depends on long-term multidisciplinary planning and continuous monitoring of the system. Physically based hydrological models provide scientific basis to simulate the past condition along with future scenario and evaluate different options and alternative strategies. Despite limitations reported by some studies, SWAT has a good potential to simulate the hydrologic conditions of Australia. Discrepancy increases when different models are used to simulate different components of a hydrological process. Being a public-domain freely available model with a wide range of applicability to simulate several components of a hydrological system, SWAT has additional advantage over other similar models. The capability of SWAT, as a single model, to simulate both water quality and quantity, including different management scenarios, makes it a more suitable tool for sustainable integrated water management and planning.

To keep the consistency between all SWAT studies in Australia, a common weather generator and soil database need to be created. More accurate parameters need to be determined specifically for Australian land use types such as eucalyptus. Addition of such tree parameters in the SWAT land use database allows more accurate estimation of water balance components. Further validation of SWAT in Australia, including its semiarid and arid zones, can make it a more viable tool under Australian conditions.

Acknowledgement This study was supported by the Charles Sturt University (CSU) Strategic Research Centre Scholarship. The first author was the recipient of this scholarship.

References

ABS (2008) Water and the Murray-Darling Basin – a statistical profile, 2000–01 to 2005–06, Australian Bureau of Statistics Cat. No. 4610.0.55.007, Canberra. http://www.abs.gov.au/ausstats/abs@.nsf/mf/4610.0.55.007. Accessed 2 Apr 2012

ABS (2013a) Australia's climate. In: Year book Australia, No. 92, 2012, Cat. No. 1301.0. Australian Bureau of Statistics, Canberra

ABS (2013b) Water account, Australia, 2011–12, Australian Bureau of Statistics Cat. No. 4610.0, Canberra. http://www.abs.gov.au/AUSSTATS/abs@.nsf/Lookup/4610.0Main+Features202011-12. Accessed 3 Jan 2014
Acreman MC, Dunbar MJ (2004) Defining environmental river flow requirements – a review. Hydrol Earth Syst Sci 8(5):861–876. doi:10.5194/hess-8-861-2004
Addams L, Boccaletti G, Kerlin M, Stuchtey M (2009) Charting our water future, economic frameworks to inform decision-making. 2030 Water Resources Group, New Delhi
Arnold JG, Fohrer N (2005) SWAT2000: current capabilities and research opportunities in applied watershed modelling. Hydrol Process 19(3):563–572. doi:10.1002/hyp.5611
Arnold JG, Srinivasan R, Muttiah RS, Williams JR (1998) Large area hydrologic modeling and assessment part I: model development. J Am Water Resour Assoc 34(1):73–89. doi:10.1111/j.1752-1688.1998.tb05961.x
Beven KJ (2012) Rainfall – runoff modelling – the primer, 2nd edn. Wiley, Chichester
Bureau of Meteorology (2005) Automatic weather stations for agricultural and other applications. http://www.bom.gov.au/inside/services_policy/pub_ag/aws/aws.shtml. Accessed 9 Jan 2014
Bureau of Meteorology (2012) Australian water resources assessment 2012. Bureau of Meteorology, Melbourne
Chiew F (2011) Climate impact on water availability. Paper presented at the SEACI Workshop, Shine Dome, Canberra, 19 July 2011
Church TM (1996) An underground route for the water cycle. Nature 380(6575):579–580
Council of Australian Governments (2004) Intergovernmental agreement on a national water initiative between the Commonwealth of Australia, and the governments of New South Wales, Victoria, Queensland, South Australia, the Australian Capital Territory and the Northern Territory. Commonwealth of Australia, Canberra
CSIRO (2011) Sustainability in Australia's arid lands. http://www.csiro.au/science/arid-land-sustainability. Accessed 10 Jan 2014
Das SK, Ng AWM, Perera BJC (2013a) Development of a SWAT model in the Yarra River catchment. Paper presented at the 20th international congress on modelling and simulation, Adelaide, Australia, December 1–6 2013
Das SK, Ng AWM, Perera BJC (2013b) Sensitivity analysis of SWAT model in the Yarra River catchment. Paper presented at the 20th international congress on modelling and simulation, Adelaide, Australia, December 1–6 2013
DFAT (2008) About Australia: Australia's environment at a glance. http://www.dfat.gov.au/facts/env_glance.html. Accessed 17 Nov 2011
Dile YT, Berndtsson R, Setegn SG (2013) Hydrological response to climate change for Gilgel Abay River, in the Lake Tana Basin – Upper Blue Nile Basin of Ethiopia. PLoS One 8(10): e79296. doi:10.1371/journal.pone.0079296
Dong W, Cui B, Liu Z, Zhang K (2013) Relative effects of human activities and climate change on the river runoff in an arid basin in northwest China. Hydrol Process. doi:10.1002/hyp.9982
Dougalla C, Rohdea K, Carrollb C, Millara G, Stevensc S (2003) An assessment of land management practices that benchmark water quality targets set at a neighbourhood catchment scale using the SWAT model. In: Post D (eds) MODSIM 2003: international congress on modelling and simulation, Jupiters Hotel and Casino, 14–17 July 2003: integrative modelling of biophysical, social and economic systems for resource management solutions : proceedings. Modelling and Simulation Society of Australia and New Zealand Inc., Canberra, , pp 1–6
Exbrayat J-F, Viney NR, Frede H-G, Breuer L (2011) Probabilistic multi-model ensemble predictions of nitrogen concentrations in river systems. Geophys Res Lett 38(12), L12401. doi:10.1029/2011GL047522
FAO (2013) Water scarcity. http://www.fao.org/nr/water/topics_scarcity.html. Accessed 6 Jan 2014
Foster SSD, Chilton PJ (2003) Groundwater: the processes and global significance of aquifer degradation. Philos Trans R Soc Lond B 358(1440):1957–1972. doi:10.1098/rstb.2003.1380
Gassman PW, Reyes MR, Green CH, Arnold JG (2007) The soil and water assessment tool: historical development. Applications, and future research directions. Trans ASABE 50(4): 1211–1250

Gassman PW, Arnold JG, White M, Srinivasan R, Reyes M, Kim NW, Chung I-M, Huang F, Griensven Av, Volk M, Abbaspour K, Watson B (2012) The worldwise use of the SWAT model: networking impacts, simulation trends, and future developments. Paper presented at the international SWAT conference and workshop, New Delhi, India, 16–20 July 2012

Githui F, Selle B, Thayalakumaran T (2012) Recharge estimation using remotely sensed evapotranspiration in an irrigated catchment in southeast Australia. Hydrol Process 26(9): 1379–1389. doi:10.1002/hyp.8274

Gleick PH (1996) Water resources. In: Schneider SH (ed) Encyclopedia of climate and weather, vol 2. Oxford University Press, New York, pp 817–823

Global Water Partnership (2013) What is IWRM? Global Water Partnership. http://www.gwp.org/The-Challenge/What-is-IWRM/. Accessed 4 Jan 2014

Hassing J, Ipsen N, Clausen TJ, Larsen H, Lindgaard-Jørgensen P (2009) Integrated water resources management (IWRM) in action. World Water Assessment Programme (WWAP), DHI Water Policy and UNEP-DHI Centre for Water and Environment, Paris

Jeong J, Kannan N, Arnold J, Glick R, Gosselink L, Srinivasan R (2010) Development and integration of sub-hourly rainfall–runoff modeling capability within a watershed model. Water Resour Manag 24(15):4505–4527. doi:10.1007/s11269-010-9670-4

Jones DA, Wang W, Fawcett R (2007) Climate data for the Australian water availability project – final milestone report. Bureau of Meteorology, Melbourne, 3001

Kollmorgen A, Little P, Hostetler S, Griffith H (2007) Water availability theme. National Perspective National Water Commission, Canberra

Kummu M, Ward PJ, Hd M, Varis O (2010) Is physical water scarcity a new phenomenon? Global assessment of water shortage over the last two millennia. Environ Res Lett 5(3):1–10. doi:10.1088/1748-9326/5/3/034006

Labadz M, Geigorescu M, Cox ME (2010) Modelling surface and shallow groundwater interactions in an ungauged subtropical catchment using the SWAT model, Elimbah Creek, Southeast Queensland, Australia. In: 19th World Congress of Soil Science, Soil Solutions for a Changing World, Brisbane, Australia, 2010

Lee L-T, Ancev T, Vervoort W (2007) Environmental and economic impacts of water scarcity and market reform on the Mooki catchment. Environmentalist 27(1):39–49. doi:10.1007/s10669-007-9011-1

Lee LY, Ancev T, Vervoort W (2012) Evaluation of environmental policies targeting irrigated agriculture: the case of the Mooki catchment, Australia. Agric Water Manag 109(0):107–116. http://dx.doi.org/10.1016/j.agwat.2012.02.011

Lévesque É, Anctil F, Van Griensven ANN, Beauchamp N (2008) Evaluation of streamflow simulation by SWAT model for two small watersheds under snowmelt and rainfall. Hydrol Sci J 53(5):961–976. doi:10.1623/hysj.53.5.961

Loucks DP (2000) Sustainable water resources management. Water Int 25(1):3–10. doi:10.1080/02508060008686793

Mango LM, Melesse AM, McClain ME, Gann D, Setegn SG (2011) Land use and climate change impacts on the hydrology of the upper Mara River Basin, Kenya: results of a modeling study to support better resource management. Hydrol Earth Syst Sci 15(7):2245–2258. doi:10.5194/hess-15-2245-2011

McKenzie NJ, Jacquier DW, Ashton LJ, Cresswell HP (2000) Estimation of soil properties using the atlas of Australian soils, CSIRO land and water technical report 11/00. CSIRO Land and Water, Canberra

MDBA (2014) Basin environment. http://www.mdba.gov.au/about-basin/basin-environment. Accessed 13 Jan 2014

Milburn T (2006) Living in a changing world. In: Lee S, Treves-Habar J (eds) Water a shared responsibility. UNESCO, Paris, pp 5–42

National Water Commission (2011) The national water initiative – securing Australia's water future: 2011 assessment. NWC, Canberra

Neitsch SL, Arnold JG, Kiniry JR, Williams JR (2011) Soil and water assessment tool theoretical documentation: version 2009, Texas Water Resources Institute Technical Report No. 406. Texas Water Resources Institute, College Station

Oelkers EH, Hering JG, Zhu C (2011) Water: is there a global crisis? Elements 7(3):157–162. doi:10.2113/gselements.7.3.157

Oki T, Kanae S (2006) Global hydrological cycles and world water resources. Science 313(5790):1068–1072. doi:10.1126/science.1128845

Pavey C, Nano C (2006) Australia's desert. Desert Wildlife of Australia. http://www.abs.gov.au/AusStats/ABS@.nsf/Previousproducts/1301.0Feature%20Article42006?opendocument&tabname=Summary&prodno=1301.0&issue=2006&num=&view=. Accessed 10 Jan 2014

Rahman K, Maringanti C, Beniston M, Widmer F, Abbaspour K, Lehmann A (2013) Streamflow modeling in a highly managed mountainous glacier watershed using SWAT: The Upper Rhone River Watershed Case in Switzerland. Water Resour Manag 27(2):323–339. doi:10.1007/s11269-012-0188-9

Ranatunga K, Nation E, Barratt D (2007) Review of water models and their application in Australia water, 2010 Technical Paper 1. Bureau of Rural Sciences, Government of Australia

Refsgaard JC, Abbott MB (1996) The role of distributed hydrological modelling in water resources management. In: Abbott MB, Refsgaard JC (eds) Distributed hydrological modelling. Kluwer Academic, Dordrecht, pp 1–16

Rosberg D, Masdsen H (2005) Concept of hydrologic modelling. In: Anderson MG, McDonnell JJ (eds) Encyclopedia of hydrological sciences, vol 1. Wiley, Chichester, pp 155–163

Saha PP, Zeleke K, Hafeez M (2013) Streamflow modeling in a fluctuant climate using SWAT: Yass River catchment in south eastern Australia. Environ Earth Sci. doi:10.1007/s12665-013-2926-6

Sing VP (1995) Watershed modeling. In: Sing VP (ed) Computer models of watershed hydrology. Water Resources Publications, Fort Collins, pp 1–22

Strauch M, Volk M (2013) SWAT plant growth modification for improved modeling of perennial vegetation in the tropics. Ecol Model 269(0):98–112. http://dx.doi.org/10.1016/j.ecolmodel.2013.08.013

Sun H, Cornish PS (2005) Estimating shallow groundwater recharge in the headwaters of the Liverpool Plains using SWAT. Hydrol Process 19(3):795–807. doi:10.1002/hyp.5617

Sun H, Cornish PS (2006) A catchment-based approach to recharge estimation in the Liverpool Plains, NSW, Australia. Aust J Agric Res 57(3):309–320. doi:10.1071/AR04015

Tang Y, Tang Q, Tian F, Zhang Z, Liu G (2013) Responses of natural runoff to recent climatic variations in the Yellow River basin, China. Hydrol Earth Syst Sci 17(11):4471–4480. doi:10.5194/hess-17-4471-2013

Trambauer P, Maskey S, Winsemius H, Werner M, Uhlenbrook S (2013) A review of continental scale hydrological models and their suitability for drought forecasting in (sub-Saharan) Africa. Phys Chem Earth Pt A/B/C 66(0):16–26. http://dx.doi.org/10.1016/j.pce.2013.07.003

Tzoraki O, Cooper D, Kjeldsen T, Nikolaidis NP, Gamvroudis C, Froebrich J, Querner E, Gallart F, Karalemas N (2013) Flood generation and classification of a semi-arid intermittent flow watershed: Evrotas river. Intl J River Basin Manag 11(1):77–92. doi:10.1080/15715124.2013.768623

UNDESA (2013) Integrated water resources management. In: International decade for action "Water for Life" 2005–20015. United Nations Department of Economic and Social Affairs. http://www.un.org/waterforlifedecade/iwrm.shtml. Accessed 24 Dec 2013

UNWATER (2009) Water in a changing world – The United Nations World Water Development Report 3. UNESCO, Paris

UNWATER (2012) Managing water under uncertainty and risk – The United Nations World Water Development report 4, vol 1. UNESCO, Paris

van Griensven A, Ndomba P, Yalew S, Kilonzo F (2012) Critical review of SWAT applications in the upper Nile basin countries. Hydrol Earth Syst Sci 16(9):3371–3381. doi:10.5194/hess-16-3371-2012

Watson BM, Ghafouri M, Selvalingam S (2003a) Application of SWAT to model the water balance of the Woady Yaloak River catchment, Australia. In: Srinivasan R, Jacobs J, Jensen, R (eds) SWAT 2003: 2nd international SWAT conference. USDA-ARS Research Lab, Temple, pp 94–110

Watson BM, Selvalingam S, Ghafouria M (2003b) Evaluation of SWAT for modelling the water balance of the Woady Yaloak River catchment, Victoria. In Post D (ed) MODSIM 2003: international congress on modelling and simulation, Jupiters Hotel and Casino, 14–17 July 2003: integrative modelling of biophysical, social and economic systems for resource management solutions: proceedings. Modelling and Simulation Society of Australia and New Zealand Inc., Canberra, pp 1–6

Watson BM, Coops N, Selvalingam S, Ghafouri M (2005a) Integration of 3-PG into SWAT to simulate the growth of evergreen forests. In Srinivasan R, Jacobs J, Day D, Abbaspour K (eds) SWAT 2005: 3rd international SWAT conference. USDA-ARS Research Lab, Temple, pp 142–152

Watson BM, Srikanthan R, Selvalingam S, Ghafouri M (2005b) Evaluation of three daily rainfall generation models for SWAT. Trans ASABE 48(5):1697–1711

Western A, McKenzie N (2004) Soil hydrological properties of Australia: user guide. CRC for Catchment Hydrology, Melbourne

Zhang X, Srinivasan R, Bosch D (2009) Calibration and uncertainty analysis of the SWAT model using genetic algorithms and Bayesian model averaging. J Hydrol 374(3–4):307–317. doi:10.1016/j.jhydrol.2009.06.023

Zyngier R, Shelly K, Baker P, Cavagnaro T (2010) Terrestrial nutrient cycling, transport and vegetation dynamics Australian Centre for Biodiversity. Monash University, Melbourne

Chapter 30
Watershed Modeling as a Tool for Sustainable Water Resources Management: SWAT Model Application in the Awash River Basin, Ethiopia

Selome M. Tessema, Shimelis Gebriye Setegn, and Ulla Mörtberg

Abstract Improving the reliability of streamflow prediction under limited data conditions is a vital step to achieve a sustainable water management system. In many areas, when planning for balancing water demands for hydropower, irrigation, and ecosystem services as well as preventing flood risk, major gaps exist on baseline information of water resources. The prediction of streamflow requires adequate understanding of the characteristics of the river basin. Awash River basin has been a subject of large-scale flooding for several years mainly due to heavy rains and inadequate water resource management. The lack of decision support tools and limitation of available data hinder research and development in the area. The main objective of this study was to characterize the hydrological components of the upper part of Awash River basin under limited data condition. The optimal approach for this purpose was considered to be statistical analysis of the time series hydrometeorological data and to adapt existing hydrological models. The physically based Soil and Water Assessment Tool (SWAT) model was successfully calibrated and validated in the watershed. The performance of the model was evaluated based on the streamflow prediction at four subbasin outlets and the main outlet of the river basin. Model validation indicated that daily streamflows were predicted reasonably which was verified by Nash-Sutcliffe values ranging from 0.55 to 0.71. The evaluations from tributary rivers indicate that the drainage area is one of the important factors that affect the direct transferring of parameter values from one watershed to another. The catchment characteristics and its different hydrological components of the water balance are discussed.

S.M. Tessema (✉) • U. Mörtberg
Division of Land and Water Resources Engineering, KTH Royal Institute of Technology, Stockholm, Sweden
e-mail: selome@kth.se

S.G. Setegn
Department of Environmental and Occupational Health and Global Water for Sustainability Program (GLOWS), Florida International University, 11200 SW 8th Street, Miami, FL 33199, USA

S.G. Setegn, M.C. Donoso (eds.), *Sustainability of Integrated Water Resources Management*, DOI 10.1007/978-3-319-12194-9_30

Keywords Hydrological characteristics • Hydrological modeling • Rainfall-runoff correlation • Streamflow prediction • Awash River basin • SWAT

30.1 Introduction

Ethiopia is known to have abundant water resources, available for hydropower and irrigation. The country also has several problems related to water resources, such as flooding, drought, and depletion of ecosystem services. The country plans for major investments in building dams for hydropower, irrigation, and flood control purposes (Ministry of Water Resources 2010). The topography of the country is rugged with clearly defined water courses, which concentrates significant flooding to the low-land flat parts of the country. For example, the Awash River basin has been a subject of large-scale flooding for several years. The impacts of flooding so far include both the tragic human life loss and extensive property damages (Emergency appeal 2006). In order to tackle water resource-related problems and to plan for balancing water demands for hydropower, irrigation, and ecosystem services as well as flood risk, realistic river discharge predictions are needed, which require hydrological characterization of the river basin. An effective way to improve the accuracy of such predictions through allocating the characteristics of the catchment is by employing hydrological models. However, in spite of the abundance of water resources in Ethiopia, there is a limited availability of hydrometeorological data that hinder research and development in the country. Hence, adapting existing hydrological models under limited data conditions is essential to understand the hydrological characteristics of the catchment. The use of these hydrological models is at the center of many research efforts for studying the nature of streamflow for different purposes, like predictions of peak discharges (e.g., Brewer et al. 1982), total volume of flood flow (e.g., Patro et al. 2009b), extent of flood (e.g., Patro et al. 2009a), or a full detail distribution of flow over space and time in a catchment (e.g., Anderson et al. 2001).

The existing hydrological simulation models can be grouped according to the runoff generation process considered in each model. When comparing models, stochastic and deterministic models are often considered to be at the top level of the classification tree, in accordance to the way they treat the randomness of hydrologic phenomena (Chow et al. 1988). Stochastic models use local hydrometric data to predict flows. These models allow for some randomness that results in different outputs and are based on the analysis of past events, in the form of rainfall and river discharge (e.g., Tesfaye et al. 2006; Anderson et al. 2007). Deterministic models generally produce a single output of runoff for a given rainfall under identical physical environment (Crawford and Linsely 1966; Abbott et al. 1986; Arnold et al. 1998; Govindaraju and Rao 2000). Recently, through ensemble hydrological modeling concept, deterministic models are able to produce ensemble outputs by including randomness either in the parameters, in the model structure, or in the forcing data (e.g., Georgakakos et al. 2004; Shrestha et al. 2005). Detailed descriptions about the types of models could be further read in Becker and Serban (1990), Refsgaard (1996), Jones (1997), Beven (2001), and Mulligan and Wainwright (2004).

In this study, two approaches were used to characterize the study area: statistical analysis of observed data and application of a conceptual hydrological model. A conceptual model is one type of a deterministic model, which can be lumped or distributed, and generally composed of mathematical descriptions of the processes of catchment response. These models represent the catchment as integrated conceptual components but also incorporate some aspects of physical processes. One of the conceptual models that are widely used for analyzing the different components of hydrology in large river basins is the Soil and Water Assessment Tool (SWAT) (Arnold et al. 1998). SWAT is a semi-distributed watershed model developed to examine the influence of topographic, land use, soil and climatic conditions on streamflow, and sediment yield. It has also been used to predict the potential for hydropower (Kusre et al. 2010) as well as the impact of land management practices on water, sediment, and agricultural chemical yields (Neitsch et al. 2005). As a physically based model, SWAT requires relatively detailed and extensive data to produce reliable predictions. It is a freely available public domain model and has been validated for different catchments in different parts of the world (Pikounis et al. 2003; Jayakrishnan et al. 2005; Ahl et al. 2008). However, there are only few applications of the SWAT model to Ethiopian conditions (Tadele and Foerch 2007; Mekonnen et al. 2009; Setegn et al. 2009a, b, 2010; Tessema et al. 2014). Few water-related research studies were conducted on the upper Awash River basin (Hailemariam 1999; Chekol 2006). The existing studies could be seen as a start of building the database for future plans of integrated water resource management.

The main aim of this study was to characterize the major hydrological components of the upper Awash River. The specific objective was to investigate the applicability of the SWAT watershed model for reliable characterization of hydrological processes of the catchment and further to assess and discuss the modeling uncertainties due to limited input data and other sources. The overall purpose of this research is to improve the baseline information for planning sustainable water resources management; balancing water demands for hydropower, irrigation, and ecosystem services; as well as mitigating flood risk.

30.2 Methodology

In order to characterize the catchments and predict the streamflow in the Awash River basin, two approaches were used: statistical analysis of the observed data and application of a hydrological model. Statistical analysis was made to get an insight into the correlation between the available rainfall and streamflow measurements at the headwaters of the Awash River basin. The application of the model involved model setup, sensitivity analysis, calibration, and uncertainty analysis. The performance of the model was evaluated by comparing the simulated flow hydrograph with the observed hydrograph and also using three model goodness-of-fit statistics methods.

30.2.1 Study Area

The upper Awash River basin is located between latitude 9° 18′ N and 8° 17′ N and longitude 37° 57′ E and 39° 4′ E with an outlet at the Hombole gauging station. The Awash River originates at an elevation of about 3000 m a.s.l. in the central part of Ethiopia. It flows to the southeast for a short distance and turns northeast along the Great Rift Valley. It joins Lake Abbe at the border of Djibouti Democratic Republic and is about 1,200 km long from its origin. The basin covers a total area of about 113,000 km^2 that includes the capital city, Addis Ababa. The present study focuses on the uppermost part of the basin with an area coverage of about 7630 km^2 (Fig. 30.1). Some of the main streams that comprise the upper Awash River basin are the Holeta, Berga, Akaki, and Melka Kuntire streams. Table 30.1 summarizes the characteristics of the sub-watersheds.

The climate of upper Awash River basin is characterized as humid to subhumid in the highlands and semiarid in the lowlands with an average annual temperature of 16–17 °C and 19–20 °C, respectively. The average annual rainfall in the highlands is about 1,200 mm and in the lowlands is about 900 mm. The elevation range in the upper Awash River basin is between 1,593 and 3,559 m a.s.l., and the length of the main stream, from the upstream to the outlet (Hombole), is approximately 310 km. About 66 % of the basin has a slope greater than 5 %. The land use of the area is dominated by agriculture with coverage of 57 % by a crop known as Eragrostis teff. The forest coverage is very minimal (about 2.5 %) compared to the rest and scattered across the basin. The dominant soil type in the area (65 %) is Eutric vertisols.

30.2.2 Available Data

The raw data used to characterize the topography and slope of the watershed was the ASTER Global Digital Elevation Model with approximately 30 m (1 arc second) posting interval. The DEM is generated and distributed by the Ministry of Economy, Trade and Industry of Japan and the National Aeronautics and Space Administration. The release year for this DEM is 2009 with an accuracy of 7–14 m and freely available in the official website of the Earth Remote Sensing Data Analysis Center of Japan (ASTER GDEM 2009).

Soil and land use data were obtained from previous work (Chekol 2006). The soil data contains a detail description of the textural, physical, and chemical properties of each soil layers for the different soil types. The properties such as soil hydrologic group, available water content, soil depth, bulk density, and

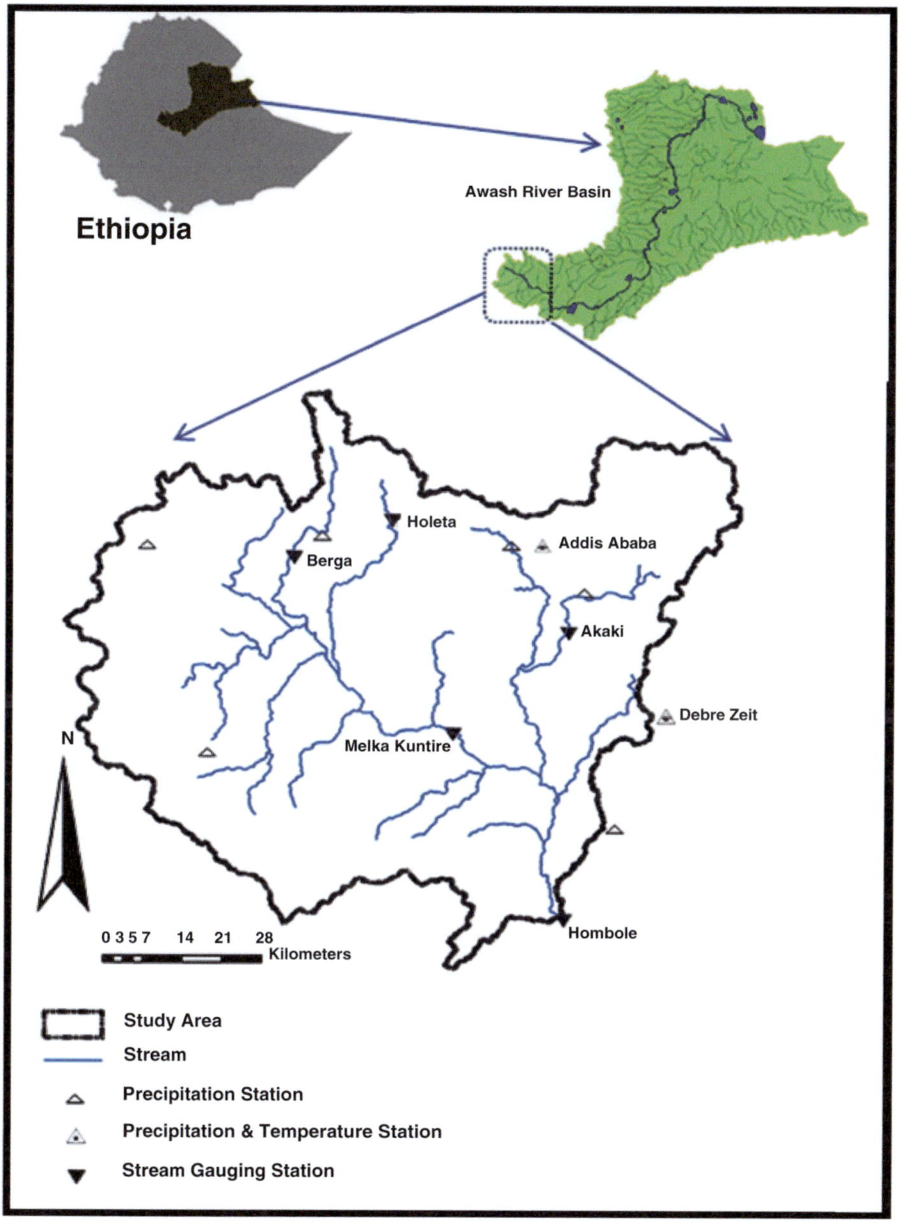

Fig. 30.1 Location map of the study area

hydraulic conductivity are among the important parameters used to determine a water budget for the soil profile and daily surface runoff. The major soil types of the study area are Eutric vertisols, Humic nitisols, Chromic luvisols, Vertic cambisol, and Lithic leptosols. The land use data are used to calculate the surface runoff and

Table 30.1 Physical characteristics of five sub-watersheds in the upper Awash River basin and the observed mean annual streamflow (mm) for 1996–2000

Characteristics		Holeta	Berga	Akaki	Melka Kuntire	Hombole
Size (km^2)		131.1	216.8	884.3	4,501.0	7,626.3
Dominant slope (%)/areal coverage (%)		>15/42.6	>15/36.5	8–15/33.4	2–5/30.8	2–5/28.9
Average elevation (m a.s.l)		2,544	2,652	2,371	2,048	2,018
Dominant soil type/depth (mm)		Luvisols/1,800	Luvisols/1,800	Vertisols/2,422	Vertisols/2,422	Vertisols/2,422
Dominant land use		Corn	Durum Wheat	Durum Wheat	Eragrostis Teff	Eragrostis Teff
Streamflow (mm)	1996	525.4	679.4	693.4	275.4	289.2
	1997	164.1	347.8	187.7	105.9	94.9
	1998	472.9	623.9	528.9	259.4	274.5
	1999	358.6	322.4	564.9	174.6	191.2
	2000	271.4	256.3	373.7	157.5	143.7

evapotranspiration in the area. The main land use types of the area are teff, durum wheat, range grasses, spring barley, summer pasture, range brush, and corn.

The climatological data used for the study were daily precipitation and maximum and minimum temperature which were obtained from the National Meteorological Agency of Ethiopia. The precipitation data were from eight measurement stations, of which six are located inside the watershed and two are nearby. Table 30.2 summarizes the annual rainfall and elevation of each station used for the study. The temperature data were obtained from two stations, of which one is located inside the watershed (Addis Ababa) and the other nearby (Debre Zeit). Daily river discharge data were obtained from the Ministry of Water Resources of Ethiopia for a total of five gauging stations (the four tributary rivers and the outlet). Table 30.1 shows the mean annual discharge for each sub-watershed. The location and spatial distribution of the stations are shown in Fig. 30.1.

30.2.3 Analysis of the Hydrometeorological Time Series Data

A statistical analysis was made on the observed time series climate and discharge data. The spatial distribution and representativeness of rainfall data in relation to observed streamflow data were assessed using a correlation coefficient method. The analysis was based on the daily data obtained from the eight rain gauges and five streamflow measurements used in the study. An additional three datasets of daily precipitation were generated, based on area-weighted method, using two stations for Akaki sub-watershed, five stations for Melka Kuntire, and eight stations for Hombole watershed. The reason for use of these additional datasets was to relate the analysis with the simulated results from the SWAT model.

30.2.4 Description of the Hydrological Model

SWAT is a physically based spatially semi-distributed watershed model developed by Arnold et al. (1998). It is a continuous time model and allows for a simulation of different physical processes in a watershed. The spatial unit for rainfall-runoff calculations is the hydrologic response unit (HRU), which is a lumped land area within a sub-watershed comprised of unique land cover, soil, slope, and management combinations. The hydrological components (like surface flow, lateral flow, groundwater flow, and evapotranspiration) are calculated for each HRU through the water balance. The cumulative total over a sub-watershed then gives the hydrological balance and main streamflow for that sub-watershed. The overall watershed hydrological balance including streamflow at the outlet of the whole watershed is then calculated from the contribution of the upstream sub-watersheds.

The hydrology part of the model is separated into two: the land phase and the routing phase of the hydrological cycle. The land phase of the hydrological cycle

Table 30.2 Annual rainfalls (mm) of eight rain gauge stations used for the study with their elevations (m a.s.l)

Year	*AAObs* (2,408 m)	*AABole* (2,354 m)	*Akaki* (2,230 m)	*Adisalem* (2,400 m)	*Debrezeit* (1,900 m)	*Ginchi* (2,290 m)	*Tulubolo* (2,100 m)	*Zukuala* (3,050 m)
1996	1,549.0	1,552.5	1,116.9	977.7	1,077.4	1,213.4	1,129.5	1,453.3
1997	952.4	885.0	777.3	1,484.2	854.4	1,027.2	651.2	920.0
1998	1,337.7	1,372.9	1,071.5	1,401.2	984.7	1,222.5	1,251.4	1,447.4
1999	947.2	946.7	955.2	846.1	921.3	1,130.3	1,085.8	1,043.6
2000	1,191.1	929.5	861.4	1,383.2	845.6	997.2	1,292.3	1,082.0

controls the amount of water that goes to the main channel from each sub-watershed, while the routing phase controls the movement of water through the channel network of the watershed to the outlet (Neitsch et al. 2005).

The hydrological cycle in the land phase as simulated by SWAT is based on the water balance equation:

$$\mathrm{SW}_t = \mathrm{SW}_o + \sum_{i=1}^{t}\left(R_{\mathrm{day}} - Q_{\mathrm{surf}} - E_a - w_{\mathrm{seep}} - Q_{gw}\right) \qquad (30.1)$$

where SW_t is the final soil water content (mm), SW_o is the initial soil water content on day i (mm), t is the time (days), R_{day} is the amount of precipitation on day i (mm), Q_{surf} is the amount of surface runoff on day i (mm), E_a is the amount of evapotranspiration on day i (mm), w_{seep} is the amount of water entering the vadose zone from the soil profile on day i (mm), and Q_{gw} is the amount of return flow on day i (mm).

The model provides two methods for the calculation of surface runoff: the Soil Conservation Service (SCS) curve number procedure (USDA-SCS 1972) and the Green & Ampt infiltration method (Green and Ampt 1911). The latter method requires sub-daily rainfall data; hence, in this study, the SCS curve number (CN) method was used. The CN is a similarity parameter ranging from 0 to 100 which serves as a representation of soil, land use, and antecedent soil water conditions for a given landscape position. The standard approach in SWAT uses CNs for various soils and land cover conditions (stored in SCS tables) that are based on average antecedent soil moisture conditions (CN2). The model incorporates two different methods for calculating the retention parameter. The first method is to allow the retention parameter to vary with soil profile water content characterized by field capacity, wilting point, and saturation water content. The second method is to allow the retention parameter to vary with the accumulated plant evapotranspiration, which makes the calculation of the daily curve number more dependent on antecedent climate. In this study, the latter method was used.

Potential evapotranspiration (PET) can be estimated by three different methods, such as Hargreaves method (Hargreaves and Samani 1985) that needs only air temperature data, Priestley-Taylor method (Priestley and Taylor 1972) that needs solar radiation and relative humidity in addition to air temperature, and Penman-Monteith method (Monteith 1965) that requires wind speed in addition to the above data. In this study, Hargreaves' method was used as it is recommended for areas where only maximum and minimum air temperature data are available (American Society of Civil Engineers 1996).

Since the curve number method was used, the amount of water that infiltrates into the soil profile is calculated as the difference between the amount of rainfall and the amount of surface runoff. This is due to the timescale of the precipitation data being daily, which makes it inconvenient to model the rate of infiltration in smaller time steps. The lateral subsurface flow, originating from the saturated zone of soil layers and contributing to the streamflow, is calculated in each layer by a kinematic storage model. Groundwater or base flow is a part of the streamflow that originates from recharged shallow aquifer through the process of percolation. In SWAT, base flow is allowed to enter the reach only if the amount of water stored in the shallow aquifer exceeds a threshold value specified by the user, while water that enters the deep aquifer is considered to be lost from the system. The rate and velocity of flow is defined by Manning's equation. Two variations of the kinematic wave model are incorporated in SWAT to route the water through the channel network; variable storage is the primary method, and Muskingum River routing is the other option. Further description of the different components of SWAT model is available in Neitsch et al. (2005).

30.2.5 Model Setup

The 30 m DEM was used to delineate the watershed and to analyze the drainage patterns of the land surface terrain. Subbasin parameters such as slope gradient and slope length of the terrain and the stream network characteristics such as channel slope, length, and width were derived from the DEM. A predefined digital stream network was used to help the delineation process to accurately capture the stream location. The stream definition was made based on the threshold area option to define the minimum size of a subbasin. The previously mentioned five stations were defined as subbasin and basin outlets in the watershed (Fig. 30.1). HRU definition was followed from the land use, soil, and slope datasets based on user-defined minimum threshold percentage of each category. In order to eliminate minor classes, based on their areal coverage, the threshold values used for this study were 8 %, 20 %, and 10 % for land use, soil, and slope, respectively. Rainfall and temperature data were obtained from the weather stations.

30.2.6 Sensitivity Analysis, Calibration, and Validation

Seven years of available data were used for the study. Calibration of the model was done for Akaki sub-watershed and Hombole watershed based on the data for the period of 1996–1998 with a prior 2-year warm-up period, 1994–1995. The warm-up or equilibration period is recommended for simulation covering a period of 5 years or less to get the hydrological cycle fully operational (Neitsch et al. 2004). The last two years' data were used for validation (1999–2000). The model was tuned by visual comparison of the observed and simulated hydrographs and analyzing the model performance statistics during the calibration period. To evaluate the model capability of simulating surface and subsurface hydrological processes, a base flow filter method developed by Arnold et al. (1995) was used to divide the total observed streamflow into surface runoff and subsurface runoff. These values were assumed to represent observed surface and subsurface runoff for this study and were compared to their equivalence of the SWAT simulated total streamflow for each year on the respective calibrated watersheds. The other three interior points at tributary rivers, Holeta, Berga, and Melka Kuntire, were used to evaluate the model performance based on the parent watershed (Hombole) calibration. In order to comprehend the ability of the model in simulating ungauged watersheds, the discharge data from the interior points were not used for the calibration process but were left for validation. In case of the Akaki sub-watershed, two scenarios were evaluated: the first was the calibrated results from using the Akaki discharge data and the second was an interior point validation for the calibrated results from using the parent watershed (Hombole) discharge data.

A *sensitivity analysis* was performed on the model to select the most sensitive parameters, out of the total 27 flow parameters included in SWAT. A sensitive parameter is one that changes the model outputs of the streamflow significantly per unit change in its value. Automated Latin-Hypercube One-factor-At-a-Time (LH-OAT) global sensitivity analysis procedure (van Griensven et al. 2006) incorporated in the model was used in the study.

The calibration of the selected sensitive parameters was undertaken in different steps. Manual calibration was the first step to obtain some level of agreement with the observed flow and also to minimize the potential range that bounds the value of each parameter. The parameters can be changed either in a lumped way, for the whole catchment, or in a distributed way, for a subbasin or HRU. The autocalibration and uncertainty analysis were done using parameter solution (ParaSol) method (van Griensven and Meixner 2007) incorporated in SWAT.

ParaSol is a multi-objective uncertainty method that is efficient in optimizing a model and providing parameter uncertainty estimates (van Griensven and Meixner 2007). It calculates objective functions (OF) based on the simulated and observed time series and aggregates the OFs into a global optimization criterion (GOC). The optimization was done by adapting the shuffled complex evolution approach for effective and efficient global minimization method (SCA-UA). The SCA-UA algorithm is a global search algorithm for the minimization of a single function that is used to deal with a maximum of 16 parameters (Duan et al. 1992). The minimized parameter space through the manual calibration was used for each parameter to reduce the running time.

The *objective function* (OF) used in this study was the sum of the squares of the residuals (SSQ) with an objective of minimizing the differences between the simulated and measured time series.

$$\mathrm{SSQ} = \sum_{i=1}^{N} \left[Q_{i,s} - Q_{i,m}\right]^2 \tag{30.2}$$

where N is the number pairs of simulated ($Q_{i,s}$) and measured ($Q_{i,m}$) streamflow values.

Since the autocalibration was based on SSQ only, which has a high priority of minimizing the large differences and often tend to force the model to underestimate the peak flows for a result of lower objective function values (Eckhardt and Arnold 2001), further tuning of parameters was made manually by comparison of the simulated and observed hydrographs.

The *validation* of the model was performed to test if the calibrated parameter set would behave consistently for the watershed using different observed datasets for different periods. In this study, the only observed data used for model validation was the measured streamflow datasets for 2 years (1999–2000).

30.2.7 Model Performance Evaluation

The performance of the model was evaluated using three model goodness-of-fit measures that are widely used in hydrology (Moriasi et al. 2007). The Nash-Sutcliffe efficiency (NSE) (Nash and Sutcliffe 1970) is a commonly used method related to the objective function of the calibration (SSQ). During the calibration process, the calculation was based on the variables within that period. When applied to the validation period, the initial variance was calculated as the sum of squares of deviation from the mean $\left(\overline{Q_m}\right)$ of the observed time series of the calibration period. The coefficient of determination (R^2) was used to characterize the amount of variance in the observed data explained by the model results.

$$\mathrm{NSE} = 1 - \frac{\sum_{i=1}^{N} \left(Q_{i,m} - Q_{i,s}\right)^2}{\sum_{i=1}^{N} \left(Q_{i,m} - \overline{Q_m}\right)^2} \tag{30.3}$$

$$R^2 = \left[\frac{\sum_{i=1}^{N} \left(Q_{i,m} - \overline{Q_m}\right) \cdot \left(Q_{i,s} - \overline{Q_s}\right)}{\sqrt{\sum_{i=1}^{N} \left(Q_{i,m} - \overline{Q_m}\right)^2 \cdot \sum_{i=1}^{N} \left(Q_{i,s} - \overline{Q_s}\right)^2}} \right]^2 \tag{30.4}$$

where $\left(\overline{Q_s}\right)$ is the mean of the simulated time series.

Percent bias (PBIAS) measures the average trend of the simulated data to be larger or smaller than their observed equivalent in the preferred time step (Gupta et al. 1999). The ideal value should be zero; however, the lower the values, the more precise the model simulations. Negative values indicate a bias toward underestimation, whereas a positive value indicates an overestimation.

$$\mathrm{PBIAS} = 100^{*} \frac{\sum_{i=1}^{N} \left(Q_{i,s} - Q_{i,m}\right)}{\sum_{i=1}^{N} \left(Q_{i,m}\right)} \tag{30.5}$$

The performance of the model can be explained based on the recommended ranges: NSE values greater than 0.75 show good model efficiency, NSE between 0.36 and 0.75 shows satisfactory performance, and value less than 0.36 is considered to be unsatisfactory (Motovilov et al. 1999; Van Liew et al. 2007; Moriasi et al. 2007). Similarly, PBIAS values greater than and equal to twenty-five percent (±25 %) are considered to be satisfactory.

30.3 Results and Discussion

30.3.1 Hydrometeorological Time Series Trend Analysis

The monthly average air temperature values, of the 10 years (1996–2006) data from the Addis Ababa and Debre Zeit stations, are summarized in Table 30.3. For the Addis Ababa station, the average temperature was 17.2 °C with the maximum and minimum temperatures being 18.9 °C and 15.8 °C, respectively. For the Debre Zeit station, the average temperature was 19.4 °C with the maximum and minimum temperatures being 21.2 °C and 17.4 °C, respectively. The maximum and minimum average temperatures for both the stations occur in May and December, respectively. However, the monthly variation was not significant as indicated by the low values of the standard deviation (average 1.1). The altitude difference, 2,408 m for Addis Ababa Observatory and 1,900 m for Debre Zeit stations, could explain the slight differences in temperatures between the corresponding months of the two stations. On an average, Debre Zeit is warmer by 2.2 °C throughout the year.

Table 30.3 Statistical summary of monthly average air temperature from two stations and area-weighted rainfall from eight stations in the upper Awash River basin

	Temperature (°C)		Area-weighted rainfall (mm)	
	Addis Ababa Observatory	Debrezeit	Mean	Standard deviation
Jan	16.7	18.5	15.7	5.62
Feb	17.9	19.8	34.7	15.10
Mar	18.5	20.7	58.0	12.11
Apr	18.7	21.1	81.9	19.59
May	18.9	21.2	87.1	14.83
Jun	17.5	20.3	154.3	36.35
Jul	16.4	19.1	271.4	28.95
Aug	16.4	19.1	284.1	37.03
Sep	16.7	19.2	135.1	17.68
Oct	16.8	18.7	29.2	8.86
Nov	16.0	17.7	8.8	3.96
Dec	15.8	17.4	6.3	3.80
Mean	17.2	19.4	97.2	
Max	18.9	21.2	284.1	37.03
Min	15.8	17.4	6.3	3.8
Standard deviation	1.03	1.20		
Total			1,166.6	

The area-weighted monthly rainfall analysis for Hombole watershed is summarized in Table 30.3. The pattern and characteristics of rainfall varies in different parts of the country mainly due to its geographic location and topography (Degefu 1987; Bekele 1997). Based on the three main rainfall regimes, the watershed is characterized by two different rainfall patterns (Bekele 1997). Out of the analyzed eight stations, three of them (Addisalem, Ginchi, and Tulubolo) fall in the regime characterized by a mono-modal rainfall pattern from March to September. The rest of the stations fall in the regime characterized by three distinct seasons: wet, small rain, and dry season. The wet season starts in June and continues up to September. The small rainy season is from February to May, while the dry season usually occurs from October to January. The standard deviation of the monthly rainfall data shows that the variation between the stations is high during the wet season (37 mm in Aug) and low in dry season (3.8 mm in Dec). The calculated area-weighted rainfall for Hombole watershed follows the mono-modal pattern, which agrees to the three stations (Addisalem, Ginchi, and Tulubolo). The mean annual rainfall was 97.2 mm with the maximum and minimum rainfall being 284.1 mm in August and 6.3 mm in December, respectively. Even though the amount of rainfall shows a distinct boundary between the two rainy seasons (wet and small rain), the separator, which is the relatively short "dry" period (Bekele 1997) in between (in this case May), resulted in a single peak rainfall pattern.

The flow regimes of Hombole watershed were analyzed using 39 years of observed streamflow data (1968–2006). The average annual runoff of the watershed is estimated at 42.97 m^3/s (177 mm). The wettest year was 1971 with a 75.99 m^3/s (313 mm) discharge, and the year 1987 with 20.88 m^3/s (86 mm) discharge was the driest. The frequency analysis shows that the exceedance probability from the mean annual streamflow value was 43 %, out of which 23 % exceeds the mean by 10 % (Fig. 30.2). The average monthly flow follows the rainfall trend which is a mono-modal curve with a peak in August. Monthly trends of rainfall, streamflow, and average temperature for Hombole watershed is presented in Fig. 30.3.

30.3.2 Spatial Distribution and Representativeness of Rainfall Data

One of the major limitations to large area hydrological modeling is the spatial variability associated with the precipitation. Even though eight rainfall gauges were used in this study, the spatial distribution in relation to the watershed area could be considered to be inadequate. According to Arnold et al. (1998), using one gauge to represent an entire sub-watershed or even by means of area-weighted methods for rainfall representation can cause considerable error in the runoff estimation. The correlation coefficient of the precipitation data and observed streamflow might help in explaining the model performance in different sub-watersheds. According to the analysis, the correlation between the observed streamflow at Akaki sub-watershed

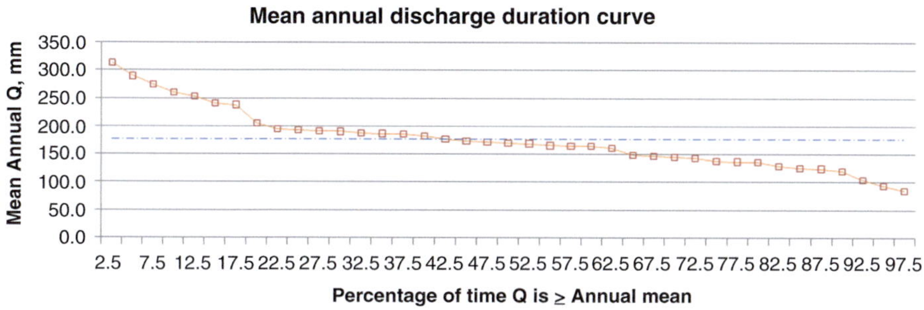

Fig. 30.2 Exceedance probability curve showing the percent of time that a given discharge equaled or exceeded the mean annual discharge of Hombole watershed through the 39 years of record (1968–2006)

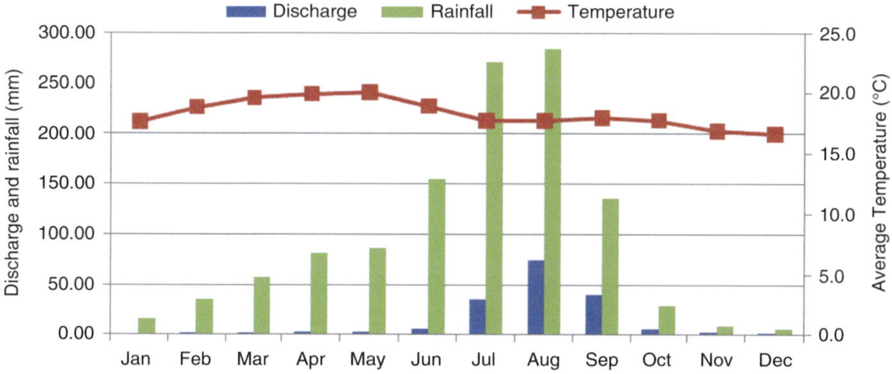

Fig. 30.3 Graphical presentation of average monthly rainfall (mm), streamflow (mm), and temperature (°C) trends of Hombole watershed

and the corresponding rainfall gauge considered by SWAT (in this case area weighted) is poor in respect of both Melka Kuntire and Hombole watersheds (Table 30.4). Holeta and Berga sub-watersheds, having only a single rainfall station, had the lowest correlation coefficient relative to the others.

30.3.3 Sensitivity Analysis

The sensitivity analysis indicated that the streamflow prediction was sensitive to variation in surface water, groundwater, and soil parameters (Table 30.5). Among the 27 hydrological parameters, the most sensitive ones were the base flow alpha factor (ALPHA_BF), initial SCS curve number II (CN2), soil evaporation compensation factor (ESCO), groundwater delay (GW_DELAY), deep aquifer percolation fraction (RCHRG_DP), threshold water depth in the shallow aquifer for “revap”

Table 30.4 Correlation coefficient between the observed rainfall and streamflow data from 1996 to 2000 (the highlighted ones are the combinations selected by SWAT)

Rainfall stations	*Holeta* Q	*Berga Q*	*Akaki Q*	*Melka Kunitre Q*	*Hombole Q*
Addis Abeba Bole	0.22	0.28	0.30	0.32	0.31
Addis Abeba Obs	0.28	0.29	0.35	0.36	0.36
Adisalem	0.17	0.24	0.18	0.35	0.32
Akaki	0.20	0.27	0.33	0.33	0.30
Debre Zeit	0.22	0.25	0.27	0.33	0.36
Ginchi	0.19	0.24	0.20	0.29	0.27
Tulubolo	0.29	0.29	0.27	0.41	0.40
Zukuala	0.17	0.16	0.21	0.25	0.25
Akaki area weighted	0.28	0.33	0.39	0.40	0.39
Hombole area weighted	0.36	0.41	0.41	0.56	0.54
Melka Kuntire area weighted	0.32	0.37	0.33	0.52	0.49

(REVAPMN), available soil water capacity (SOL_AWC), and saturated hydraulic conductivity (SOL_K). The sensitivity of each parameters for that particular watershed depends upon the topography, the soil types, and the land use/land cover types of the watershed that significantly impact the infiltration of water into the soil, movement of surface water and groundwater, and other hydrological components.

30.3.4 Model Calibration

The calibration range and the final fitted values for the sensitive parameters are summarized in Table 30.5. The base flow recession constants from filter techniques (Arnold et al. 1995) were used as the initial estimates for the base flow alpha factor parameter, which was 0.0424 for Akaki and 0.0193 for Hombole. The base flow recession constant is a direct index of groundwater flow response to changes in recharge (Smedema et al. 2004). The values may vary from 0.05 to 0.10 day^{-1} for land with slow response to 5–10 day^{-1} for land with a rapid response. In SWAT model, the possible parameter space for the base flow recession constant ranges

Table 30.5 Selected flow-sensitive SWAT model parameters with their default value, calibration range, and final estimate using the ParaSol method

		Calibration range		Final estimate	
Parameter name	Default value	Akaki	Hombole	Akaki	Hombole
CN2	[a]	±25 %	−20–0 %	−15.4 %	−17.5 %
ALPHA_BF	0.048	0–1	0–0.05	1.0	0.049
GW_DELAY	31	0–500	0–10	2	1
RCHRG_DP	0.05	0–1	0.05–1	0.001	0.305
REVAPMN	1	0–500	1–10	100	1
ESCO	0.95	0–1	0.5–0.95	0.7	0.65
SOL_AWC	[b]	±25 %	−20–5 %	−4.8 %	−14 %
SOL_K	[b]	± 25 %	−20–5 %	−6 %	−10.3 %
SURLAG	4	1–24	0–4	2	0.5

[a]SWAT-driven parameter values
[b]Model input data

between 0 and 1. The calibration result shows that the alpha factor parameter holds the topmost value from the preferred range in both the stations: Akaki (1.0) and Hombole (0.049). This might explain their difference in response to recharge. According to Smedema et al. (2004), the base flow recession constant tends to increase as the basins become smaller. On the other hand, the parameter that controls the evaporation from the shallow aquifer (REVAPMN) is much larger for Akaki (100) than for Hombole (1). These significant ranges in parameter values could substantiate the spatial heterogeneity that exists between the watersheds (Van Liew et al. 2007).

Peak flows: The predicted values of daily peak flows compared to the observed streamflows differed between the two watersheds (Fig. 30.4). For Akaki sub-watershed, underestimation of the daily peak values was observed during the entire wet season, while some overestimations occurred during the small rainy season. For Hombole watershed, the simulated flow showed overestimation on some of the daily peak flows in both seasons.

The comparison between the observed and simulated streamflow hydrographs indicated that there is a good agreement between the observed and simulated discharges of the calibrated model, which could be verified by higher values of the performance evaluators (Table 30.7). For daily data, the NSE values for Akaki and Hombole watersheds were 0.53 and 0.72, respectively. Similarly, the corresponding R^2 values were 0.82 and 0.97, while the PBIAS values were −2.86 and 2.28. The negative and positive values of the PBIAS indicate the underestimation and overestimation of the total simulated flow, respectively. According to the recommended ranges of model goodness-of-fit measures, the performance of the model is satisfactory for both the watersheds (Motovilov et al. 1999; Van Liew et al. 2007; Moriasi et al. 2007).

Analysis of the monthly data indicated a trend similar to that of the daily time step. The NSE values were 0.82 and 0.97 for Akaki and Hombole watersheds, respectively. Likewise, the R^2 values were 0.83 and 0.97, while the corresponding

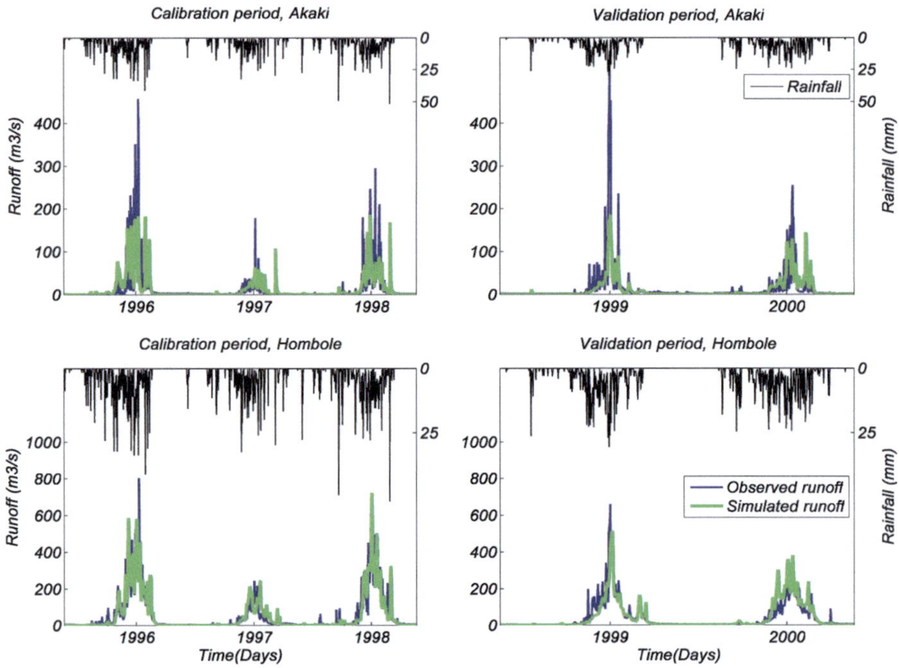

Fig. 30.4 Daily observed and simulated hydrographs for Akaki sub-watershed (*top*) and Hombole watershed (*bottom*) during the calibration (1996–1998) and validation (1999–2000) periods

PBIAS values were −2.62 and 2.24. According to the performance ranges, the model showed good performance in predicting the monthly streamflow/discharges.

The difference between the two watersheds in model performance could be due to the forcing data. The correlation between the rainfall and runoff for Akaki watershed was poorer than for Hombole (Table 30.4). One of the main sources of model uncertainties is acknowledged (Refsgaard and Storm 1996; Beven 2001) to be errors in the input variables such as rainfall and temperature. Hence, the poor model performance in Akaki sub-watershed could be due to poor quality of the gauged climatic variables as well as the very coarse spatial distribution of weather stations in the sub-watershed.

30.3.5 Model Validation

After a successful calibration of the model for prediction of streamflow, the final calibrated parameter values were incorporated into the SWAT model for validation. Validation involves a comparison of model results with an independent dataset with no further adjustments. For this study, an independent precipitation and temperature data and streamflow dataset (1999–2000) were used. The result indicated that the daily and monthly simulated hydrographs were in good agreement with the observed flows (Fig. 30.4).

The validation period analysis of the model performance showed that for Akaki watershed, the NSE was 0.55 and 0.82 on daily and monthly time steps, respectively (Table 30.7). Likewise, the R^2 and PBIAS were 0.59 and −23.38 on daily basis and 0.86 and −23.27 on monthly time steps. Even though, the model performance was satisfactory, the PBIAS showed an underestimation of the observed total streamflow in a much larger quantity than the calibration period. For Hombole watershed, the NSE was 0.71 and 0.91 for daily and monthly time steps, respectively. Similarly, the R^2 was 0.74 and 0.92, while the PBIAS was 3.61 and 3.55 for daily and monthly time steps, respectively.

Peak flows: The daily peak flows showed some variation for this period. For Akaki sub-watershed, the underestimation of the simulation that was observed during the calibration period continued in both the years (Fig. 30.4). While for Hombole watershed in the year 1999 it showed an underestimation, especially in the rising limb, in the year 2000, the simulation resulted in an overestimation. The magnitude of the underestimation in year 1999 was exceptional for Hombole when compared to all the other years. This might be due to the various trends between the different rainfall stations that could affect the area-weighted rainfall on which the simulation is dependent (Table 30.2).

Generally, as shown in Table 30.6, the model simulated with an overall trend of underestimation of the subsurface runoff in the relatively dry years (1997 and 1999) for both Akaki and Hombole watersheds. An exception was the 1997 simulation for Hombole watershed, which was a less pronounced overestimation in comparison to the overestimation of the surface runoff. The overall performance, both during calibration and validation period, might indicate that the model performs relatively well during wet years.

30.3.6 Hydrological Simulations at Tributary Rivers

The model performance on interior points, based on a parent watershed calibration, is summarized in Table 30.8. The sub-watersheds in the table are listed in the order of the increasing drainage area. A noticeable trend shown in the table was an improvement in performance, on both the NSE and R^2 values, that was directly proportional with the size of the sub-watersheds. The performance was poor for the smaller interior points, in particular for Holeta and Berga. The possible explanation could be the insufficient capacity of smaller basins to dampen out input signals and consequent input errors as discussed in Reed et al. (2004) and Shrestha et al. (2005). The coefficient of variation for the daily streamflow data showed that the smaller sub-watersheds exhibited more variability than the larger ones, which could affect the accuracy of the simulation (Fig. 30.5). According to Hirpa et al. (2010), the statistical analysis on daily flow data shows that the degree of multifractality of river flow decreases with increasing watershed area, which could point out the

Table 30.6 Observed and simulated annual runoff volume for Akaki and Hombole watersheds with the corresponding area-weighted rainfall during the calibration (1996–1998) and validation (1999–2000) periods

	Annual rainfall (mm)	Surface runoff (mm)		Subsurface runoff (mm)		Total runoff (mm)		Difference (%) between predictions and observations in:		
Year	Observed	Observed	Simulated	Observed	Simulated	Observed	Simulated	Surface runoff	Subsurface runoff	Total runoff
Results for ***Akaki***										
1996	1,350.2	558.6	518.3	134.8	143.6	693.4	662.0	−7.2	6.5	−4.5
1997	871.8	112.2	155.6	75.5	66.5	187.7	222.0	38.7	−11.9	18.3
1998	1,215.3	409.6	360.9	119.3	124.8	528.9	485.7	−11.9	4.6	−8.2
1999	950.9	440.8	243.4	124.1	80.9	564.9	324.4	−44.8	−34.8	−42.6
2000	1,039.4	271.5	284.2	102.2	110.4	373.7	394.5	4.7	8.0	5.6
Results for ***Hombole***										
1996	1,207.3	165.3	164.4	123.9	122.9	289.2	287.3	−0.5	−0.8	−0.7
1997	952.9	52.0	66.8	42.8	49.6	94.9	116.4	28.5	15.9	22.7
1998	1,284.7	169.0	158.8	105.4	111.1	274.5	269.9	−6.0	5.4	−1.7
1999	996.5	112.7	99.5	78.5	61.5	191.2	161.1	−11.7	−21.7	−15.7
2000	1,147.9	77.8	97.2	65.9	88.7	143.7	185.8	24.9	34.6	29.3

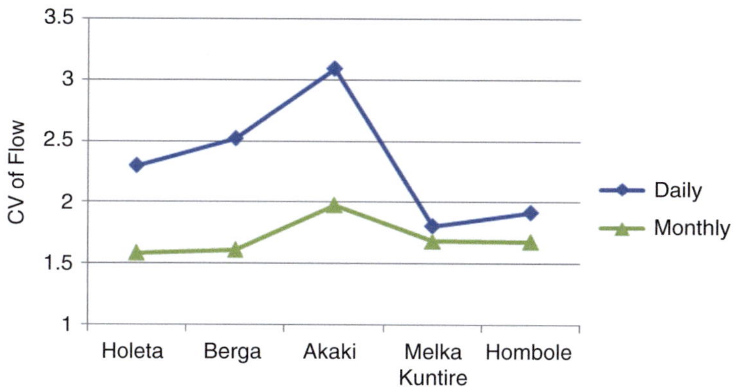

Fig. 30.5 Coefficients of variation (CV) of observed daily and monthly streamflow from 1996 to 2000 for each watershed

similarity between larger basins in preserving the different intensities of flow fluctuations. The study concludes that the watershed area is an important factor that controls the long memory of river flow fluctuations. Hence, the direct transferring of parameter values should be based on catchment characteristic analysis that considers the effects of the watershed area. On the other hand, the monthly variation showed that the catchments could, to some extent, stabilize the input signals at this timescale (Fig. 30.5). The streamflow response seemed to incorporate the influence of the catchment characteristics and damp the fluctuations of the rainfall, which could explain the better performance of the model at a monthly timescale. The uncertainties that come from the spatial distribution of rainfall is also another possible reason as illustrated by the coefficient of correlation analysis (Table 30.4).

The performance of the model at Akaki sub-watershed was analyzed from the two scenarios (Tables 30.7 and 30.8): the first one is the performance based on the calibration of the model using the streamflow data of Akaki, and the second one is the performance of the model at Akaki as an interior point. The comparison, based on all the three model evaluators (NSE, R^2 and PBIAS), showed that the model performed better in the first case (Table 30.7) for a daily timescale. The monthly timescale comparison indicated a relatively poor performance in the first scenario during the calibration period, while it showed better performance during the validation period. The PBIAS values indicate a better performance in all the years for the first scenario. Even though the performance of the model mostly improved by using the streamflow data for calibration, the results also indicate the possibility of using the model for ungauged interior points having similar catchment characteristics as that of a gauged watershed.

Table 30.7 Model performance statistics for prediction of streamflow in Akaki and Hombole watersheds in comparison with the uncalibrated model simulation during the calibration period (1996–1998)

		Akaki sub-watershed		Hombole watershed	
Evaluator	Period	Daily	Monthly	Daily	Monthly
NSE	Uncalibrated	0.25	0.78	−1.59	−0.44
	Calibration	0.53	0.82	0.72	0.97
	Validation	0.55	0.82	0.71	0.91
R^2	Uncalibrated	0.35	0.79	0.54	0.91
	Calibration	0.53	0.83	0.75	0.97
	Validation	0.59	0.86	0.74	0.92
PBIAS	Uncalibrated	10.42	10.79	117.31	117.26
	Calibration	−2.86	−2.62	2.28	2.24
	Validation	−23.38	−23.26	3.61	3.55

30.3.7 Water Balance of Upper Awash River Basin

The major components of water budget were the total amount of precipitation (RF) falling on the watershed, followed by actual evapotranspiration (ET), surface runoff (SurQ), groundwater flow (GWQ) originating from the shallow aquifer, percolation to the deep aquifer (D_Aq), lateral flow (LATQ) originating from the soil profile, soil water storage (ΔSW), and transmission losses (TransL) that occur due to leaching through the streambed of tributary channels. The water balance components of Hombole watershed are shown in Fig. 30.6 for each of the simulated years. The results showed that the loss of water through ET was dominant in all the years. The amount of ET loss was between 59 and 73 % of the total annual precipitation, with the maximum percentage (73 %) occurring in the driest year (1997). The foremost contributor to the water yield was the surface runoff with an exception in the year 1999 when the groundwater flow was greater than the surface runoff by about 1.2 mm.

The yearly soil water storage (including the vadose zone) resulted in a surplus with a more pronounced amount in the relatively dry years (1997 and 1999), although an exceptional deficit occurred in the year 1996 (Fig. 30.6). The yearly trend of all the water balance components, as shown in Fig 30.6, pronounced that the change in storage is inversely correlated to the other components. For example, the increase in the amount of surface runoff was followed by a decrease in soil water storage in relative to the previous year. This reversed relation of the soil water storage was also observed with respect to the other water budget components such as precipitation and actual evapotranspiration. This might further explain the overall interrelation of the water budget components. For example, as in the case of the year 1998, as the amount of precipitation increased, the capacity of the soil for further infiltration reduced, which results in an increment of surface runoff and evaporation from bare soil.

Table 30.8 Model performance statistics for prediction of streamflow at four interior points (Holeta, Berga, Akaki, and Melka Kunitre) based on a parent watershed calibration (Hombole) in comparison with the uncalibrated model simulation during the calibration period (1996–1998)

		Holeta		Berga		Akaki		Melka Kuntire	
Evaluator	Period	Daily	Monthly	Daily	Monthly	Daily	Monthly	Daily	Monthly
NSE	Uncalibrated	−2.26	−0.79	−0.58	0.49	0.25	0.78	−2.49	−0.76
	Calibration	−1.00	−0.20	0.01	0.59	0.46	0.83	0.63	0.79
	Validation	−0.38	0.29	−0.13	0.45	0.37	0.63	0.53	0.76
R^2	Uncalibrated	0.06	0.33	0.12	0.61	0.35	0.79	0.52	0.88
	Calibration	0.10	0.39	0.21	0.64	0.47	0.87	0.67	0.89
	Validation	0.05	0.37	0.08	0.38	0.43	0.73	0.55	0.82
PBIAS	Uncalibrated	49.93	49.27	9.45	9.23	10.42	10.79	120.75	120.59
	Calibration	21.36	20.89	−8.87	−9.05	−17.77	−16.21	−33.37	−33.44
	Validation	−17.46	−17.37	3.21	3.58	−38.17	−36.91	−32.85	−33.01

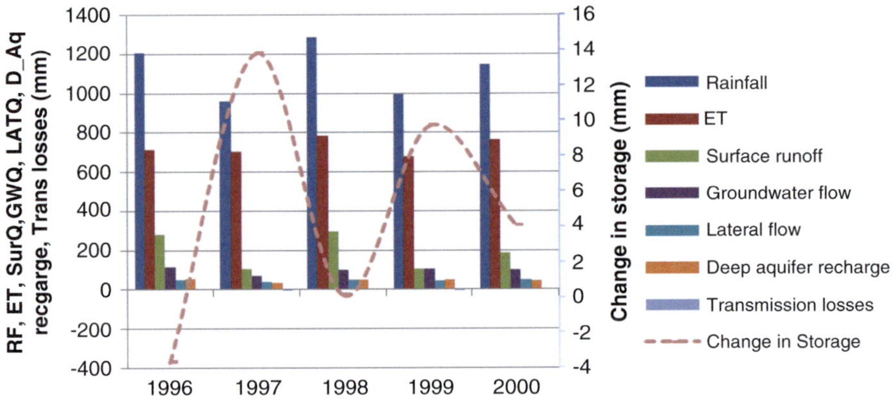

Fig. 30.6 Water balance components of Hombole watershed with the corresponding average annual values (mm) for each year of the simulation period

30.3.8 Model Prediction Performance and Uncertainties

The overall performance of the model depends on different factors. The forcing input data (rainfall and temperature) and the streamflow observations used to calibrate and validate the model outputs could be mentioned as the major uncertainty sources. Apart from data quality, in a place like the Awash River Basin with an altitude ranging from 250 to 3,600 m a.s.l., the extrapolation of rainfall from a distant gauge is an obvious source of bias. The limited ability for rain gauges to capture the localized events that contribute to surface runoff is one of the problems in using the point rainfall data. This might add some prediction uncertainty.

The main limitation in using a distributed hydrological model is the large number of parameters that need to be optimized in order to reach an acceptable prediction of the hydrological processes. Even though some of the parameters are measurable in the field, it is not practicable to get error-free measurements. The scale of the measured parameters that usually does not match the model element or discretization scale, which is much larger (Beven 2001), also adds to the limitations. During the calibration process, the values of the parameters for best fit depend on the initial values, in this case the SWAT default parameter values designed for catchments in the United States. The autocalibration method used in this study gave only one optimum set of parameter as the best parameter values. Since this process takes an extended time of running before giving any result, minimizing the range of this value (potential parameter space), based on the manual calibration, for each calibrated parameter was essential. The existence of multiple optimal parameter sets that can give a good prediction of streamflow is usually possible, which is explained in the concept of equifinality of model structures and parameters (Beven 1993, 2001; White and Chaubey 2005). For Akaki sub-watershed, the potential default parameter space was used during the autocalibration of all the sensitive parameters, while for Hombole, a range was defined based on the manual calibration. This might explain the significant difference in the alpha factor and REVAPMN and RCHRG_DP parameters between the two watersheds.

Even though the SWAT model is distributed based on HRU units, the parameterization was done in a more lumped way, mainly on subbasin scales, during the calibration process. The total number of HRUs was 399, and this large number made it difficult to calibrate each of them separately. Apart from the spatial variation, the temporal variation is also important for areas characterized by a distinct seasonal trend. The calibration process for the curve number is especially sensitive for seasonal variation, in that the performance of the model would be improved at the expense of the peak flows and vice versa.

In this study, the Hargreaves equation (Hargreaves and Samani 1985) was used to estimate the potential evapotranspiration (PET) due to the limited availability of data. This is advantageous in areas like Ethiopia, where it is unlikely to get a good quality of solar radiation, humidity, and wind data. The two main drawbacks of the Hargreaves equation are that it overestimates in humid regions and underestimates in arid areas and the associated error occurs when used for shorter time periods and intervals of less than a week (Jensen et al. 1990; Droogers and Allen 2002). Hargreaves' method also does not include the effect of wind on evapotranspiration. In cases where the wind is a predominating factor, the method can introduce some errors (Setegn et al. 2009a).

30.4 Conclusions

The comparison between the observed and simulated streamflow indicated that there was a good agreement between the observed and simulated discharge of the calibrated model, which was verified by higher values of coefficient of determination (R^2) and Nash-Sutcliffe efficiency (NSE) and good agreement in the hydrographs. The model evaluation statistics for streamflows gave acceptable results which were verified by NSE and $R^2 > 0.36$ and PBIAS ± 25 %.

The analysis on the streamflow measurements from the five stations showed that the flow regime of each sub-watershed differed according to their catchment characteristics like drainage area, slope, land use, and soil type. The coefficient of variation of the streamflows on a daily scale showed high variability in smaller basins; however, on a monthly scale, the variation was more or less similar for all watersheds. The 39 years of record of streamflow at the outlet showed that the mean annual discharge exceeded about 43 % of the time.

The hydrological water balance analysis showed that surface runoff is an important component of the total discharge within the watershed that contributed more than the groundwater flow. More than 63 % of the water losses from the annual precipitation were through the evapotranspiration process. Even though the model performed in a satisfactory level, the performance level should not be generalized equally for all purposes. The daily results are much more important than the monthly if the simulation is for flood analyses. By contrast, for hydropower and irrigation purposes, the monthly results could help for allocating and planning. The calibrated model can be used to analyze the effects of climate and land use change and different management scenarios.

References

Abbott MB, Bathurst JC, Cunge JA, O'Connel PE, Rasmussen J (1986) An introduction to the European hydrological system-system hydrologique European, "SHE", 1: history and philosophy of a physically based, distributed modelling system. J Hydrol 87:45–59

Ahl RS, Woods SW, Zuuring HR (2008) Hydrologic calibration and validation of SWAT in a snow-dominated rocky mountain watershed, Montana, U.S.A. J Am Water Resour Assoc (JAWRA) 44(6):1411–1430

American Society of Civil Engineers (1996) Hydrology handbook. ASCE manuals and reports on engineering practice No. 28. American Society of Civil Engineers (ASCE), New York

Anderson J, Refsgaard JC, Jensen KH (2001) Distributed hydrological modeling of the Senegal River Basin-model construction and validation. J Hydrol 247:200–214

Anderson PL, Tesfaye YG, Meerschaert MM (2007) Fourier-PARMA models and their application to river flows. J Hydrol Eng 12(5):462–472

Arnold JG, Allen PM, Muttiah R, Bernardt G (1995) Automated base flow separation and recession analysis techniques. Ground Water 33(6):1010–1018

Arnold JG, Sirinivasan R, Muttiah RS, William JR (1998) Large area hydrologic modeling and assessment. Part 1, model development. J Am Water Resour Assoc (JAWRA) 34:73–89

ASTER GDEM (2009) ASTER global digital elevation model. Earth Remote Sensing Data Analysis Center (ERSDAC). http://gdem.ersdac.jspacesystems.or.jp/. Accessed 15 Apr 2010

Becker A, Serban P (1990) Hydrological models for water-resources system design and operation. WMO Oper Hydrol Rep Geneva 34:34–80

Bekele F (1997) Ethiopian use of ENSO information in its seasonal forecasts. Internet J Afr Stud 2. http://www.bradford.ac.uk/research-old/ijas/ijasno2/ijasno2.html

Beven KJ (1993) Prophecy, reality and uncertainty in distributed hydrological modeling. Adv Water Resour 16:41–51

Beven KJ (2001) Rainfall-runoff modeling. The primer. Wiley, Chichester

Brewer MS, Lee R, Helvey JD (1982) Predicting peak stream flow from an undisturbed watershed in the Central Appalachians. J Am Water Resour Assoc (JAWRA) 18(5):755–759

Chekol DA (2006) Modeling of hydrology and soil erosion of Upper Awash River Basin. Dissertation, University of Bonn

Chow VT, Maidment DR, Mays LW (1988) Applied hydrology. McGraw-Hill, New York

Crawford NH, Linsely RK (1966) Digital simulation in hydrology: the Stanford watershed model IV. Technical report No. 39, Stanford University, Palo Alto, CA

Degefu W (1987) Some aspects of meteorological drought in Ethiopia. In: Glantz M (ed) Drought and hunger in Africa: denying famine a future. Cambridge University Press, Cambridge, pp 23–26

Droogers P, Allen RG (2002) Estimating reference evapotranspiration under inaccurate data conditions. Irrig Drain Syst 16:33–45

Duan QY, Sorooshian S, Gupta V (1992) Effective and efficient global optimization for conceptual rainfall-runoff models. Water Resour Res 28(4):1015–1031

Eckhardt K, Arnold JG (2001) Automatic calibration of a distributed catchment model. J Hydrol 251:103–109

Emergency Appeal (2006) Ethiopia: floods. Preliminary appeal no. MDRET003, 18 August 2006. International Federation of Red Cross and Red Crescent Societies

Georgakakos KP, Seo D-J, Gupta H, Schaake J, Butts MB (2004) Towards the characterization of stream flow simulation uncertainty through multimodal ensembles. J Hydrol 298:222–241

Govindaraju RS, Rao AR (eds) (2000) Artificial neural networks in hydrology. Kluwer Academic, Dordrecht

Green WH, Ampt GA (1911) Studies on soil physics (Part I): the flow of air and water through soils. J Agric Sci 4:1–24

Gupta HV, Sorooshian S, Yapo PO (1999) Status of automatic calibration for hydrologic models: comparison with multilevel expert calibration. J Hydrol Eng 4(2):135–143

Hailemariam K (1999) Impact of climate change on the water resources of Awash River Basin, Ethiopia. Clim Res 12:91–96

Hargreaves GH, Samani ZA (1985) Reference crop evapotranspiration from temperature. Appl Eng Agric 1(2):96–99

Hirpa FA, Gebremichael M, Over TM (2010) River flow fluctuation analysis: effect of watershed area. Water Resour Res 46, W12529. doi:10.1029/2009WR009000

Jayakrishnan R, Srinivasan R, Santhi C, Arnold JG (2005) Advances in the application of the SWAT model for water resources management. Hydrol Process 19:749–762

Jensen ME, Burman RD, Allen RG (eds) (1990) Evapotranspiration and irrigation water requirements. ASCE manuals and reports on engineering practice No. 70. The Society, New York

Jones JAA (1997) Global hydrology: processes, resources and environmental management. Addison Wesley Longman Limited, Essex

Kusre BC, Baruah DC, Bordoloi PK, Patra SC (2010) Assessment of hydropower potential using GIS and hydrological modeling technique in Kopili River basin in Assam (India). Appl Energy 87:298–309. doi:10.1016/j.apenergy.2009.07.019

Mekonnen MA, Wörman A, Dargahi B, Gebeyehu A (2009) Hydrological modelling of Ethiopian catchments using limited data. Hydrol Process 23:3401–3408

Ministry of Water Resources (2010) Projects and programs: dams and hydropower. Ministry of Water and Energy, Addis Ababa, Ethiopia. http://www.mowr.gov.et/index.php?pagenum=4.3&pagehgt=1000px. Accessed 08 Feb 2011

Monteith JL (1965) Evaporation and the environment. In: The state and movement of water in living organisms. In: XIXth symposium on the society of experimental biology, Cambridge University press, Swansea, pp 205–234

Moriasi DN, Arnold JG, Van Liew MW, Bingner RL, Harmel RD, Veith TL (2007) Model evaluation guidelines for systematic quantification of accuracy in watershed simulations. Trans ASABE 50(3):885–900

Motovilov YG, Gottschalk L, England K, Rodhe A (1999) Validation of distributed hydrological model against spatial observations. Agric For Meteorol 98:257–277

Mulligan M, Wainwright J (2004) Modelling and model building. In: Wainwright J, Mulligan M (eds) Environmental modeling: finding simplicity in complexity. Wiley, Chichester, pp 7–73

Nash JE, Sutcliffe IV (1970) River flow forecasting through conceptual models: Part I – a discussion of principles. J Hydrol 10(3):282–290

Neitsch SL, Arnold JG, Kiniry JR, Srinivasan R, Williams JR (2004) Soil and water assessment tool input/output file documentation: version 2005. USDA Agricultural Research Service and Texas A&M Blackland Research Center, Temple

Neitsch SL, Arnold JG, Kiniry JR, Williams JR (2005) Soil and water assessment tool theoretical documentation: version 2005. USDA Agricultural Research Service and Texas A&M Blackland Research Center, Temple

Patro S, Chatterjee C, Mohanty S, Raghuwanshi NS (2009a) Flood inundation modeling using MIKE FLOOD and remote sensing data. J Indian Soc Remote Sens 37:107–118

Patro S, Chatterjee C, Mohanty S, Raghuwanshi NS (2009b) Hydrodynamic modeling of a large flood-prone river system in India with limited data. Hydrol Process 23:2774–2791

Pikounis M, Varanou E, Baltas E, Dassaklis A, Mimikou M (2003) Application of the SWAT model in the Pinios River Basin under different land-use scenarios. Global Nest Int J 5(2): 71–79

Priestley CHB, Taylor RJ (1972) On the assessment of surface heat flux and evaporation using large-scale parameters. Mon Weather Rev 100:81–92

Reed S, Koren V, Smith M, Zhang Z, Moreda F (2004) Overall distributed model intercomparison project results. J Hydrol 298:27–60

Refsgaard JC (1996) Terminology, modeling protocol and classification of hydrological model codes. In: Abbott MB, Refsgaard JC (eds) Distributed hydrological modelling. Kluwer Academic, Dordrecht, pp 17–39

Refsgaard JC, Storm B (1996) Construction, calibration and validation of hydrological models. In: Abbott MB, Refsgaard JC (eds) Distributed hydrological modelling. Kluwer Academic, Dordrecht, pp 41–54

Setegn SG, Srinivasan R, Melesse AM, Dargahi B (2009a) SWAT model application and prediction uncertainty analysis in the Lake Tana Basin, Ethiopia. Hydrol Process 24(3):357–367

Setegn SG, Srinivasan R, Dargahi B, Melesse AM (2009b) Spatial delineation of soil erosion vulnerability in the Lake Tana Basin, Ethiopia. Hydrol Process 23(26):3738–3750

Setegn SG, Dargahi B, Srinivasan R, Melesse AM (2010) Modeling of sediment yield from Anjeni-Gauged watershed, Ethiopia using SWAT model. J Am Water Resour Assoc (JAWRA) 46(3):514–526

Shrestha RK, Sayama T, Tachikawa Y, Takara K (2005) Use of disaggregated rainfall data for distributed hydrological modeling in Yodo River basin. Annuals of the Disaster Prevention Research Institute, Kyoto University, No. 48 B

Smedema LK, Vlotman WF, Rycroft DW (2004) Modern land drainage: planning, design and management of agricultural drainage systems. Taylor & Francis Group Plc, London

Tadele K, Foerch G (2007) Impacts of land use/cover dynamics on stream flow: the case of Hare watershed, Ethiopia. In: The proceedings of the 4th international SWAT conference UNESCO-IHE, Delft, The Netherlands, 02–06 July 1978

Tesfaye YG, Meerschaert MM, Anderson PL (2006) Identification of periodic autoregressive moving average models and their application to the modeling of river flows. Water Resour Res 42, W01419. doi:10.1029/2004WR003772

Tessema SM, Lyon SW, Setegn SG, Mörtberg U (2014) Effects of different retention parameter estimation methods on the prediction of surface runoff using the SCS curve number method. Water Resour Management 28:3241–3254

USDA Soil Conservation Service (SCS) (1972) National engineering handbook section 4 hydrology. USDA, Washington, DC

van Griensven A, Meixner T (2007) A global and efficient multi-objective auto-calibration and uncertainty estimation method for water quality catchment models. J Hydroinf 9(4):277–291

van Griensven A, Meixner T, Grunwald S, Bishop T, Diluzio M, Srinivasan R (2006) A global sensitivity analysis method for the parameters of multi-variable catchment models. J Hydrol 324:10–23

Van Liew MW, Veith TL, Bosch DD, Arnold JG (2007) Suitability of SWAT for the conservation effects assessment project: comparison on USDA agricultural research service watersheds. J Hydrol Eng 12(2):173–189

White KL, Chaubey I (2005) Sensitivity analysis, calibration, and validation for a multisite and multivariable SWAT model. J Am Water Resour Assoc (JAWRA) 41(5):1077–1089

Index

S.G. Setegn, M.C. Donoso (eds.), *Sustainability of Integrated Water Resources Management*, DOI 10.1007/978-3-319-12194-9

MIX
Papier aus verantwortungsvollen Quellen
Paper from responsible sources
FSC® C105338

If you have any concerns about our products, you can contact us on
ProductSafety@springernature.com

In case Publisher is established outside the EU, the EU authorized representative is:
Springer Nature Customer Service Center GmbH
Europaplatz 3, 69115 Heidelberg, Germany

Printed by Libri Plureos GmbH
in Hamburg, Germany